Petroleum Refining Technology

W0193077

Petroleum Refining Technology

Indra Deo Mall

PhD, FIE (I), LMIIChE, LM IPPTA

Professor
Department of Chemical Engineering
Indian Institute of Technology–Roorkee
Roorkee 247667
Uttrakhand

CBS

CBS Publishers & Distributors Pvt Ltd

New Delhi • Bengaluru • Chennai • Kochi • Mumbai • Pune
Hyderabad • Kolkata • Nagpur • Patna • Vijayawada

ISBN: 978-81-239-2543-1

Copyright © Author and Publisher

First Edition: 2015

Published by Satish Kumar Jain and produced by Varun Jain for
CBS Publishers & Distributors Pvt Ltd
4819/XI Prahlad Street, 24 Ansari Road, Daryaganj, New Delhi 110 002, India.
Ph: 23289259, 23266861, 23266867   Website: www.cbspd.com
Fax: 011-23243014   e-mail: delhi@cbspd.com; cbspubs@airtelmail.in.
Corporate Office: 204 FIE, Industrial Area, Patparganj, Delhi 110 092
Ph: 4934 4934   Fax: 4934 4935   e-mail: publishing@cbspd.com; publicity@cbspd.com

Branches

- **Bengaluru:** Seema House 2975, 17th Cross, K.R. Road,
  Banasankari 2nd Stage, Bengaluru 560 070, Karnataka
  Ph: +91-80-26771678/79   Fax: +91-80-26771680   e-mail: bangalore@cbspd.com
- **Chennai:** 7, Subbaraya Street, Shenoy Nagar, Chennai 600 030, Tamil Nadu
  Ph: +91-44-42032115   Fax: +91-44-42032115   e-mail: chennai@cbspd.com
- **Kochi:** 36/14 Kalluvilakam, Lissie Hospital Road, Kochi 682 018, Kerala
  Ph: +91-484-4059061-65   Fax: +91-484-4059065   e-mail: kochi@cbspd.com
- **Mumbai:** 83-C, Dr E Moses Road, Worli, Mumbai-400018, Maharashtra
  Ph: +91-22-24902340/41   Fax: +91-22-24902342   e-mail: mumbai@cbspd.com
- **Pune:** Bhuruk Prestige, Sr. No. 52/12/2+1+3/2 Narhe, Haveli
  (Near Katraj-Dehu Road Bypass), Pune 411 041, Maharashtra
  Ph: +91-20-64704058/59, 32392277   Fax: +91-20-24300160   e-mail: pune@cbspd.com

Representatives

- **Hyderabad**   0-9885175004
- **Nagpur**   0-9021734563
- **Vijayawada**   0-9000660880

- **Kolkata**   0-9831437309, 0-9051152362
- **Patna**   0-9334159340

Printed at Repro India

Foreword

Petroleum refining industry provides basic feedstocks for a large number of chemical industries and thus plays a crucial role in meeting the basic needs of mankind. The developing nations have been seeing modernization and rapid expansion of petroleum industry. The refining industry, today, faces various challenges which include the wide variation of crude oil quality, fluctuating costs, stringent environmental standards, improved quality of products and minimization of the bottom of the barrel through heavy residue utilization.

Besides giving a brief overview of occurrence, exploration and drilling of petroleum, this book *Petroleum Refining Technology* covers in-depth characterization aspects of raw materials as well as products, various unit processes of importance in the refining industry, hydrogen production and management as well as integration of a refinery with a petrochemical complex.

The author, Prof ID Mall, has indeed put in a herculean effort in compiling various facts of petroleum refining technology. This book will be highly useful to students, researchers, faculty and practising engineers in the industries.

I wish the author all success in his endeavour.

Dr SJ Chopra
Chancellor
University of Petroleum and
Energy Studies
Dehradun

and

Former CMD
Engineers India Limited
New Delhi

Preface

Oil and natural gas are the important part of everyday life and their availability has changed the whole economy by providing basic needs of mankind in the form of fuel, petrochemicals, feedstock for fertilizer and energy for power sector. Now the world economy runs on oil and natural gas. Processing of oil and natural gas for producing fuels and other value-added products has become very important activity in the modern society. Availability of LNG (liquefied natural gas) has further improved the environment. Recent development on technology of natural gas to liquids (GTL) will further improve the availability of fuel to transportation sector.

Petroleum refinery is one of the important sectors of economy and playing a vital role in the industrialization, urbanization and meeting basic needs of mankind through supplying energy for domestic, industrial and transport sectors, and as feedstocks for fertilizers, synthetic fiber, synthetic rubber, polymers, intermediates, explosives, agrochemicals, dyes, paints, etc. Globally it processes more materials than any other industry. With expected population of 7.6 billion by 2020, huge increase in demand for transportation fuels, electricity, and many other consumer products made from petrochemical route is expected.

Petroleum refining technology is one of the important courses in chemical and petroleum engineering, and petrochemical courses at BTech, dual degree and MTech programs. The book contains 26 chapters. Chapters 1 and 2 deal with an overview of petroleum refining, Indian and global refining industries. Chapter 3 describes crude oil and natural gas origin, occurrence, exploration, drilling and processing. Chapter 4 describes crude oil and natural gas characteristics and evaluation while Chapter 5 deals with crude oil desalting and distillation. Chapter 6 deals with thermal cracking and reactions, visbreaking, coking. Chapter 7 deals with catalytic cracking process and technological development in FCC and cracking catalyst. Chapter 8 describes hydrocracking processes. Chapter 9 deals with catalytic reforming while Chapter 10 describes about alkylation, isomerization and polymerization. Chapter 11 describes about lubricating oil production while Chapters 12 and 13 describe about bitumen and grease production respectively. Chapter 14 describes about hydroprocessing, sulfur removal and sulfur recovery. Chapter 15 describes about natural gas processing, transportation and its use as petrochemical feedstock. Chapter 16 deals with oxygenates while Chapter 17 describes about biofuels. Production of hydrogen and hydrogen management has been discussed in Chapter 18. Chapter 19 highlights about petroleum and petrochemical integration and major petrochemical processes—steam cracking, aromatic production, LAB manufacture, FCC gases and petrochemical feedstock. Chapter 20 deals with petroleum product quality analysis, specification and standards. Chapter 21 describes the various separation processes in petroleum industry and technological industry. Chapter 22 deals with energy consumption and energy conservation measures in petroleum refinery industry. Chapter 23 describes the corrosion problem in petroleum industry while Chapter 24 describes about environmental management in petroleum refinery. Chapter 25 describes safety and hazard management in petroleum processing. Chapter 26 describes about the storage, blending and transportation of crude, gas and petroleum products.

In glossary of terms, a list of important terms used in chapters is given. In the appendix, abbreviations, physicochemical properties of hydrocarbons, list of petroleum and petrochemical industries, research and consultancy organizations and chemical, petroleum and petrochemical industries are given.

Indra Deo Mall

Acknowledgments

I would like to express my sincere thanks to Dr SJ Chopra, Chancellor, University of Petroleum and Energy Studies, Dehradun, for his encouragement, inspiration and valuable comments and Foreword. I am greatly indebted to Dr SN Upadhyaya, Former Director, Institute of Technology, Banaras Hindu University, who has always been a source of inspiration and encouragement. Mr PD Bagri, Former Executive Vice-President, Orient Paper Mills, for his kind support during my professional career. I would like to offer my sincere gratitude to Dr Prem Vrat, Former Director, Indian Institute of Technology, Roorkee, and Former Vice-Chancellor, UP Technical University. I would like to offer my sincere gratitude to Prof MM Sharma and Dr RA Mashelkar, who are source of inspiration to chemical engineers. I would also like to pay my sincere gratitude to all my teachers at IT, BHU, Varanasi. I am thankful to my friend Prof IM Mishra, Professor, Chemical Engineering, IIT–Roorkee, for his help and support. Thanks are also due to Dr Vimal Chanda Srivastava, for his support and sincere efforts.

I wish to pay a rich tribute to late Shri Dhiru Bhai Ambani, a great visionary, who has brought India on the world's petroleum and petrochemical map and revolutionized petroleum and petrochemical sector.

I wish to pay a rich tribute to late Prof AP Mall, who had been a source of inspiration and moral support.

I would like to offer my sincere gratitude to *Hydrocarbon Processing, Hydrocarbon Asia, Oil and Natural Gas Journal, Chemical Engineering Progress, Research and Industry,* Texas, USA, M/S Reliance Industries Ltd, Mumbai; Indian Oil Corporation Ltd, Panipat Refinery; Gas Authority of India Ltd, New Delhi; Indian Petrochemical Corporation Ltd (now Reliance Industries), Vadodara, for providing important technical information and plant photographs. Mr RK Ghosh, Director, Refineries, Indian Oil Corporation; Mr MM Paniappan, Former Head, Technical Cell, IOC, Panipat Refinery; Mr Sanjeev Singh, Executive Director, IOCL, Paradip Refinery; Dr RP Verma, Former Executive Director, IOC R and D Center, Faridabad; Dr SN Kaul, Former Acting Director, Neeri; Mr RK Ghosh, Director, Refineries, IOCL; Mr Sanjeev Singh, Former Executive Director, IOC, Panipat and Paradip Refineries, now Director, Refineries, IUCL; Mr A Basu, CMD, CPCL; Mr DK Karwal, DGM, IOC, Panipat Refinery, for their help. I am grateful to Mr Sushil Kumar, President, Reliance Industries, and Executive Director, IPCL, Vadodara (now Reliance Industries).

I would like to express my sincere gratitude to Mr Chandrakant Thakur, for his untiring efforts in preparation of manuscript for second edition and preparing the process flow diagrams. I also acknowledge the contribution of my student V Subbaramaiah and some of my other students Anang Swapnesh, Nitin Kumar,

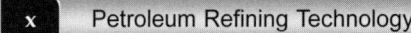

Seema, Bhawna Bajpai, Priyanka Sanjay Gautam, Shialendra Vikky Anand for their contribution.

Last but not the least, I sincerely acknowledge the patience, support, and understanding of my wife Indira, for standing behind me during the ups and down of my life. I also acknowledge the support of Vishal and Pallavi, Kavim, Gunjan and Sanjay, Priyansh, Shagun.

Indra Deo Mall

Contents

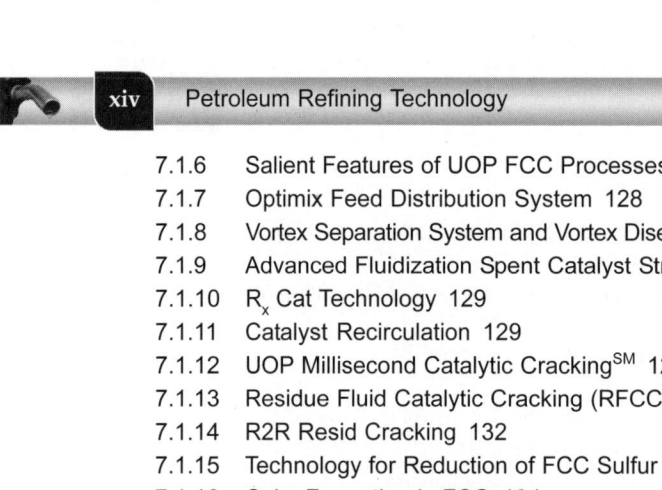

20. Product Quality Analysis, Specification and Standards   364–387

Introduction

Oil and natural gas are the important part of everyday life and their availability has changed the whole economy. It is playing a vital role in the industrialization and urbanization and meeting basic needs of mankind through supplying energy for domestic, industrial, transport sectors and as feedstocks for fertilizers, synthetic fiber, synthetic rubber, polymers, intermediates, explosives, agrochemicals, dyes, paints, etc. Now the world economy runs on oil and natural gas. Processing of oil and natural gas for producing fuels and other value added products has become very important activity in the modern petroleum industry. Availability of liquefied natural gas (LNG) has further improved the environment. Recent development of technology of natural gas to liquids (GTL) will further improve the availability of fuel to transportation sector. With expected population of 7.6 billion by 2020, demand for transportation fuels, electricity, feedstocks for petrochemical feedstock and other consumer products derived from petroleum and petrochemical route is also expected to increase.

Petroleum, in some form or other is not new. The use of petroleum and its derivatives has been reported since the pre-Christian times and was known largely through use in many of the older civilizations. Petroleum and its derivatives such as asphalt, bitumen have been known and used for almost 6000 years [Speight, 1999]. First documented use of asphalt has been reported in 3800 BC. The kerosene lamp invented in 1854, ultimately created the first large scale demand for petroleum. Modern petroleum refining began in 1859 with discovery of petroleum in Pennsylvania and subsequent commercialization [Speight, 1999]. About 150 years ago, in the hills of northwestern Pennsylvania, Edwin Drake was the man who struck South America's first oil well [PetroFed, 2006]. The exploration of petroleum originated in the latter part of the nineteenth century. The Second World War gave rise to fast scientific and technological development to produce more distillate products from bottom of the barrel. The beginning of American petroleum industry was with the discovery of oil in 1859 by Edwin Drake, near Titusville, Pennsylvania. There has been continuous innovation in refinery technology over the last century largely in order to meet the increasing requirements of gasoline and diesel for automobile, petrochemical feedstock for chemical and intermediate, synthetic fiber, polymer, rubber and consumer goods, and more environmentally friendly processes and products [www.allbusiness.com]. There has been continuous upgradation in the process of technology, catalyst, energy and environmental management to meet the consumer's demand for better and different types of products and meeting the challenges of global environmental concern.

As a result of various technological developments and rise in the demand of petroleum and petrochemical products, today's refinery is highly integrated and complex, producing fuels, lubricants and broad range of petrochemicals. Future refinery will not only be a refinery to produce fossil fuel, but also will be a petrochemical refinery to produce value added petrochemicals. Distinct periods in the evolution of refinery are—the separation and thermal period, catalytic period, quantitative reaction engineering period and era of environmental friendly technologies and cleaner fuels.

Driving forces for changes in the refining technology has been the increasing demand for clean, distillate transportation fuels, higher crude oil costs especially for light, sweet crude and declining heavy residual fuel-oil markets [Kovac et al., 2006]. Another major development during recent years has been the use of biomass based renewable fuels; biogasoline and biodiesel.

Though the hydrocarbon industry is on an upcycle, refineries which are the core of this industry face as many challenges as the increasing options to increase refinery margins. Pressures towards cleaner products, differing qualities of crude with shift towards heavier and dirty crudes and now volatile and high crude prices have made the refinery's talk challenging. Increasing product yields, getting value added products and improving product mix flexibilities through integration with petrochemicals are now necessary though this existed for some time now [Rohatgi, 2008].

A standalone refinery configuration in the present context will have (i) flexibility to handle heavy and sour crude oils, with suitable provisions to process opportunity crudes; (ii) adequate capacity of secondary conversion and residue upgrading facilities to ensure high overall yields of light and middle distillates and (iii) appropriate upgrading facilities (hydrodesulfurization, CCR reforming, isomerization, alkylation, etc.) to meet product quality complying with Euro IV or better specifications [Joshi, 2009].

1.1   PROCESSES INVOLVED IN PETROLEUM REFINING

A typical refinery consists of wide variety of processes for producing various products which are being used as fuel, source of energy and feedstock. Summary of the processes involved in petroleum refining is given in Table 1.1 and Fig. 1.1. Product profile of a typical refinery is given in Table 1.2 and Fig. 1.2.

Table 1.1 Summary of the processes involved in petroleum refineries

Process	Description
Desalting	Desalting process is used for removal of the salts like chlorides of Ca, Mg, Na, and other impurities as they cause fouling and results in corrosion of various equipments and acts as catalyst poisons. Two stage desalting is done.

Contd.

Table 1.1 Summary of the processes involved in petroleum refineries *(Contd.)*

Process	Description
Atmospheric distillation	The crude oil is first desalted to remove the impurities. Deesalted crude oil is heated in a heat exchanger and furnace to about 400 °C and fed to atmospheric distillation column and separated into its various fractions like light gases, gasoline, naphtha, kerosene, light fuel oil, diesel oils, gas oil, heavy residue.
Vacuum distillation	Distillation of the residue from atmospheric distillation column is done in vacuum distillation column to make it suitable for catalytic cracking or hydrocracking, or manufacturing of lubricating oils, and a residue (which is blown or further, distilled to form bitumen).
Thermal cracking/ visbreaking	Thermal cracking/visbreaking cracking involves cracking of atmospheric and vacuum distillation column heavy residue products at high temperature without use of any catalyst to produce lighter fuel products and other heavier fractions.
Coking and delayed coking	Coking is basically a thermal cracking process where cracking of the heavy residue is done primarily to produce more value added products such as gasoline and diesel along with production of coke and to reduce low-value residual fuel oils.
Catalytic cracking	Catalytic cracking is used to break larger hydrocarbon molecules into light gases, LPG, gasoline, naphtha, diesel, light and heavy cycle oil using catalyst. Coke is also formed due to cracking.
Hydrocracking	Hydrocracking converts the heavier feedstocks (fractions that are more difficult to crack and cannot be cracked effectively in catalytic cracking units) into gasoline, jet fuels, diesel fuels, low-pressure gas, and low-sulfur fuel oil.
Catalytic reforming	Catalytic reforming is used to process primarily low octane straight run naphtha into high octane fuel. It is also used for the production of aromatics.
Hydrotreating and hydroprocessing	Hydrotreating and hydroprocessing are used to remove impurities such as sulfur, nitrogen, oxygen, and trace metal impurities which are poisonous to catalyst.
Alkylation	Alkylation is one of the important conversion process to produce high octane gasoline from alkylation of isobutane and olefinobtained from FCC. Earlier sulfuric acid, HF was used. Now, solid zeolite base catalyst have replaced the conventional sulfuric acidand hydrofluoric acid catalyst.
Isomerization	Isomerization is used to improve octane number by altering the arrangement of molecule without adding or removing anything from original molecule. Typically, paraffins (C_4H_{10} and C_5H_{12} from CDU) are converted to isoparaffins having a much higher octane. Isomerization is used for improving the octane number of low octane naphtha.

Contd.

Table 1.1 Summary of the processes involved in petroleum refineries *(Contd.)*

Process	Description
Polymerization	Polymerization is a oligomerization process. The process is used to convert propane and butane to high octane gasoline blending components.
Solvent extraction	Solvent extraction is being increasingly used in refinery and petrochemical industry. Solvent extraction is used for removal of non-aromatics from aromatics and to dissolve and remove aromatics from lube oil feedstock which improves viscosity, oxidation resistance, colors, and gum formation. Solvent extraction is also used for separation of many close boiling hydrocarbons in refinery and petrochemical complexes.
Dewaxing	Dewaxing is a process of removing wax from lube oil, generally after deasphalting and solvent extraction so as to produce lubricants with low pour point and to recover wax for further processing.
Propane deasphalting	Deasphalting is a solvent extraction process for removing asphalt or resins from viscous hydrocarbons from vacuum column to produce feedstock for lube oil refining or catalytic cracking processes.
Merox extraction	This process is used for removal of mercaptans or for conversion of mercaptans into disulfide using a catalyst. The process is a combination of Merox extraction and sweetening.
Hydrodesulfurization	The process involves removal/reduction of the sulfur content of fuel. The separation of reaction gases is done in a two-stage separator. The stripper bottom is the finished product. The separator gases are then sent to amine wash for removal of H_2S.

Table 1.2 Products from a typical refinery

Product	End-uses
Light gases containing methane and hydrogen	Fuel and hydrogen in hydroprocessing
LPG	Domestic fuel
Ethylene from FCC gas	Ethyl benzene
C_4 and C_5 gases from FCC	MTBE, TAME, isobutylene and other petrochemical feedstock
Propylene from FCC	Feedstock for propylene and other petrochemicals
Gasoline	Transport fuel
Naphtha	Feedstock for catalytic reforming, feedstock for petrochemicals such as ethylene, propylene and fertilizers, etc. and as fuel in power plants.

Contd.

Table 1.2 Products from a typical refinery *(Contd.)*

Product	End-uses
Reformate	High octane gasoline, feedstock for petrochemicals such as ethylene benzene, benzene, toluene, p-xylene, o-xylene, etc.
MTBE, TAME	Oxygenate for reformulated gasoline
Jet/aviation turbine fuel/superior kerosene oil	Aviation and domestic fuel
High speed diesel	Fuel for transport sector, agriculture and captive power generation
Light diesel oil	Fuel for small industrial units, start up fuel for power generation
Sulfur	Feedstock for sulfuric acid, pharmaceuticals
Petroleum coke	Fuel for power plants and cement plants carbon electrode, synthesis gas through gasification
Kerosene	Domestic fuel, feedstock for LAB, jet fuel
Lubricating oil	Lubricants and grease
Petroleum wax	Packing, adhesive, cosmetics, rubber, candles, chewing gum, inks, plastics, rubber
Bitumen	Surfacing of roads
Other minor products (benzene, toluene, MTO, LABFS, CBFS, paraffin wax, etc.)	Petrochemical feedstock for value added products.

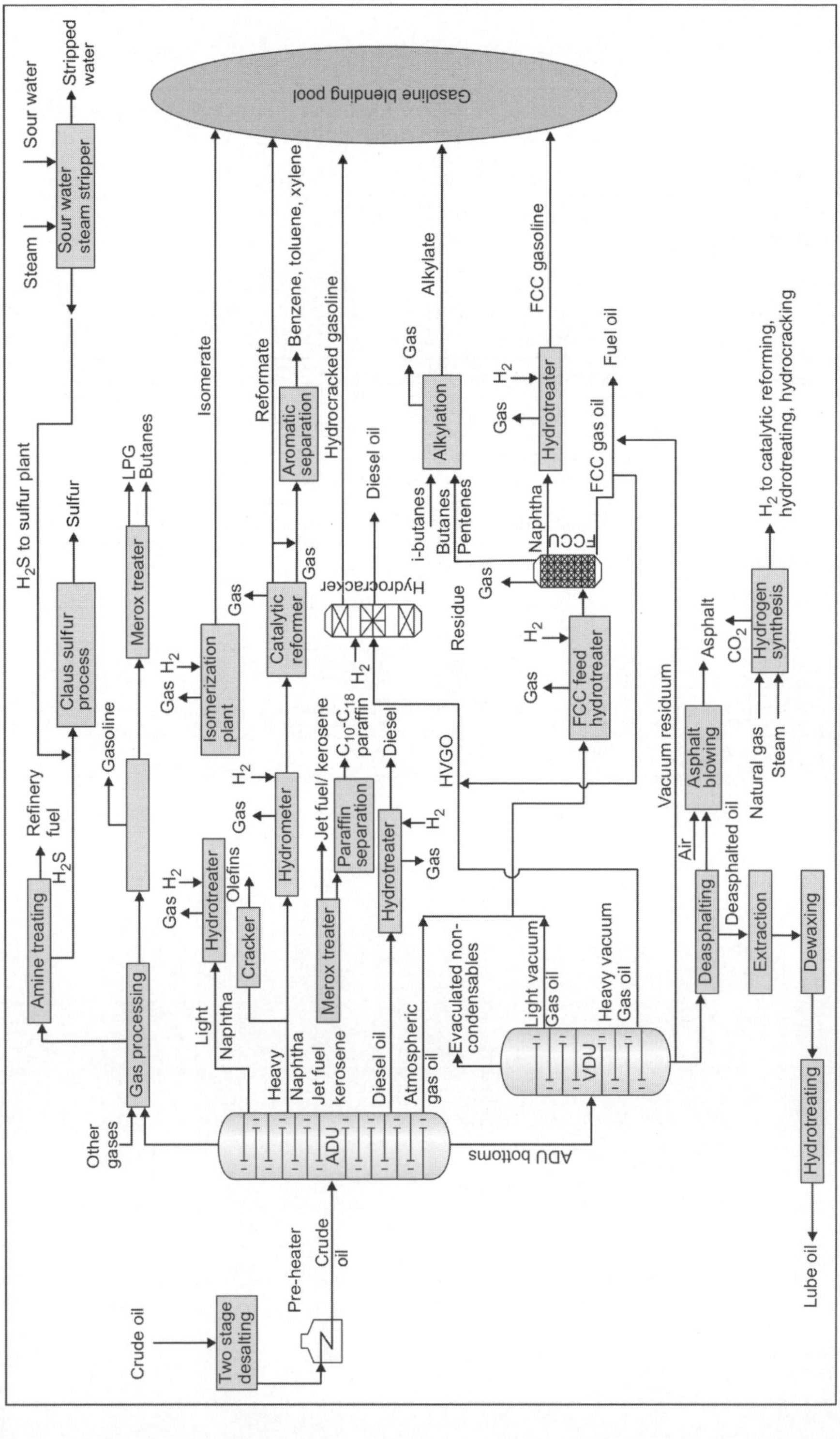

Fig. 1.1 General petroleum refining flow

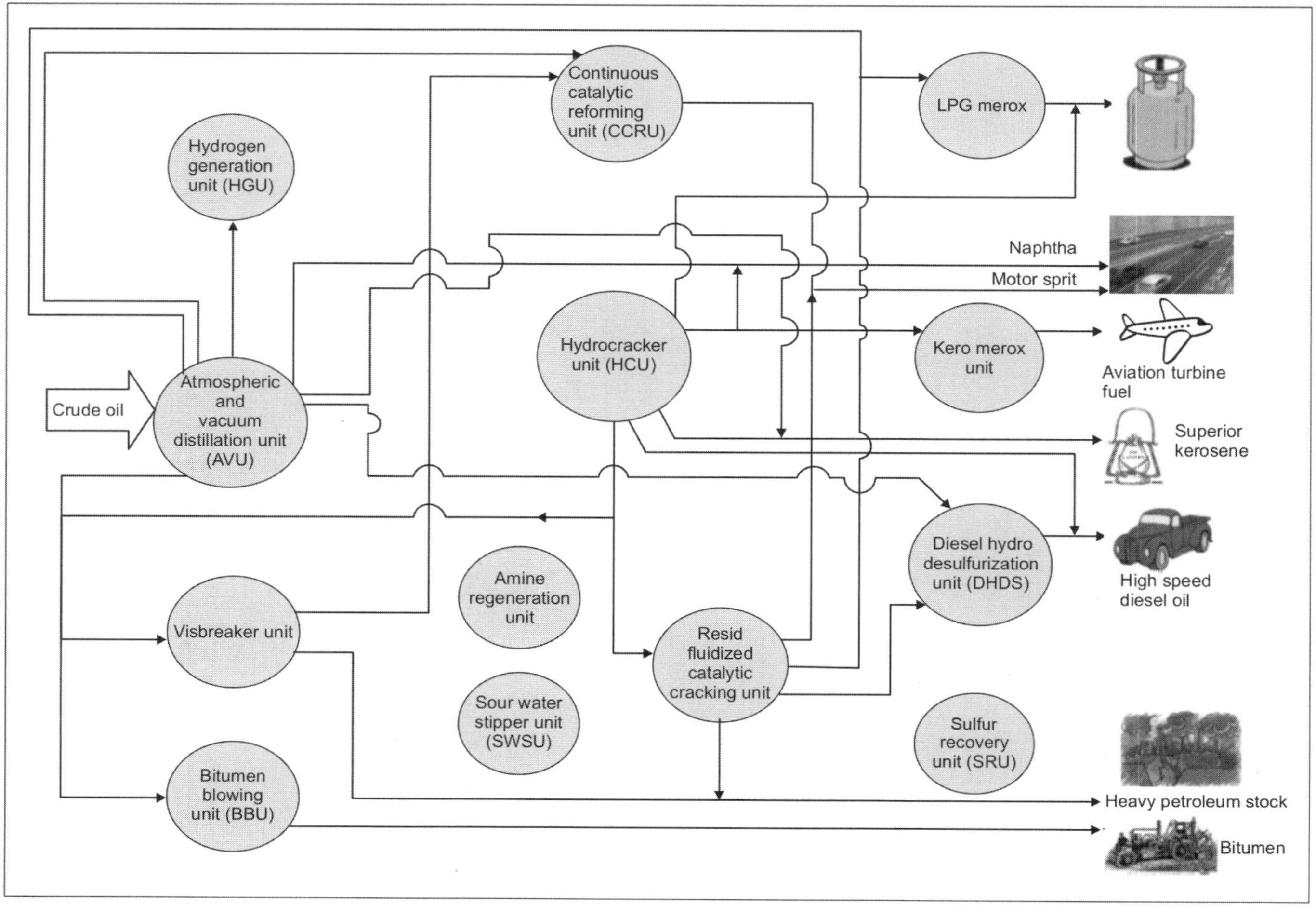

Fig. 1.2 General flow and product profile of a typical refinery

1.2   WORLD PETROLEUM INDUSTRY

Global demand for petroleum for petroleum products is expected to increase to 90 million barrels/day by 2010 and is projected to touch 105 million bpd by 2020 [Chemical industry Annual-January 2007]. World oil demand is expected to grow by 1.2–1.4 million barrels a day on every year for the next 15 years and 60% of that demands growth will be from Asia and more so from China and India [Acharya, 2007]. Growing demand of transportation fuel due to rising population, a turbulent supply market of crude oil, lack of spare production capacity and continuous turmoil in several oil producing capacities has been responsible for increasing crude oil prices [Acharya, 2007]. Global refinery capacity is given in Table 1.3. World crude oil production is given in Table 1.4. Total World Consumption* of oil is given in Table 1.5. Total world refinery throughputs is given in Table 1.6. Total world production* of natural gas is given in Table 1.7. Total world production* of natural is given in Table 1.8.

Table 1.3 Global refinery capacities (thousand barrels daily)

Year	2002	2008	2011*
North America	20143	21086	21382
S. and Cent. America	6296	6658	6590
Europe and Eurasia	25037	24704	24570
Middle East	6915	7603	8011
Africa	3228	3151	3317
Asia Pacific	22451	26123	29135
Total world	84070	89324	93004

*Atmospheric distillation capacity on a calendar-day basis.
Source: BPL energy data 2012 Statistical Review of World Energy available online at www.bp.com/statisticalreview2004.

Table 1.4 Total world production* of oil (million tons)

Year	2006	2008	2010	2011	2012
Crude oil production	3963.3	3991.8	3977.8	4018.8	4118.9

*Includes crude oil, shale oil, oil sands and NGLs (natural gas liquids—the liquid content of natural gas where this is recovered separately). Excludes liquid fuels from other sources such as coal derivatives.
Sources: Annual Report, 2012–13; Indian Petroleum and Natural Gas Statistics, Govt. of India, New Delhi.
Courtesy: BP 2012 Statistical Review of World Energy www.bp.com/statisticalreview2012.

Table 1.5 Total world consumption* of oil (million tons)

Year	2006	2008	2010	2011	2012
World total	3950.1	3994.8	4038.2	4081.4	4130.5

*Inland demand plus international aviation and marine bunkers and refinery fuel and loss.
Source: Annual Report 2012–13, Indian Petroleum and Natural Gas Statistics, Govt. of India, New Delhi.
Courtesy: BP 2012 Statistical Review of World Energy www.bp.com/statisticalreview2012.

Table 1.6 Total world refinery throughputs (million tons per year)

Year	2006	2008	2010	2011	2012
Refinery capacity	4330.8	4430.8	4556.1	4575.6	4593.2
Refinery throughputs	3699.8	3720.5	3737.5	3760.6	3784.2

Source: Annual Report 2012–13, Indian Petroleum and Natural Gas Statistics, Govt. of India, New Delhi.
Courtesy: BP 2012 Statistical Review of World Energy available online at www.bp.com/statisticalreview2012.

Table 1.7 Total world production* of natural gas (trillion cubic feet)

Year	2006	2008	2010	2011	2012
Natural gas	6405.0	6534.0	6925.2	7360.9	6614.1

*Excluding gas flared or recycled.
Source: Annual Report 2012–13, Indian Petroleum and Natural Gas Statistics, Govt. of India, New Delhi.
Courtesy: BP 2012 Statistical Review of World Energy available online at www.bp.com/statisticalreview2012.
Review of World Energy available online at www.bp.com/statisticalreview2012 [www.poweron.ch].

Table 1.8 Total world consumption of natural gas

Trillion cubic meters

Year	2006	2008	2010	2011
Billion cubic meters	2824.3	3005.1	3153.1	3222.9

Courtesy: BP 2012 Statistical Review of World Energy www.bp.com/statisticalreview2012 [www.poweron.ch].

The US geological survey has estimated that world may be endowed with resources of nearly 5.25 trillion barrels (720 billion tons of oil and gas, of which nearly half has been proven). Most of the world oil reserves known today (85%) are concentrated in ten and almost all (except the US) are suppliers to the world and are likely to remain major suppliers in future also [Butola, 2007].

There has been continuous development in production of oil and gas economically. Advances notably in horizontal drilling and fracturing have made tight gas, shale gas, light tight oil production much more economically and environmentally attractive. Some of the major advances in oil and gas production are [Dawe, 2001; Alvarado and Manrique, 2010]:

- Seismic exploration, computer intensive data processing and interpretation: 3-D seismic
- Innovation in onshore and offshore drilling, coiled tubing drilling
- Improved downhole separation of water and gas
- Shared earth model
- Advances in drilling fluid: Polymer free fracturing fluid, surfactant based fracturing fluids
- Horizontal drilling and hydraulic fracturing
- Deep water extraction
- Improved or enhance oil recovery (EOR) techniques: Thermal methods, polymer flooding, CO_2 EOR.

Increasing energy demand and stringent environmental regulations are the major driving force for new investments in refining technologies. Inorder to meet the rising product demand of light transportation fuels and declining availability of light sweet crude oils, the refining industry has two options: Either to build new grassroots refineries to accommodate heavy crude oils or to process in the existing refineries. Natural gas which accounts for 22% of the world energy mix is rapidly becoming the most exciting resource of energy. Its share in the world consumption is expected to increase to 28% by 2025 [Singh et al., 2005].

Global petroleum demand is expected to average 2.2% during next 10 years. Table 1.9 shows the World's Top Petroleum Refineries.

Table 1.9 World top petroleum refinery

Rank Jan 1, 2009	Company	Crude capacity b/cd[1]
1	Exxon Mobil Corp	5,632,000
2	Royal Dutch Shell PLC	4,599,000
3	Sinopec	3,611,000
4	BP PLC	3,328,000
5	ConocoPhillips	2,696,000
6	Petroleos de Venezuela SA	2,678,000
7	Total SA	2,738,000

Contd.

Table 1.9 World top petroleum refinery *(Contd.)*

Rank Jan 1, 2009	Company	Crude capacity b/cd[1]
8	Valero Energy Corp	2,596,000
9	China National Petroleum Corp	2,440,000
10	Saudi Aramco	2,433,000
11	Petroleo Brasileiro SA	1,997,000
12	Chevron Corp[2]	1,997,000
13	Petroleos Maxicanos	1,703,000
14	National Iranian Oil Co	1,451,000
15	Nippon + Oil Co Ltd	1,317,000
16	Rosneft	1,293,000
17	OAO Lukoil	1,217,000
18	Reosol YPF SA	1,105,000
19	Kuwait National Petroleum Co	1,085,000
20	Marathon Oil Corp	1,016,000
21	Pertamina	993,000
22	Agip Petroli SPA	904,000
23	Sunoco Inc	880,000
24	Flint Hills Resources	817,000
25	SK Corp	817,000
26	Indian Oil Corp Ltd	777,000

Source: Oil and Gas Journal December 22, 2008.

1.3 INDIAN PETROLEUM INDUSTRY

Indian petroleum industry has made tremendous growth during last six decades with only one refinery with just a 0.25 million tons refining capacity at the time of independence. Next phase of development was with setting up of three major refineries with assistance of International oil companies Shell, Caltex and Esso. India's present refining capacity is 149 million tons with 17 public sector and two private sector refineries. Growth of Indian refining industry and India petroleum industry statistics are given in Tables 1.10 and 1.11 respectively.

Table 1.10 Growth of Indian refining industry

Year	Growth
1866	Oil discovery at Nahorpung.
1882	Assam Railway and Trading Corp (AR and TC) acquired rights for oil exploration.
1889	Oil production started at Digboi.

Contd.

Table 1.10 Growth of Indian refining industry *(Contd.)*

Year	Growth
1893	Right were granted to Assam Oil syndicate to start refining and first refinery started at Margheritta, Assam.
1899	Digboi Refinery commissioned supplanting above refinery with a capacity of 500 barrel.
1921	Burmah Oil Corporation (BOC) started oil exploration.
1931	Crude oil production; 250,000 tons.
1947–57	Multinational oil companies Burmah Shell, Esso Stanvac setup two refineries at Mumbai and Caltex setup at Visakhapatnam. Exploration of oil in Upper Assam at Naharkatia and Moran by BOC.
1954	Indian oil exploration with Russian geologists.
1956	Formation of Oil and Natural Gas Commission and production of crude oil and gas. Discovery of Cambay Oil field.
1958	Public sector Indian Refineries Ltd (IRL) came into oil field (to create refining and pipeline infrastructure).
1959	Oil India Ltd formed as joint venture between Government of India and Burmah Oil Company.
1962	First public sector refinery started at Guwahati by Indian Oil Refinery Ltd.
1962	Indian Oil Blending Ltd (IOBL) was formed for manufacture of lube oil and greases (joint venture between Indian Oil Company and Mobil Petroleum Corp).
1964	IRL and Indian Oil Company merged to form Indian Oil Corporation Ltd.
1965	Engineers India Ltd started.
1967–77	Discovery of oil and gas at Dabka, Santhal and Balol in Gujarat, Amguri, Charali; and Borholla in Assam, and Baramura in Tripura.
1977–87	Discovery of oil in large quantity at Bombay High.
1974	IOBL became part of IOCL.
1977–87	Discoveries of oil and gas in the Bombay offshore area (Panna field, Bassein field Mumbai offshore, Heera, Ratna, South and Mid Tapi). Oil and gas reserves discovered at Gandhar and Dahej (Cambay basin).
1981	Oil and Gas found in Krishna-Godavari-Cauvery basin.
1995	Oil was found in Cachar (Assam) and Nagaland.
1998	Assam Oil Co vested in IOCL as Assam Oil Division.
1999	First joint sector refinery MRPL commissioned.
2000	Panipat Refinery of Indian Oil Corporation commissioned. Reliance Petroleum Refinery commissioned at Jamnagar, Gujarat. Numaligarh refinery in Assam commissioned.
2006	Indian Oil's first linear alkyl benzene plant commissioned at IOC Vadodara India.

Contd.

Table 1.10 Growth of Indian refining industry *(Contd.)*

Year	Growth
2006	Indian Oil's first terephthalic acid plant commissioned at IOC Panipat refinery
2006	ONGC entered in Petroleum and refinery sector by acquiring MRPL
2008	First crude oil production from deep water in Krishna Godavari basin
2009	Reliance commissioned its 27 billion refinery
2009	ONGC to start mega petrochemical complex at Dahej, Gujarat to be commissioned in 2012
2009	HPCL Mittal energy to setup petrochemical hub to produce 400 tpd polypropylene at Ludhiana
2009–10	IOC Panipat Refinery commissioned naphtha cracker, polyethylene and polypropylene plant
2012	HPCL Bhatinda refinery commissioned
2011	BPCL Bina refinery commissioned
2014	Chennai petroleum corporation to double capacity to 3,90,000 bpd

Table 1.11 Growth of Indian petroleum industry at a glance

Item	Unit	2006–07	2008–09	2010–11	2011–12	2012–13*
1	2	3	4	5	6	7
1. Reserves: **Balance recoverable**						
i. Crude oil	Mn. tons	725.38	773.29	757.44	760	758
ii. Natural gas	Bn. cub. mtr.	1054.52	1115.26	1241.00	1330	1355
2. Consumption						
i. Crude oil (in terms of refinery crude throughput)	Mn. tons	146.55	160.77	206.15	204	219
ii. Petroleum products (excl RBF)	"	120.75	13.60	141.79	148	157
3. Production						
i. Crude oil	"	33.99	33.51	37.71	38	38
ii. Petroleum products	"	135.26	150.52	190.36	203	218
iii. LPG from natural gas	"	2.09	2.16	2.17	NA	NA

Contd.

Table 1.11 Growth of Indian petroleum industry at a glance *(Contd.)*

Item	Unit	2006–07	2008–09	2010–11	2011–12	2012–13*
1	2	3	4	5	6	7
4. Imports and exports						
i. Gross imports:						
(a) Qty: Crude oil	Mn. tons	111.50	132.78	163.59	172	185
LNG	"	6.81	8.06	8.95	10	10
Pol. products	"	17.66	18.52	17.34	16	16
Total (a)	"	**135.97**	**159.36**	**189.88**	**197**	**210**
(b) Value: Crude oil	₹ Billion	2190.29	3481.49	4559.09	6722	7847
LNG	"	56.50	95.48	127.19	204	282
Pol products	"	411.60	608.46	558.12	681	684
Total (b)	"	**2658.39**	**4185.43**	**5244.40**	**7607**	**8813**
ii. Exports:						
(a) Qty:						
Pol. products	Mn. tons	33.62	38.90	59.13	61	63
(b) Value:						
Pol. products	₹ Billion	810.94	1220.64	1961.12	2846	3201
iii. Net imports:						
(a) Qty: Crude oil	Mn. tons	111.50	132.78	163.59	172	185
Pol. products	"	–15.96	–20.38	–41.80	–45	–48
Total (a)	"	**102.35**	**120.46**	**130.75**	**136**	**147**
(b) Value: Crude oil	₹ Billion	2190.29	3481.49	4559.09	6722	7847
Pol. products	"	–399.34	–612.18	–1403.00	–2166	–2517
Total (b)	"	**1847.45**	**2964.79**	**3283.28**	**4557**	**5330**
iv. Unit value of crude oil imports (gross)	₹/MT	19643.50	26220.98	27868.29	39144	42461
5. India's total exports	₹ Billion	5717.79	8407.55	11188.23	14660	16353
6. Pol. imports as percentage of India's total exports						
i. Gross imports	%	46.49	49.78	46.87	NA	NA
ii. Net imports	%	14.18	14.52	17.53	19	20

Contd.

Table 1.11 Growth of Indian petroleum industry at a glance *(Contd.)*

Item	Unit	2006–07	2008–09	2010–11	2011–12	2012–13*
1	2	3	4	5	6	7
7. Contribution of oil sector to center/state resources						
i. Royalty from crude oil	₹ Billion	58.57	71.55	91.21	159	181
ii. Royalty from gas	"	10.75	16.24	25.27	33	39
iii. Oil development cess	"	71.77	68.87	83.14	81	159
iv. Excise and custom duties	"	718.93	705.57	1026.17	952	986
v. Sales tax	"	539.49	633.49	807.09	1004	1114
vi. Dividend	"	115.27	98.61	NA	130	141
8. Natural gas						
i. Gross production	Bn. Cub. Mtr.	31.75	32.85	52.22	48	41

* Provisional: As on 1st April initial year; NA: Not Available.

Source: Annual report ministry of petroleum and natural gas 2012, Public Sector Undertakings/ DGCI & S, Kolkata , Ministry of Finance [petroleum.nic.in].

1.4   MAJOR TECHNOLOGICAL DEVELOPMENT IN OIL AND NATURAL GAS PRODUCTION

Oil and gas industry is technology oriented and is highly cost intensive. However, due to technological development it has been possible to enhance the oil and gas production and supply fuels and feedstock for world economy. Development and application of new advanced technology coupled with development of computational skill has played an important role in finding and developing new oil and gas resources and developing economically large oil and gas deposits offshore. The three-dimensional (3D) visualization center allows the look of subsurface and explores what is there. Returns on refining assets have fallen to inadequate levels due to low growth for major refined products, poor upgrading margins, increased competition and there is need for refinery/petrochemical integration, process technology developments. Major factors influencing petroleum refining industry are rising crude oil prices; growing demand of petroleum products in developing Asia and Asia pacific, decline in crude oil production in some of the countries.

1.5   MAJOR PRODUCTS DEMAND, QUALITY SCENARIO

With increasing demand of light distillates for motor gasoline and petrochemical feedstocks and changing scenario in the environmental and clean fuel requirement, there has been continuous upgradation in the technology and future decisions on technology adoption will be governed by the environmental consideration. Historical and future trends in gasoline and diesel specifications [Absi-Halabi et al., 1997] and Indian Fuel Quality Trends in Future for Petrol and diesel in India are given in Tables 1.12 and 1.13.

Table 1.12 Historical and future trends on global and Indian gasoline quality trends

Period	Specifications
Prior to 1994	Lead phase out Lower RVP
1995–2000	Reformulated gasoline Zero lead Benzene: 1% max. Lower aromatics: 25 %vol max. Oxygen: 25 %wt min. RVP: 7.2/8.1 psi max.
Beyond 2000	Reformulated gasoline phase II Lower sulfur content: 30% Benzene: 0.8% max. Aromatics: 22.0% max. RVP: 7.0% Olefins: 4% Oxygen: 2%

Gasoline specification in India

	BS II	BS III	BSIV
Sulfur (%wt max.)	0.05	0.015	0.005
Benzene [%vol max. (Metros)]	5 (3)	1	1
Aromatics (%vol max.)	–	42	35
Olefins (%vol max.)	–	21 (i8@)	V

Table 1.13 Historical and future trends on global and Indian diesel quality trends

Period	Specifications	
Prior to 1993	Sulfur: Cetane:	1.0%, 0.2% (Sweden: 0.001–0.05% from 1991) No limit (Sweden: 47–50% from 1991)
1993–2000	Aromatics: Sulfur: Cetane:	No limits (Sweden: 5–25% from 1991) 0.2–0.05% 40–46 min.

Contd.

Table 1.13 Historical and future trends on global and Indian diesel quality trends *(Contd.)*

Period	Specifications
Beyond 2000	Aromatics: 35% max (California 10% max.) Sulfur: Further reduction to Sweden levels Cetane: Further increase 40–46 minutes Aromatics: Further reduction Volatility: Increase over present level

Source: Absi-Halabi, 1997 Hydrocarbon Processing.

Diesel specification in India

	BS II	BS III	BS IV
Cetane number	48	51	51
Sulfur (%wt max.)	0.05	0.035	0.005
Distillation (°C at 95%)	370	360	360
Polycyclic aromatics (%wt max.)	–	11	11

1.6   EMERGING CRUDE OIL AND NATURAL GAS RESOURCES

Opportunity crude, tar sands or oil sands, shale oil, shale gas are emerging crude oil and natural gas resources and posing big challenge to hydrocarbon industry for their effective and economical utilization. Oil sands are heavy black viscous oil and consist of clay sand, heavy black viscous oil and posses big challenge in extraction, processing and further downstream processing. The processing of oil sands derived feed stock because very nature of heavy fed stocks such as bitumen presents inherent challenge to refineries [Ovchinnikov et al., 2013]. Huge amount of tar sands are available, however, all are not recoverable. Largest deposit exist in Alberta (Canada) and Venezuela. In the US, tar sand is present. Processing of opportunity crude poses big challenge because of high sulfur, metal content and total acid number, fewer lighter components, a high density and viscosity, high gel-asphalt content, high salts and heavy metal content. Shale gas and shale oil has become important source of energy and petrochemical feedstock. With the development of horizontal drilling and hydrofracturing viability of shale gas and shale oil/tight oil has increased. However, high metal contents in shale crude especially iron create number of problems [Graf et al., 2014]. Shale oil is generally light paraffinic and sweet and low in nickel and vanadium and refinery processing shale oil will have higher light Cu [Bryden et al., 2014]. Tight oil is also opportunity crude and poses problem in further processing. Shale liquids production (crude oil and condensate) is currently the profit driver for the upstream drilling business [Conzalez, 2013]. There is growing pervasiveness of oil sands as a vibal feedstock for global oil processing. Processing of crude derived from oil sands because of their heavy nature and processing of these crude is accomplished by hydrocracking or coking together with other subsequent hydrotreating steps [Ovchinnikov et al., 2013].

1.7   PETROLEUM AND PETROCHEMICAL INTERFACE

Though the hydrocarbon industry is on an upcycle, refineries which are the core of this industry face as many challenges as the increasing options to increase refinery margins through integration of crackers, aromatic production recovery of FCC gases. Petroleum and petrochemical industries are closely interlinked and petrochemical industry is totally dependent on petroleum industry for feedstock. Fast growth of petroleum industry has given real impetuous to growth of Indian chemical industry where the sales is US \$30 billion due to availability of petroleum feedstock for further downstream processing to commodity and value added products. With increasing demand of petroleum feedstock and petrochemicals some of the refineries have gone for integrated refinery and petrochemical complex in India. Today's refinery is a highly integrated and produces a broad range of petrochemicals in addition to traditional fuels, lubricants and other petroleum products. Maximizing value addition to refinery streams is given in Tables 1.14 and 1.15. Figures 1.3, 1.4 and 1.5 show typical integration scenario of petroleum refinery with petrochemical industry. Integration of petroleum and petrochemicals has been discussed in detail in Chapter 19.

In the changing scenario, petroleum refining and petrochemical and fertilizer production integration will be of vital importance for maximizing the use of by-products and improving the overall economy of a petroleum refinery.

Table 1.14 Maximizing value addition to refinery streams

Streams	Utilization
Fuel gas	H_2
FCC: • Ethylene • Propylene • Butylene	Ethyl benzene → Styrene Cumene, iso-propanol Methyl ethyl ketone, MTBE, xylenes
C_3	Propylene + H_2
C_4	Discussed separately
LPG	BTX
C_5	TAME
Light naphtha	LPG, BTX
Heavy naphtha	Aromatics
Kerosene	n-paraffins → LAB
LCO (FCC unit)	Mixed naphthalenes
Coker kerosene	α-olefins

Table 1.15 Maximizing value in addition to refinery streams

Refinery	Petrochemical	Fertilizer complex
Fuel gas	H$_2$	Hydrogen
Naphtha	Olefins, aromatics, synthesis gas	Synthesis gas, ammonia, urea, nitric acid, methanol
		Caprolactam, melamine, dimethyl formamide
Kerosene	Linear alkyl benzene	
Gasoil	Synthesis gas and synthesis gas derivatives	Synthesis gas, ammonia, urea, nitric acid, methanol, melamine, dimethyl formamide
Petrocoke	Synthesis gas and synthesis gas derivatives, hydrogen	Synthesis gas, ammonia

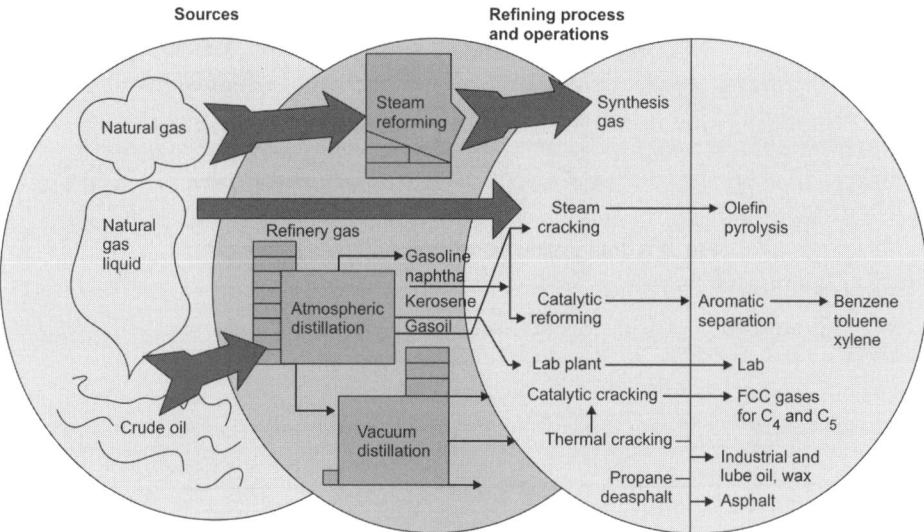

Fig. 1.3 Petroleum and petrochemical interface

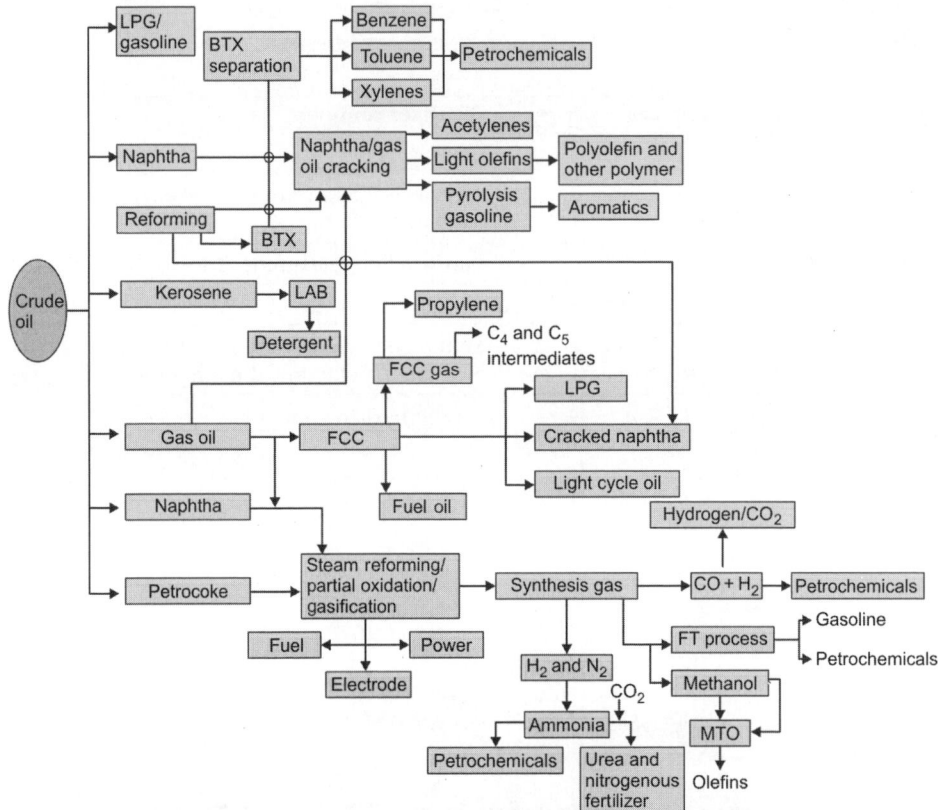

Fig. 1.4 Integration of refinery and petrochemical

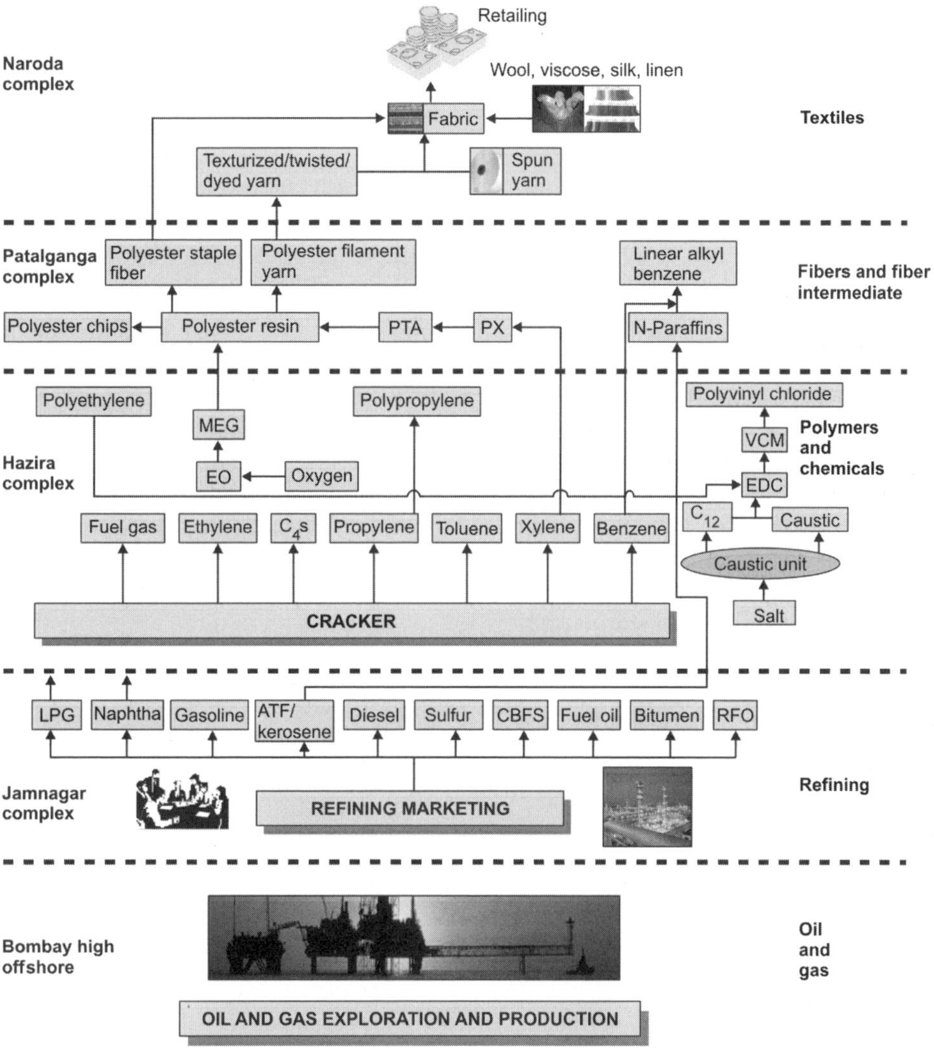

Retailing

Naroda
complex

Wool, viscose, silk, linen

Fabric

Textiles

Texturized/twisted/
dyed yarn

Spun
yarn

Patalganga
complex

| Polyester staple fiber | Polyester filament yarn | | | | Linear alkyl benzene | Fibers and fiber intermediate |

Polyester chips | Polyester resin ← PTA ← PX ← | N-Paraffins

Polyethylene

Polypropylene

Polyvinyl chloride

MEG

VCM

Polymers
and
chemicals

Hazira
complex

EO ← Oxygen

EDC

Fuel gas | Ethylene | C₄s | Propylene | Toluene | Xylene | Benzene | C₁₂ | Caustic

Caustic unit

CRACKER

Salt

LPG | Naphtha | Gasoline | ATF/ kerosene | Diesel | Sulfur | CBFS | Fuel oil | Bitumen | RFO

Jamnagar
complex

REFINING MARKETING

Refining

Bombay high
offshore

Oil
and
gas

OIL AND GAS EXPLORATION AND PRODUCTION

Fig. 1.5 Typical integration of a refinery and petrochemical
Courtesy: Reliance Industries

1.8 GAS TO LIQUID TECHNOLOGY (GTL)

Although GTL technology has been in development since the start of 1900s, but its large scale, worldwide application remains limited and it was the only past decade which has brought renewed vigor to the field and GTL technology will play an increasingly significant role as crude oil resources are depleting [Jamieson et al., 2007]. GTL technology natural gas is converted into liquid fuels and future refinery is expected to be GTL based refinery. A total of 10,00,000 bbl/day capacity of GTL plants are being considered for construction. GTL technology is mature and the cost per barrel is competitive [HC, 2007]. Figure 1.6 shows the concept of a GTL based refinery. GTL technology produces cleaner fuel with reduced sulfur dioxide and CO emission.

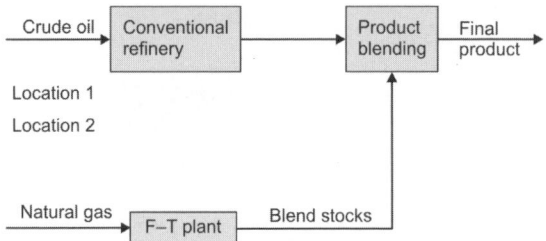

Fig. 1.6 Fischer-Tropsch blendstocks from gas based plants

Source: Chopra, 2003.

1.9   EMERGING SCENARIO IN PETROLEUM REFINING

Rising energy requirements, increased costs of crude oil and shift in crude mix towards heavier and sour crude have played major role in the choice of refinery configuration. Future growth of the refineries is driven by freedom of flexibility of choice in selection of crude oils to be processed, increased demand of refined products confirming to Euro III and Euro IV norms, shift to lighter distillates, more diesel cuts and other more value added product, petrocoke gasification, integration of refinery with petrochemical to have more value added products, compilation of stringent environmental standards.

Development of gasification technology and FT synthesis offer another potential linkage between refinery, gas/petrochemicals. Future emerging scenario in petroleum refining is given in Fig. 1.7.

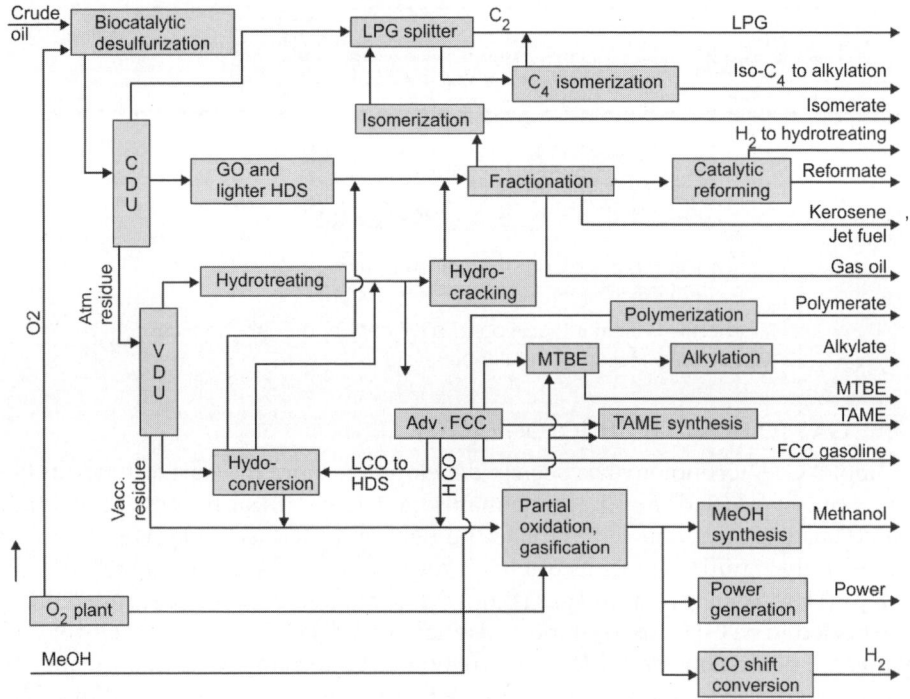

Fig. 1.7 Future refinery

Source: Chopra, 2003.

1.10 REFINERY CHALLENGES

Refinery operation is energy intensive operation. Due to to increasing demand of crude oil and petroleum products, refinery margin has decreased and need innovation technologies. Major influencing factors are shaping refinery design and operation which include access to crudes, location to markets, size and complexity of facilities, market demand, cost control and improving environmental performance and compliance of stringent environmental and safety regulations, fuel regulations. Now refinery will have to go for better sulfur management to meet the requirement of ultralow sulfur specification. Major challenges are:

- Increasing products demand due to increasing population
- Demand for better products quality with low sulfur emissions
- New environmental regulations
- Decreasing quality of feedstocks
- Increasing "s" (> 1.5%), decreasing api gravity (< 0.8) (heavier crude)
- Increasing level of microconstituents
- Growing demand of petroleum products
- Decline in domestic crude oil production
- Gradually increasing heavier feedstock
- Investment in billions of dollars for environmental protection
- Low profit margins.

Refining industry has under gone a major transformation in the last decade due to stringent regulatory standards, market requirement, fluctuating crude prices, tighter regulation on product quality and enforcement various fuel norms for reducing emissions, shifting trends towards heavier and opportunity crude [Lageteau et al., 2012]. Worldwide refiners are upgrading and modernizing refinery for increasing flexibility, improving energy efficiency, meeting environmental standards and regulation , increasing yields of value added products, integration of refineries with petrochemicals. Some of the key challenges for petroleum industry is given in Fig. 1.8 [Sarkar, 2014].

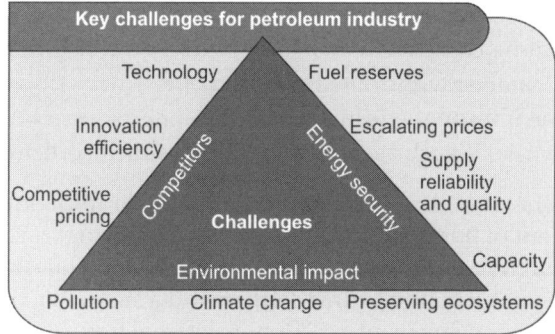

Fig. 1.8 Key challenges for refinery

Source: Sarkar, 2014.

1.11 PROCESSING OF OPPORTUNITY CRUDE OIL AND SHALE OILS AND SHALE LIQUID

There has been continuous change in the crude quality from light to heavier crude and now high TAN opportunity crude. Changing crude oil quality is given in Table 1.16.

Table 1.16 Changing world crude oil quality

Properties	1985	1990	1999	2010
Sulfur (%wt)	1.14	1.12	1.41	1.51
API gravity	32.7	32.6	32.2	31.8
"S" in residue (%wt)	3.07	3.26	3.91	4.0
Properties	1985	1990	1999	2010
Sulfur (%wt)	1.14	1.12	1.41	1.51
API gravity	32.7	32.6	32.2	31.8
"S" in residue (%wt)	3.07	3.26	3.91	4.0

Source: Samanti, 2012; RK "Refining Challenges and Trends" 6th Summer School on "Petroleum Refining and Petrochemicals" June 6, 2012, Organized by IOCL, New Delhi.

Producing clean fuel or high value petrochemical products from opportunity crude is big challenge. Opportunity crude are characterized by high sulfur content, metal content and total acid number, fewer lighter components, a high density and viscosity, high gel-asphalt content, high salts and heavy metal content which give rise to corrosion in equipment and concerns with product quality, environmental protection and processibilty [Qing, 2010]. Processing of heavy crude is big challenge to refinery. Various problems associated are [Venkatata Rammana et al., 2010].

- Desalter performance and associated problem
- Corrosion related problem in various equipments
- Catalyst life may be exhausted at a faster rate
- Drop in FCC and hydrocracker conversion
- Throughput reduction due to design limitations
- Increased operating cost due to increased hydrogen consumption
- Increased FCC catalyst withdrawal requirements
- Increased cetane improver additive consumption
- Product quality issues such as density of diesel, sulfur content of fuel oil.

Production of shale oils and shale gas has increased rapidly during recent time due to advancement of horizontal drilling and hydraulic fracturing. Shale oils are light, sweet with high paraffinic content and low acidity. Shale oil is characterized by low asphaltenes content, low sulfur content and a significant molecular weight distribution of the paraffinic content. Quality of shale oil is highly variable and processing requires innovative solutions [Wright and Sandu, 2005]. Shale oil quality is also varying in color, sulfur content from place to place. Although availability of shale oil have has increased during recent years, however some of the problem are inherently present due to high paraffinic content and pose problems in transportation, storage and refining. Shale liquids production (crude oil and condensate) is currently the profit driver for the upstream drilling business [Gonzalez, 2013]. Shale gas sector has rapidly evolved during last decade. Availability of shale gas/liquids have changed now the petrochemical feed stock scenario and also the economics in the United States. India has also huge shale deposit in basins of Camby basin, Assam-Arakan, Gujarat, Rajasthan, Tmax, coastal area in KG and Cauvery basins, Indogangetic plain [Tulsi Das, 2011]. Some of the major applications are in ethane based steam crackers, propane dehydrogenation, gas to liquid plants, methanol and aromatics production [Gonzalez, 2013]. Availability of shale gas with new horizontal fracturing and drilling has made shale gas business more profitable by reducing cost of natural gas [Fig. 1.9]. Summary of the potential impact of shale gas on hydrocarbon processing industry is given in Fig. 1.10 [Gunaseelan and Thundyli, july, 2013].

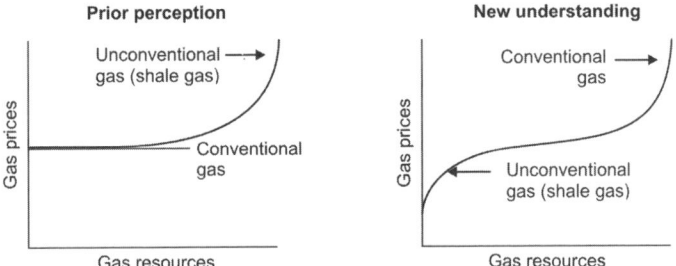

Fig. 1.9 Shale gas economics changing perception

Source: Tulsi, 2011.

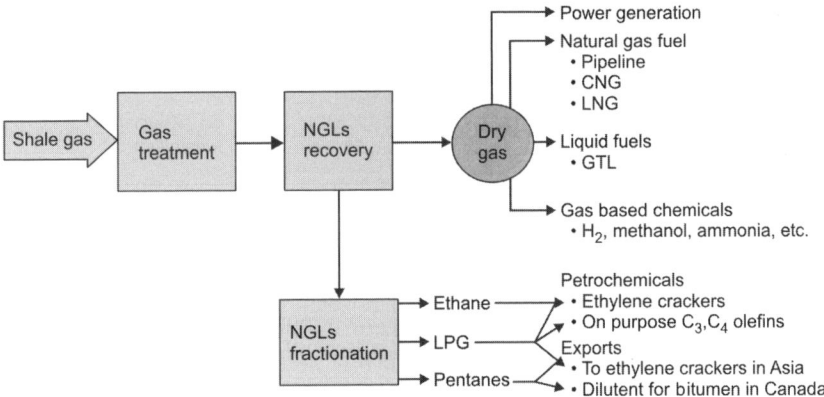

Fig. 1.10 Summary of potential impacts of shale gas hydrocarbon processing Industry

1.12 CARBON FOOTPRINTS IN PETROLEUM INDUSTRY

Petroleum refining is facing new challenge because of the strict green house gas regulations to reduce CO_2 emissions into the atmosphere. In a typical refinery 4–10% (depending on the refinery complexity) of the energy equivalent in the incoming crude is consumed as fuel and are significant source of CO_2 emissions. Worldwide refinery operations are responsible for about 1000 million tons annual CO_2 emissions which is about 6% of total manmade emissions worldwide [Vardarajan et al., 2009]. CO_2 emission increase with heavier feedstocks, cleaner fuels, conversion and complexicity [Turrowicz, 2011]. Although the CO_2 is emitted from various processes of refinery, however, major source of CO_2 emissions in a refinery are fire heaters, FCC and hydrogen production, captive power plants. Typical CO_2 emission distribution in a refinery is given in Fig. 1.11 [Turowicz, 2011].

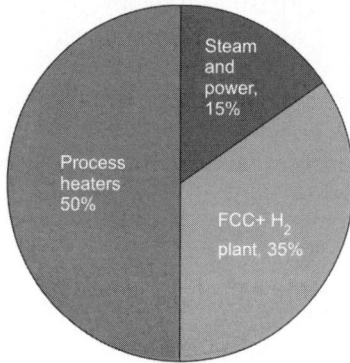

Fig. 1.11 Typical CO_2 distribution

Some of the opportunities in refinery to reduce CO_2 emissions are in the area:
- Improving energy efficiency
- Fuel substitution—petroleum fuel to natural gas
- Crude substitution
- Recovery and utilization of waste gas in refinery facilities
- Hydrogen production and management
- Alternate treatment processes including bioprocesses
- Formulations of the products with non-oil components/additives, use of biofuels
- Gasification.

REFERENCES

1. Absi-Halabi M, Stamislaus A and Qabazard H. Trends in Catalysis Research to Meet Future Refinery Needs. Hydrocarbon Processing, Feb, 1997, p45 [web.upmf-grenoble.fr].
2. Annual Report, 2006–07 (Public Sector Undertakings/DGSCI, Kolkata, Ministry of Finance).
3. Alvarado V, Manrique E. Enhanced Oil Recovery: Update review. Energies, 2010 p1529.
4. Benoit B and Zurlo J. Overcoming the Challenges of Tight/Shale Oil Refining. Processing Shale feedstocks, 2014, p37.
5. BPL. Energy Data 2004. Statistical Review of World Energy available online at www.bp.com/statisticalreview 2004 [www.poweron.ch].
6. Bryden K, Habbib Jr ET and Topete O. Catalytic Solution for Processing Shale Oils in the FCC. Processing Shale feedstocks, 2014, p25.
7. Butola RS. Oil Time for a Bolder Strategy 2007. Hindu, p163.

8. Chemical industry Annual, Jan 2007, P83.

9. Chopra SJ. Refinery for Future. QIP Short-term Course on Advances in Hydrocarbon Engineering, June 23–July 4, 2003, IIT Roorkee, India.

10. Ford J. Drilling Operations in Modern Petroleum Technology Vol 1, edited by Dawe RA, John Wiley & Sons Ltd, New York, 2001.

11. Gonzalez R. Shale Gas Feeds Petrochemical Expansion. Processing Shale feedstocks 2013, p17.

12. Gunaseelan P and Thundylijuly M. Shale Hydrocarbons Ushering an Industrial Renaissance in North America, Gas processing special supplement to hydrocarbon processing. July 2013, p19.

13. HC, 2007. Hydrocarbon Processing, Jan 2007, p21.

14. Jamleson A, McManus G and Mackenzie W. GTL Production will Partially Ease Regional Diesel. Naphtha Imbalances, Oil and Gas Journal, March 19, 2007, p49.

15. Joshi MK. Refining The Challenge for Green Technologies. Petrofed, January–March 2009, Vol 7, Issue 1, p3.

16. Kovac M, Movik G and Elliott JD. Upgrade Refinery Residuals into Value-Added Products. Hydrocarbon Processing, June 2006, p57.

17. Leargeteau D, Ross J, Laborde M and Wisdom L. Challenges and Opportunities of 10 ppm Sulfur Gasoline: Part 1. Petroleum Technology, Quarterly, Q3, 2012.

18. Oil and Gas Journal, Dec 22, 2008.

19. Petrofed, October–December, 2006, Vol 4, Issue 4, p9.

20. Ovchinnikov M, Ginestra J, Rauschining D, Gillespie B and Carlson K. Oil Sands Derived Feed Processing.

21. PTQ, 2013. Petroleum Technology, Quarterly, Q3, 2013, p71.

22. Qing W, 2010. Processing High TAN Crude: Part 1. PTQ, Q4, 2010, p35.

23. Ovchinnikov M, Ginestra J, Rauschning D, Gillespie and Carlson K. Oil Sands Derived Feed Processing Petroleum Technology, Quarterly, Q3, 2013.

24. Rohatgi M. Refining Scenario—The Road Ahead. Chemical Industry Digest, Annual, Jan 2008, p109.

25. Samanti RK. Refining Challenges and Trends. 6th Summer School on Petroleum Refining and Petrochemicals, June 6, 2012, Organized by IOCL, New Delhi at IIPM Gurgaon.

26. Sarkar. Invited Talk National Conference on Innovation and Development of Chemical Technology, 28th Feb 2014, IP University, New Delhi.

27. Singh H. Hydrocarbon Asia, Sep/Oct 2005, p34.

28. Speight JG. The Chemistry and Technology of Petroleum, 3rd edition, Marcel Dekker, 1999, p1.

29. Turrowicz M. Refinery Trends and Challenges, IOC L Summer School, IPM, Gurgaon, June 7, 201.

30. Das Tulsi. Shale Gas. Journal of Petrotech, January–March 2011, p35.

31. Vardarajan K, Handa SK and Goyal GD. Lovraj Kumar Memorial Annual Workshop 2009. Managing Carbon Footprints in the Process Industry, Organized By EEIL.

32. Venkatata Rammana U, Abhiram C and Roy G. Opportunity Crude Oil Processing and Process Challenges.

33. Wright B and Sandu C. Processing Shale Oils Require Innovative Solutions. Hydrocarbon Processing, July 2013, Vol 92, No. 7, 51.

Indian Petroleum Industry

Although Indian petroleum industries have revolutionized our life and are providing the major basic needs of rapidly growing, expanding and highly technical civilization as a source of energy for domestic, industrial, transport sectors and as feedstock for fertilizers, synthetic fibers, synthetic rubbers, polymers, intermediates, explosives, agrochemicals, dyes, paints, etc. Hydrocarbon sector is also playing an important role in meeting around 45% of India's primary energy requirements [CEW, 2009]. Share of future energy supply for crude and gas requirements is given in Table 2.1

Table 2.1 Total energy requirements (MTOE)

Year	2011–12	2016–17	2021–22	2026–27	2031–32
Hydro	12	18	23	29	35
Nuclear	17	31	45	71	98
Coal	283	375	521	706	937
Oil	186	241	311	410	548
Natural gas	48	64	97	135	197
Total	546	729	997	1351	1815

Sources: IEP Report, Page 28, Table 2.10, Annual Report 2012–13, Indian Petroleum and Natural Gas Statistics, Govt. of India, New Delhi.

2.1 PROFILE OF INDIAN PETROLEUM INDUSTRY

The Indian petroleum industry is one of the oldest in the world; oil having struck at Makum, near Margeritta in upper Assam in 1867. The petroleum industry in India stands out as an example of the strides made by the country in its march towards economic self-reliance. At independence in 1947, international companies controlled the industry. Indigenous expertize was scarce, if not non-existent. Today 50 years later, the industry is almost entirely in the public domain with skill and technical know-how comparable to the highest international standards.

The search of oil in India began in 1866, since then there has been continuous search. Burmah Oil Company (BOC) earlier looked oil exploration. Large sizes of oil fields were discovered by BOC in Naharkatiya in 1953 and 1956. Although indigenous production of crude oil went up from 0.25 million tons in 1948 to 0.39 million tons in 1956, real growth in the crude oil exploration and production began only during 1957–67 when Oil and Natural Gas Commission was setup in August 1956 and Oil India, a joint venture of Government of India and Burmah Oil Company, was setup in 1959 with objectives of planning, promoting and implementing programs for exploration and exploitation of petroleum resources throughout the country. Since then continuous efforts have been made to intensify the production of oil and gas in the country. Various on land and offshore exploratory drilling and drilling of crude oil were started in various parts of country which include Jammu & Kashmir, Punjab, Assam, Rajasthan, Gujarat, Western Coast, Bombay high and Krishna-Godavari basins. The new revolutionary licensing policy will impart and impetuous to crude oil exploration in the country. Crude oil production during 2010–11 was 37.71 million tons while the gas production was 52.22 billion cubic meters. A number of measures is being taken to bridge the growing gap between indigenous crude production and import of crude. Several measures have been taken to enhance hydrocarbon reserves and its production in the country. This includes: Development of new fields, additional development of existing field, scaling up and implementation of enhanced oil recovery schemes, indenting on the services of international experts, improved maintenance of reservoir health, conduct of 3D seismic surveys for precise reservoir delineation, application and optimization of artificial lift operations, stimulation of wells, optimization and redistribution of water injection, infill drilling.

In the first four rounds of NELP (New Exploration Licensing Policy), 90 blocks covering 9.0 lakh sq km have been signed and total area under exploration has grow up to 12,40,000 sq km. Today as much 74% of area under exploration belongs to the NELP blocks. The exploration investments in three phases are about ₹ 19,050 crore, which is expected to substantially increase in case of discoveries of hydrocarbon [Annual Report, 2012]. Exploration under NELP has shown positive results with discoveries made in Krishna-Godavari deep water and in Cambay on land, including significant gas discovery in Krishna-Godavari basin. Under the Coalbed methane policy [www.indianstocks.com], eight blocks have been awarded in the states of Jharkhand, Madhya Pradesh and West Bengal for exploration and production of coalbed methane, which is an environment friendly, non-conventional source of gas. Profile of Indian Petroleum Industry is shown in Table 2.2 [Annual Report, 2011–12]. Coalbed methane which is an environment friendly clean fuel is similar to conventional natural gas. To give impetuous to its exploration and production, Government of India has formulated a CBM policy. Contracts with PSUs/private sector for thirteen blocks on nomination basis have been signed. Measures have been taken by the government to intensify exploration and enhance hydrocarbon reserves. This includes: Developments of new fields, additional development of existing fields, implementation of enhanced oil recovery schemes, resource to specialized technology, enlisting the services of international experts, encouraging participation of private and joint venture companies in the exploration program and activation of the NELP [www.gasandoil.com]. Today ONGC Reliance,

Essar and Videocon are looking for oil along with multinationals like Premier Oil, Cairn Energy and Shell, Stumberger [www.franchiseindia.com]. With the participation of private sector, oil and gas exploration and drilling has changed completely. Import and export of petroleum is mentioned in Table 2.3.

Table 2.2   Profile of Indian petroleum industry

Parameters	Unit	2002–03	2005–06	2006–07	2007–08	2008–09	2009–10	2010–11
Production (onshore and offshore)								
Crude oil	'000' tons	33044	32190	33988	34118	33508	33690	37712
Natural gas	Million cubic meters	31389	32202	31747	32417	32845	47496	62222
Production of petroleum products								
Light distillates	'000' tons	28619	32427	38104	40111	40222	51197	55197
Middle distillates	"	55937	64432	71225	76649	80309	93790	99776
Heavy ends	"	19584	22891	25931	28170	29985	34782	35391
Total		104140	119750	135260	144930	150516	179769	190364
Consumption of petroleum products								
Light distillates	'000' tons	31755	33662	37076	38557	39878	39086	41433
Middle distillates	"	52065	54423	57595	62823	66378	71198	74949
Heavy ends	"	20306	25129	26078	27568	27343	27911	25402
Total		**104126**	**113213**	**120749**	**128948**	**133599**	**138195**	**141784**

Table 2.3   Import and export of petroleum on reserves

Parameters		2002–03	2003–04	2006–07	2007–08	2008–09	2009–10	2010–11
Import of crude oil and petroleum products								
Total import	'000' tons	89217	98435	129162	144133	151300	173922	180958
Export of petroleum products								
Total export	'000' tons	10289	14620	33624	40779	38902	50974	59133

Contd.

Table 2.3   Import and export of petroleum on reserves *(Contd.)*

Parameters		2002–03	2003–04	2006–07	2007–08	2008–09	2009–10	2010–11
Reserves								
Crude oil	Million tons	741	733	725	770	773	775	758
Natural gas	Billion cubic meter	751	854	1055	1090	1115	1149	1242

Source: Petroleum Planning and Analysis Cell, New Delhi, Annual Report, 2007–08, 2011–12.

The refining capacity has been gradually increased over the years by setting up of new refineries in the country as well as by expanding the capacity of the existing refineries in order to meet the growing demand of petroleum products [www.petrosilicon.com]. During the past five decades, there has been significant increase in crude production from 33.69 to 37.71 million tons from 2009–10 to 2010–11 and in refining capacity from 183.386–187.386 million tons per annum from 1/4/2010–1/4/2011. Real breakthrough in the petroleum refining industry has been in the setting up of the Reliance Refinery Project at Jamnagar with one of the largest grassroots refineries in the world with more than 27 million tons refining capacity. IOC refineries, BPCL, HPCL have already gone for expansion of their existing plants.

The demand of crude oil is expected to touch 235 million tons and that of gas up to 326.14 mmscmd by the year 2012 as per 11th plan [CEW, 2009]. In the optimistic scenario the oil production is expected to increase steadily to 79 million tons per annum by 2031–32. The gas production is also expected to increase to 46 bcm by 2031–32. Structure of petroleum refining industry is described in Fig. 2.1.

2.2 STRUCTURE

There are 21 refineries (seventeen in public sectors, one in joint sector and three in private sector) in the country with a capacity of about 193.386 million tons. Domestic production of crude oil is 34 million tons. A new grassroots refinery and the expansion of existing refineries will enhance the refining capacity to about 234.066 million tons (excluding export oriented units by the end of 11th plan). **India Natural Gas Supply and Demand Imbalance** is given in Fig. 2.2 [Hydrocarbon Vision, 2020].

At industry level, the major players are:
   a. Exploration: ONGC, OIL, RIL/JVs
   b. Refining: IOCL, BPCL, HPCL, ONGC, RIL
   c. Crude pipelines: IOCL, ONGC, OIL
   d. Product pipelines: IOCL, BPCL, HPCL and Petronet
   e. Gas pipelines: GAIL, ONGC
   f. New projects: IOCL, RIL, HPCL, BPCL, ONGC, ESSAR.

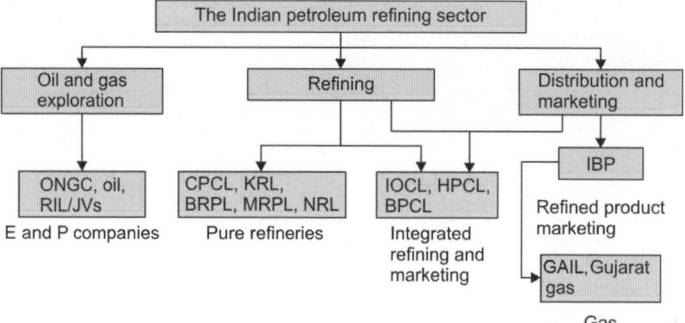

Fig. 2.1 Structure of Indian petroleum industry

Source: Reliance, 2002/2003.

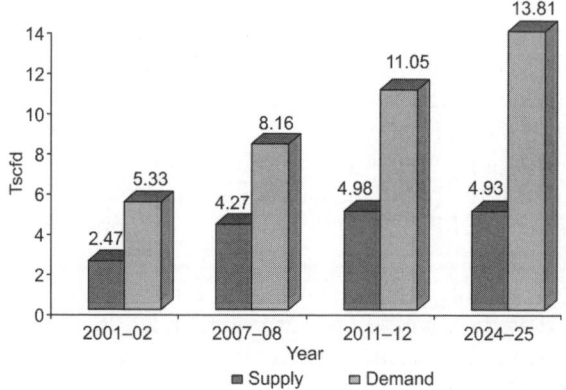

Fig. 2.2 India natural gas supply and demand imbalance

Source: Hydrocarbon Vision, 2020.

2.3   PETROLEUM REFINING

Primary commercial energy consumption in India is given in Table 2.4. Installed capacity and refinery crude throughput is given in Fig. 2.3 and Table 2.5. Oil and gas reserves in India are given in Table 2.6. Number of wells and metreage drilled is given in Table 2.7. Exploratory and development drilling in India is given in Table 2.8. Industrywise offtakes of natural gas in India is given in Table 2.9. CNG activities in India are given in Table 2.10. Existing petroleum product pipelines in the country are given in Table 2.11. Crude oil and natural gas reserves in India are given in Table 2.12. Production of crude oil and natural gas in India is given in Table 2.13. Production of petroleum products in India is given in Table 2.14. Consumption of petroleum products in India is given in Table 2.15. Import and exports of crude oil and petroleum products in India is given in Table 2.16. Proposed refineries in India are given in Table 2.17.

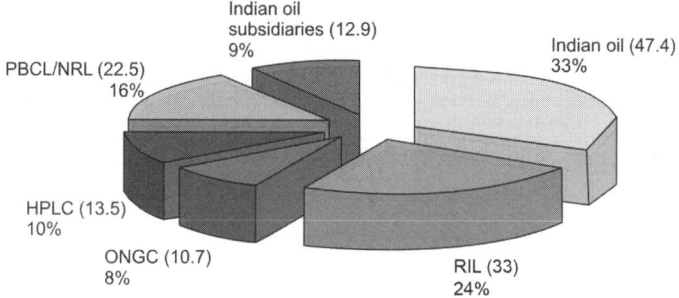

Fig. 2.3 Current Indian refining capacity
(as on 1.4.2006) figures in brackets: Refinery capacity in MMTP

Table 2.4 Primary commercial energy consumption in India

Source	Unit	1990–91	2004–05	2005–06	2006–07	2007–08	2008–09	2009–10*
1	2	3	4	7	8	9	8	9
1. Pol. products (incl. RBF)	MMT	57.75	120.17	119.76	135.26	144.93	150.52	179.77
2. Natural gas (net)	BCM	12.77	30.78	32.20	31.75	32.417	32.845	47.50
3. Coal	MMT	211.73	382.61	407.04	430.83	457.08	492.76	532.06
4. Lignite	MMT	13.77	30.34	30.06	31.29	33.98	32.42	34.07
5. Electricity (incl. non-utilities)	Bn. KWH	289.40	665.80	698.40	752.50	813.10	840.80	909.50

* Provisional. MMT: Million metric tons; BCM: Billion cubic meters.
Note: Data reflects despatches of coal / lignite (incl. stock differential).
Source: Ministry of Finance/ Economic Survey. Annual report, 2007–2008, 2011–2012, Ministry of Petroleum and Natural Gas, India.

Table 2.5 Installed capacity and refinery crude throughput

('000' tons)

	Installed capacity as on						
	1/4/2012	1/4/2013	2005–06	2007–08	2009–10	2011–12	2012–13
a. Public sector	120066	120066	96946	112541	112117	120895	120303
IOC, Guwahati, Assam	1000	1000	864	920	1078	1058	956
IOC, Barauni, Bihar	6000	6000	5553	5634	6184	5730	6344
IOC, Koyali, Gujarat	13700	13700	11543	13714	13206	14253	13155

Contd.

Table 2.5 Installed capacity and refinery crude throughput *(Contd.)*

('000' tons)

	Installed capacity as on		2005–06	2007–08	2009–10	2011–12	2012–13
	1/4/2012	1/4/2013	2005–06	2007–08	2009–10	2011–12	2012–13
b. Public sector	**120066**	**120066**	**96946**	**112541**	**112117**	**120895**	**120303**
IOC, Haldia, West Bengal	7500	7500	5502	5715	5686	8072	7490
IOC, Mathura, Uttar Pradesh	8000	8000	7938	8033	8107	8202	8561
IOC, Digboi, Assam	650	650	615	564	600	622	660
IOC, Panipat, Haryana	15000	15000	6507	12821	136115	15496	15126
IOC, Bongaigaon, Assam	2350	2350	2356	2020	2220	2188	2356
Total IOCL	**54200**	**54200**	**40878**	**49421**	**50696**	**55621**	**54649**
BPCL, Mumbai, Maharashtra	12000	12000	10298	12746	12516	13355	13077
BPCL, Kochi, Kerala	9500	9500	6939	8134	7875	9472	10105
Total BPCL	**21500**	**21500**	**17237**	**20880**	**20391**	**22828**	**23183**
HPCL, Mumbai, Maharashtra	6500	6500	6249	7409	6965	7506	7748
HPCL, Visakh, Andhra Pradesh	8300	8300	7980	9409	8796	8682	8028
Total HPCL	**14800**	**14800**	**14229**	**16818**	**15761**	**16189**	**15777**
CPCL, Manali, Tamil Nadu	10500	10500	9680	9802	9580	9953	9105
CPCL, Narimanam, Tamil Nadu	1000	1000	682	464	517	611	640
Total CPCL	**11500**	**11500**	**10362**	**10266**	**10097**	**10565**	**9745**
NRL, Numaligarh, Assam	3000	3000	2133	2568	2619	2825	2478
ONGC, Tatipaka, Andhra pradesh	66	66	93	63	55	69	57
MRPL, Mangalore, Karnataka	15000	15000	12014	12525	12498	12798	14415

Contd.

Table 2.5 Installed capacity and refinery crude throughput *(Contd.)*

('000' tons)

	Installed capacity as on		2005–06	2007–08	2009–10	2011–12	2012–13
	1/4/2012	1/4/2013					
c. Private sector	78000	80000	33163	43562	80651	81179	88273
RPL, Jamnagar, Gujarat	33000	33000	343090	36616	34415	32497	32613
RPL (SEZ), Jamnagar, Gujarat	27000	27000	–	–	32735	35186	35892
Essar Oil Ltd, Vadinar	18000	20000	–	6631	13501	13496	19769
Total (a + b)	198066	200066	130109	156103	192768	202074	208576

Notes: 1. CPCL and BRPL are subsidiaries of IOC.
    2. KRL and NRL are subsidiaries of BPCL.
    3. MRPL is subsidiary of ONGC.
Source: Annual Report, 2007–2008, 2011–2012, 2012–2013, Ministry of Petroleum and Natural Gas, India.

Table 2.6 Oil and gas reserves in India

AREA	1990	2005	2006	2007	2008	2009	2010	2011
1	2	3	4	5	6	7	8	9
Crude oil (Million metric tons)								
Onshore	307	376	387	357	404	406	403	403
Offshore	432	410	369	368	367	368	371	354
Total	739	786	756	725	770	773	775	757
Natural gas (Billion cubic meters)								
Onshore	229	340	330	270	304	328	334	394
Offshore	457	761	745	785	786	788	815	824
Total	686	1101	1075	1055	1090	1115	1149	1241

Note: The oil and natural gas reserves (proved and indicated) data relate to 1st January for the year 1990 and thereafter 1st April of each year.
Source: ONGC, OIL and DGH [Annual Annual Report, 2007–2008, 2011–2012, Ministry of Petroleum and Natural Gas, India].

Table 2.7 Number of wells and metreage drilled

Area	2005–06		2006–07		2007–08		2008–09		2009–10	
	Wells	Metre-age	Wells	Metre-age	Wells	Metre-age	Wells	Metre-age	Wells	Metre-age
	(No.)	(000)	(No.)	(000)	(No.)	(000)	(No.)	(000)	(No.)	(000)
1	2	3	4	5	6	7	8	9	10	11
a. Onshore										
Assam	58	225	48	190	67	227	80	206	63	259
Gujarat	124	215	128	223	156	255	167	296	199	335
Others	45	119	55	128	48	132	53	14	64	173
Total (a)	227	559	231	541	271	614	300	636	346	768
b. Offshore										
Bombay high	51	171	66	201	88	258	81	252	82	251
Others	23	71								
Total (b)	74	242	66	201	88	258	81	252	82	251
Grand total (a + b)	301	801	297	742	359	872	381	888	428	1019

* Provisional   **Source:** Public Sector Undertakings.

Source: Annual Report, 2007–2008, 2011–12, Ministry of Petroleum and Natural Gas, India.

Table 2.8 Exploratory and development drilling in India

Area	2005–06		2006–07		2007–08		2008–09		2009–10	
	Wells	Metre-age	Wells	Metre-age	Wells	Metre-age	Wells	Metre-age	Wells	Metre-age
	(No.)	(000)	(No.)	(000)	(No.)	(000)	(No.)	(000)	(No.)	(000)
1	2	3	4	5	6	7	8	9	10	11
a. Exploratory										
Onshore	82	236	74	187	79	209	90	243	110	298
Offshore	32	106	22	76	27	84	32	97	34	128
Total (a)	114	342	96	263	106	293	122	340	144	426
b. Development										
Onshore	145	322	157	354	192	405	210	393	236	470
Offshore	42	137	44	125	61	174	49	155	48	123
Total (b)	187	259	201	479	251	579	259	548	284	593
Grand total (a + b)	301	801	297	742	359	872	381	888	428	1019

Source: Annual Report, 2011–2012, Ministry of Petroleum and Natural Gas, India.

Table 2.9 Industrywise offtakes of natural gas in India

(Million cubic meter)

Industry	2005–06	2007–08	2008–09	2009–10	2010–11	2011–12	2012–13
1	2	3	4	5	6	7	8
a. Energy purposes:							
1. Power generation	11878	12037	12603	21365	23583	18912	12849
2. Industrial fuel	3780	3323	5912	2322	999	1127	1139
3. Tea plantation	151	160	154	167	193	175	182
4. Domestic fuel	75	38	102	246	1584	1913	1996
5. Captive use/LPG	5048	1804	1885	5433	5770	6343	5961
Shrinkage							
6. Others	1120	1324	1535	1838	6551	5759	3224
Total (a)	22052	18686	22191	31371	38680	34229	25311
b. Non-energy purposes:							
1. Fertilizer industry	7762	9823	9082	13168	10444	10406	10702
2. Petrochemicals	1175	1432	1105	1264	470	576	437
4. Others**	36	638	611	703	1424	1508	1950
Total (b)	8973	11893	10798	15135	12338	12490	13089
Grand total (a + b)	31025	30579	32989	46506	51018	46719	38400

** Excludes offtakes of natural gas by ONGC.
Note: Excludes gas supplied to IGL, MGL, Bhagyanagar Gas, TNGCL, BMC and GGCL and sponge iron use.
Source: ONGC, OIL, DGH and GAIL.

Table 2.10 CNG activities in India as on 1/4/2006

Item	Delhi	Maharashtra			Gujarat				Grand
	Delhi	#	Pune	Panval	Vadodara	Surat	Ankleshwar	Total	total
	IGL	MGL	MNGL	GAIL	GAIL	GGCL	GGCL		
1	2	3	4	5	6	7	8	7	8
Station type									
Mother	99	12	3	0	2	3	1	6	120
Online	65	127	–	0	1	27	4	32	224
Daughter	47	10	11	1	2	4	2	8	77
Booster									
Daughter	2	0	0	0	0	0	1	1	3
Total	213	149	14	1	5	34	8	47	424

Contd.

Table 2.10 CNG activities in India as on 1/4/2006 *(Contd.)*

Item	Delhi	Maharashtra			Gujarat				Grand
	Delhi	#	Pune	Panval	Vadodara	Surat	Ankleshwar	Total	Total
	IGL	MGL	MNGL	GAIL	GAIL	GGCL	GGCL		
1	2	3	4	5	6	7	8	7	8
No. of vehicles									
Cars$	281802	66552	1399	38	984	70502	8465	79951	429742
Autos	121854	140533	12991	163	2327	50448	10630	63405	338966
LCV/ RTVs	5389	2043	–	0	0	2994	146	3140	10552
Buses*	16655	3720	115	2	72	584	62	718	21210
Others/ Phatphat Sewa	5681	1188	–	0	0	0	0	0	6869
Total	**431381**	**214036**	**14505**	**203**	**3383**	**124528**	**19303**	**147214**	**807339**
Price & (₹/kg)	29.00	31.84	37.5	41.0	32.05	32.25	35.25		
Average consumption									
TPD	1706.06	940.37	35.70	0.44	21.89	216.23	34.92		
MMSCMD	2.24	1.31	0.05	0	0.02	0.31	0.05		
MMSCMD: Million standard cubic meters per day					TPD: Tons per day				

#: Mumbai/Thane/Mira Bhayander. NA: Not Available.
$: Includes taxies.
*: Buses are registered at Gandhinagar, but taking CNG at Vadodara CNG Station.
&: Price at Daughter Booster Station / Mother Station.

Sources:	Conversion factors:
1. IGL: Indraprastha Gas Ltd.	1 kg = 1.313 SCM of gas
2. MGL: Mahanagar Gas Ltd.	1 kg = 1.41 SCM of gas
3. GAIL: GAIL (India) Ltd.	1 kg = 1.32 SCM of gas
4. GGCL: Gujarat Gas Company Ltd.	

Note: Earlier, the Daughter Station was being fed from Mother Compressor at Vaghodia. However, an online station was commissioned for supply of CNG at GSFC, Vadodara in the month of November, 2005 and it has been upgraded to mother cum online station from 1st April, 2006.

Sources: Annual Report, 2007–2008, 2011–12, Ministry of Petroleum and Natural Gas, India.

Table 2.11 Existing petroleum product pipelines in the country

Oil Co.	Name of the pipeline	Capacity* (MMTPA) 1/4/2011	Length (km) 1/4/2011	Throughput ('000' tons)			
				2006–07	2007–08	2008–09	2009–10
1	2	3	4	5	6	7	8
Product pipelines							
IOC	Barauni-Patna-Kanpur	5.30	745	4050	3870	3580	4000
IOC	Guwahati-Siliguri	1.40	435	1230	1310	1510	1800
IOC	Haldia-Barauni	1.30	525	960	940	820	
IOC	Haldia-Mourigram-Rajbandh	1.40	277	1900	1580	1290	2500
IOC	Koyali-Ahmedabad	1.10	116	540	610	540	500
IOC	Mathura-Jalandhar	3.50	434				2600
IOC	Panipat-Delhi (Somipat Meerut Branchline)		182				700
IOC	Mathura-Delhi	3.70	258				2500
IOC	Panipat-Bhatinda	1.50	219	1290	1440	1510	1500
IOC	Mathura-Tundla	1.20	77	250	240	280	300
IOC	Koyali-Viramgam-Sidhpur-Sanganer	4.10	1056	3010	3180	3240	3000
IOC	Panipat-Rewari	1.50	155	1240	1380	1420	1500
IOC	Chennai-Trichi-Madurai	1.80	683	820	1370	1690	1900
IOC	Koyali-Dahej		103	180	900	800	1000
IOC	Chennai ATF	0.2	95		0	30	200
IOC	Chennai-Bangalore	1.50	290				0
HPCL	Mumbai-Pune	3.70	508	3240	3360	3370	3400
HPCL	Vizag-Vijayawada-Secunderabad	5.40	572	3500	3850	3880	3800
HPCL	Mundra-Delhi	3.80	1054		620	3390	4800
BPCL	Mumbai-Manmad-Mangliya-Piyala-Bijwasan	11.80	1389	3790	8010	10490	11200
OIL	Numaligarh-Siliguri	1.70	654				800
PCCK	Kochi-Coimbatore-Karur	3.30	293	–	1360	1560	1500

Contd.

Table 2.11 Existing petroleum product pipelines in the country *(Contd.)*

Oil Co.	Name of the pipeline	Capacity*	Length	Throughput ('000' tons)			
		(MMTPA)	(km)	2006–07	2007–08	2008–09	2009–10
		1/4/2011	1/4/2011				
1	2	3	4	5	6	7	8
Product pipelines							
PMHB	Mangalore-Hasan-Bangalore	2.10	362	1440	2141	2450	2500
Subtotal		66.8	10747	27440	36161	41850	52100
LPG pipelines							
IOC	Panipat-Jalandhar	0.70	274			600	400
GAIL	Jamnagar-Loni	2.50	1250	2030	2232	2090	2400
GAIL	Vizag-Vijayawada-Secunderabad	0.7	300	460	522	650	700
Subtotal		3.90	1824	2490	2754	3340	3500
Crude pipelines							
OIL	Duliajan-Bongaigaon-Barauni	8.40	1193	6000	6070	6220	6200
IOC	Salaya-Mathura-Panipat	21.00	1870	21910	21680	22650	22400
IOC	Haldia-Barauni/Pradip barauni	11.00	1302	7060	7240	8680	11600
IOC	Mundra-Panipat	8.40	1194	3640	6940	6910	7400
ONGC	Mumbai High-Uran-Trunk Pipeline	15.63	204				8600
ONGC	Heera-Uran-Trunk Pipeline	11.50	81				5530
ONGC	Kalol-Nawagam-Koyali	8.54	141				5170
ONGC	MHN-NGM Trunk line	2.26	77				2180
ONGC	CTF, Ank to Koyali Oil Pipeline (AKCL)	2.00	98				2090
ONGC	Lakwa-Moran oil line	1.50	18				430
ONGC	Geleki-Jorhat oil line	1.50	48				650
ONGC	NRM to CPCL	0.74	4				240
ONGC	KSP-WGGS to TPK Refinery	0.08	14				30

Contd.

Table 2.11 Existing petroleum product pipelines in the country *(Contd.)*

Oil Co.	Name of the pipeline	Capacity*	Length	Throughput ('000' tons)			
		(MMTPA)	(km)	2006–07	2007–08	2008–09	2009–10
		1/4/2011	1/4/2011				
1	2	3	4	5	6	7	8
Crude pipelines							
ONGC	GMAA EPT to S. Yanam Unloading Terminal	0.15	4				120
IOC	Haldia-Barauni	11.00	1302	7060	7240	8680	11600
Subtotal		**92.7**		**6250**	**38430**	**41930**	**44460**	**72640**
Grand Total		**163.4**		**18821**	**68360**	**80845**	**89650**	**128240**

* Includes Kurukshetra-Roorkee-Nazibabad branch lines.

Source: Petroleum Planning and Analysis Cell [Annual Report, 2007–2008, 2011–2012, Ministry of Petroleum and Natural Gas, India].

Table 2.12 Crude oil and natural gas reserves in India

Parameters	Unit	2005–06	2006–07	2007–08	2008–09	2009–10	2010–11	2011–12	2012–13
Reserves									
1. Crude oil	Million tons	756	725	770	773	775	757	760	758
2. Natural gas	Billion cubic meter	1075	1055	1090	1115	1149	1278	1330	1355

Source: Annual Report, 2012–2013, Ministry of Petroleum and Natural Gas, India.

Table 2.13 Production of crude oil and natural gas in India

Parameters	Unit	2005–06	2006–07	2007–08	2008–09	2009–10	2010–11	2011–12	2012–13
Production									
1. Crude oil	Million tons	32.19	33.99	34.12	33.51	33.69	37.68	38.09	37.86
2. Natural gas	Billion cubic meter	32.20	31.75	32.42	32.85	47.50	52.22	47.56	40.68

Source: Annual Report, 2012–2013, Ministry of Petroleum and Natural Gas, India.

Table 2.14 Production of petroleum products in India

Parameters	Unit	1990–91	2001–02	2002–03	2005–06	2006–07	2007–08	2008–09	2009–10
Production of petroleum products	'000' tons								
1. Light distillates		10,023	26539	28619	32427	38104	40111	40222	51197
2. Middle distillates		26,344	54409	55937	64432	71225	76649	80309	93790
3. Heavy ends		12,195	19056	19584	22891	25931	28170	29985	34782
Total		48,562	100004	104140	119750	135260	144930	150516	179769

Source: Annual Report, 2006–07, 2011–12.

Table 2.15 Consumption of petroleum products in India

Parameters	Unit	1990–91	2001–02	2002–03	2005–06	2006–07	2007–08	2008–09	2009–10
Consumption of petroleum products	'000' tons								
1. Light distillates		9,801	29618	31755	33662	37076	38556	39878	39086
2. Middle distillates		33,106	51439	52065	54423	57595	62823	66378	71198
3. Heavy ends		12,128	19375	20306	25129	26078	27567	27343	27912
Total		55,035	100432	104126	113213	120749	128946	133599	138196

Source: Annual Report, 2006–07; Annual Report, 2011–2012; Ministry of Petroleum and Natural Gas, India.

Table 2.16 Import and exports of crude oil and petroleum products in India

Parameters	Unit	2000–01	2005–06	2006–07	2007–08	2008–09	2009–10
Total import of crude oil	'000' tons	74100	99409	115502	121672	132775	159259
	₹ crore	65932	171702	219029	272669	348149	375378
Total import of petroleum	'000' tons	9270	13440	17660	22461	18524	14662
Products	₹ crore	12093	27972	41160	61000	60846	33754
Total export of petroleum	'000' tons	8370	23461	33624	40779	38902	50974
Products	₹ crore	7672	49974	81094	110789	122066	144037

Source: Annual Report, 2007–08; Annual Reports, 2011–2012; Ministry of Petroleum and Natural Gas, India.

Table 2.17 Proposed refinery projects

Name	Capacity, million tons	Location
Bina Refienry	6	Bina (Madhya Pradesh)
Eastern India Refinery	6	Abhaychandrapur (Odisha)
Bhatinda Refinery	9	Bhantida (Punjab)
Paradip Refinery	15	Paradip (Odisha)
Cuddalore Refinery	6	Cuddalore (Tamil Nadu)
Essar Oil Refinery	12	Jamnagar (Gujarat)
Hinduja Refinery	2 (lube refinery)	Haridaspur (Odisha)

2.4 FACTORS INFLUENCING PETROLEUM REFINING INDUSTRY

Factors influencing petroleum refining industry are crude oil prices (worldwide oil prices reached a 5 year high in 1996 are now declining), growing demand of petroleum products (developing Asia, Asia Pacific), decline in domestic crude oil production (USA, India, etc.), gradually increasing heavier feedstock, investment in billions of dollars for environmental protection, low profit margins. Major challenges which Indian refineries are likely to face are: Increasing products demand, demand for better products quality and new environmental regulations, decreasing quality of feedstocks, increasing 'S' (> 1.5%), decreasing API gravity (< 0.8) (heavier crude), increasing level of microconstituents, value addition.

Key issues facing Indian Refineries are low margins and unacceptable profitability, requirements posed by changes in product specifications resulting from environmental consideration; focus on improving bottom-line of existing assets. Expected trends in refining in India are higher conversion, better yield and improved efficiency to generate more value from each barrel of crude oil, reforming of naphtha, process to increase LPG production, conversion process like hydrocracker to convert heavy ends into middle distillates, residue upgradation, hydrodesulfurization, improvements in lube oil production, hydrogenation process to improve stability, an alternative additive for octane improvement which is as effective as TEL, fuel additives to improve stability and undue deposit forming tendencies.

Future vision of Indian refining industry is environmentally sound, energy efficient with lower energy intensity, safe and easy to operate, operation with minimum inventory, sustainable, valuable and profitable development, complete synergy between refineries and product consumers.

2.5 CONSUMPTION OF THE PETROLEUM PRODUCTS

Petroleum products are consumed in two ways: For energy production (85–90%) and for non-energy production (10–15%). Energy products include: LPG, gasoline, ATF/SKO, high speed diesel oil, low speed diesel oil and fuel oil/LSHS. Non-energy products include:

- Naphtha: Fertilizer / petrochemical feedstock
- Lubes: Engines and machines
- Waxes: Food packaging and others
- Bitumens: Highways, airfields
- Petroleum coke: Metallurgical industry
- Solvents and others.

The changing scenario of oil consumption is affecting the demand projections of the distillates. Demand projection and percentage growth of total petroleum products are given in Table 2.18. Infact, this distribution is increasingly turning towards motor gasoline, LPG, kerosene, diesel and towards petrochemical feedstock at the expense of heavy ends. Demand projection distillatewise is given in Table 2.19.

Table 2.18 Demand projection of total petroleum products (million tons)

Year	Demand projections	Growth (%)
2000–01	92.71	5.8
2005–06	118.38	5.0
2010–11	148.99	4.7
2015–16	189.22	4.9
2020–21	242.30	5.1

Table 2.19 Demand projection—distillatewise ('000' tons)

Year	Light distillate	Middle distillate	Heavy distillate	Total ends
2000–01	20,379	55,231	17,102	92,712
2005–06	27,803	70,245	20,330	1,18,378
2010–11	35,147	89,728	24,118	1,48,993
2015–16	45,262	1,15,171	28,788	1,89,221
2020–21	59,246	1,48,459	34,593	2,42,298

Distillatewise consumption pattern of petroleum products by sector is given in Table 2.20. Supply demand scenario of petroleum products is given in Table 2.21.

Table 2.20 Distillatewise pattern of consumption ('000' tons)

Year	Light distillate	Middle distillate	Heavy distillate	Total ends
2000–01	21.98	59.57	18.45	100.0
2005–06	23.49	59.34	17.17	100.0
2010–11	23.59	60.22	16.19	100.0
2015–16	23.92	60.96	15.12	100.0
2020–21	24.45	61.27	14.28	100.0

Table 2.21 Supply demand—petroleum products (million tons)

Year	Demand without meeting gas deficit	Demand with meeting gas deficit	Estimated refining capacity	Estimated crude requirement
1998–1999	91	103	69	69
2001–2002	11	138	129	122
2006–2007	148	179*	167	173
2011–2012	195	95**	184	190
2024–2025	368	368	358	364

*Assuming 15 MTPA of LNG import by 2007.

**Assuming that by 2012, adequate gas is available through imports and domestic sources.

Source: Report of the subgroup on development of refining, marketing, transportation and infrastructure requirements (1999).

2.6 HYDROCARBON VISION 2020

The gap between supply and availability of crude oil, petroleum products as well as gas from indigenous sources is projected to increase requiring increasing emphasis on exploration and production sector. The expert team setup by the Government of India to work on "Indian Hydrocarbon Vision, 2020" has estimated the availability of around 336 million tons of liquid petroleum products two decades from now, while assuming a demand of 300 million tons for liquid products. A significant part of the capacity will come from a single refinery of Reliance Petroleum Ltd (50 million tons), two green field projects of 15 million tons each at Deogarh/Ratnagiri and Krishnapatnam and projects of Indian Oil Corporation Ltd (133.74 million tons), Bharat Petroleum Corporation Ltd (37 million tons), Hindustan Petroleum Corporation Ltd (34.2 million tons) and Essar Oil with 18 million tons. In order to meet the requirement of crude oil and product supply various new pipelines projects have been projected by expert team on "Indian Hydrocarbon Vision, 2020".

Table 2.22 highlights the projects that will add to additional refining capacity in coming years. With all this and the ongoing expansion projects completed, India would have accomplished a refining capacity of 250 MMTPA gradually by the end of 11th Five-Year Plan (2007–12).

Table 2.22 Projects to add to refining capacity (MMTPA)

Company	Location	Capacity
IOCL	Paradip Refinery (new)	15.0
IOCL	Panipat Refinery (expansion)	3.0
BPCL	Bina Refinery (new)	6.0
BPCL	Allahabad Refinery (new)	7.0
HPCL	Bhatinda (Punjab) Refinery (new)	9.0

Contd.

Table 2.22 Projects to add to refining capacity (MMTPA) *(Contd.)*

Company	Location	Capacity
ONGC/MRPL	Mangalore Refinery (expansion)	15.0
ONGC	Balmer (Rajasthan Refinery (new)	7.5
BPCL/KRL	Kochi Refinery (expansion)	3.0
	Kakkinada Refinery (new)	5.0
RIL	Jamnagar 2nd Refinery (new)	30.0
ESSAR	Vadinar Refinery (new)	14.0
Total		**114.5**

2.7 EMERGING INDIAN SCENARIO OF PRODUCT QUALITY

Emerging Indian scenario of product quality is:

- Increased production of motor gasoline
- Demand for better quality products (fuels and lubricants) meeting new environmental regulations
- Lube base stocks (high VI, high oxidation stability, low pour point, low volatility)
- Fuel oil (low sulfur)
- Hydroprocessing (HC-fuel and lube, MHC, HDT, etc.)
- Reformulated fuels, e.g. RFG and synthetic lubes
- Additives
- Redefinition of catalytic reforming technology
- Selective dehydrogenation
- Hydroisomerization
- Hydrocatalytic isodewaxing (diesel/lubes)
- Alkylation
- Development of environment friendly solid acid catalysts
- More hydroprocessing units to consume heavier feedstock for production of value product having better quality stability and low sulfur
- Extensive R & D efforts towards technology adsorption, adaptation, development/improvement.

Comparative consumption of key petroleum products (current pattern, %wt) is given in Table 22.3.

Table 2.23 Comparative consumption of key petroleum products (current pattern, %wt)

Components	World	USA	India
LPG	9.4	11.2	10.2
Gasoline	25.7	43.8	17.6 (A)

Contd.

Table 2.23 Comparative consumption of key petroleum products (current pattern, %wt) *(Contd.)*

Components	World	USA	India
Jet fuel	5.8	8.4	2.9
Kerosene	2.4	0.4	9.5
Gas oil	27.3	19.6	39.9
Fuel oil	13.3	4.1	12.0 (B)
Others	16.1	12.5	7.9
Total	**100**	**100**	**100**

Note: (A) Includes LSHS.

Chronology of diesel specification developed in India is 1460 given in Table 2.24. Typical process units and HSD blending components in Indian refineries is given in Table 2.25. Indian and European diesel key specifications are given in Table 2.26. Comparison of Indian and European gasoline key specifications is given in Table 2.27.

Table 2.24 Chronology of diesel specification developed in India (IS 1460)

Parameter	1974	1980	1995	2000	2005 (EUIII)
Cetane number (min.)	42	42	45	48	51
Distillation (°C max.)					
85% (Vol.)	–	–	–	350	–
90% (Vol.)	366	366	366	–	–
95% (Vol.)	–	–	–	370	360
Sulfur (%wt max.)	1.0	1.0	1.0 (0.25)	0.25 (0.05)	0.05 (0.035)
PAH (%vol max.)	–	–	–	–	11
Total sediments (mg/100 ml)	1.0 [1]	1.0 [1]	1.6 [2]	1.6 [2]	2.5 [3]

[1]: DEF 2000–16; [2]: UOP 413; [3]: ASTM D-2274.

Table 2.25 Typical process units and HSD blending components in Indian refineries

S. No.	Process unit	HSD component	Remarks
1	FCC/RFCC	TCO	Low cetane
2	Crude	SRGO	Crude specific
3	HCU	Gas oil	High cetane, low sulfur
4	DCU, VBU	Kero/gas oil	Low cetane
5	DCU, VBU	Cracked naphtha	Low RON, high olefins, diolefins, high sulfur

Table 2.26 Indian and European diesel key specifications

Property	Indian			European	
	1996 IS-1460	2000	2005	ECE 1993	ECE 2000
Density (kg/m²)	820–880	820–860	820–845	820–860	820–845
Cetane number (min.)	45	45–48	50–51	49	51
Sulfur, [%wt (max.)]	1.0	0.25/0.05	0.05	0.05	0.035
Polyaromatics (% max.)	NS		10.00		11.0
T95 (°C max.)	360 (T90)		360	370	360

NS: Not specified.

Table 2.27 Comparison of Indian and European gasoline key specifications

Property	Indian			European	
	1996 IS-2796	2000 IS-2796	2005 Proposed	ECE 1993	ECE 2000
Density, kg/m²	–	700–750		725–780	
Distillation					
E-70 (%vol)	10–40	10–40			
E-100 (%vol)	40–70	40–70	45–70	40	46
Octane number	87/NS	–	89/79	95/85	
RON/MON	(82 AKI)	(84 AKI)	91/81 and 93/83		
Aromatics [%V/V (max.)]	NS	NS	40		45
Olefins [%V/V (max.)]	NS	NS	25		18
Benzene [%V/V (max.)]	NS	NS	1.0		2.0
RVP, KPA	35–70	35–60	–	35–70	
Sulfur [%wt (max.)]	0.2	0.05	0.03	0.05	0.02

NS: Not specified, AKI: Anti-knock index.

2.8   BIOFUEL IN INDIA

With rapid growth of transport sector in India there has been continuous increase in oil import which is about 70% of the total refining capacity. Although India is one of the major sugar producing countries in the world, however, due to constraint of supply from sugar industry, the Government of India had decided to supply 5% ethanol-blended petrol (E-5) in some states. So far biodiesel is concerned, unlike countries like Malaysia, Thailand, use of edible oil is not option for biodiesel production and main focus in India has been on the use of non-edible seeds. There is scope for cultivation of non-edible oil seeds especially Jatropha. Planning Commission has proposed a National Mission on biodiesel and Jatropha curcas, which includes large scale plantation, collection of seeds and setting up of plants for producing biodiesel. Projected demand for petrol, diesel and biofuel requirements in India is given in Tables 2.28 and 2.29.

Table 2.28 Petrol demand scenario

Year	Petrol demand (million tons)	Ethanol blending requirement (in metric tons)		
		@ 5%	@ 10%	@ 20%
2006–2007	10.07	0.50	1.01	2.01
2011–2012	12.85	0.64	1.29	2.57
2016–2017	16.40	0.82	1.64	3.28

Source: Planning Commission, 2003.

Table 2.29 Biodiesel demand scenario

Year	Diesel demand (million tons)	Biodiesel blending requirement (in metric tons)		
		@ 5%	@ 10%	@ 20%
2006–2007	52.32	2.62	5.23	10.46
2011–2012	66.91	3.35	6.69	13.38
2016–2017	83.58	4.18	8.36	16.72

Source: Planning Commission, 2003.

REFERENCES

1. Annual Report, 2007–2008, 2011–2012, Ministry of Petroleum and Natural Gas, India.
2. www.indianstocks.com [CEW, Dec 2002, p108].
3. CEW, 2009. Outlook for Indian Oil and Gas Sector. Chemical Engineering World, 24, Jan 2009.
4. IEP Report, p39–41, Table 2.10; Annual Report, 2012–13, Indian Petroleum and Natural Gas Statistics, Govt. of India, New Delhi.
5. Reliance Review of energy Markets, Dec 2002/Aug 2003.
6. Planning Commission, 2003.
7. Report of the Subgroup on Development of Refining, Marketing, Transportation and Infrastructure Requirements, 1999.
8. Chopra SJ. Refinery for Future. QIP Short-term Course on Advances in Hydrocarbon Engineering, June 23–July 4, 2003, IIT Roorkee, India.

3

Crude Oil and Natural Gas Origin, Occurrence, Exploration, Drilling and Processing

The world petroleum derived from the Latin words *petro* and *oleum* and the greek words petra—a rock and elation—oil means literally rock oil and refers to hydrocarbons that occur widely in the sedimentary rocks in the form of gas, liquids, semisolids or solid [Speight, 1999]. Crude oil is believed to have formed over millions of years. Two types of theories have been proposed to explain the formation—biogenic theory and abiogenic theory. According to biogenic proposes, oil is formed from the preserved remains of prehistoric zooplankton and algae which have been settled to the sea bottom in large quantities under the anoxic condition [www.commoditiescontrol.com]. Abiogenic theory proposes that hydrocarbons of purely geological origin exist in the planet [http://en.wikipedia.org/wiki/petroleum].

Oil and gas production includes exploration, drilling, extraction, and treatment. A typical flow diagram showing hydrocarbon chain from exploration to dispensing is given in Fig. 3.1 [KMPIE, 2006]. During last three decades there has been continuous development in oil and gas drilling and exploration.

3.1 OIL AND NATURAL GAS ORIGIN, OCCURRENCE

Oil and natural gas are believed to be formed from the prehistoric plant and animals which settled into the sea along with sand, silt and rocks in hundreds of millions of years ago. The preserved organic matter undergoes three phases of alteration-diagenesis, catagenesis, and metagenesis. The dead organic matter, due to temperature and pressure conditions deep beneath earth is transformed into organic form called kerogen. Kerogen is precursor for petroleum and is a heteroatomic macromolecular material, insoluble organic solid of variable composition. Three major types of kerogen have been identified: Type I kerogen (algal), type II kerogen (lptinic) and type III kerogen (humic). Different forms of kerogen are formed because of the different kinds of debris deposited in the sediment and the conditions that prevail in the sediment over geological time [Speight, 1999]. Types I and II kerogen tend to occur in marine environments while type III kerogen is produced from lignin and higher woody products and occur in continental (fluvial and deltaic). The formation of petroleum from kerogen is postulated to involve the following path [Speight, 1999]. Origin of oil and gas is shown in Fig. 3.2.

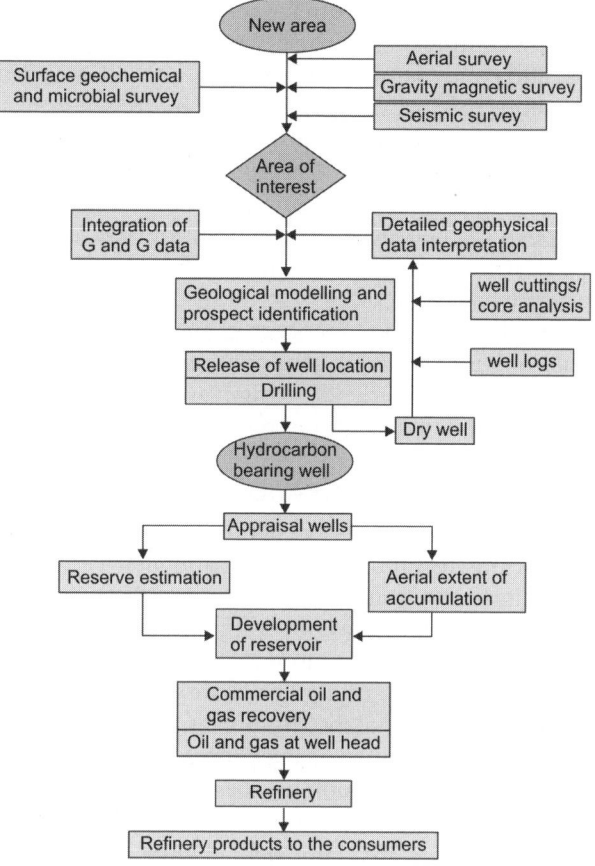

Fig. 3.1 Hydrocarbon chain from exploration to dispensing

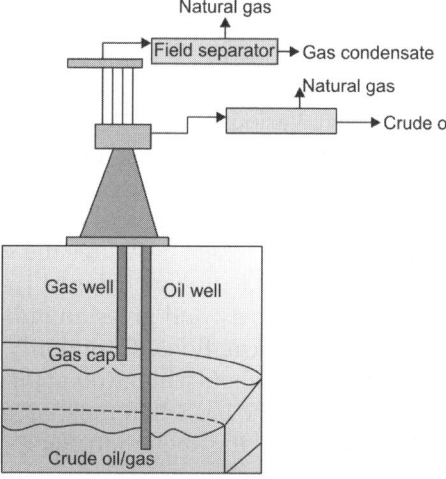

Fig. 3.2 Oil and gas

Although etymologically, petroleum means rock oil or mineral oil, crude oil is a naturally occurring complex mixture of three phases—gaseous (natural gases), liquid (crude oils) and solid/semi-solid (bitumen, asphalt, tars and pitches). It is a complex heterogeneous mixture of compounds of hydrogen from simple structure to most complex structures such as porphyrins, asphaltenes, resins, polyaromatic structures, etc. [Bhatnagar, 1996]. Various hydrocarbons present in the crude oil are paraffins, olefins, naphthenes, aromatics, diolefins, acetylene, etc. Crude oil is believed to have been formed by transformation of organic matter and marine organisms deep below the surface of the earth under the conditions that existed in that layer of the earth, has taken place in millions of years. Various processes that have contributed to the formation of organic matter to petroleum deposits are chemical, bacterial or radioactive or combination of these. All rocks containing deposits of petroleum are of sedimentary origin and can be classified—fragmental (classic), chemical (precipitated) and miscellaneous.

3.2 OIL AND NATURAL GAS EXPLORATION

Oil and natural gas are commonly associated with one another in the rocks of the earth's crust and various techniques of explorations are common as both occur under similar geological conditions [Sell and Dosset, 1958]. Oil and gas exploration is one of the most critical parts of the crude oil production as finding of oil and gas was largely a matter of luck and huge money is involved in oil exploration. Oil and gas exploration requires understanding of a basin, understanding of the nature of rocks, their succession, their spatial distribution and the structural attributes along with understanding the importance of the processes of hydrocarbon generation, migration, accumulation and alteration in a basin.

Oil exploration tools are based on fundamental variables in the earth's physical conditions like gravity change, magnetic field change, time change, and electrical resistance [www.rigsworld.com]. Various methods of oil exploration are: Direct indications, geological methods and geophysical methods.

Geological exploration is based on mapping both surface and sub-surface. Various techniques of geological exploration are gravimetric method and magnetic method and seismic method. Various types of mappings are topographical mapping, photo geology and gelological mapping. Subsurface maps are obtained from well logs, core drilling and startigraphy method. Well logs are representation of some rock property or property versus depth. Various methods of well are—sample log, drilling time log, electric logs, radioactivity logs and caliper logs [Gatlin, 1960].

Earlier torsion balance and similar instruments use earth's gravitational field and the way the field varied according to differences in mass distribution near the earth's surface. Presently, gravimeter which measures the variations in the earth gravitational field has supplemented torsion balance [www.tshaonline.org]. Gravimeter detects differences in gravity and gives an indication of the location and density of underground rock formations. Commonly used gravity meters consist of a weighted boom that pivots about a hinge point [Speight, 1999]. Magnetic method of oil exploration works on the basis that most oil occurs in sedimentary rocks that are nonmagnetic. Igneous and metamorphic rock rarely contains oil and is highly magnetized. The magnetometer is especially designed magnetic compass and detects minute differences in the magnetic properties of rock formations [www.tshaonline.org]. The field balance and the airborne magnetometer are types of field balance and the airborne magnetometer and measure the slight difference in magnetism in rocks. Seismic method uses the refraction and reflection of sound waves to identify subsurface formations. In seismic method variation in speed of the transmission of elastic earth waves or sound waves through different geological structures is measured by time.

Startigraphy exploration consists of in establishing correlations between wells, matching fossils, strata, rock hardness or softness, and electrical and radioactivity data to determine the origin, composition, distribution and succession of rock strata [www.tshaonline.org]. The seismic surveys and the exploratory wells provide a lot of information on the volumes of oil and gas present in the field, their distribution in the reservoir and expected production [www.bpes.com].

Recent developments in the seismic surveys are three-dimensional seismic imaging technology and four-dimensional time lapse seismic visualization. Seismic techniques and satellite imaging have facilitated the discovery of promising new natural gas reservoirs. These have nearly doubled the success rate of new field [bulk.resource.org]. Three-dimensional seismic imaging technologies are based on bouncing of acoustic or electrical vibrations of underground surfaces and generate massive amounts of data which can be plotted on a virtual map that allows explorers to identify areas where commercial quantities of oil and natural gas may have accumulated [www.classroom-energy.com] [http://www.api.org/ehs/performance/explore/moreexplorationproduction.cfm]. Four-dimensional time lapse seismic visualization can capture three-dimensional seismic data over time, providing a motion picture view of the flow patterns of oil and natural gas underground.

3.3   OIL AND NATURAL GAS RESERVOIR

A reservoir is porous permeable subsoil formation containing a natural, individual and separate accumulation of hydrocarbons (oil and/or gas) [Coss, 1993]. Reservoir rocks are made up of sandstone and/or carbonates (99% of the total). Reservoir rocks must possess fluid holding and fluid transmitting capacity [Speight, 1999]. Cap rock acts as a seal to prevent the escape of oil and gas from the reservoir rock and has a far lower permeability than a reservoir rock [Jenner,1986].

As the source rocks become buried under more sediment, the pressure rises and the hydrocarbons are very slowly squeezed from the source rocks into neighboring porous rocks, such as sand stones. Reservoir can have any combination of oil and gas: Oil with no gas, gas with no oil or both oil and gas together. Gas being less dense rises to the top of the reservoir [www.bpes.com]. The reservoir rock is porous and permeable, and the structure is bounded by impermeable barriers which trap the hydrocarbons [Coss, 1993]. The distribution of the fluids in a reservoir rock is dependent on the densities of the fluids and the detained capillary properties of the rock [Jenner,1986]. Most of the oil reservoirs contain three fluids: Gas, oil and water. Reservoir estimates are commonly classified as proven, unproven, probable and possible and undiscovered.

Oil reservoirs differ greatly from one another in shape, size, in formation of characteristics and in oil properties. All nearby wells drilled to the same reservoir can have different depths by formation, characteristics and holes sizes.

The existence of reservoirs impregnated with hydrocarbons indicates the presence of trap capable of stopping the hydrocarbons from migrating. A trap is an area bounded by a barrier lying upwards from the flow and is provided by a layer of impermeable rocks called cap rock. Traps may be classified as: Structural traps, stratigraphic traps and combination traps [Coss, 1993]. Structural traps are formed by deformation of the earth's crust by either faulting or folding. Stratigraphic traps are formed by changes in lithology, generally a disappearance of the containing bed or porosity. Combination traps are having both structural and stratigraphic features [www.pete.metu.edu.tr] [Galtin,1960]. Various properties used for characterization of reservoir rocks are porosity, permeability and saturations. Porosity is the storage capacity of the rock while permeability is the rate of flow of fluid through the rock.

Porosity = Pore volume of the rock/Total volume rock.

Specific or absolute permeability of a rock is the ability of the rock to allow a fluid with which it is saturated to flow through its pores. Permeability is expressed in term of Darcy. Darcy is the permeability that allows a fluid of 1 centipoise viscosity to flow at a velocity of 1 cm/s for pressure drop of 1 atm/cm.

The term reserves are applied to the recoverable volumes that appear producible.

3.4   WELL LOGGING

This is a method used for determining saturation. A well log is essentially a instrument for reservoir assessments and is the recording of a characteristics of the formations intersected by a bore hole, as function of depth. Various types of well loggings are drill's log, sampler logs, mud logs, electric logs, radioactivity logs and miscellaneous logs [Gatlin, 1960]. Amongst various well loggings, electric logging and radioactive logging are commonly used. Electric logging was developed by Marcel Schlumberger and was introduced to the United States in 1929 [Gatlin, 1960].

Electric logging is the recording, in uncased sections of a bore hole, of the resistivity of the subsurface formations, generally along with spontaneous potentials generated in the bore hole.

Resistivity = RA/L

R = Resistance of conductor

A = Cross-sectional area of conductor

L = Length of conductor.

Commonly used resistivity devices are macrodevices and microdevices.

Electric logs must be run in open hole to avoid short circuits. However, this restriction is not in case of radioactivity logging as radioactivity logging may be run in either open or cased holes. Radioactivity logging may be gamma ray well logging or neutron well logging. Gamma ray and neutron curves indicate natural and artificially produced radiation intensities within well.

3.5   DRILLING AND PRODUCTION

After the oil and gas exploration the next step in oil production is drillings of an exploratory well which is very expensive. Surface environment of oil and gas can be categorized as either onshore (land rigs) or offshore. There has been continuous development in the drilling technologies. Earlier oil and natural gas wells were drilled vertically called vertical drilling. However, new directional and horizontal drilling technologies allow drills to deviate from the vertical plane to horizontal or beyond [www.classroom-energy.com] [API: July, 2007]. Various advances in directional drilling now permit multilateral drilling, where multiple offshoots of a single well borne radiate in different directions and can contact resources at different depths [www.classroom-energy.com].

3.5.1 Onshore Drilling

Onshore rigs are highly portable and can be easily shifted from one site to another, Land rigs can be classified on the basis of maximum drilling depth (MDD): Light duty (MDD 1000–1500 m), medium duty (MDD 1200–300 m), heavy duty (MDD 3500–5000 m) and very heavy duty (MDD 5500–7500 m+). Onshore or land rigs are highly portable and can be easily shifted from one site to another as the work at one site is finished. Major systems of a land oil rig are: Power system; mechanical system: Hoisting system, turntable; rotating equipment: Swivel, kelly, rotary table, drill string, drill bits; casing: Large diameter concrete pipe; circulation system; pumps, drilling mud; derrick: Support structure that holds the drilling apparatus; blowout preventer: High pressure valves.

3.5.2 Offshore Drilling

In nearly every corner of the globe, thousands of offshore installations are producing oil and gas in varying water depths. First offshore well drilled in 1897 near Summerland, California. The offshore rigs are further classified based on drilling platform as floating and fixed. Offshore rigs are barge rigs, submergible rigs, and jack up or self-elevating rigs, semi-submergible rigs and drill ships structure rigs. Onshore rigs are the cantilevered mast or jack knife derrick type. Mobile jack-up type of onshore rags have three legs for stability and are used in shallow water up to depth of 100 m. Platform type rigs are permanently fixed to sea floor. Another type of offshore rig is mobile semi-submersible type which flat on pontoons and are held in position with tensioned chain.

3.6   OIL AND GAS DRILLING

Various steps involved in drilling and production of oil and gas are preparation of drilling, drilling, extraction and processing of oil and gas. There are two basic methods of drilling of oil: Cable tool method employing percussion and the rotary method employing rotary and grinding action.

3.6.1 Rotary Drilling

In rotary drilling, a bit used to cut the formation is attached to steel pipe called drill pipe. The bit is lowered to the bottom of the hole. The pipe is rotated from the surface by means of rotary table, through which it inserted a square or hexagonal piece of pipe called kelly. The turning action of the rotary table is applied to the kelly, which in turn rotates the drill pipe and the drilling bit. [www.blueflameenergy.com].

In the rotary drilling system, the rotary rig drills a hole by rotating a column of hollow steel pipes in the bottom of which a bit which carves, chisels or grinds a hole depending on the type of the bit used. Three principal types of bits are used in rotary drilling operations.

- Drag or fish tail bits
- Rolling cutter bits
- Diamond bits.

Fluid circulation is very important feature of rotary drilling. As the bit continues with drilling, the cuttings need to be removed. To remove these cuttings a fluid is circulated which brings the cuttings to the surface. Drilling fluid called drilling mud is pumped through the pipe and out through the bit and returned to the surface via the pipe. The basic functions of drilling fluids are [Jenner, 1984; Singh 2006; Ford, 2001]:

- To cool and lubricate the drilling bit and the drilling strings
- To remove rocks cuttings continuously out of bore hole and bring them on the surface
- To maintain the required hydrostatic pressure on the bore hole to counter formation pressure
- To prevent squeezing or caving of formations and to plaster the sides of the bore hole
- To allow the acquisition of formation data
- To minimize damage to any potential production zone.

Drilling mud carries the small bits of rock (cuttings) from the drilling process to the surface so that they can be removed and fills the well-bore with fluid to equalize pressure and prevent water or other fluids in underground formations from flowing into the well-bore during drilling [www.spe.org].

Various types of drilling fluids are water based drilling fluid and oil based drilling fluids. Various water based drilling fluids are fresh water, brackish or seawater, salt saturated, inhibited, surfactant, polymer, aerated. Various oil based drilling fluids are either oil-base or water-in-oil emulsion completing a well and preparing for the production of oil involves insertion of a casing which comprises one or more strings of tubing and which is carried out in part during drilling [Speight, 1999].

Direction drilling also called horizontal drilling, deviated, or slant drilling is the process of drilling and indirect path to reservoir that cannot be reached directly beneath the drilling site [pickensplan.ning.com]. Advances in directional drilling now permit multilateral drilling, where multiple offshoots of a single well-borne radiate in different directions and can contact resources at different depths [www.classroom-energy.com].

3.6.2 Cable Tool Drilling

Cable tool drilling achieves penetration by pulverizing the rock at the bottom of the hole by rising and dropping heavy bit and taking out as mud. Cable tool drilling was the first drilling method used to reach oil and water deposits. Cable tool drilling is an absolute method and rarely used these days. Now, this method has been replaced by rotary method.

In cable drilling, drilling is accomplished by lowering a wire line or cable into the hole. On the end of the line is a heavy chisel-shaped piece of steel called the drilling bit. An up and down motion is applied to the line at the surface. This churning action chips small pieces of rock from the formations.

No significant amount of fluid is circulated in the hole. The only fluid in the hole is what has come from the formations being penetrated.

A distinct disadvantage of cable tool drilling is that when high pressure oil and gas formations are encountered, there is no fluid in the hole to control them resulting in frequent blowout [www.most.gov.mm].

3.6.3 Hydraulic fracturing and Horizontal drilling

Hydraulic fracturing is used to increase or restore the rate of fluid flow within the shale reservoir. The fractures are created by pumping large volume of fluid at high pressure. Horizontal drilling creates maximum bore hole surface area in contact with the shale (Fig. 3.3).

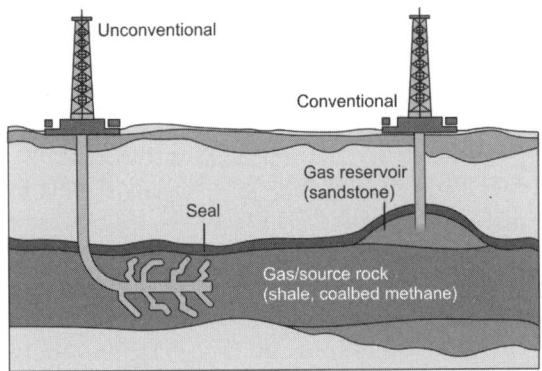

Fig. 3.3 Conventional vs shale gas from horizontal drilling

Source: Tulsi Das, 2011. www.dteenergy.com.

3.6.4 Rigs

Various types of rigs are: Swamp barge, jacket with tender, jack-up, semi-submersible, fixed platform and drill ship.

3.6.5 Casing

The major functions of casing are:

- Contains formation pressures and prevents fracturing of the upper and weaker zone
- Prevents the hole from caving in
- Confines production to the well bore
- Provides an anchor for the surface equipment and artificial lift equipment
- Separates the formations behind the pipe and limits the production to the selected zone.

Various types of casings are given below.

- *Conductor pipe:* Outermost casing of diameter in the range of 30–42 inches and of 20–50 ft long. Installed to prevent excessive caving around the sides of the hole.
- *Surface casing:* Protects fresh water formations from contamination by oil, gas or salt water from the deeper production formations. Provides a place to attach the blowout preventers (BOP) 2000 ft long.

- *Intermediate casing:* Protects the hole against loss of circulation in shallow formations.
- *Liner strings:* Runs from the bottom of the previous string to the bottom of the open hole.
- *Production casing:* Isolates the oil and gas from undesirable fluids in the producing formation serves as the protective housing for the equipments used in the well.

3.7 PRODUCTION CONCEPTS

3.7.1 Dissolved Gas Drive

Oil is displaced when the gas is liberated from solution in the oil. This happens when production reduces the pressure in the formation.

3.7.2 Gas Cap Drive

The agent is free gas cap originally present and overlying the oil bearing zone. The reduced pressure results in the gas cap expanding. As the gas cap expands downwards and invades the oil zone down—structure it drives the oil towards the regions of reduced pressure—the producing wells.

3.7.3 Water Drive

Water from adjacent aquifers encroaches into the oil bearing portion of the reservoir. In response to the reduced pressure from the well-bore, the water flows in the direction of the reduced pressure, displaces the oil from the porous rock, and drives the oil ahead of it towards the well.

3.8 WELL COMPLETION AND OIL EXTRACTION

Completing a well and preparing for production of oil after removal of drilling rig, involves insertion of a casing part of which is carried out during drilling. casing provides a permanent wall to borehole. Casing is provide to prevent caving of bore hole, contamination of water, exclude water from the producing formations, provide an adequate means of controlling well pressures [Speight, 1999]. Next steps are installation of an assembly of valves known as Christmas tree above a master valve at the casing head [Speight, 1999].

3.9 ENHANCED OIL RECOVERY (EOR)

Crude oil production from well include three steps: primary, secondary and tertiary recovery process. Tertiary recovery process is called enhanced oil recovery step. Recovery of oil during conventional primary phase of production is poor. Inorder to improve oily recovery water is injected in secondary recovery. The process is called water flooding. However, in order to enhance oil recovery tertiary oil recovery is being practiced. Some of the enhanced oil recovery are thermal methods using thermal recovery (steam injection), chemical enhanced oil recovery, polymer flooding, gas injection using natural gas, nitrogen or CO_2, high pressure air injection. CO_2 EOR getting importance as EOR and potentially as a sequestration strategy in recent year [Alvarado and Manrique, 2010] moves through formation mixing with oil droplets expanding them and moving them to producing wells CO_2.

3.10 OIL AND NATURAL GAS PROCESSING

Flow diagram for processing of oil and gas from well is given in Fig. 3.4 [Ravindranath and Habibula, 1992]. The process involves stabilization of oil with associated gas, separation of oil from water in separator, final transportation of oil to desalting plant. Gas with associated condensate is sent to slug catcher unit for separation of gas from condensate (NGL). Condensate is further fractionated to get LPG and heavier fraction NGL which can be fractionated to kerosene, diesel, etc. The natural gas after sweetening can be supplied as feedstock for fertilizer plant.

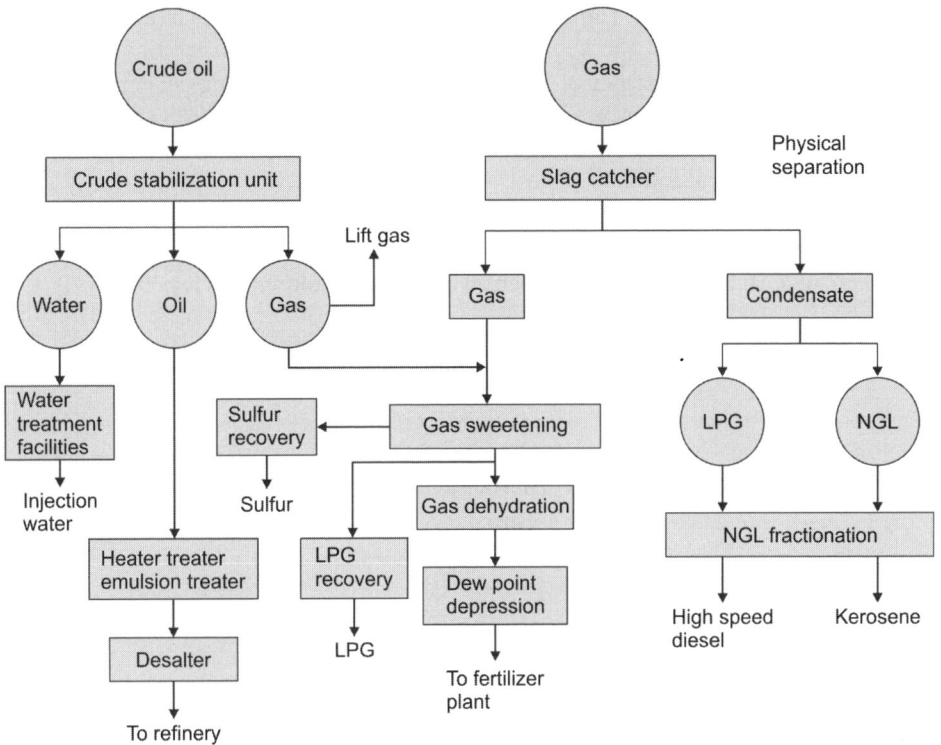

Fig. 3.4 Oil and gas processing

Source: Ravindranath and Habibula, 1992.

REFERENCES

1. AlvaradoV and Manrique E. Enhanced Oil Recovery: Update Review. Energies, 2010, p1529.

2. API, 2007. http://www.api.org/ehs/performance/explore/moreexploration production.cfm.

3. Bhatnagar AK. Analytical Developments in Crude Oil and Product Characterization. Challenges in Crude Oil Evaluation, edited by Nagpal HM, Tata McGraw Hill Publishinhg Co Ltd, New Delhi, 1996, p1.

4. Coss R. Basics of Reservoir Engineering, Editions Technip, 1993. Jenner JW. Modern Petroleum Technology, edited by Hobson GD, Part 1, John Willey & Sons, New York, 1984, p123.

5. Galtin C. Petroleum Engineering Drilling and Well Completions, Prentice Hall, INC Englewood Cliffs, NJ, 1960.

6. Jenner JW. Modern Petroleum Technology, edited by Hobson GD, Part 1, John Wiley & Sons, New York, 1984, p123.

7. October 2002—EPA issues the publication, Exemption of Oil and Gas Exploration and Production Wastes from Federal Hazardous Waste Regulations.

8. Ravindranath K and Habibula M. Hydrocarbon Condensate Fractionation in Oil and Gas Processing Complex. Chemical Engineering World, Vol 27, No. 10, Oct 1992, p43.

9. Sell G and Dossett HA. Handbook of the Petroleum Industry, George Newness, London, p158.

10. Singh RJ. Chemical Journey of Oil: An Economic and Technological Perspective in Relation to Oilfield Process Chemicals. Industry Digest, Aug 2006, p49.

11. KMPIE, 2006. An Economic and Technological Perspective in Relation to Oil Field Process. Chemical Industry Digest, Aug 2006, p49.

12. Speight JG. The Chemistry and Technology of Petroleum, Marcel Decker Inc, New York, 1999.

13. Das Tulsi. Shale Gas. Journal of Petrotech, January–March, 2011, p35.

4

Crude Oil and Natural Gas Characteristics and Evaluation

Crude is a complex mixture of hydrocarbon-rich fluids and includes three phases: Gaseous (natural gas), liquid (crude oil) and solids or semisolid (bitumen and asphalt). Natural gas can be an associated or a non-associated natural gas. Associated gas is closely connected with crude oil and is separated during initial stages of processing of crude oil after it is extracted from well and during fractionation. Non-associated gas or natural gas predominantly from kerogen III is separately generated.

4.1 COMPOSITION OF CRUDE OIL

Crude oil is a complex mixture of hydrocarbons predominantly organic carbon hydrogen which may include from 1 to 60 carbon atoms and non-hydrocarbons containing nitrogen, sulfur and oxygen in significant amounts and smaller quality of trace metals like nickel, vanadium and other elements [Borchardt, 2006]. Consistency of crude oil vary from source to source and average crude oil contains about 84% carbon, 14% hydrogen, 1–3% sulfur, and each of nitrogen, oxygen, metals and salts [gisceu.net]. [OSHA technical Manual, Section IV Chapter 2 [http://www.osha.gov/dts/osta/otm/otm iv/otmiv2.html]]. Crude oil contains three major groups of hydrocarbon: Paraffins, aromatics and naphthenes. Composition of petroleum is given in Tables 4.1 and 4.2.

4.1.1 Hydrocarbons

Table 4.1 Composition of petroleum

Hydrogen family	Distinguishing characteristics	Major hydrocarbons	Remarks
Paraffins (alkanes) general formula— C_nH_{2n+2}	Straight carbon chain	Methane, ethane, propane, butane, pentane, hexane	Paraffin content normally decreases with increasing molecular weight of the hydrocarbon. These are present in gases and paraffin wax. Boiling point increases as the number of carbon atoms increases. With number of carbons 25–40, paraffin becomes

Contd.

Table 4.1   Composition of petroleum *(Contd.)*

Hydrogen family	Distinguishing characteristics	Major hydrocarbons	Remarks
			waxy, paraffinic crude may contain up to 20–50% paraffins. Paraffinic feedstocks are preferred for kerosene.
Isoparaffins (isoalkanes)	Branched carbon chain	Isobutane, isopentane, neopentane, isooctane	The number of possible isomers increases as in geometric progression as the number of carbon atoms increases. These are present in the heavier fractions of crude and have higher octane number than paraffins.
Olefins (alkenes) general formula— C_nH_{2n}	One pair of carbon atoms	Ethylene, propylene	Olefins are not present in crude oil, but are formed during the processes. Undesirable in the finished product because of their high reactivity. Low molecular weight olefins have good antiknock properties.
Naphthenes general formula— C_nH_{2n+2}	5 or 6 carbon atoms in ring	Cyclopentane, cyclohexane and their alkylated derivatives.	The average crude oil contains about 50% by weight naphthenes. Naphthenes are modestly good components of gasoline.
Aromatics	6 carbon atoms in ring with three around linkages.	Benzene, toluene, xylene, ethyl-benzene, cumene, naphthalene	Aromatics are not desirable in kerosene and lubricating oil. Benzene is carcinogenic and hence undesirable part of gasoline. In general, aromatic increases with increasing molecular weight. Polynuclears which are very complex in nature are present in heavier fractions of crude.

Table 4.2   Non-hydrocarbon products

Non-hydrocarbons	Compounds	Remarks
Sulfur compounds	Hydrogen sulfide, mercaptans, thiophenes	Undesirable due to foul odor 0.5–7%. Heavier crude oil contains higher sulfur.
Nitrogen compounds	Quinotine, pyridine, pyrrole, indole, carbazole	Presence of nitrogen compounds in gasoline and kerosene degrade the color of product on exposure to sunlight. They may cause gum formation normally less than 0.2.

Contd.

Table 4.2 Non-hydrocarbon products *(Contd.)*

Non-hydrocarbons	Compounds	Remarks
Oxygen compounds	Naphthenic acids, phenols, ketones and carboxylic acids	Content traces to 2%. These acids cause corrosion problem at various stages of processing and pollution problem.
Trace metals	Nickel, iron and vanadium	Presents in small quantities and undesirable part.
Salts	Inorganic salts: Sodium chloride, magnesium chloride, and calcium chloride	These salts causes corrosion problem at various stages of operation and must be remove before processing.
Naphthenic acid		Results in corrosion of equipment

Source: OSHA Technical Manual, Section IV, Chapter 2, http://www.osha.gov/dts/osta/otm/otm iv/htm.

4.1.2 Paraffins [Alkanes]

Straight chain C_2H_{2n+2}

Methane	(CH_4)	
Ethane	(C_2H_6)	CH_3CH_3
Propane	(C_3H_8)	$CH_3CH_2CH_3$
Butane	(C_4H_{10})	$CH_3CH_2CH_2CH_3$
Pentane	(C_5H_{12})	$CH_3CH_2CH_2CH_2CH_3$
Hexane	(C_6H_{14})	$CH_3CH_2CH_2CH_2CH_2CH_3$
Heptane	(C_7H_{16})	$CH_3CH_2CH_2CH_2CH_2CH_2CH_3$
Octane	(C_8H_{18})	$CH_3CH_2CH_2CH_2CH_2CH_2CH_2CH_3$
Nonane	(C_9H_{20})	$CH_3CH_2CH_2CH_2\ CH_2CH_2CH_2CH_2CH_3$
Decane	$(C_{10}H_{22})$	$CH_3CH_2CH_2CH_2\ CH_2CH_2CH_2CH_2CH_2CH_3$

4.1.3 Isoparaffins

Isoalkanes branched carbon chain C_2H_{2n+2}, n = 4 or more number of isomers increase in geometric proportion of branched carbon chain.

Isobutene (C_4H_{10})

$$CH_3$$
$$|$$
$$CH_3CHCH_3$$

Isopentane (C_5H_{12})

$$CH_3$$
$$|$$
$$CH_3CHCH_2CH_3$$

Neopentane (C_5H_{12})

$$CH_3$$
$$|$$
$$CH_3CCH_3$$
$$|$$
$$CH_3$$

4.1.4 Olefins (Alkenes)

C_2H_{2n}

Ethylene (C_2H_4)	$CH_2=CH_2$
Propylene (C_3H_6)	$CH_2=CHCH_3$
Butylenes (C_4H_8)	$CH_2=CHCH_2CH_3$

Naphthenes

Saturated ring compounds (cycloparaffins)

Cyclopentane (C_5H_{10})

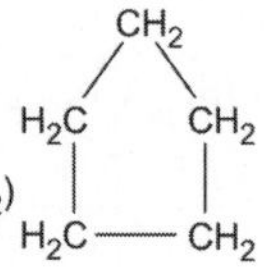

$$
\begin{array}{ccc}
 & CH_2 & \\
H_2C & & CH_2 \\
| & & | \\
H_2C & \!\!\!-\!\!\! & CH_2
\end{array}
$$

Methylcyclopentane (C_6H_{12})

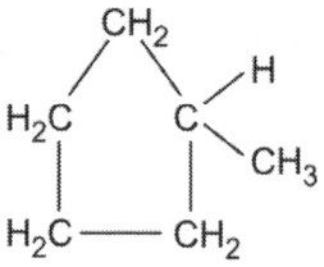

$$
\begin{array}{ccc}
 & CH_2 & \\
H_2C & & C\!\!\diagdown^{H}_{CH_3} \\
| & & \\
H_2C & \!\!\!-\!\!\! & CH_2
\end{array}
$$

Cyclohexane (C_6H_{12})

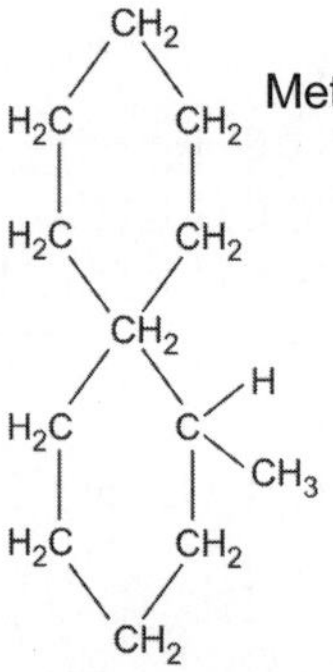

Methylcyclohexane (C_7H_{14})

$$
\begin{array}{ccc}
 & CH_2 & \\
H_2C & & CH_2 \\
H_2C & & CH_2 \\
 & CH_2 &
\end{array}
$$

$$
\begin{array}{ccc}
 & CH_2 & \\
H_2C & & C\!\!\diagdown^{H}_{CH_3} \\
H_2C & & CH_2 \\
 & CH_2 &
\end{array}
$$

1, 2-dimethylcyclohexane (C_8H_{20})

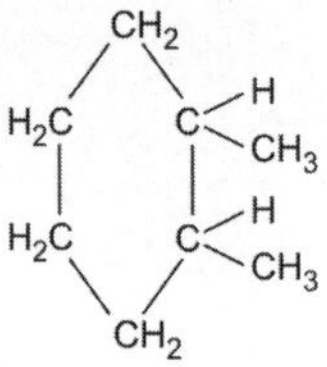

$$
\begin{array}{ccc}
 & CH_2 & \\
H_2C & & C\!\!\diagdown^{H}_{CH_3} \\
H_2C & & C\!\!\diagdown^{H}_{CH_3} \\
 & CH_2 &
\end{array}
$$

4.1.5 Aromatics

Aromatics cyclic compound containing benzene ring

Benzene   Toluene   Ethyl benzene

o-xylene   p-xylene   m-xylene   Naphthalene

4.2   CRUDE OIL EVALUATION

Crude oil exhibits wide variation in composition and properties from one field to another, one change in the crude quality has multifarious effects that include changes in the product quality, change in the processing scheme, throughput, and material of construction, economics and effluent quality. Crude oils range consistency from water to tar-like solids, clear to black. Crude oils may be classified as paraffinic, intermediate and naphthenic base and aromatic. Crude oil evaluation is one of the important aspects in the refinery operation. Crude oil characterization has become further important as now wide variation in crude quality starting from conventional crude to now opportunity crude with high TAN, shale oil, oil sands which vary widely in quality. Crude oil evaluation and chemical analysis in petroleum industry gives valuable support to refinery operation through:

- Exploration of downstream utilization
- Crude oil evaluation to assess the general characteristics of crude oil for further processing
- Characterizing and tracking, contaminants in opportunity crudes
- Evaluation of the various intermediate products, various unit processes in the refinery and characterization of the petroleum products
- Process selection and development
- Additive performance evaluation
- Catalyst screening and evaluation
- Development and production and characterization of lubricants
- Product development/value addition
- Evaluation of feedstocks for petrochemical feed
- Development of property, structure and correlations
- Monitoring of environmental pollutants.

General information required in case of crude oil are: Base and general properties of crude oil, presence of impurities like sulfur, chloride, metals, nitrogen, moisture, oxygen, etc.; operating and design data; yield pattern of light medium and heavy distillate; properties of various distillates; hydrocarbon distribution in middle and heavy distillates. Crude oil evaluation data is utilized by [http://www.petrotechnical.com/pti/crudechar.html].

Planning of upstreaming	: Determination of the economic viability of new fields
Organizations of supply	: To assign crude value for individual grades
Refinery operations	: Scheduling crude receipts and determine product yields
Modelling and optimization	: Optimizing refinery crude slates
Research and development	: Design of equipment and process planning [www.petrotechintel.com]

Distillation and contaminant data ultimately in refinery improves the accuracy of various refinery operations and decision making [Golden et al., Nov 1995, 1]. The assessment of the quality of product is of equal importance. Some of the important parameters are specific gravity, smoke point, pour point, viscosity, freezing point, stability test, corrosion tests, flammability tests, acidity and alkalinity, ash, sediment, wax, sulfur, inorganic constituents and contaminants, organic constituents, RON and MON tests, etc. Table 4.3 shows requirement of various parameters for evaluation of crude oil, fuel, lubricating oil, petrochemical feedstocks and product quality.

Some of the major analytical instruments used are gas chromatograph, HPLC, surface air and pore volume analyzer, TBP distillation unit (ASTMD2892), high temperature simulated distillation, FIA apparatus, atomic absorption spectrophotometer, UV/visible spectrophotometer, inductively coupled plasma atomic emission spectroscopy (ICP-AES), MSCP, wiped film evaporators potentiograph, wickbold sulfur apparatus, automatic distillation apparatus, vacuum distillation apparatus, etc. Some of the engines/ATF special tests include CFR engine, WSTM, jet fuel thermal oxidation test, conradson-ASTM D189/microcarbon residue-ASTM D4530, asphaltene IP-143, ASTM d3279, ASTM D4124.

Rheological characteristics of crude oil from different sources and petroleum products are important for development of drag reducers and flow additives. Instrumented pig is a hightech pipeline probe to monitor the health of crude and product pipeline.

Table 4.3   Requirement of various parameters for evaluation of crude oil, lube base oil

Product	Information required
Crude oil and lube base crude oil	Base of the crude, TBP assay, yield data, specific gravity, API, sulfur, RVP, pour point, wax, viscosity, water, salt, acidity, alkalinity, TBP yield, LPG potential, trace metals

4.2.1 Density and API Gravity

Density data is required for weight to volume and vice versa calculation, checking consistency of crude oil, control of refinery operation and give rough estimation of crude oil whether it will give higher or lower middle distillate. Specific gravity is influenced by the chemical composition of petroleum. API gravity is the one major criterion used for evaluation of crude oil and other hydrocarbon products. Lighter crude has higher API gravity while heavier hydrocarbon has lower API gravity. API gravity of lighter crude oil may be of the order of 45 whereas in heavier asphaltenes API is 10–12.

1. Density = Mass/Volume
2. Specific gravity = Density of substance/Density of reference substance
3. API gravity = – 131.5
4. Crude oil can be classified according to specific gravity in following four major categories [Chatila, 1995].
   - Light crude   Sp. gr. d^{15}_4 < 0.825
   - Medium crudes   0.825   Sp. gr. d^{15}_4 < 0.875
   - Heavy crude   0.875   Sp. gr. d^{15}_4 < 1.000
   - Extraheavy crude   Sp. gr. d^{15}_4 > 1.000
5. Density and relative density data of light hydrocarbons, including LPG, are useful information required for transportation, storage and regulatory purposes.
6. API of mixture can be calculated from the API of the individual component.
   $$(API)_{Mix} = (API)_a\,W_a + (API)_b\,W_b + (API)_c\,W_c$$

4.2.2 Base of the Crude Oil

For characterization of the crude oil base-paraffinic/intermediate/naphthenic and for measure of the aromaticity, various parameters used are characterization factor, BMCI, VGC.

4.2.2.1   Characterization Factor

$K = /$sp. gr. at 15.6/15.6
   T_B = Mean average boiling point in rankin
   Paraffinic base k > 12.1
   Intermediate base k = 11.5–12.1
   Naphthenic k = 11.5
   Aromatics k = 9.8–12.0

4.2.2.2   BMCI (Bureau of Mines Correlation Index)

BMCI = 48640/°K + 473.7 g – 456.8
   K = Average boiling point in °K, g = Specific gravity at 15.6/15.6 °C
   BMCI value:
   Paraffinic < 15
   Intermediate = 15–50
   Naphthenic > 50

4.2.2.3   Viscosity Gravity Correlation (VGC)

$$\text{VGC} = \frac{10G - 1.0752\log(V-38)}{10 - \log(V.\,38)}$$

G is sp. gravity and V is saybolt universal viscosity

Paraffinic base: 0.80–0.83

Intermediate base: 0.83–0.88

Naphthenic base: 0.88–0.95

4.2.2.4   Viscosity Index

Viscosity index (VI) = 100 × (L–U)/(L–H)

   L = Viscosity at 100 °F of reference oil with VI = 0

   H = Viscosity at 100 °F of reference oil with VI = 100

   U = Viscosity at 100 °F of the base oil

   H = Viscosity at 210 °F of the base oil

Viscosity index is also a measure of base of oil. Naphthenic base oil have viscosity index between 40 and 80 with low pour point while paraffinic base oil have higher viscosity index (> 95) with high pour point.

4.2.3 TBP Assay

It is done for generating distillation data and for study of variations of some key properties throughout the distillation range and gives an almost exact picture of a crude petroleum by measuring the boiling points of the components of the crude. As complete component-by-component analysis of crude oil sample is not practical, the composition can be approximated by a true boiling point (TBP) distillation.

TBP is determined by fractionating the crude oil into nearly 50 cuts having a very narrow distillation intervals which allow them to be considered as fictitious pure hydrocarbons whose boiling points are equal to the arithmetic average of the initial and final boiling points, while other physical characteristics being average properties measured for each cut [Chatila, 1995] (Fig. 4.1).

4.2.4 Advance Laboratory Distillation

Advance laboratory distillation data are important in evaluation of crude distillation, product molecular distillation, high temperature simulated distillation (HTSD), ASTM D2892 distillation, microdistillation, high vacuum potstill distillation unit (ASTM D5236).

4.2.4.1   Molecular Distillation

Molecular distillation is an advance separation process. It explores high vacuum, operation at reduced temperatures, and low exposition of material at the operating temperature.

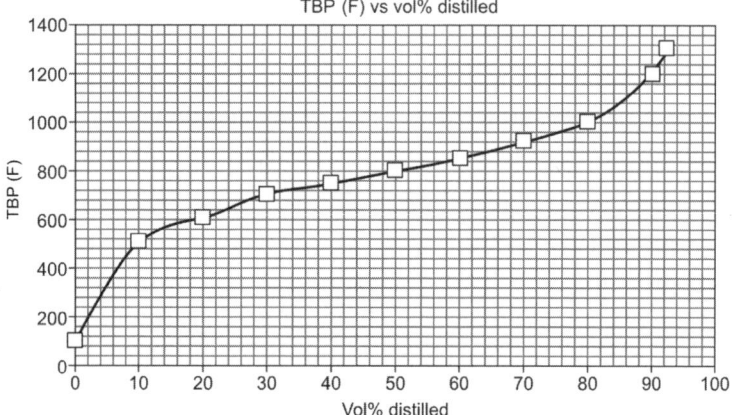

Fig. 4.1 Typical TBP curve

Source: Banik, 2006.

4.2.4.2　High Temperature Simulated Distillation (HTSD)

Characterization of hydrocarbon by HTSD is a relatively new method of characterization of petroleum products and has wide range of applications. Major application of HTSD included crude oil characterization, reduced crude, FCC feeds, products, slurries, hydrotreater feed and products, VGO and vacuum residues [home.earthlink.net].

4.2.4.3　Automated TBP Distillation Unit (ASTM D2892)

ASTM D2892 distillation method is used for evaluating distillation yield curves of crude oils and crude products. This involves batch distillation of the sample at prescribed conditions of boil up rate, reflux ratio and still pressure.

4.2.5　Viscosity

Viscosity indicates the relative mobility of various crude oils and is a measure of the resistance to flow liquid. Temperature has a marked effect on viscosity. Viscosity is related to complexity of its molecule and fractions having low boiling have a low viscosity and those having high boiling point have a high viscosity. Redwood viscometer and Saybolt viscometer are commonly used for viscosity determination.

4.2.5.1　Kinematic Viscosity

Kinematic viscosity is the ratio of absolute viscosity to density. Viscosity data is of primary importance in the design and selection of wide range of petroleum products and storage and transportation system.

4.2.5.2　Saybolt Viscometer

It determines the time required to flow through a calibrated orifice under precisely controlled condition.

Kinematic viscosity and saybolt viscosity can be correlated by the expression [Speight, 1999].

Kinematic viscosity = a × Saybolt s + b / Saybolt s

where a and b are constants.

4.2.6 Water, Sediments and Salts

Water, sediments and salts are present in very small quantity and cause irregular behavior in the distillation and cause blocking and fouling of heat exchangers, catalyst poisoning and result in corrosion, erosion and deposits. Presence of salts also favors coke formation due to formation of hot spots. Salt in the form of chlorides of sodium, calcium and magnesium are commonly present in the crude.

4.2.6.1 Water and Sediment

Water is present in the crude partly in solution and partly in the form of a stable emulsion. Presence of sediments which are entrained in the crude are highly undesirable and may be in the range of 0.1–2.0%. Water content is determined by Dean and Stack. Sediment and water is determined by centrifuging a mixture of crude oil and toluene. Salt content is determined by titrating the water extract with KCNS/AgNO$_3$ sediments in crude oil and fuel oil is also determined by extraction with toluene as per ASTM D473.

4.2.6.2 Salt

Presence of salt in the crude oil has harmful affect on various stages of processing. Some of the major problems associated with presence of salt are: Irregular behavior in distillation, corrosion, increased ammonia consumption and product contamination. Salt content is determined by measuring the conductivity of a solution of crude oil in polar solvent when subjected to an alternating electrical current and is obtained by comparison of the resulting conductance to a calibration curve of known salt mixtures. Electronic salt determinator measures the salt content, conductance and temperature of crude oil according to ASTM D3230 [www.koehlerinstrument.com].

4.2.6.3 Trace Metals

Metals present in the crude poisoned the catalyst. Analysis of trace metals in petroleum distillates by inductively—coupled plasma atomic emission spectro-photometer (ICP-AES) enables more consistent low detection limits.

4.2.7 Reid Vapor Pressure and Light End Analysis

Reid vapor pressure is a measure of the vapor pressure of petroleum and petroleum products and indicates the relative percentage of gaseous and lighter hydrocarbons present in crude oil and the fuel. Vapor pressure is a critical factor in the handling and performance of liquids petroleum and liquefied gas (LPG) products. The vapor pressure of gasoline is subject to environmental regulations.

4.2.8 Pour Point

Pour point is an important parameter which gives an idea of the lowest temperature at which a crude oil and fuel can be used and approximated indication of its pump ability. The problems associated with handling of waxy crude oils are related to the precipitation of wax and subsequent gelling of crude oil when it is cooled from high reservoir temperature to ambient condition during its production and transportation. Pour point is an indicator of the lowest temperature of utility for petroleum products.

For estimating the relative amount of wax present in the crude oil, pour point is the lowest temperature at which movement of the oil is observed. Oils used for lubricants must have a low pour point. Pour points of crude oils generally vary in the range from –60 to +30 °C.

4.2.9 Identification and Structural Group Analysis [Speight, 1999]

The crude oil is a complex mixture of saturated hydrocarbons, saturated hetero-compounds, and aromatic hydrocarbons, olefinic hydrocarbons and aromatic heterocompounds. With the advancement of the instrumental analysis technique like chromatography and spectroscopic methods, now it has been possible to go in depth study of this identification and structural group analysis. Some of the measure analytical instruments used are gas chromatography, ion exchange chromatography, simulated distillation by gas chromatography, absorption chromatography, gel permeation chromatography, high-performance liquid chromatography and supercritical fluid chromatography. The application of spectroscopic methods include infrared spectroscopy, nuclear magnetic resonance spectroscopy, mass spectroscopy, electron spin resonance, X-ray diffraction, inductively coupled plasma emission spectroscopy, X-ray absorption spectroscopy and atomic absorption spectrophotometer. Methods used for analysis of metal and other petroleum products are given in Table 4.4.

Table 4.4 Analytical methods used in petroleum refinery

Types of analysis	Analytical methods/instruments
Content of various elements in petroleum products, metal analysis	X-ray fluorence (XRF), inductively coupled argon plasma (ICAP), atomic absorption spectrometry, X-ray diffraction (XRD), thermogravimetric analysis (TGA), CHNS analyzer
Total sulfur	XRF
Metal mapping	ICAP
PONA (paraffins, olefins, naphthenes and aromatics benzene)	Nuclear magnetic resonance (NMR), high pressure liquid chromatography (HPLC)
Oxidation and thermal stability	TGA
Detailed component analysis	GC, HPLC, GEL permeation chromatograph (GPC), thin layer chromatography (TLC), supercritical fluid chromatography (SFC), ion exchange analysis
Structural group analysis	NMR, mass spectrometry, electron spin resonance, X-ray diffraction.
Molecular weight and molecular weight distribution	Vapor pressure osmometry, boiling point elevation, size exclusion chromatography, NMR, mass spectrometry, capillary GC

Contd.

Table 4.4   Analytical methods used in petroleum refinery *(Contd.)*

Types of analysis	Analytical methods/instruments
Oxygenate analysis	GC, HPLC, NMR, GC-MS
Trace and ultratrace level analysis	ICAP, AAS, XRF, ICP-Ms
Oxidation and thermal stability	DSC, TGA
Additive identification	Total number NMR, IR, HPLC, GPC and Ms
Characteristics of lubricants (base of oil lubricants, metals, additive compound, additive metals, physicochemical properties)	IR, NMR, GC, ICAP, XRF, DRES, XRD
Characterization of asphaltenes	NMR/XRD

4.3   QUALITY OF CRUDE OILS PROCESSED IN INDIA

The total capacity of crude oil production in India is not adequate to meet the country's demand of crude oil and petroleum products. In order to meet the total requirement, about 7 million tonnes of crude is being imported. The various sources of indigenous and imported crude oils are given in Table 4.5.

Table 4.5   Various sources of indigenous and imported crude oils

Indigenous crude oil	Assam crude, North Gujarat and Ankleshwar crude, Bombay High, Satellite field, KG basin, Ravva crude, Cavery Basin crude.
Imported crude oil	Arab mix, Lavan blend, Upper Zkem, Iran mix, Dubai, Kuwait, Suez mix, Zeit bay, Arab medium, Quaiboe, Miri light, Bonny light.

Characteristics of various crude oils are given in Tables 4.6–4.11. Yield of the various products also varies with the type of crude being processed. The change of crude oil quality has multifarious effect, which includes change of product pattern, change in processing scheme, throughput, material of construction, economics and effluent quality (Mishra and Unikrishnan, 1996). Typical yield pattern of some of the crude oils is given in Table 4.12.

Table 4.6   Characteristics of various crude oils

Sources of indigenous crude	Salient features
Assam Crude, Naharkhotia/ Moran ONGC, Lawkwa, Rudrasagar	31° API, sulfur 0.3%, pour point +30 °C, high aromatics, total distillate yield 65%, 27°API, sulfur 0.3%, high aromatics, distillate yield 57%.
Ankleshwar crude	48° API, sulfur 0.1%, pour point +18 °C, distillate yield 80–82% (light distillates 24%, middle distillate 47%), wax content 9.9%, total sulfur 0.02%, light paraffinic crude.

Contd.

Table 4.6 Characteristics of various crude oils *(Contd.)*

Sources of indigenous crude	Salient features
North Gujarat crude	28° API, sulfur content 0.1%, pour point +27 °C, distillate yield low 33–35%, high organic acidity, low yields of naphtha and moderate value of research octane number. Kerosene and gas oil yields are relatively low.
Bombay High crude	38° API, sulfur 0.2%, pour point +30 °C, distillate yield 65–70% (light distillate 245, middle distillates 46%), high aromatics
Narimanam crude	46° API, sulfur 0.1%, pour point 3 °C, distillate yield 80%, intermediate base and low wax content resulting in low pour point. Sulfur content is very low. Quite low yields of VGO and residual stocks. Octane number of naphtha is of moderate order.
KG Basin Ravva crude	36° API, sulfur 0.1%, pour point +30 °C, distillate yield 61%.

Table 4.7 Typical properties of various types of crude oils

Crude properties	Kuwait	Upper zakum	Arab mix	Basrah light
API gravity	31.4	33.1	30.8	33.7
Sp. gravity (60 °F)	0.868	0.8591	0.8712	0.8559
Sulfur (%wt)	2.52	2	2.4	1.95
Nitrogen (%wt)	0.12	0.0973	0.137	0.095
Pour point (°C)	–15	–27	–15	–15
Vanadium (ppm)	30	8.77	22	18
Nickel (ppm)	8	6.9	7.2	5
Con. carbon (%wt)	5.3	4.42	5.44	4.18
Ash content (%wt)	0.06	–	–	0.01
Salt (lbs/1000 bbls)	3	–	5	1
Kv @ 37.8 CST	9.78	–	9.41	6.5

Source: IOC Panipat Refinery.

Table 4.8   Typical properties of various types of crude oils

Crude properties	Dubai	Suez mix	Lower zuluf	Arab med zuluf	Mandji blend
API gravity	31.1	31.9	40.6	31.1	30.1
Sp. gravity (60 °F)	0.8699	0.8654	0.8216	0.8696	0.875
Sulfur (%wt)	2	1.52	1.05	2.48	1.11
Nitrogen (%wt)	0.17	0.22	0.0525	0.137	–
Pour point (°C)	–9	3	–15	–33	9
Vanadium (ppm)	42	41	0.36	43	65
Nickel (ppm)	14	25	0.3	13	55
Con. carbon (%wt)	4.62	3.6	1.56	5.05	4.7
Ash content (%wt)	–	–	0	–	0.05
Salt (lbs/1000 bbls)	–	–	1	5	–
Kv @ 37.8 CST	7.48	8.74	2.9	11.01	17.9

Source: IOC Panipat Refinery.

Table 4.9   Typical properties of various types of crude oils

Crude properties	Bombay High	Quai-Boe	Brent Blend	Brega	ES Sider	Zaire	Bonny Lt	Escrands	Planca	Ekofisk	Seria Lt
API gravity	39.2	35.8	38.6	40.4	37.0	31.7	35.3	36.4	40.1	43.4	36.2
Sp. gravity (60 °F)	0.82	0.84	0.83	0.82	0.84	0.87	0.85	0.84	0.82	0.811	0.84
Sulfur (%wt)	0.15	0.12	0.29	0.21	0.45	0.13	0.11	0.12	0.11	0.14	0.07
Nitrogen (%wt)	0.015	0.054	0.09	0.1	0.14	–	–	0.095	0.08	–	0.02
Pour point (°C)	9	9	9	0	9	27	3	6	-3	–12	–
Vanadium (ppm)	0.1	0.3	2.5	2.7	1.7	1.5	0.4	0.4	0.1	0.11	0.3
Nickel (ppm)	1.4	3.3	0.8	3.5	4.8	17.8	1.2	4.3	0.1	1.65	0.5
Carbon (%wt)	1.2	0.92	1.69	1.67	3	3.58	1	1.3	0.9	1.24	0.3
Ash content (%wt)	0.01	0	0	0	–	–	0.01	–	–	–	–
Salt (lbs/1000 bbls)	15	9.2	2.8	–	–	–	12	2	–	–	15
Kv @ 37.8 CST	3.75	3.57	3.64	3.58	5.56	5.56	3.49	3.51	3.09	3.14	2.69

Courtesy: IOC Panipat Refinery.

Table 4.10 Typical characteristics of Assam crude parameters analysis

Characteristics	Unit	Range
Density at 15 °C	g/cc	0.8781–8839
Water content	%vol	4.3–7.2
Sediment	%wt	0.0018–0.002
Salt content	%wt	0.008–0.01

Table 4.11 Typical characteristics of the crude

Characteristics	Unit	Range
Density at 15 °C	g/cc	0.8824
API gravity		28.86
Water content	%vol	3.4
BS and W	%vol	4.0
Pour point	°C	+18
Salt content	%wt	0.007
Kv at 40 °C	C_5+	6.4
Kv at 50 °C		4.6
RCR	%wt	2.2
Sulfur	%wt	0.22
ASTM distillation		
IBP	°C	71
5%	°C	103
10%	°C	129
20%	°C	185
30%	°C	251
40%	°C	288
Rec. at 300 °C	5 vol	43.0

Source: BRPL, Assam.

Table 4.12 Typical yield pattern of some of the crude oils

Product	Bombay High	Quai-Boe	Brent Blend	Brega	ES Sider	Zaire	Bonny Lt	Escrands	Planca	Ekofisk	Seria Lt.
LPG	2.9	2.6	3.2	2.8	3.3	2.0	2.7	2.5	2.6	2.9	2.1
MS U/L	10.0	10.0	10.0	10.0	10.0	10.0	10.0	10.0	10.0	10.0	10.0
Naphtha	11.1	7.3	11.7	12.1	1.7	1.7	8.9	8.7	12.9	21.9	11.8
ATF	3.7	3.7	3.7	3.7	3.7	3.7	3.7	3.7	3.7	3.7	3.7
SKO	8.3	8.3	8.3	8.3	8.3	8.3	8.3	8.3	8.3	8.3	8.3
HSD	51.3	54.4	46.7	46.9	46.4	46.4	54.0	54.2	49.5	37.8	53.5
FO	0.0	0.0	0.0	0.0	0.0	0.0	0.0	0.0	0.0	0.0	0.0
HPS	3.6	4.6	7.3	7.1	18.7	18.7	3.3	3.5	3.8	6.4	1.7
Bitumen	0.0	0.0	0.0	0.0	0.0	0.0	0.0	0.0	0.0	0.0	0.0
Sulfur	0.4	0.4	0.4	0.4	0.5	0.5	0.4	0.4	0.5	0.3	0.2
F and L	8.7	8.7	8.7	8.7	8.7	8.7	8.7	8.7	8.7	8.7	8.7
Total	**100.0**	**100.0**	**100.0**	**100.0**	**100.0**	**100.0**	**100.0**	**100.0**	**100.0**	**100.0**	**100.0**

Source: IOC Panipat Refinery.

REFERENCES

1. Banik S. Distillation Options in Refinery. Petrotech Society's Summer School Program on Advances in Petroleum Refining Industry, July 3–8, 2006.

2. Borchardt JK. In Kirk Othmer Encyclopedia of Chemical Technology, 5th edition, Volume 18, Willey Interscience, 2006.

3. Golden SW, Craft S and Villalanti DC. Refinery Analytical Techniques Optimize Unit Operations Performance. Hydrocarbon Processing, Nov 1995, p1.

4. OSHA Technical Manual, Section IV, Chapter 2, http://www.osha.gov/dts/osta/otm/otm iv/otm iv 2.html.

5. http:// www.petrotechnical.com/pti/crudechar.html.

6. Chatila SG. Evaluation of Crude Oils in IFP Petroleum Refining, Part 1: Crude Oil Petroleum Products, Process Flow Sheets; Edition Technip, 1995, p315.

7. Speight JG. The Chemistry and Technology of Petroleum, 3rd Edition. Marcel Dekker Inc, New York, 1999, p300.

8. Petrotech Society's Summer School Program on Advances in Petroleum Refining Industry, July 3–8, 2006.

9. Mishra AK and Unnikrishnan A. Overview of the Quality of Crude Oils Processed in India. Challenges in Crude Oil Evaluation, edited by Nagpal JM, Tata McGraw Hill Publishing Company Limited, New Delhi.

5

Crude Oil Distillation

Crude oil is a complex mixture of wide variety of hydrocarbons with a varying composition and characteristic. Crude oil distillation is the first major operation in the refinery for separation of the crude oil into diffraction fraction which is having wide range of applications in energy and petrochemical, fertilizer sector. The various steps involved are desalting, atmospheric distillation and vacuum distillation.

The crude oil always contains salts, solids, trace metals and water which cause corrosion and fouling which are different stages of refinery operation. Entrained water droplets in crude oil contains entrained water droplet's salt. The salt forms corrosive acid during processing and, therefore, is detrimental to process equipments. In order to transport and processing of crude oil in subsequent operation starting from crude oil distillation, catalytic reforming, fluid catalytic cracking, hydrocracking without causing fouling and corrosion removal of salt is essential. Pretreatment of crude oil is one of the important steps in the refinery operation.

5.1 PRETREATMENT OF CRUDE OILS

Crude oil coming from the ground is mixed with variety of substances like gases, water, dirt (minerals), etc. Impurities in the crude oil are either oleophobic or oleophilic. Oleophobic impurities include salt, mainly chloride and impurities of Na, K, Ca and Mg, sediments such as salt, sand, mud, iron oxide, iron sulfide, etc. and water present as soluble emulsified and/or finely dispersed water. Oleophilic impurities are soluble and are sulfur compounds, organometallic compounds, Ni, V, Fe and As, etc., naphthenic acids and nitrogen compounds. Oleophobic impurities are removed during pretreatment of the crude oil. Pretreatment takes place in two ways in crude oil processing.

1. Field separation
2. Crude desalting

5.1.1 Field Separation

For removal of the gases, water and dirt that accompany crude oil coming from the ground, field separation is the first step. Processing of crude oil is accomplished in the site of the oil wells in the field itself. The field separator consists of a large vessel, which gives a quieting zone to permit gravity separation of three phases: Gases, crude oil and water (with entrained dirt) [Hoffman, 1997].

5.1.2 Crude Desalting

Crude oil contains varying amounts of inorganic salts (NaCl, $CaCl_2$, $MgCl_2$ and so on). The presence of these salts cause difficulty during crude oil processing resulting in corrosion, plugging and fouling of equipment, and poisoning of catalyst in subsequent processing units [Bai, 2012]. Excessive sodium accelerates furnace coking and excessive coking may occur when sodium salt is between 15 ppm and 30 ppm. Switching to heavier crudes, desalter operations suffer resulting in poor desalting efficiency requiring multistage desalting. In order to mitigate the effects resulting from salt, crude oil desalting is done.

Crude desalting consist of water washing. This operation is performed at the refinery site to get additional crude oil clean up. The crude oil coming from field separator will continue to have some water/brine and dirt entrained with it. These impurities are to be further removed during desalting process to avoid corrosion and catalyst poisoning. Water washing removes much of the water-soluble minerals and entrained solids. There are two types of desalting: Single stage and multistage desalting.

Commercial crudes, salt contents 10–200 ppb, earlier 10–20 ppb were considered satisfactorily low. However, many refiners now aim at 5 ppb or less (1–2 ppb) which is not possible through single stage desalting; hence two-stage desalting is required. Single stage desalting with water recycle is usually justified, if salt content in crude is less than 40 ppb. Two-stage desalting involves dehydration followed by desalting. Double stage desalting is better for residuum hydrotreating. Fuel oil quality is better.

5.2 CRUDE OIL PROCESSING

The crude oil processing section in refinery consist of following sections:
- Crude oil desalting
- Atmospheric distillation section
- Stabilizer section
- Naphtha splitter and caustic wash section
- Mineral turpentine oil splitter
- Vacuum distillation column
- LPG amine and caustic wash section
- Sour fuel gas amine treatment section.

5.2.1 Desalter

The electric desalting process is commonly used in the refineries. The process consists of:
- Heating of crude oil
- Mixing of the crude with sufficient DM water ahead of the desalter to dissolve the salts present in the crude
- Electric desalting unit where very high voltage of about 18,000 volts is applied.

The raw crude oil is first preheated to reduce the viscosity which helps in efficient mixing and separation. High electric field results in vibration and disturbance of the emulsion resulting in breaking of emulsion droplets which assume polarity and travel to and from towards the electrodes. Due to changing polarity of the electrodes, the water particles travel up and down; collide with each other and coalesce. This results in formation of big droplets which settle at the bottom of the desalter. Figure 5.1 illustrates the process of electrical desalter.

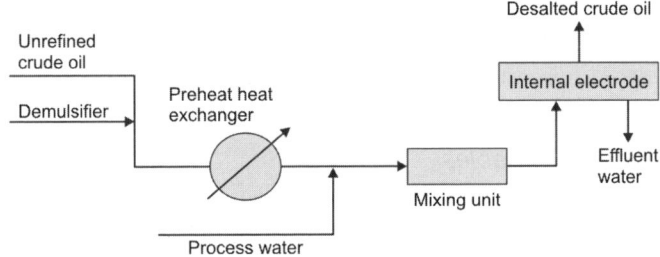

Fig. 5.1 Electrical desalting

The crude oil and water after mixing is processed to desalter through a valve across which a controlled pressure drop is maintained which causes emulsification of the oil and water. Desalting and dewatering of crude oil is one of the key processes for removal of undesirable contaminants to avoid the corrosion and removal of contaminants which may be deterrent in further downstream processing of crude oil and fouling of downstream equipments and deactivating the catalyst.

Sometimes surfactants are added when crude contains large amount of solids. Desalter effluent is a major source of contaminated waste water and may be in the range of 2–3 gallons of waste water per barrel of crude oil feed. Demulsifier dosing improves separation between the crude oil and emulsion and reduces water carry-over along with crude [Mandal, 2005].

Several chemical treatment programs have been developed to improve the overall performance of desalting system. The most common chemical treatment includes the application of an oil emulsion breaking chemical injected into crude oil upstream of desalter [Kremer and Bieber, 2008].

For some crude blends, destabilization of asphaltenes in the crude charge can cause excessive emulsion build up, with eventual water carryover and oil carry-under problems. Aphaltene stabilizer which prevents asphaltenes from accumulating in the emulsion, thus reducing their concentration in the desalter effluent water [Kremer and Bieber, 2008]. Typical operating condition of a desalter unit is given in Table 5.1.

Table 5.1 Typical operating condition of a desalter

	Arabian light	Arabian heavy
Crude inlet flow (MT/hour)	1.280	1.324
Crude inlet temperature (°C)	128	128
Operating pressure [kg/cm^2 (gauge)]	258.4	25.7
Desalting water (5–6% of crude flow rate) (MT/hour)	74.4	74.5

Contd.

Table 5.1   Typical operating condition of a desalter *(Contd.)*

	Arabian light	Arabian heavy
Mixing valve pressure drop [kg/cm^2 (gauge)]	1.5–2.0	0.5–2.0
Demulsifying chemical (3 ppm) (kg/hour)	3.84	3.97
Current (amp)	65.47	65.47

5.2.1.1   Operating Variables

Some of the major operating variables in desalter are water injection rate and pressure drop, oil-water interface level, mixing energy, desalter temperature, desalter pressure, demulsifier injection rate and voltage.

Water injection rate is limited to 3–5% of the crude flow rate. Although, increasing water addition rate results in better washing and higher removal, however, water injection rate above 4.5% may lead to higher carryover of oil resulting in higher loss of oil [Mandal, 2005]. Desalter performance is generally improved by increasing the wash water rate which increases water droplets. The droplets are closer together and coalescing of small droplets into larger ones is facilitated. Increasing the wash water rate also dilutes the concentration of stabilizing molecules oil/water interface, which reduces the coalescing of droplets [Kremer and Bieber, 2008].

Optimum mixing is important for removal of contaminants. In lack of enough mixing, all of contaminants will be not removed. If there is too much mixing, then very small water droplets are formed, producing an emulsion which cannot be fully resolved in the desalter vessel. Overmixing results in high BS and W in the desalted crude oil and in poor salt removal [Kremer and Bieber, 2008].

The mixing energy (as measured in psi drop across the mixing device, or ΔP) varies from a low of 1 psi up to the refinery that consistently runs 19–22 psi [Kremer and Bieber, 2008]. The higher the differential pressures across the mixing valve, the better the mixing. Higher desalter pressure is maintained in order to prevent vaporization of crude. Too large pressure is avoided as it causes excessive emulsification, poor separation of water and oil. Oil–water interface level is monitored as too high level results in carryover of water with crude [Mandal, 2005]. Improved desalting efficiency can be achieved by using some emulsion breaking chemicals which produce a more rapid resolution of the emulsion in the desalted crude [Kremer and Bieber, 2008].

Optimum temperature is around 157 °C. Lower temperature results in higher viscosity resulting in slower separation of oil and water. Higher temperature results in increase in conductivity which will lead to drop in grid voltage. Raising the temperature improves the oil separation in the desalter. This is due to the decrease in viscosity of the hydrocarbon due to the increase in temperature. However, there is negative impacts due to raising temperature as (i) asphaltenes can become unstable as the oil–water interface to stabilize the emulsion and cause oil under carry; (ii) high BS and W in the desalted crude and (iii) higher conductivity resulting into excessive current draw and potential loss of grids [Kremer and Bieber, 2008].

Optimum voltage of 3,000 volts/inch distance between the electrodes is maintained. High voltage will increase operating cost while lower voltage will lead to poor demulsification.

5.2.1.2 Advances in Desalter

A new development in desalting process is membrane filtration by VSEP industrial scale ultrafiltration. The permeate contains less than 1 mg/L of total solid. Different types of desalters are now reported for improving the desalting efficiency. Hanover offers a range of crude treatment system is based on mechanical/thermal and electrostatic coalescing technology.

5.2.1.3 Gastech Dehydrator and Desalting

Electrostatic AC/DC dual wave dehydrator/desalting design utilizes both alternating and direct current through multiple plates to promote enhanced water extraction and level reduction. Gastech electrostatic heater/treater consists of horizontal heater–treater combines the advantages of emulsion breaking heat with electrostatic dehydration in a single vessel package [http://www.gastecheng.com/ desalter.html].

Various factors affecting desalter efficiency are mixing efficiency, inlet temperature, location, electrostatic field type and intensity and electrode design.

5.2.1.4 Preflash Section

The desalted crude after preheating goes to preflash section where the flashed vapors are separated and directly go to the primary crude splitter.

- Crude API   :   28.5
- TAN in crude :   0.5 mg KOH/g
- Sulfur   :   2.3 %wt
- Salt content   :   43 ptb
- Viscosity   :   18 cSt @ 40 °C

5.2.1.5 Problems in Desalter

Commonly problems encountered during desalting operation are [Mandal, 2005]:
- High oil carryover along with brine
- No improvement in desalter performance even after demulsifier dosing
- Crude flow fluctuation during normal operation from ex-desalter
- Very high crude loss along with brine during crude change over
- Frequent tripping or high current in desalter transformer during low density with high BS and W crude oil processing.

5.3   CRUDE OIL DISTILLATION

Desalted crude flows to atmospheric and vacuum distillation through crude pre-flashing section. Atmospheric and vacuum distillation column is the main primary separation processes producing various straight run products, e.g. gasoline to lube oils/vacuum gas oils (VGO). These products, particularly the light and middle distillates, i.e. gasoline, kerosene and diesel are more in demand than their direct availability from crude oils, all over the world.

Crude oil distillation consists of atmospheric and vacuum distillation. Distillation process is shown in Fig. 5.2. The heavier fraction of crude oil obtained from atmospheric column requires high temperature. In order to avoid cracking at higher temperature, the heavier fractions are fractionated under vacuum. Typical operating parameter for atmospheric distillation column is given in Table 5.2. Various streams from atmospheric and vacuum distillation column are given in Table 5.3.

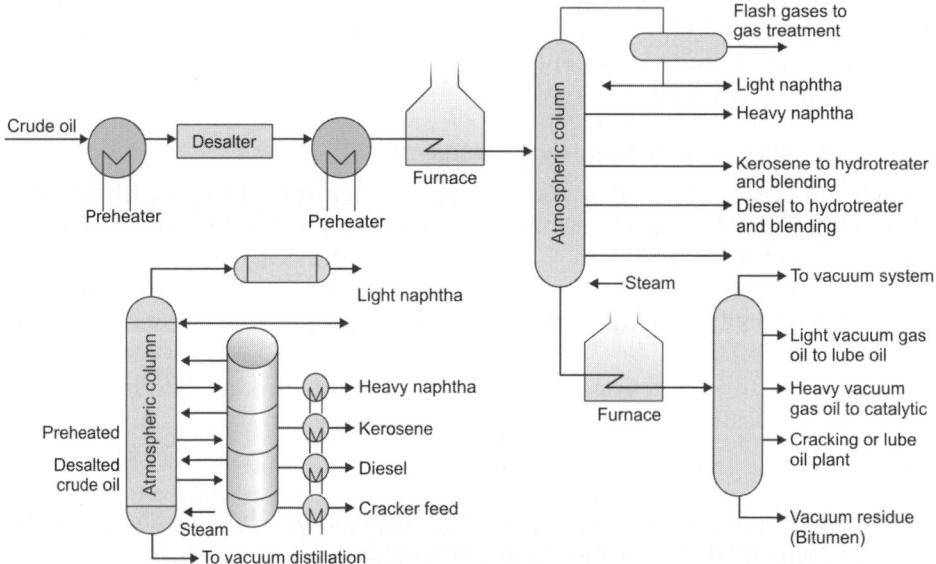

Fig. 5.2 Crude oil distillation

Table 5.2 Typical operating parameter for atmospheric crude distillation column

Crude heater	Arab mix	Upper zakum
Feed (m³/hour)	553.4	
Crude rate (each pass m³/hour)	182	180
Inlet temperature (°C)	287	291.3
Transfer temperature (°C)	374.5	379
Crude column	**Arab mix**	**Upper zakum**
Flash zone temperature	367.5	370.7
Flash zone pressure (kg/cm² g)	2.18	2.74
Column top temperature	114.9	113
Column top pressure	1.7	2.3
Steam to column (kg/hour)	6,650	7,200
Column bottom temperature	361.5	363.5
Column bottom pressure	2.21	2.81
Overflash	33.7	

Contd.

Table 5.2 Typical operating parameter for atmospheric crude distillation column *(Contd.)*

Draw off temperature (°C)	Arab mix	Upper zakum
Heavy naphtha	156.5	160.2
Kerosene	204.7	203.8
GO	283.9	276.5
EGO	340.8	341.3
Atm. column bottom temperature	361.5	363.5
Heavy naphtha (HN) stripper	**Arab mix**	**Upper zakum**
Outlet temperature (°C)	183.8	192.6
Reboiler outlet temperature (°C)	188.1	198
HN stripper vapor temperature (°C)	170.3	174.4
Go stripper	**Arab mix**	**Upper zakum**
Outlet temperature (°C)	273	266.5
Stripping steam (kg/hour)	1170	1650
Go vapor return temperature (°C)	279.6	272.7
Kero stripper	**Arab mix**	**Upper zakum**
Outlet temperature (°C)	190	189.4
Stripping steam (kg/hour)	2050	1570
Kero stripper vapor temperature (°C)	198.1	197.4
Naphtha stabilizer	**Arab mix**	**Upper zakum**
Feed temperature (°C)	120	120
Column top temperature (°C)	62.4	62.3
Column top pressure (kg/cm² g)	7.9	7.9
Column bottom temperature (°C)	165.2	148.4
Naphtha splitter	**Arab mix**	**Upper zakum**
Feed temperature (°C)	116.4	96.2
Column top temperature (°C)	93.8	83.2
Column top pressure (kg/cm² g)	1.2	1.2
Column bottom temperature (°C)	143.2	143.3
Column tray no. 1 (bottom temperature) (°C)	146	146.4
EGO stripper	**Arab mix**	**Upper zakum**
Outlet temperature (°C)	330	331.8
Stripping steam (kg/hour)	240	220
EGO vapor return temperature (°C)	337	338.2

Courtesy: CPCL, Chennai.

Table 5.3   Various streams from atmospheric and vacuum distillation column

Column	Fraction	Temperature (°C)	Carbon range	Uses
Atmospheric column	Fuel gases	> 40	C_1–C_2	Fuel
	LPG		C_3–C_4	Domestic fuel
	Straight run gasoline	20–90	C_6–C_{10}	Gasoline pool
	Naphtha (medium and heavy)	130–180	C_6–C_{10}	Catalytic reforming and aromatic plant feedstock steam cracker, synthesis gas manufacture
	Kerosene	150–270	C_{11}–C_{12}	Aviation turbine fuel, domestic fuel, LAB feedstock (paraffin source)
	Light gas oil	230–320	C_{13}–C_{17}	High speed diesel component
	Heavy gas oil	320–380	C_{18}–C_{25}	High speed diesel component
Vacuum column	Light vacuum gas oil	370–425	C_{18}–C_{25}	Feed to FCC/HCU
	Heavy vacuum gas oil	425–550	C_{26}–C_{38}	Feed to FCC/HCU
	Vacuum slop	550–560		RFCCU feed
	Vacuum residue	> 560	> C_{38}	Bitumen/visbreaker feed

5.3.1   Atmospheric Crude Distillation

5.3.1.1   Fired Heater

The fired heaters consist of radiant and convection section. Radiant section consists of the burners and forms the combustion chamber. Radiant section houses the burners and forms the combustion chamber or fire box. Convection section receives heat from the hot flue gases leaving the radiant section and is, therefore, placed above the radiant section. It contains rows of tubes for superheating steam and heating oil. Thermal oxidation of the most of the crude oils occurs at temperature around 340–370 °C. Skin temperature of tubes is limited to 550 °C.

5.3.1.2   Air Preheater Section

Air preheater recovers the waste heat from fuel gases coming from the fired heaters.

5.3.1.3   Atmospheric Distillation Column

The preheated and partially vaporized crude oil from the fired heaters enters the flash section column. The flashed hydrocarbon vapours from the flash section are fractionated into different sections above the flash zone. Non-flashed fraction from the flash zone moves downward as reduced crude oil. Four fractions are separated in the atmospheric column. The overhead vapors are condensed. The condensed overhead product is separated as hydrocarbon and water, in the reflux drum from which water is withdrawn and sent to sour water section. The unstabilized naphtha containing fuel gas, LPG and naphtha is partially refluxed and sent to naphtha stabilizer section. Typical temperature at furnace coil outlet is in the range of 310–340 °C.

Three other products separated in atmospheric column are heavy naphtha, kerosene and gas oil. These fractions are withdrawn from different sections of the column and in some of the refinery two side streams of gas oil are withdrawn as light gas oil stream and heavy gas oil stream.

Heavy naphtha is the first side stream withdrawn from the atmospheric column.

Kerosene is withdrawn as second side stream from the atmospheric column which is used from—manufacture of aviation turbine fuel and partly superior kerosene. Kerosene is first desulfurized in kerosene hydrodesulfurization section and then send to storage tank.

Gas oil is withdrawn as other side stream from the atmospheric column which is used as diesel blending stock is first desulfurized in the diesel hydro-desulfurization section.

Various streams drawn from a typical atmospheric column are: Overhead stream, top circulating reflux, heavy naphtha draws off, kerosene circulating reflux, kerosene draws off, diesel circulating reflux, diesel draws off, gas oil draws and reduced crude.

5.3.1.4   Naphtha Stabilizer Section

Unstabilized naphtha from atmospheric column is pumped to the naphtha stabilizer section for separating LPG as top fraction and naphtha as bottom. Stabilizer overheads are condensed and flow to reflux drum from where part of the hydrocarbon is refluxed and balance LPG goes to caustic and amine treatment unit sweetening. Stabilized naphtha from the stabilizer section goes to naphtha splitter for separating into light naphtha, medium naphtha and heavy naphtha.

5.3.1.5   Naphtha Splitter

In the naphtha splitter column naphtha is split into naphtha overhead and bottom product stream of different boiling points.

5.3.1.6   Gas Plant

LPG from lighter component of naphtha stabilizer component is separated in gas plant. Propane may be also separated, if necessary for deasphalting unit. A typical gas plant may consist of deethanizer, amine washing and depropanizer.

5.3.2 Vacuum Distillation

The bottom product also called reduced crude oil from the atmospheric column, is fractionated in the vacuum column. Distillation under vacuum permits fractionation at lower temperature which avoids cracking of the reduced crude oil and coking of the furnace tube. Reduced crude oil from the atmospheric column is first preheated in a vacuum furnace and introduced in the flash zone of the vacuum column. The flash zone temperature and pressure are around 380–400 °C and 115–120 mm Hg.

Vacuum is maintained by a three-stage steam ejector system. Steam is injected in the flash zone which further reduced the partial pressure of the hydrocarbons. The vaporized portion of the hydrocarbon in the flash zone is fractionated into various fractions and withdrawn as four side stream. The hydrocarbon vapours from various side streams are condensed. The various fractions are slop oil, light vacuum gas oil, heavy vacuum gas oil, and vacuum residue. Vacuum in the column is maintained by two- or three-stage ejector system with surface condenser. Typical yield of Merey and Arabian light crude oil is given in Tables 5.4 and 5.5 respectively. Typical operating condition of vacuum distillation column is given in Table 5.6.

CRUDE DISTILLATION

Table 5.4   Yields: Typical for Merey crude oil

Crude unit products	%wt	°API	Pour (°F)
Overhead and naphtha	6.2	58.0	–
Kerosene	4.5	41.4	–85
Diesel	18.0	30.0	–10
Gas oil	3.9	24.0	20
Light temperature vacuum gas oil	2.6	23.4	35
Heavy vacuum gas oil	10.9	19.5	85
Vacuum bottoms	53.9	5.8	(120)*
Total	100.00	8.7	85

Licensor: Foster Wheeler USA Corp.

Source: Hydrocarbon Processing, Nov 1998, p66.

Table 5.5   Yields: Typical for Arabian light crude

Products		%wt
Gas	C_1–C_4	0.7
Gasoline	C_5–150 °C	15.2
Kerosene	150–250 °C	17.4
Gas oil (GO)	250–350 °C	18.3
VGO	350–370 °C	3.6
Waxy distillate (WD)	370–575 °C	28.8
Residue	575 °C+	16.0

Licensor: Shell International Oil Products BV.

Source: Hydrocarbon Processing, Nov 1998, p67.

Table 5.6 Typical operating condition for vacuum distillation unit

Estimated coil outlet temperature of vacuum heater for Arab mix crude is 413.5 °C and Upper zakum crude is 411.5 °C

Vacuum heater	AM	UZ
Pass flow m^3/hour	82.2	73.2
Inlet temperature (°C)	361.5	363.5
Transfer temperature (°C)	413.5	411.5
Vacuum column		
Flash zone temperature (°C)	398.3	393.5
Flash zone pressure (mm HgA)	39	39
Bottom temperature (°C)	350	350
OVHD temperature (°C)	60	60
OVHD pressure (mm HgA)	20	20
Vacuum diesel draw off temperature (°C)	114.2	113.8
Vacuum diesel CR flow (m^3/hour)	149.4	156.5
Vacuum diesel CR temperature (°C)	50	50
Vacuum diesel IR (m^3/hour)	30.3	30.8
GO draw off temperature (°C)	210.4	210.8
LVGO draw off temperature (°C)	243.5	244.1
LVGO CR flow rate (m^3/hour)	93.7	93.5
LVGO CR temperature (°C)	177.9	178.6
LVGO internal reflux rate (m^3/hour)	0.6	0.6
HVGO draw off temperature (°C)	309.6	310.7
HVGO CR flow rate (m^3/hour)	373.8	350.6
HVGO CR temperature (°C)	243.2	244.4
Wash oil slop + Distillate draw off temperature (°C)	376.5	374
VR quench return temperature	259.6	239.4
VR quench flow rate	56.7	28.1

5.3.3 Effect of Crude Characteristics

Crude characteristics show considerable impact on performance of crude distillation and product distribution. Effect of crude characteristics on performance of crude oil distillation and product characteristics is given in Table 5.7.

Table 5.7   Effect of crude characteristics on performance of crude distillation

Characteristics	Effect
API	API is a measure of "heaviness" or "lightness". API = (141.5/density) – 131.5 API > 30, light crude API < 28, heavy crude
Viscosity	Viscosity is a measure of resistance to flow and is highly dependent on temperature. It is an important parameter for effective desalting as high viscosity crudes need high temperatures. There is a limit for temperature in desalters.
UOP K (characterization factor)	Characterization factor is an important characteristics of crude oil and it is a measure of paraffinity and aromaticity of crude. High UOP K is desired for high conversion in FCC. Aromatic molecules cannot be cracked in FCC. They will simply take ride through the plant [www.slideshare.net].
Total acid number (TAN)	Total acid number is a measure of naphthenic acid contents in crude. High TAN crude leads to corrosion in various sections of the unit large number of NA species are present in crude. All nap. acids are not corrosive. Latest research indicates that TAN is not a complete corrosion index. TAN with 2.5 may corrode at higher rate than TAN with say 6.
Bottom sediments and water (BS and W)	Bottom sediments and water (BS and W) is a measure of water and water dissolved substances like mud, sand and sludge. Lower the BS and W—the higher the reliability of the unit. BS and W is one of the major pointers for corrosive materials in crude.
Sulfur	Sulfur content is a measure of "sourness" and "sweetness" of crude and affects the product quality. It is removed from hydrotreaters by reacting with H_2 and recovered as elemental sulfur in sulfur recovery unit [www.slideshare.net]. S > 2.5, sour crude; S < 2, sweet crude
Total salts	Total salts is a measure of contaminant in crude and cause overhead corrosion or foul-up exchangers by settling and scaling. It is removed from desalters by washing and settling.
VGO metals	VGO metals are the measure of metals content in VGO fraction. Ni and V are known as poisons of VGO hydrotreater catalyst. Metals in VGO are controlled by controlling wash rate in slop wax section of vacuum column.

5.3.4 Operating Variables in Crude Oil Distillation Column

Some of the major operating variables in atmospheric column are crude charge heater outlet temperature, crude column top temperature, circulating reflux, column top pressure, product withdrwal temperature,oil reflux flows side stripper steam flow rates and product flow rate. Major vatriables in the vacuum column are crude temperature, flash zone tmperature, flash zone pressure, residuum bottoms temperature, residence time, slop wax reflux flow rate and heavy vacuum gas.

5.3.5 Chemical Injection

Presence of naphthenic acid, sulfur compounds and salt possesses major threat to various equipments in crude oil distillation unit. For protection of atmospheric tower overheads system which is more prone to corrosion, chemical and corrosion inhibitor facilities have been provided to reduce and minimize corrosion problem. Chemical and inhibitor injection facilities involve injection of caustic solution into crude and combination of ammonia and corrosion inhibitor into overhead vapor/reflux lines of atmospheric distillation column. Details of various types of corrosion problems, corrosion remediation measures are given in Chapter 23.

5.4 PROCESSING OF OPPORTUNITY CRUDE OIL

Producing clean fuel or high value petrochemical products from opportunity crude is a big challenge. Opportunity crudes are characterized by high sulfur content, metal content and total acid number, fewer lighter components, a high density and viscosity, high gel-asphalt content, high salts and heavy metal content which give rise to corrosion in equipment and concern with product quality, environmental protection and processibility [Qing, 2010]. Processing of heavy crude is a big challenge to refinery. Strong incentives exist in the low margin, highly competitive petroleum refining business to process low cost opportunity crudes [Eric-Vetters and Clarida, 2013]. With the use of more and more opportunity crudes there has been continuous problem in crude oil desalting. Desalter upset due to increase in desalter current resulting in decline grid voltage, decreasing emulsion breaking efficiency and oil under carry, water carryover, salt slippage and overhead corrosion results with stable emulsion which is difficult to break [Srinivasan et al., 2013].

Various problems associated with the utilization of opportunity crudes are [Venkatata Rammana et al., 2010]:

- Desalter performance and associated problem
- Corrosion related various equipment problems
- Catalyst life may be exhausted at a faster rate
- Drop in FCC and hydrocracker conversion
- Throughput reduction due to design limitations
- Increased operating cost due to increased hydrogen consumption
- Increased FCC catalyst withdrawal requirements
- Increased cetane improver additive consumption
- Product quality issues such as density of diesel, sulfur content of fuel oil.

Processing Options for Opportunity Crudes

Processing of opportunity crude possesses challenges and needs careful analysis of the crude type and its impact on the existing refinery.

- Metallurgy upgradation to process high TAN crude
- Upgradation of desalting, CDU and VDU
- Upgrading for heavy oil processing through blending with conventional lighter crude
- Upgradation of hydrocracker performance by increasing severity
- Catalyst improvement to handle high contaminants

REFERENCES

1. Bai ZS. Crude Oil Desalting Using Hydrocyclones. Chinese Journal of Chemical Engineering, 20 (2): 212–219 (2012).

2. Eric-Vetters and Clarida D. Maintaining Reliability When Processing Opportunity Crudes. Petroleum Technology, Quarterly, Q4, 2013, 59.

3. http://www.gastecheng.com/desalter.html.

4. Hoffman HL. Petroleum and Its Product. Riegel's Handbook of Industrial Chemistry, 9th edition, edited by James A Kent, 1997, p480.

5. Hydrocarbon Processing, Nov 1998, p66–67.

6. Kremer L and Bieber S. Rethink Desalting Strategies When Handling Heavy Feedstocks. Hydrocarbon Processing, Sep 2008, p113.

7. Mandal KK. Improve Desalter Control. Hydrocarbon Processing, April 2005, p77.

8. Qing W. Processing High TAN Crude, Part 1. Petroleum Technology, Quarterly, Q4, 2010, p35.

9. Srinivasan V, Subbramaniyam M and Shah P. Processing Strategies for Metallic and High Acid Crudes. Petroleum Technology, Quarterly, Q3, 2013, p51.

10. Venkatata Rammana U, Abhiram C and Roy G. Opportunity Crude Oil Processing and Process Challenges. Compendium 16th Refinery Technology Meet, organized by Center for High Technology and Indian Oil Corporation Limited, Feb 17–19, 2011, p33.

6

Thermal Cracking

Thermal cracking process for upgradation of heavy residue has been used since long and still it is playing an important role in the modern refinery through upgradation of heavy residue and improving the economics of the refinery through production of lighter distillate and other valuable products like low value fuel gas and petroleum coke. Although, petroleum coke was first made by North Western Pennsylvanian in the 1860s using cracking, however, real breakthrough in the thermal cracking process was with development of first cracker by William Burton and first used in 1913.

Heavy residues are the mixture molecules consisting of an oil phase and an asphaltene phase in physical equilibrium with each other in colloidal form. Asphaltenes are high molecular weight, relatively high atomicity molecules containing high levels of metals. During thermal cracking, the long molecules, thus depleting the oil phase in the residue. Asphaltene cracking is the most difficult component to process and asphaltenes in the feed remain unaffected, additional asphaltenes may be formed by secondary polymerization reactions. At a certain, asphaltenes are disturbed and precipitated. At this stage of conversion, the product residue becomes unstable.

Development of Cracking Processes

Year	Process
1861	Thermal cracking
1910	Batch thermal cracking
1912–13	Burton cracking
1914–22	Continuous cracking process

In a thermal cracking process, the heavy residue is cracked at higher temperature resulting in random distribution of molecular sizes. Various thermal cracking processes and process conditions are given in Table 6.1. The viscosity reduction of a feedstock increases with increased conversion (yield of gas and naphtha) up to a certain level, where maximum viscosity reduction is obtained. Increase in conversion beyond this value leads to decrease in viscosity reduction. There is a limiting conversion up to which a stable product can be obtained. Conversion beyond this leads to an unstable product which is undesirable. The conversion at which viscosity inversion takes place, may be different for each feedstock and needs to be established for each feedstock.

Asphaltenes, aromatic and paraffin content in the feedstock affect the limiting conversion for a stable product in the following manner:

- Higher asphaltenes lead to lower conversion
- Higher aromatics lead to higher conversion
- Higher paraffins lead to lower conversion.

Table 6.1   Various thermal cracking processes and process conditions

Process	Process conditions
Visbreaking	Mild thermal cracking (low severity) and mild (470–500 °C) heating at 50–200 psig improve the viscosity of fuel oil, low conversion (10%) to 221 °C, residence time 1–3 min, heated coil or drum
Hydrovisbreaking	In hydrovisbreaking, cracking is done in presence of hydrogen under pressure, with hydrogen donor or under hydrogen pressure with catalyst.
Delayed coking	Operates in semi-batch mode, moderate (482–515 °C) heating at 90 psig, soak drums (451–482 °C), coke walls, coked until drum solid, coke (removed hydraulically) 20–40% on feed, yield 221 °C, 30%
Fluid coking	Severe (510–520 °C) heating at 10 psig, oil contact refractory coke bed fluidized with steam—even heating, higher yield of light ends (< Cs), less coke yield
Flexicoking	Flexicoking is an extended form of fluid coking and uses a gasifier, temperature from 830–1000 °C
Gasification	Cracking of heavy residue into gaseous products and operate at temperature higher than 1,000 °C

Source: Rassev, 2003; Sloan, 1991; Patil, 2009; Rana et al., 2007.

Thermal cracking: The thermal cracking process may be medium, high and ultra-high (cracking with higher temperature and with very short residence time).

6.1   THERMAL CRACKING REACTIONS

Thermal reactions involve primary and secondary reactions. Primary reactions result in the decomposition of large molecules into small molecules. In secondary reactions some of the primary products interact to form higher molecular weight materials [Speight, 1999]. Thermal cracking reaction is a free radical chain reaction. Free radical is formed by carbon bond rupture which is favored by high temperature [Egloff, 1937]. Breaking of the bonds results predominantly in an increase in small molecules. Breaking of the bonds also leads to free radical formation that can also result in subsequent polymerization or condensation [McGrath, 2001].

Free Radical Mechanism

Initiation

$$C_6H_{14} \rightarrow C_2H_5 + C_4H_9$$

Propagation

$$C_2H_5 + C_6H_{14} \, \alpha \, C_2H_6 + C_6H_{13}$$
$$C_4H_9 + C_6H_{14} \rightarrow C_4H_{10} + C_6H_{13}$$
$$C_4H_9 \rightarrow C_3H_6 + CH_3$$
$$C_6H_{13} \rightarrow C_4H_8 + C_2H_5 \text{ (many other products)}$$

Termination

During the cracking operation, some cokes are usually formed. Coke is the end product of thermalization reaction in which two large olefin molecules combine to form an even larger olefinic molecule.

$$C_{10}H_{21}\text{-}CH = CH_2 + CH_2 = CH\text{-}C_{10}H_{21} \rightarrow C_{10}H_{21}\text{-}CH = CH\text{-}CH_2\text{-}CH_2\text{-}C_{10}H_{21}$$

Higher boiling petroleum stock $\longrightarrow$ Lower boiling products

Free radical chain reaction:

Free radical + hydrocarbon $\longrightarrow$ Stable end product

6.2   VISBREAKING

Visbreaking remains an important, relatively inexpensive bottom-of-the barrel upgrading processing in many areas of the world [Sieli, 1998]. With increasing use of heavier crude, there has been continuous upgradation in the visbreaking technology. The recent combination of the visbreaking technologies of Foster Wheeler and UOP and the addition of new innovative coil visbreaker design features have given the coil process a competitive advantage over the traditional soaker visbreaker [Sieli, 1998]. Higher conversion versions of visbreaking are available from companies including Shell Global Solutions and Foster Wheeler/UOP (Auaconversion^{TM}) while still achieving a stable fuel oil product [Phillips, 2003].

The vacuum gas oil (VGO) content in vacuum residue (VR) and its severity affect the stability of unconverted visbreaker residue. Decreasing VGO content in VR with high aromatic content did not affect conversion, but improve the VBR stability [Stratiev et al., 2008].

For upgrading of heavy feedstock the visbreaking process emerges as a cheapest refining tool [Tondon et al., 2007]. Visbreaking is essentially a mild thermal cracking operation at mild conditions and is based on thermal cracking of heavy residue operated at lower severity than thermal cracker. The purpose of visbreaking is to produce lower viscosity fuel oil. Cracking of the heavy feedstock takes place in the furnace coil alone or a combination of furnace and soaker. Long chain molecules in heavy feedstock are broken into short molecules thereby leading to a viscosity reduction of feedstock. The cracked product contains gas, naphtha, gas oil and furnace oil, the composition of which will depend upon the type of feedstock processed. A typical yield pattern may be gas 1–2%, naphtha 2–3%, gas oil 5–7% and furnace oil 90–92%.

Visbreaking is essentially a mild thermal cracking operation at mild conditions wherein long chain molecules in heavy feedstock are broken into short molecules thereby leading to a viscosity reduction of feedstock. During the cracking gas, naphtha and gas oil are produced as by-products.

A given conversion in visbreaker can be achieved by two ways:

- High temperature, low residence time cracking: Coil visbreaking
- Low temperature, high residence time cracking: Soaker visbreaking.

Some of the visbreaking technologies are given in Table 6.2.

Visbreaking Conditions

- Inlet temperature: 305–325 °C (15–40 bar)
- Exit: 480–500 °C (2–10 bar)
- With soaking 440–460 °C (5–15 bar)

Table 6.2   Various visbreaking technologies

Foster Wheeler/UOP coil type visbreaking [Negin and Van Tine, 2004; Refining processes, 2008]	Cracking of preheated vacuum residue takes place in furnace coil where partial vaporization and mixed cracking takes place, which is fractionated. Both coil and soaker type visbreaking processes are available.
Tervahl process (IFP)	The unit consists of heat recovery section, heater soaker and stabilization section.
Shell soaker visbreaking process	In shell soaker visbreaking, the preheated vacuum residue is charged to visbreaker and from there, the product goes to soaker. Cracking takes place in both visbreaker and soaker. Lower fuel requirement, increased heater run length, higher conversion and better viscosity reduction.
UOP aqua conversion process	Vacuum residue is heated to high temperature in heater and soaker (which may be optional) where thermal cracking takes place and characterized by uniform degree of conversion, high in tube velocities.
Hydrovisbreaking [Patil, 2009]	Hydrovisbreaking process takes place in presence of hydrogen which enhanced both yield and stability of liquid products. The process can be operated under both hydrogen pressure and other with hydrogen donors.

6.2.1 Coil Visbreaker

In coil visbreaking process the desired cracking is achieved in the furnace at high temperature in the furnace coil itself. The desired conversion is achieved by maintaining short residence time of 1–2 minutes in the reaction zone of the coil and further the reaction is stop by adiabatic flashing using relief valve and quenching with cold oil. The temperature of reaction products of cracking are subsequently quenched and distilled in a downstream fractionator.

6.2.2 Soaker Visbreakers

The visbreaker thermal cracking unit consists of thermal cracker furnace, a visbreaker furnace, a flash tower and fractionation column. The furnace operates at a lower outlet temperature where 20–25% conversion takes place only, while major conversion takes place in the soaker. Soaker is essentially a vertical cylinder with height to diameter ratio of about 6 and the residence time in the soaker is about 10–20 minutes. Soaker results in saving of energy because of lower temperature with less coke tendency and larger gas oil yield. The products from soaker drum are quenched and distilled in the downstream fractionator.

Advantages of Soaker Visbreaking

Lower capital expenditure (10–15% lower)
- Smaller furnace
- Less waste heat recovery equipment
- Lower pressure drop through the furnace.

Lower fuel consumption
- 30–35: Less direct fuel
- 15% less on net fuel, 0.2% on feed.

Better or more selective yield

Improved operation
- Longer run length, e.g. 1 year
- High on stream time: 330SD/A
- Less susceptible to upsets
- Better process control
- Less waste heat steam.

6.2.3 Visbreaker Furnace

The visbreaker furnaces consist of a bottom radiation and top convection zone. In the convection section in addition to the coils for feed heating, there are MP and LP steam superheating coils as well as MP steam generation coils to improve the thermal efficiency. As furnace tube exposed to high temperature preferred material is 9% Cr–1% Mo.

Advantages

- Fifteen percent reduction in fuel oil
- Larger running time between two decoking operations. Coke deposit rate 3–4 times slower than in conventional units.

  Better selectivity towards gas and gasoline productivity.

Process Description

The preheated visbreaker feed from the preheat exchanger is charged to the heater at about 325 °C and pressure of about 55 kg/cm². The heater is essentially a furnace. The visbreaker furnace consist of a bottom radiation and top convection zone in the convection section in addition to the coils for feed heating, there are MP and LP steam superheating coils as well as MP steam generation coils to improve the thermal efficiency. The feed is first preheated in the convection zone of the furnace before entering the radiation zone of the furnace. The furnace skin temperature is measured at various parts of the radiation zone. Maximum allowable tube skin temperature is around 650 °C. The feed from the furnace is sent to the bottom of the soaker drum. Soaking drum is a reactor in downstream and the visbreaker which lengthen the feed residence time allowing the furnace to operate at lower temperature. The thermal cracking which is initiated in the furnace is completed in the soaker drum. Residence time of about 20 minutes is maintained in the soaker. The operating temperature is maintained such as to reach desired conversion. The cracked products from the cracker are quenched with gas oil to maintain a temperature of 425 °C which stop further reaction and prevent production of an unstable bottom product. The quenched product from the visbreaker unit is sent to fractionator where it enters to the flash zone of the column at about 420 °C. The various products from the column are gases, gasoline, kerosene, gas oil and the residue. The overhead products of the column containing gases and gasoline sent to separation and stabilization section for separating into LPG and cracked gasoline. Part of the visbreaker tar is injected into the column bottom to avoid coking at the bottom section.

Typical products from visbreaking process are: Gases, visbreaker naphtha (C_5–166 °C), light visbreaker gas oil (166–343 °C), heavy visbreaker gas oil (343–471 °C) and visbreaker tar (470 °C).

6.2.4　Decoking of the Furnace Tubes

Decoking of the furnace tube is done by steam air which involves burning of the coke deposits on the internal walls of the furnace coils.

Operating Variables

Variables in visbreaker are feed rate, furnace transfer temperature, visbreaker tar quench to transfer line, fractionation pressure, fractionation top temperature, circulation, reflux flow, visbreaker tar quench to fractionator bottom, visbreaker tar quench to visbreaker tar stripper bottom, stabilizer temperature and pressure. Temperature pressure and residence time are the most critical variables so increasing any one of these will result in increase in overall severity [Negin and Van Tine, 2004].

Typical cracking reaction taking in a visbreaker may be splitting of C–C bond, condensation of cyclic molecules to polyaromatics and condensation of cyclic molecules, formation of sulfur compounds, H_2S, thiophenes, mercaptans and phenol.

Maximum conversion in visbreaking can be obtained by keeping the proper process in severity for any processed feed (controlling heater outlet temperature) and use of antifoulants/anticoke chemicals. Increase in run length of 50–75% without a reduction in severity or conversion [Vivirz and Respini, 2014].

Feed rate: Reduction in feed rate will result increase in residence time in the furnace coil while increasing flow will result decrease in residence time. Decrease in feed flow rate will lead in high severity condition and result in lower viscosity and pour point. Increased flow rate will result in high viscosity and pour point due to low severity.

Furnace transfer temperature (FTT): In case of high FTT, the cracking increases resulting in more gas, naphtha, and gas oil with low viscosity of tar, increased coke formation and reduced run length. The temperature is normally kept in the range of 480–490 °C depending on type of feedstock and type of visbreaker unit. Visbreaker tar quench reduces the temperature in transfer line and stop further cracking.

6.2.5 Hydrovisbreaking

In hydrovisbreaking, cracking is done in presence of hydrogen. Three versions of hydrovisbreaking are available—under hydrogen pressure, with hydrogen donor and under hydrogen pressure with catalyst in suspension [Raseev, 2003; Patil, 2009]. Residue feed preheated in preheat exchangers before entering the visbreaker heater where it is heated to visbreaker temperature. The high efficiency heater is also utilized to superheat stripping steam. Heater effluent is sent to the soaker drum where most of the thermal cracking and viscosity reduction takes place under controlled conditions. Soaker drum effluent is flashed and then quenched in the fractionator. Heat integration is maximized in order to minimize fuel consumption. The flashed vapors can be fractionated into gas, gasoline, gas oil and visbreaker residue [www.cbi.com]. The gas oil fraction can be included with the visbreaker effluent. It is also possible to obtain a heavy vacuum gas oil fraction by adding a vacuum flasher downstream of the fractionator (Fig. 6.1).

Cutter stocks, such as light cycle oil or heavy atmospheric gas oil, may be added to the visbreaker residue/gas oil mixture to meet the desired fuel oil specification. Process has been described in Figs 6.2 and 6.3. Visbreaking process is also operated without soaker (Fig. 6.3). Yield of light Arabian feed by visbreaking process is given in Tables 6.3 and 6.4.

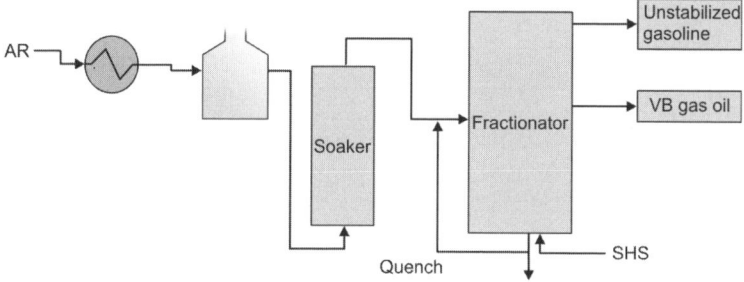

Fig. 6.1 Visbreaking with soaker

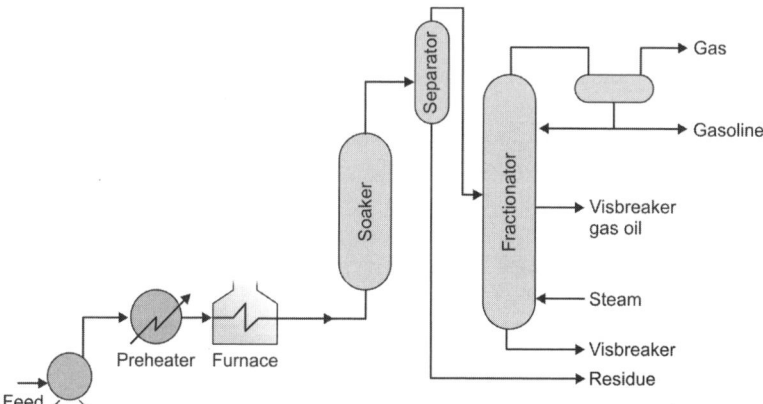

Fig. 6.2 Visbreaking with soaking

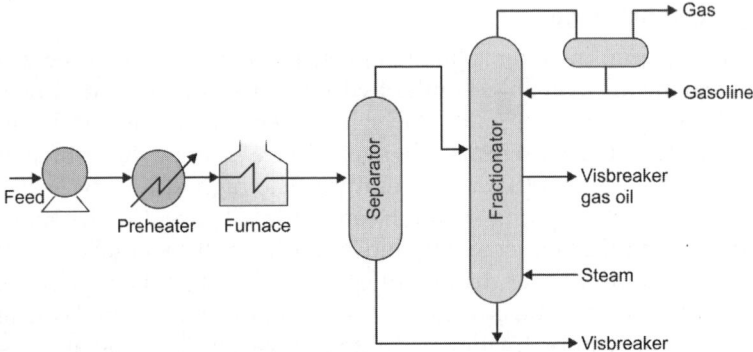

Fig. 6.3 Visbreaking without soaking

Visbreaking

Table 6.3   Yield: Light Arabian atmospheric and vacuum residues

Feed source type [www.cbi.com]	Light Arabian atm. resid	Light arabian vac. resid
Gravity (°API)	15.9	7.1
Sulfur (%wt)	3.0	4.0
Concarbon (%wt)	8.5	20.3
Viscosity, CKS @ 130 °F	150	30,000
CKS @ 210 °F	25	900
Products (%wt)		
Gas	3.1	2.4
Naphtha (C_5–330 °F)	7.9	6.0
Gas oil (166–316 °C)	14.5	15.5
Visbreaking resid (6 °C)	74.5	76.1

Source: Hydrocarbon Processing, Nov 1998, p67.

Table 6.4   Yield: Light Arabian product

	T	H
Feed/product visc (ratio @ 50 °C)	20–50	50–100
Product pour point (°C)	< –5	< –10
Desulfurization (%)	5–10	10–20
Conradson carbon	Slight increase	Slight decrease

Source: Hydrocarbon Processing, Nov 1998, p67.

6.3   COKING

Coking is a very severe form of thermal cracking and converts the heaviest low value residue to valuable distillates and petroleum coke. It is relatively severe cracking operation to convert residual oil. Product represents the complete conversion of petroleum residues to coke and lighter products. Recycle is used further to convert heavy distillate fractions to lighter products.

The reactions taking place in the coking are thermal cracking, condensation and polymerization reactions which occur in both sequentially and simultaneously [Hughes, 2004]. Various types of coking processes are delayed coking, fluid coking and flexi coking. Various commercial coking processes are mentioned in Table 6.5. During this process residue is generated, yield of this residue is mentioned in Tables 6.6 to 6.8.

Table 6.5   Various commercial coking processes

Process	Process description
Foster Wheeler (FW)/UOP delayed coking	In this process, feed is charged to the fractionator where it combines with recycle and is pumped to coker furnace where partial vaporization and mild cracking takes place. The vapor—liquid mixture from the furnace goes to coker drum where further cracking takes place. The overhead product from the coker goes to fractionator for separation into gas, naphtha, light and heavy gas oils. Heater inlet temperature　　　　: 482–510 °C Coker drum pressure, psig　　　　: 15–100 psig Recycle ratio, vol/vol feed%　　: 0–100
Kellogg Brown and root delayed coking	In this process, hot residual oil is fed to the base of fractionator where it mixes with the condensed recycle. The combine feed is heated in the furnace to suitable temperature to cause subsequent formation of coke in the coke drum. The coke drum overhead vapors flow to fractionator where they are separated to wet gas, naphtha, light and heavy gas oils and recycle. ["Sasol picks ABB Lummus delayed coker. Sasol Industries Pty Ltd contract]. Heater inlet temperature　　　　: 482–510 °C Coker drum pressure, psig　　　　: 15–90 psig Recycle ratio, vol/vol feed %　　: 0–100 [www.cbi.com]
Exxon flexicoking	It is a continuous process and involves thermal cracking of the feed in a fluidized bed system. This process contains an additional step of gasification of the coke produced at 870 °C. The gases leaving the gasifier are low calorific value fuel gases at 800–1,500 kcal/m³ (4,200 to 5,000 kJ/std m³) and is burnt in the furnace or power plants. It can be applied to wide variety of feedstocks.

Contd.

Table 6.5   Various commercial coking processes *(Contd.)*

Process	Process description
Exxon fluid coking	In fluid coking, 15–25% of the coke is burnt with air to satisfy the process heat requirements. Remaining coke is withdrawn. Residue is coked by spraying into a fluidized bed of hot, fine coke particles. This allows higher temperature with shorter contact time than delayed coking resulting in increased light and medium hydrocarbons with less cake generation. Typical fluid coking process is shown in Fig. 6.4.
ABBS Lummus global delayed coking	This process is similar to UOP process. The feedstock is fed to the bottom of the fractionator where it mixes with condensed recycle. The combined stream is pumped to the coker heater where desired coking temperature is achieved before entering the coker drum. Overhead vapor from the coker drum is fractionated into overhead gas, naphtha, gas oils and the recycle that joins the feed. Heater inlet temperature : 482–510 °C Coker drum pressure, psig : 15–90 psig Recycle ratio, vol/vol feed % : 0–100
ConocoPhillips delayed coking	Fresh feed combined with recycle distillate is pumped to the coker through furnace. Either high pressure steam or boiler feed water is injected into each of the furnace which helps in maintaining the optimum temperature and residence time in the furnace. The coker overhead product is quenched and fed to fractionator where it is fractionated to overhead gas, naphtha, light and heavy gas oils. Major benefits: maximum coking capacity, higher liquid yields, improved safety, greater operability and reliability. Coil out let temperature = 920–945 °C.
SYDEC™ delayed coking process licensed by Foster Wheeler	The selective yield delayed coking (SYDEC) is a low pressure; low recycle design for maximum liquid yields. Preheated fresh feed along with recycle stream is charged to coker heater. Steam is injected in the heater coil to maintain required minimum velocity and residence time and to suppress coke. Vapor liquid from the furnace enters coke drum where coke is formed along with light hydrocarbons which are fractionated.

Sources: Refining Processes, 1998; HC, 1994; Hughes et al., 2004; Feintuch and Negin, 2004; UOP, 2005; (www.uop.com).

Table 6.6 Yields of various residues

Feedstock	Middle east vac. residue	Vacuum residue of hydrotreated bottoms	Coal tar pitch
Gravity (°API)	7.4	1.3	−11.0
Sulfur (%wt)	4.2	2.3	0.5
Conradson carbon (%wt)	20.0	27.6	–
Products (%wt)			
Gas + LPG	7.9	9.0	3.9
Naphtha	12.6	11.1	–
Gas oils	50.8	44.0	31.0
Coke	28.7	35.9	65.1
Installation: More than 55 units			
Licensor: ABB Lummus Global Inc.			

Source: Hydrocarbon Processing, Nov 1998, p62 [www.cbi.com].

Yields: Example, Arabian heavy 1,0500 F + vacuum resid, 3.2 °API, 28.5 %wt concarbon, 5.6 %wt sulfur, 975 °F recycle cutpoint mode:

Table 6.7 Yields of Arabian heavy residue

Light ends (%wt)	12.9
C_5–221 °C (%wt)	14.4
221–343 °C (%wt)	10.2
343–525 °C (%wt)	27.3
For fluid coking: Net coke yields (tons/bbl)	0.05
Installation: Eighteen units built (over 530 kbpd capacity)	
Licensor: Exxon Research and Engineering Co	

Source: Hydrocarbon Processing, Nov 1998, p62.

Table 6.8 Yields of various residues

Feed, sources	Venezuela	N. Africa	–
Type	Vac. Residue	Vac. Residue	Decant oil
Gravity (°API)	2.6	15.2	–0.7
Sulfur (%wt)	4.4	0.7	0.5
Conradson carbon (%wt)	23.3	16.7	–
Operation products (%wt)	Max dist.	Anode coke	Needle coke
Gas	8.7	7.7	9.8
Naphtha	10.0	19.9	8.4
Gas oils	50.3	46.0	41.6
Coke	31.0	26.4	40.2
Installation: More than 58,000 tpd of fuel, anode and needle coke.			
Licensor: Foster Wheeler USA Corp/UOP LLC			

Source: Refining Processes, 1998, p63.

6.3.1 Delayed Coking

Delayed coking is the most effective process to decarbonize and demetallize heavy petroleum residues. It can provide 20–40% of the downstream hydroprocessing feedstocks [Elliot, 2003, fwc.com]. A relatively low cost resid-upgrading option delayed coking is favored as a relatively low cost resid-upgrading option [Bansal et al., 1994]. Some of the major technological developments in the delayed coking are [Phillips and Liu Fang, 2003]:

- Development of automated coke drum unheading devices, allowing to carry out the decoking procedure safely
- Understanding of process parameters affecting yields, coker product qualities and coke qualities
- Design and operation of major equipments items, in particular coke drums (allowing shorter coking cycle) and delayed coker heater (online spalling/decoking and minimization of coking in furnace tubes [www.coursehero.com]

The process involves thermal conversion of vacuum residue or other hydrocarbon residue resulting in fuel gas, LPG, naphtha, gas oil and coke. Delayed coking consists of thermal cracking of heavy residue in empty drum where deposition of coke takes place. The product yield and quality depends on the typed feedstock processed. The process is batch-continuous process. The process remains a preferred residue upgradation option because of its ability to handle heaviest, contaminated crude. Globally, one-third of the installed residue upgrading plant is by delayed coking [Phillips and Liu Fang. 2003, www.coursehero.com].

Typical delayed coking consists of a furnace to preheat the feed, coking drums where the fractionation of the product takes place. The feed stream is switched between two drums while one drum is in operation and another drum is switched off for decoking. The feed is first preheated in furnace where desired coking temperature is achieved and fed to the coking drums normally installed in pairs where the cracking reaction takes place and the coke is deposited in the bottom of the reactor. The reactions involved in delayed coking are partial vaporization and partial cracking, cracking of two vapor phases in the coke drum and successive cracking used polymerization of liquid phase resulting in formation of coke in the drum. The temperature in the tubular reactor is maintained just lower than the intense formation of coke begins. This is achieved by injection of water or steam delay the formation of coke. The coke drums overhead vapor flows to the fractionating column where they are separated into overhead stream containing wet gas LPG and naphtha and two side gas oil streams. Typical coke drum consists of about 16–18 hours decoking. Figure 6.4 illustrates the process of coking.

Recycled stream from the fractionating column combines with the fresh feed in the bottom of the column and is further preheated in coke heaters and flows to the coke drums is shown in Fig. 6.5. When a coke drum is filled, the heated streams from the coke are sent to the other drum. Some of the commercial delayed coking processes are FW delayed coking process, FW/UOP delayd coking, Kllong Brown and Rot delayed coking and ConocoPhillips delayed coking.

Optimizing coker operation is crucial. Effective use of coker recycle can improve the yield and the coker—liquid products quality. Recycle is produced by the direct condensation of the distillate tail in coke drum vapors. Zero recycle and ultralow recycle operations for fuel grade coking can effectively optimize coke integration in a clean fuel programmed [fwc.com]. Impact of reducing recycle ratio is: Lower coke formation, higher liquid yield [Elliot, 2003]. Zero recycle minimizes coke and maximizes liquid product yields. Characteristics of feed and product is given in Table 6.9.

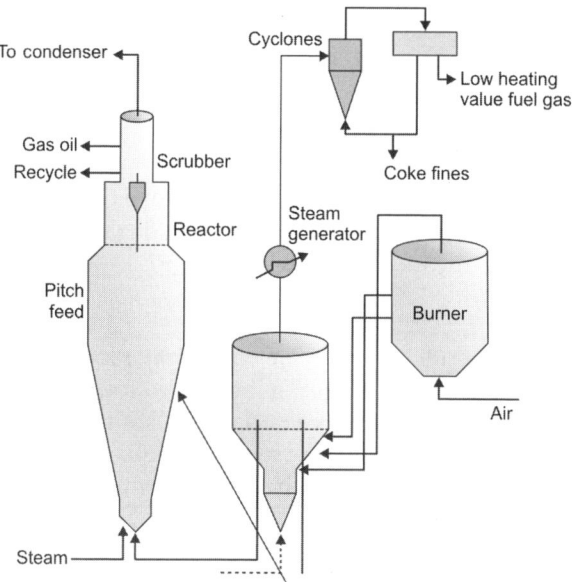

Fig. 6.4 Fluid coking

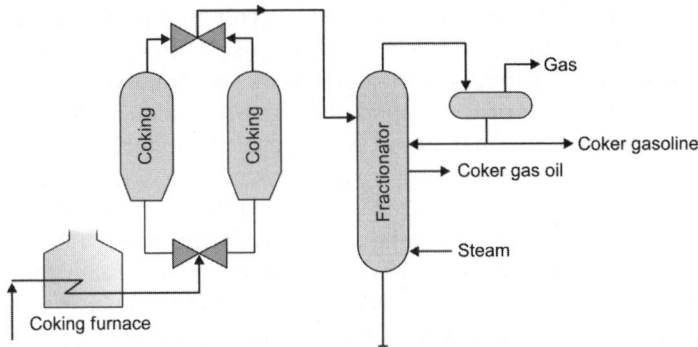

Fig. 6.5 Process flowchart of coking

Typical Operating Condition in Delayed Coking

Table 6.9   Feed and product characteristics

	Mol. Wt.	Density	Sulfur composition	Viscosity at 50 °C	Pour point
Feed	540.31	0.9916	4.06	–	2.2
Gas $(C_1–C_4)$	348	–	–	–	–
C_5 177 °C	100.2	0.723	1.2	–	–
117–357	210	0.8816	2.6	2.6	–
357 + Cal.	500	1.008	4.3	2100	+21

Reactions

The cracking reactions taking place in delayed coking are complex. Three distinct reactions occur [UOP 4223–46]:

- Partial vaporization and mild cracking of the feed as it passes through the coker's furnace
- Cracking of the vapor as it passes through the coke drum
- Successive cracking and polymerization of liquid trapped in the coking drum until it is converted to vapor and coke [uop.com].

Coker furnace

Coker furnace is a most critical part of the coking operation and provides all heat required in the process. The furnace may be two to four pass with tube of high alloy. Multiple burners are fitted in the radiant wall of the furnace opposite from the tubes and fired vertically during course of time coke is deposited in the furnace resulting in higher pressure drop and high tube skin temperature. Optimum furnace design criteria are [Elliot, 1996]:

- High in-tube velocities result in maximum heat transfer coefficient
- Minimum residence time in the furnace especially above the cracking temperature threshold

- A constantly rising temperature gradient
- Optimum flux rate with minimum practicable maldistribution based on peripheral tube surface
- Symmetrical piping and coil arrangement within furnace enclosure
- Multiple steam injection for each heater pass [coking.com].

Operating variables

Several key variables affect delayed coker yields and economics which include feedstocks quality, coker temperature, recycle rate and coke drum pressure [Guthrie, 1960; Bansal et al., 1994].

Feedstock variables: Characterization factor, degree of reduction, conradson carbon, sulfur content, metallic constituents.

Process variables: Time, temperature, pressure, recycle feedstock ratio, coke removal factors.

Engineering variables: Batch, semi-continuous or continuous, capacity and size factors, coke removal equipment, coke handling, storage and transportation [Mantell, 1976].

Recycled stream from the fractionating column combines with the fresh feed in the bottom of the column and is further preheated in coke heaters and flows to the coke drums. When a coke drum is filled, the heated streams from the coke are sent to the other drum.

Temperature: Higher temperature results in more vaporization of the inlet material causing low coke yield. Furnace outlet temperature of 485–505 °C is maintained. High temperature results in hard coke while coke is soften when too low temperature is maintained. Typically, temperature is varied by changing furnace outlet temperature [Bansal et al., 1994].

Pressure: Higher pressure results in increase in coke and gas yields which is undesirable as basic objective is to improve the yield of distillation with less coke. Reducing coke drum pressure reduces coke make and increases liquid product yield earlier coke drum operating pressure. Cokers are now being designed and revamped to operate at pressure as low as 15 psig [Bansal et al., 1994].

Recycle ratio: Higher recycle ratio results in higher coke and gas yields. Lower recycle ratio is always desirable for higher yield of liquid product. The unit's throughput ratio (TPR) is defined as the volumetric ratio of the furnace charge to fresh feed. As TPR is reduced coke make decreases while C_5^+ liquid yield increases. However, heavy coker gas oil contaminants, metals and sulfur increase with decreasing TPR. Although conventional cokers are designed with TPRs of 1.10–1.15, new cokers are being now designed with TPRs of 1.05 or lower to maximize liquid yields [Bansal et al., 1994].

Decoking

During the delayed coking process the coke level is build up in the drum. After the coke reaches a predetermined level in one drum, the flow is diverted to another drum for operation. There are two methods of decoking—mechanical and hydraulic. Various steps involved in decoking operation are: Steam stripping of the uncracked residual hydrocarbons, cooling of the coke drum with water, draining of water, decoking, and pressure checking, and heating of the cold coking drums for use.

6.4   COKE FORMATION MECHANISM, PROPERTIES AND STRUCTURE

Cokers are defined by types of coke produced such as: Fuel grade, anode grade and needle grade.

Mechanism of coke formation

There are two mechanisms of coke formation in the cocking process:

- The colloidal suspension of the asphaltenes and resin compounds is distorted, resulting in precipitation of highly cross-linked structure of amorphous coke. The compounds are also subjected to cleavage of the aliphatic groups.

  Polymerization and condensation of free aromatic radicals, grouping a large number of these compounds to such a degree that dense high grade coke is eventually formed [Sawarker et al., 2007].

  Coke yield = 1.6 CCR (conradson carbon residue)

  Gas yield = 7.8 + 0.144 × CCR

  Naphtha yield = 11.29 + 0.343 CCR

  Gas yield = 110 – Coke – Gas yield – Naphtha yield

- Low feedstock characterization factor and high carbon residue increase coke yield and quality of gas oil end-point

- Weight of sulfur in the coke will be greater than in the feedstock.

- Bulks of metal present in the feedstock will concentrate in the coke.

6.5   PETROLEUM COKE

Petroleum coke is carbonaceous solid derived from coker. Various types of petroleum cokes are produced in the quality of which it depends upon the type of coker, feedstock and operating conditions. Petroleum coke can be classified as fuel grade coke or regular, anode grade coke or needle grade coke. Sponge coke is most common type of coke, while needle coke can be made only from special feedstock. Major application of petroleum coke is as fuel, electrode, metallurgical use, gasification. Petroleum coke can be used along with coking coals in metallurgical processes. Newer application of the delayed coking coke is in the gasification to produce low-Btu gas or syn-gas.

6.6   NEEDLE COKE

Needle coke is a premium grade petroleum coke, used in manufacturing of graphite electrodes for the arc furnace in the metallurgy industry [petrotech2009.org]. Needle coke is hard and dense mass formed with different structures of carbon threads. Needle coke is highly crystalline and can provide properties needed for manufacturing graphite electrode. It can withstand temperatures as high as 2,800 °C. Some of the important properties of needle coke are:

- Higher real density (more than 2.13 g/cc)

- Low sulfur (less than 0.5 %wt) and ash content (less than 0.5 %wt) and lower coefficient of expansion (CTE) (less than 1.0×10^{-6} cm/cm/C for petroleum grade)

- Low CTE allows the electrode to operate with high temperature gradient without spalling the hot tip.

Sulfur present in needle coke causes problems with puffing, swelling or blatting in the green electrodes during graphitization affecting the density and mechanical strength of the electrode [Indian Oil technologies Ltd, http://www.iocltech.com/coke_faq.html]. Needle coke is produced from low sulfur highly aromatic feedstocks. It is used for manufacturing of graphite electrodes used in electric arc furnace. While producing high value needle coke, the feedstock must be highly aromatic and have low asphaltene, sulfur and ash contents [petrotech2009.org].

6.7 COKE PROPERTY IMPROVEMENT

Producing high quality needle coke from aromatic tars with no asphaltenes or very high quality anode grade coke paraffinic, low sulfur residues commonly require using of high recycle operations especially for the former [Elliot, 2003, fwc.com]. Table 6.10 presents comparisons of typical fuel coke and anode coke operations [Elliot, 2003].

Table 6.10 Comparison of typical fuel coke and anode coke operations

Operation		Fuel grade coke	Anode coke
Feed type		Heavy, high sulfur VR	Light, sweet VR
Operating conditions		Low pressure	Moderate pressure
Fuel gas	Iv% FOE	5.11	6.76
C_3/C_4 LPG	Iv%	7.31	8.15
Naphtha	Iv%	19.87	13.92
Lighter coker gas oil	Iv%	24.49	33.00
Heavy coker gas oil	Iv%	29.22	25.74
Coke	%wt	30.51	29.92

Source: Elliot, 2003
Courtesy: Hydrocarbon Processing.

6.8 IOC NEEDLE COKE TECHNOLOGY [INDIAN OIL TECH/2003/002]

IOC has developed needle coke technology and producing needle coke in its Guwahati refinery. The technology is primarily focused on production of needle coke in an existing delayed coker unit using available feed stream without any cost of pretreatment. Formation of needle coke requires specific feedstocks, special coking and also special calcination conditions. Typical yield of needle coke is 16–25%.

Needle coke properties are strongly dependent on the feedstock characteristics. Sulfur, metals and other impurities of the feed, unless separated, would be concentrated in the coke and their presence in the needle coke adversely affects its price. Requirement of petroleum coke is mentioned in Table 6.11.

Process conditions: Coil outlet temperature : 470–510 °C
Recycle ratio : 0.2–0.8 hours
Cycle time : 18–36 hours
Coke drum pressure : 1.5–9.5 kg/cm^2

Petroleum coke calcining: Petroleum calcining is a process of making an anode grade coke.

Calcined petroleum coke IS 8502: 1994 [www.indianoilcorp.com].

Table 6.11 Requirements for petroleum coke (Clauses 3.2, 3.3 and 3.3.3)

Sl No.	Characteristic	Requirement for				Method of test	
		Raw petroleum coke		Calcined petroleum coke		Annex	IS 14
		Grade A	Grade B	Grade A	Grade B		
1	2	3	4	5	6	7	8
	Moisture content (as received) (percent by mass max.)	10				–	P:13
	Moisture content (after initial drying for raw petroleum coke only), percent also special calcinations conditions (by mass max.) [www.indianoilcorp.com]	2.0		0.1		–	do
	Ash, percent by mass (max.)	0.45		0.5			P:12
	Volatile matter (percent by mass, max.)	11.0		0.4		–	P:13
Density							
	Vibrated bulk (g/cm^2)			To be reported		–	P:13
	Real* (g/cm^2, min.)			2.03			P:13 and P:1
	Fixed carbon (% by mass, min.)	85.0		99.0		A	
	Sulfur total (% by mass, max.)	1.25	2.5	1.25	2.5		P:33
Trace metals							
	Silicon (Si) (% by mass, max.)	To be reported		0.05			P:131
	Iron (Fe) (% by mass, max.)	do		0.04			P:127
	Vanadium* (V), (% by mass)	do		0.03			P:79
	Nickel (Ni) (percent by mass)	do		To be reported			P:128

* For graphite industry a higher real density and low vanadium content product is required; the limits for this may be settled between the purchaser and the supplier.

6.9   TECHNOLOGICAL DEVELOPMENT IN DELAYED COKING

Some of the major technological developments in the delayed coking are [Phillips and Liu, 2003]:

- Development of automated coke drum unheading devices, allowing the operator to carry out the decoking procedure safely
- Understanding of process parameters affecting yields, coker product qualities and coke qualities
- Design and operation of major equipment items, in particular coke drums allowing shorter coking cycle and delayed coker heater (online spalling/decoking and minimization of coking in furnace tubes [www.coursehero.com].

REFERENCES

1. Akbar M and Geelen H. Visbreaker Uses Soaker Drum. Hydrocarbon Processing, May 1981, p81.
2. Bansal BB, Gentry AR and Moretta JC. Improve Your Coking Process. Hydrocarbon processing, Feb 1994, p63.
3. Catalaa KA, Karrs MS and Sieli G. Advances in Delayed Coking Heat Transfer Equipment.
4. Egloff G. The Reaction of Pure Hydrocarbons. Reinhold, 1937.
5. Elliot JD. Optimize Coker Operations. Hydrocarbon Processing, Sep 2003, p85.
6. Elliot JH. How Petroleum Delayed Coke Forms in a Drum, Foster Wheeler Corporation, 1996.
7. Feintuch HM and Negin KM. FW Delayed Coking process in Handbook of Petroleum Refining Processes, 3rd edition, edited by Meyers RA, McGraw Hill Companies, 2004.
8. Guuthrie VB. Petroleum Products Handbooks, McGraw Hill, New York, 1960.
9. Hughes GC, Michelle I, Wohlgenant MI and Doerksen BJ. ConocoPhillips Delayed Coking Process in Visbreaking and Coking, In Handbook of Petroleum Refining Processes, 3rd edition, edited by Meyers RA, 2004.
10. Indian Oil Tech/2003/002. Indian Oil Technologies Ltd., Faridabad, India.
11. Mantell Charles L. Petroleum Based Carbon. ACS Symposium Ser 211976: …
12. McGrath M. Thermal Cracking, Visbreaking and Coking. In Modern Petroleum Technology Vol. 2. Downstream, edited by Lucas AG, John Willey & Sons, 2001.
13. Negin KM and Van Tine FM. FW/UOP Visbreaking Process. Handbook of Petroleum Refining Processes, 3rd edition, edited by Meyers RA, 2004.
14. Patil A. Improve Refinery Visbreaking Process for Higher Yields of Valuable Middle Distillates. Chemical Industry Digest, Aug 2009, p69.
15. Phillips G and Fang L. Hydrocarbon Engineering, Sep 2003, www.fwc.com. fwc.com.
16. Rana MS, Samano V, Ancheyta and Diaz JAI. A Review of Recent Advances on Process Technologies for Upgrading of Heavy Oils and Residues. Fuel, June 2007, p1216.
17. Rassev S. Industrial Implementation of Thermal Processes. In Thermal and Catalytic Processes in Petroleum Refining, Marcel Dekker, 2003, p137.

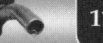

18. Refining Processes, 1998, 2008; Hydrocarbon Processing, Nov 1998; Hydrocarbon Processing, Sep 2008.

19. Sawarker AN, Pandit AB, Samant SD and Joshi JB. Petroleum Residue Upgrading Via Delayed Coking: A Review. Canadian Journal of Chemical Engineering, 85, 2007, p1.

20. Sieli GM. Visbreaker—The Next Generation, Foster Wheeler Publication, 1998.

21. Sloan ND. Process Heavier Crude Blends. Hydrocarbon Processing, Nov 1991, p99.

22. Speight JG. The Chemistry and Technology of Petroleum, 3rd edition, Marcel Dekker Inc, 1999.

23. Stratiev D, Kirilov K, Belchev and Petkov P. How do Feedstocks Affect Visbreaker Operations. Hydrocarbon Processing, June, 2008, p105.

24. Tondon D, Dang GS and Garg MO. Visbreaking: A Flexible Process to Reduce the Pour Point of Heavy Crude Oils. Journal of Petrotech Society, June 2009, p44.

25. Virzi M and Respini. Increasing Conversion and Run Length in a Visbreaker. Petroleum Technology, Quarterly, Q1, 2014, p123.

26. www.coursehero.com.

7
Catalytic Cracking

Distillation and thermal cracking were main processes in 1900s for production of gasoline, kerosene, diesel, fuel oil and lubricants. The first catalytic cracking plant was commissioned in 1936 based on Houdry Process using fixed bed catalytic cracking process. This initiated a new era in refining industry. Major advantages of catalytic cracking over thermal cracking are (i) lower C_1 and C_2 hydrocarbons with higher yield of C_3 and C_4 hydrocarbons, (ii) lower cracking temperature resulting in lower diolefins in product with improved oxidation stability of the gasoline fractions, (iii) higher octane ratings of gasoline [Walther, 2005] and (iv) higher cracking reaction rate [Irion and Neuwirth, 2003]. Major breakthrough in the catalytic cracking process was development of zeolite catalysts which demonstrated vastly superior activity, gasoline selectivity, and stability characteristics compared to original amorphous silica alumina catalyst [Hemler and Smith, 2004]. Major cracking catalyst development is given in Table 7.1. Some of the historical developments in the catalytic cracking process are fixed bed process (1936–1941), moving bed process (1941–1955), fluidized bed processes (1942 onwards). Commercially used catalytic cracking processes are fluidized catalytic cracking (FCC), residue fluidized bed cracking (RFBC), deep catalytic cracking (DCC) and hydrocracking. Hydrocracking process has been discussed in Chapter 8.

7.1 FLUID CATALYTIC CRACKING

Fluid catalytic cracking (FCC) is the most important cost effective refinery process to upgrade low cost feedstocks into transportation fuels [Yen Yung, 2010]. FCC converts low value crude oil fraction into a variety of higher value products [www.touchbriefings.com] which include gasoline, diesel, heating oil and valuable gases containing LPG, propylene and C_4 and C_5 gases. FCC units are versatile and can be operated in three main modes: Maximizing middle distillate, gasoline and olefins respectively by means of the adequate combination of various parameters like catalyst type, catalyst to oil ratio, rise of outlet temperature and recycle of fractionators bottom. The FCC naphtha accounts for about 45% of the gasoline pool. Each FCC operation is unique as a result of difference in feedstock quality, hardware, refining objectives and limitations [Salbilla, 1999]. FCC process plays a vital role in meeting two main global demands: Reducing fuel oil production and increasing propylene supply from the refinery [Ross, 2005]. Now modern FCC units can take a wide variety of feedstocks and can adjust operating conditions to maximize production of gasoline, middle distillate (LCO) or light olefins to meet different markets [www.nt.ntnu.no] [Chen,Ye-mon, 2006]. FCC gasoline contributes to the gasoline pool with high percentage of olefins [Nee et al., 2006], sulfur and benzene. Refinery FCC process also produces one-third of global propylene supply [Long et al., 2011].

Major breakthrough has been achieved as the result of innovations in catalyst formulations which resulted in higher gasoline selectivity, large coke selectivity, improved stability [Leiby, 1992] *in situ* control of emission. The first catalyst was amorphous silica—alumina which was followed by spray dried microspheroidal catalysts. Alumina content was 10–14% in first generation FCC catalyst that was further increased to 25–30% [Leiby, 1992].

Table 7.1   Milestones in cracking catalyst development

Year	Catalyst development
1915	Aluminum catalyst
1936	Activated natural clays
1940	The first catalyst of synthetic silica alumina (Houdry process)
1942	First commercial production of FCC catalyst
1946	First time use of microspherodial catalysts
1959	The synthesis and commercialization of Y zeolites
1962	First time use of zeolite catalyst (Mobil invented catalyst)
1964	Development of ultrastable catalyst (USY) and of those promoted with rare earths (REY)
1974	Introduction Pt combustion promoter
1974	Additives for the fixation of SO_2 in the regenerator and its elimination as H_2S
1975	Catalyst passivation against nickel
1978	Catalyst passivation against vanadium poisoning
1981	Al-sol bound catalysts (Davison)
1982	Commercialization of spinel sox additive (arco)
1983	Performance improvement by treating the catalyst with $(NH_3)_2SiF_6$ [Raseev. "Theoretical Basis of Catalytic Cracking", Thermal and Catalytic Pro…]
1984	Introduction of ZSM-5 type additive
1985	Maxofin FCC process
1988	Silicon enrichment by means of siliconeszzzakio882
1990	Introduction of NI tolérant matrix technologies
1990s	Rare earth-free stabilized zeolite-Y-catalyst
1992	Commercialization of V-trapping technology (Davison)

Contd.

Table 7.1   Milestones in cracking catalyst development *(Contd.)*

Year	Catalyst development
1995	Introduction of additives for gasoline sulfur reduction
1996	Introduction of CSSN/Z-17 high activity zeolites (Davison)
1997	Commercialization of Nox control additive technologies (Davison)
2001	Introduction of sulfur reduction catalyst
2003	Introduction of Z-28 and Z-30 high activity zeolites (Davison)
2004	Introduction of tunable reactive matrics (Davison) [JRD Nee. "Fluid Catalytic Cracking (FCC), Catalysts and Additives", Kirk-O...]
2010	Zero and low rare earth FCC catalysts (RepLaCeR series)

Source: Nee et al., 2006; Raseev, 2003; Srikantharajah, 2010.

Due to rising demand of gasoline and LPG transportation sector and propylene for petrochemical sector, now the refineries are operated in distillate mode, LPG mode or propylene mode. Another development has been the newer Resid FCC process which has become one of the most attractive ways for upgrading economically atmospheric or vacuum residues [Ross, 2005]. Some of the major constraints during recent years have been the use of heavier FCC feedstocks which make feed vaporization more difficult.

Although common for 60 years there has been continuous upgradation in FCC technology in process design, subsystem hardware, catalyst with newer highly active zeolites, higher regenerator temperature to achieve complete catalyst regeneration, and processing of heavier feedstock.

FCC is a worldwide refinery conversion process for production of high octane gasoline motor fuel. However, higher world demand of polypropylene makes FCC based petrochemical plants promising. FCCU accounts for 32% of worldwide petrochemical polypropylene production. The value of FCC derived petrochemical products is much than that of FCC derived petroleum products. Majority of FCC units produce naphtha as the primary product for blending into the gasoline pool, however, the light olefins from FCC can be used for motor fuel alkylation, oxygenates or petrochemical [Shorey et al., 1999]. There are about 400 FCC units operating worldwide [Hydrocarbon Processing, November, 1999]. FCC naphtha makes up 30–40% of the gasoline in a typical refinery pool. Presently, in India there are 13 FCC units with total design capacity of 17.55 million tons are operating in different refineries [Ghosh, 2002].

FCC in various regions is given in Table 7.2. Presently, we have 715 million tons FCC capacity. World FCC capacity by region is given in Table 7.3 [Walther, 2003]. Various FCC products and their uses are given in Table 7.4.

Light olefin yield can be increased in FCC processes by the use of ultrastabilized zeolite Y (USY)-based catalyst, use of medium pore zeolite (ZSM-5) as additive, increasing the severity and operating FCCU in overcracking regime [Shanker and Badoni, 2001]. Major landmarks in the history of FCC are introduction of zeolite catalyst, ultrastable-Y zeolite; switch over from bed cracking to riser cracking and introduction of number of additives for boosting gasoline octane, yield of light olefins, SO_x control, NO_x control and CO combustion promotion. In addition major improvements which have taken place are two-stage generation, catalyst coolers, improved feed distribution, quick separation of catalyst from the products at the top of the riser, improved stripper design, cold wall design, etc. [Ghosh, 2002].

Table 7.2   FCC units in various regions

Name of the country	No. of FCC units
America	183
Europe and Africa	79
FSO	26
Middle east	9
India	13
PRC	43
Rest of Asia	49
Total	**402**

Table 7.3   Worldwide FCC capacity by region

Region	Share of FCC (%)
North America	48
Western Europe	18
Asia	15
Others	19

Table 7.4   Various products and their uses

Product	Composition and uses
Light gases	Primarily H_2, C_1 and C_2s, ethylene can be recovered
LPG	C_3s and C_4s containing light olefins suitable for alkylations
Gasoline	C_5^+ high octane component for gasoline pool or light fuel
Light cycle oil (LCO)	Blend component for diesel or light fuel
Heavy cycle oil (HCO)	Fuel oil or cutter oil
Clarified oil	Carbon black feedstock
Coke	Used in regenerator to provide reactor heat demand

In 1962, Exxon Mobil introduced a new class of crystalline synthetic zeolite FCC catalysts that rapidly dominated the catalyst market [Leiby, 1992]. Cracking catalysts have undergone many revolutionary changes. Various FCC technologies are given in Table 7.5 [Hydrocarbon Processing Petroleum Refining Processes; 1999, 2008]. Another development in FCC is residue fluid catalytic cracking which is now fast growing version of conventional FCC process. The economics of RFCC unit is mainly governed by feed quality (CCR, metals, sulfur), product quality and environmental impact. New FCC process called fexible dual riser fluid catalytic cracking, can polymerize gasoline olefin and increase propylene significantly [Wang et al., 2003]. The majority of FCC units yield 4–6% propylene whereas ultra-high ZSM-5 based catalyst additives has the potential to produce up to 15% propylene [hcasia.safan.com]. Davison polymerization, a new technology that has exceeded 20% propylene in the riser pilot plants [Nee, 2003]. A typical FCC reactor is shown in Fig. 7.1.

Table 7.5 FCC technologies

Process technology licensers	Process description	Remarks
ABB Lummus Global Inc. FCC	The process consists of advanced reaction system, high efficiency catalyst stripper and a mechanically robust single stage fluidized bed regenerator. The feedstock may be virgin hydrotreated gas oil, but it may also include lube oil extract, coke gas oil and residue.	Product: High-octane gasoline, light olefins and distillate. High propylene yield in selective component cracking.
Kellogg Browns Root FCC	The converter is polymerized at to efficiently combine Kellog's proven orthoflow features with Mobil's advance designed features. Feedstock is gas oil and residues.	Product: Light olefins, high-octane gasoline and distillate. More than 120 units (> 2.5 million bpd fresh feed).
Exxon Research and Engg Co. Exxon Flexi Cracking III R FCC	This converts high boiling hydrocarbons like residual gas oil, lube extracts and/or asphalt into petrochemicals, LPG, blend stocks for high octane gasoline, distillates and fuel oil. The system consists of high efficiency closed coupled riser cyclone, feed injection system attachments, improved stripper design. Typical yield of Flexi cracking III R FCC is given in Table 7.6.	Product: Light olefins for alkylation polymerization or etherification, LPG, petrochemical intermediates, high-octane gasoline, distillate, fuel oil.
Stone and Webster Engineering Corp./Institute Fruncaij die Petrol	This unit selectively converts gas oil feedstock into high-octane gasoline, distillate and C_3–C_4 olefins by catalytic and selective cracking in a short contact time riser.	Product: High-octane gasoline, distillate and C_3–C_4 olefins.

Contd.

Table 7.5   FCC technologies *(Contd.)*

Process technology licensers	Process description	Remarks
ABB Lummus Global Inc. FCC	Selective conversion of feedstock which may be virgin or hydrotreated gas oil. It may also include lube oil extract, coke gas oil and residue. The process consists of advance reaction system, high efficiency catalyst stripper and a mechanically robust, single stage fast fluidized bed regenerator.	13 installation. Products: High-octane gasoline, light olefins and distillate. High propylene yield in selective component cracking.
Mobil Kellogg Brown and Root FCC	This unit is used for conversion of gas oils and residues to high value products. The unit consists of the efficient flexible orthoflow catalytic cracking process.	More than 120 installations. Products: Light olefins, high-octane gasoline and distillate.
UOP LLCFCC	Selectively convert gas oil and residue feedstock into higher value products using FCC/RFCC/Petro FCC/MSCC process. FCC: Standard feeds and some residue up to 4 %wt (conradson carbon) RFCC: Highly contaminated and/or residue quality feeds. Petro FCC: VGO, HVGO and hydrocracked gas oil. MSCC: All feeds are suitable	Products: Light olefins for alkylation polymerization, or etherification, petrochemical intermediates, high-octane gasoline, distillate and fuel oil more than 120 units.
UOPR2R resid Cracking	The process can convert feeds that are difficult to process, having high metal contents (up to 50 ppm) or high conradson carbon (8–10%). Improved catalyst riser, catalyst regeneration in two stages.	Improved gasoline and gas oil yields, improved regeneration and reduced catalyst consumption has been reported.
UOP RxCat	The Rx Cat technology provides the ability to increase both conversion and selectivity by recycling a portion of carbonized catalyst back to riser [uop.com].	Increased conversion, increased gasoline yield, decreased dry gas yield, increased propylene yield.
High severity FCC(HS-FCC	The main objective of high severity FCC process is to produce significantly more propylene and high quality gasoline. The special features of this process include a	Increased propylene production

Contd.

Table 7.5   FCC technologies *(Contd.)*

Process technology licensers	Process description	Remarks
	down-flow reactor, high reaction temperature, short contact time, and high catalyst to oil ratio [ajse.kfupm.edu.sa].	
MIP technology (maximizing iso-paraffins and clean gasoline and propylene)	The conventional FCC riser, which is essentially a straight pipe and does not have applicable means to provide two distinct zones for cracking reactions, i.e. the high temperature zone for cracking reactions and the low temperature zone for converting reaction hydrogen transfer, alkylation and isomerization MIP technology has two zone risers to optimally accommodate the above two desired FCC reactions.	Improved gasoline quality with higher octane number and reduced olefins, sulfur and benzene

Source: Refining Processes, 1998; Ghosh, 2002; Niccum et al., 2001; Murcia, 1992; Long et al., 2011.

Table 7.6   Typical yield of Exxon flexicracking III-R fluid catalytic cracking

	Residue feed mogas operation	VGO + lube extracts distillate operation	VGO feed mogas operation
Feed			
Gravity (°API)	22.9	22.2	25.4
Con. carbon (%wt)	3.9	0.7	0.4
Quality	80% atm. residue (hydrocarbon)	20% lube extracts	50% TBP–794 °F
Product yields			
Naphtha (lv% ff)	78.2	40.6	77.6
(C_4/FBP)	(C_4/430 °C)	(C_4/260 °F)	(C_4/430 °F)
Mid Dist. (lv% ff)	13.7	49.5	19.2
(IBP/FBP)	(430/645 °F)	(260/745 °F)	(430/629 °F)

Source: Refining, 1998.
Courtesy: Hydrocarbon Processing, Nov 1998, p67.

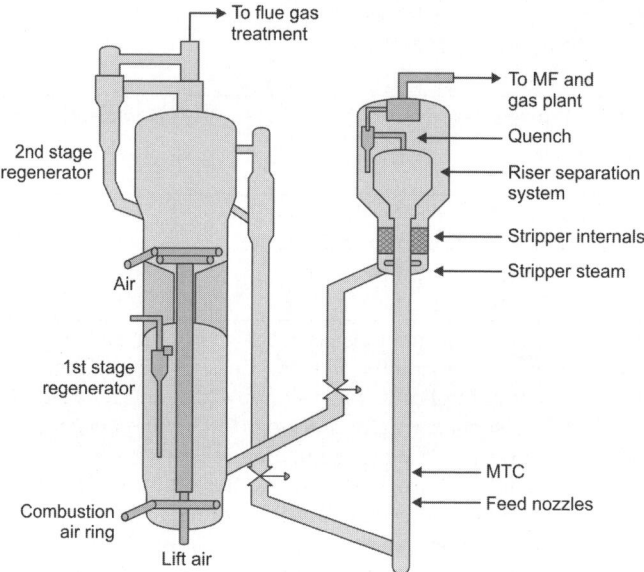

Fig. 7.1 Fluidized bed catalytic cracking reactor

7.1.1 Description of FCC Process

FCC unit can be divided into two sections: (i) reactor and regenerator system and (ii) fractionation section which operates in an integrated manner. Typical block diagram for a FCC process is given in Fig. 7.2. Modern FCC process employs riser cracking and zeolite catalyst. Typical feed specifications of FCC are given in Tables 7.7 and 7.8.

The feed to the unit along with the recycle streams is preheated to a temperature of 365–370 °C and enters the riser where it comes in contact with hot regenerated catalyst at a temperature of about 630–650 °C. The hot catalyst vaporizes the feed and the cracking reaction takes place in the riser. This vaporization fluidizes the catalyst up the riser to the disengaging T-section at the top of the riser. The cracked hydrocarbon exit from the reactor through a cyclone to the fractionation column where FCC gases including LPG and gasoline are removed and overheated as vapor. The extruded catalyst is collected in the cyclone and returned to the reactor through the dip leg. The spent catalyst (which is deposited on it) flows through the regenerator, where the coke is burnt off. The heat of combustion increases the catalyst temperature and supplies heat to the feed going to the riser. The regenerated catalyst is withdrawn from bottom of regenerator and fed to the bottom of the riser reactor. The regenerator slide valve controls the catalyst circulation rate. Schematic diagrams of FCC unit are shown in Figs. 7.2 and 7.3.

In the fractionators section, the reactor vapors are fractionated into recycle gas oil, which are released to the riser for further cracking, and converted into products—clarified slurry, heavy cycle oil, cracked light cycle oil, unstabilized gasoline and wet gas. Unstabilized gasoline separated from the lighter products is sent to the debutanizer for separation of C_3 and C_4 as overhead products. LPG is amine and Merox treated. Main column bottom is sent to slurry settler where the

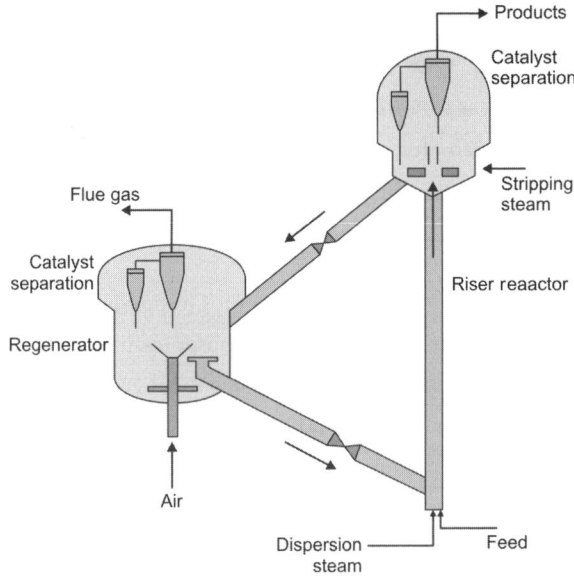

Fig. 7.2 Schematic diagram of a typical FCC unit

Table 7.7 Typical feed specifications of FCC

Feedstock quality specifications	Unconverted OHCU bottoms	Vacuum gas oil (VGO) 100% UZ	VR (AM)	VB naphtha
	ASTM D1160			D86
IBP	337	284	506	31 (1%)
10%	394	402	580	58
30%	437	431	628	85
50%	473	454	685	108
70%	511	478	765	128
90%	548	515	865	148
FBP	562	532	992.3	155 (98%)

Courtesy: CPCL, Chennai.

Table 7.8 Typical process condition in FCC

Operating parameters

Description	Design Case
Raw oil feed (at heater f-01 inlet)	114 sm^3/hour
Furnace f-01 inlet temperature	291°C
Reactor feed temperature	371°C
Reactor vapor temperature	549 °C
Level in main column bottom	50–60 %

Contd.

Table 7.8 Typical process condition in FCC *(Contd.)*

Reactor-regenerator process conditions

Reactor-regenerator process conditions	Parameter	Design case
1.	Raw oil temperature (°C)	371
2.	Combined feed ratio	1.0
3.	%wt steam to feed distributor (estimate)	1.0
4.	%wt steam to lift gas distributor (estimate)	1.5
5.	%wt sponge absorber off gas to lift gas distributor	0
6.	Catalyst to oil wt ratio	7.87
7.	Reactor temperature (°C)	549
8.	Reactor pressure (kg/cm²/g)	2.11
9.	Regenerator temperature (°C)	677
10.	Air to coke weight ratio	13.46
11.	Conversion as produced 380 °F (193 °C) @ 90 %vol	91.5

Courtesy: CPCL, Chennai.

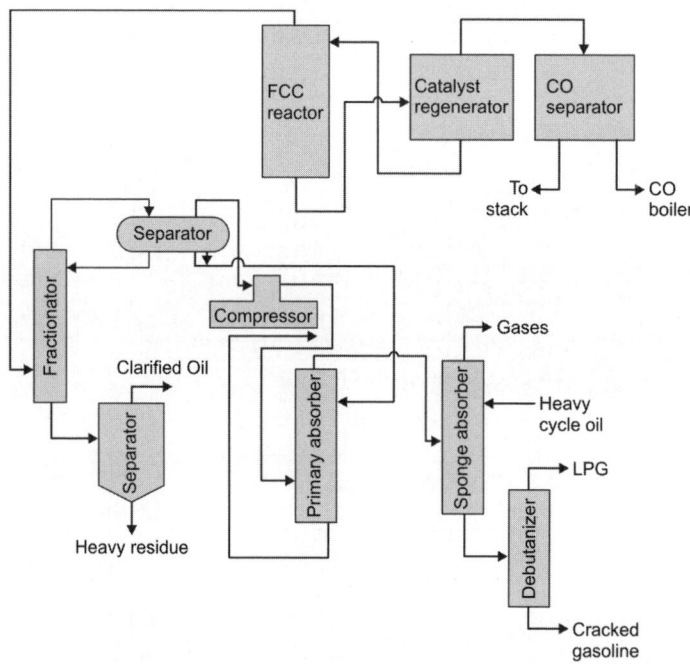

Fig. 7.3 Schematic diagram of a Typical FCC Unit

settled slurry containing catalyst is recycled. The water and gasoline vapors are withdrawn from the top of the fractionator and further separated into stripper and absorber. The function of the stripper–absorber section is to separate the main fractionator overhead product into a dispropanized fuel gas product and a liquid feed to the debutanizer of low ethane content. The fractionator overhead product enters the stripper–absorber section as both, a liquid stream and a vapor stream [Eriksson, 1993]. The various products from stripper–absorber section are: LPG, light naphtha, C_3 and C_4 gases and heavy naphtha. Important variables for spent catalyst stripper performance are temperature, pressure, steam rate, steam distribution, catalyst distribution, diameter and height of the stripper. The coke on the catalyst is drawn off in the regenerator. The forming rate depends on temperature, pressure, O_2 concentration and superficial air velocity, residence time of catalyst and air in the regenerator and presence of CO promoter additive. Unstabilized gasoline from the bottom of the primary absorber unit is sent to stripper where light components are removed, and then it proceeds to debutanizer where C_3 and C_4 are removed as overhead LPG. LPG is amine and Merox treated to remove sulfur compound and then sent to LPG pool as a feedstock to propylene recovery unit.

Propylene is recovered from cracked LPG which is basically a mixture of propane, propylene, butane and butylenes. First a mixture of propane and propylene is obtained from splitter which is further fed to a splitter from where propylene is obtained as separate stream.

7.1.2 Gas Concentration Unit

The function of this unit is to separate gas, LPG and stabilized gasoline from unstabilized gasoline, and wet gases obtained from main fractionator. The unit consists of compressor section, rectifier absorption, debutanizer and C_3/C_4 splitter. In the rectified absorber, the hydrocarbon vapors are absorbed in lean oil. The butanizer separate C_3/C_4 from the cracked gasoline. LPG from debutanizer is pretreated for H_2S removal in two amine absorbers in which countercurrent treatment with diethanol amine (DEA) takes place.

The LPG is further treated in Merox unit where the low molecular weight mercaptan is removed using NaOH. The mercaptan-rich caustic along with Merox catalyst is routed to regeneration unit where mercaptans are oxidized to disulfides. The disulfides are subsequently separated from solvent by coalescing, gravity settling and decanting.

7.1.3 Chemistry of Cracking and Process Variables

7.1.3.1 Chemistry of Cracking

The reactions involved in catalytic cracking process are complex and a large number of thermal and catalytic reactions take place. Catalytic reactions are heterogeneous in nature and involve short lived reaction intermediates of the carbocation type. Catalytic reactions involve isomerization, beta scission, hydrogen transfer, dehydrogenation and various condensation reactions [Bonifay and Marcilly, 2001]. Chemical reactions taking place in catalytic cracking process are numerous and complex. Both thermal and catalytic reactions as well as many side reactions such as hydrogen transfer progress simultaneously [Yen, 2006]. Heavy molecules crack according to three mechanisms [Lieby, 1992].

- Cracking on the external zeolite surface
- Thermal cracking
- Matrix cracking.

Thermal cracking is non-selective and governed by free radical mechanism with a very little skeleton isomerization, while catalytic cracking is selective cracking producing more of C_3–C_6 hydrocarbons with extensive isomerization and is governed by ionic mechanism. Maximum gasoline yield increases with the paraffinity of the feedstock and the conversion of feed in FCC increases with the reactor temperature and with the catalytic activity [Sarkar, 1998]. The main products of catalytic cracking are in C_4–C_{12} range. The degree of branching is high and little butadiene is formed. FCC catalyst system is complex and is a mixture of functional components [Avidan, 1992] which include Y-zeolite as main component and a large number of additives such as Ni passivator, SO_x emission control additive, vanadium passivator, CO combustion promoters and octane enhancing additives. All these additives find extensive applications on increasing the end operation and profitability. Main products from FCC processes are light gases: LPG, gasoline, light cycle oil (LCO), heavy cycle oil (HCO), clarified oil (CLO) or slurry for fuel oil (SLO) coke.

Some of the major reactions taking place are cracking of alkanes and alkenes cracking, alkenes polymerization, cycloalkanes ring opening and cyclization of alkenes, dealkylation of alkyl cyclanes, dehydrogenation of cyclohexanes, dealkylation of alkyl—aromatics, dealkylation of polycyclic hydrocarbons, cracking of sulfur and nitrogen containing compounds [Raseev, 2003].

7.1.3.2 Major Operating Variables

There are two major FCC operating variables [Bonifay and Marcilly, 2001]:

1. *Independent variables:* Reaction temperature, preheat temperature, recycle flow rate, catalyst activity and selectivity, regenerator combustion mode—total or partial.

2. *Dependent variables reactor temperature and catalyst circulation:* Regenerator temperature, catalyst circulation rate, conversion and combustion air flow rate.

Higher reaction temperature results in increase in catalyst circulation, regenerator temperature, conversion, production of C_4, C_3 and dry gas, gasoline octane number and olefin content, LCO aromacity, coke yield [Bonifay and Marcilly, 1998]. A higher preheating temperature with a constant reaction temperature cause a decrease in catalyst/feed ratio, coke yield, conversion and increases regenerator temperature. Increase in the reactor temperature results in higher gasoline yield. This can be achieved by increasing the catalyst circulation rate [Leuenberge, 1988]. The regenerator flow temperature is getting higher to achieve more complete catalyst regeneration [www.touchbriefings.com]. With increasing temperature, the proportion of cycloalkanes dehydrogenation and thermal cracking reactions increased resulting in increase of the unsaturated and aromatic character of the gasoline [Raseev, 2003].

There are three main controls in FCC that can convert more gas oil to gasoline and lighter products: Catalyst activity, catalyst circulation rate and reactor temperature. By increasing circulation rate and reactor temperature, maximum gasoline yield can be achieved and obtained. Catalyst activity also plays an important role. Too low activity results in lower gasoline activity while too high activity will overcrack gasoline. Catalyst activity is maintained at the highest target level consistent with good gasoline activity [Leuenberger, 1988].

Effect of Feed Properties

Feed characteristic is the most important factor in FCC operation as it effects on the heat balance, ultimate cracking, severity and inherent crackability of molecular structures. As feed hydrogen content increases gasoline yield and LPG yields decline. Hydrotreating FCC feed can increase FCC by 8–12% [Campagna et al., 2001]. Hydrogen content of the vacuum gas oils can vary widely in hydrogen content from under 12 %wt for aromatic feeds to over 14% for paraffinic gas oils. In FCC, cracking of paraffins occurs readily while cracking of aromatic ring does not occur. More aromatic feeds will result in poor FCC yields. High hydrogen content gives higher FCC yield in terms of gasoline and LPG and lower coke C_2 minus yield increase significantly as the hydrogen content of the feed declines while coke yields increase slightly [Campagna et al., 2001/02]. In FCC, feed quality is critical because of its impact on the heat balance and the ultimate cracking severity, in addition to inherent crackability of the molecular structures. Hydrotreating of FCC feed is a viable option to produce ultralow sulfur gasoline and diesel fuels [Campagna et al., 2001/02]. Feed FBP: Effect of feed FBP on FCC performance is given in Table 7.9.

Table 7.9   Effect of feed FBP on FCC performance

Feed	BH (400–550 °C)	BH (400–620 °C)
Reaction temperature (°C)	510	510
W/F (min.)	0.70	0.70
216 conversion	65.77	59.69
Yields (%wt)		
H_2	0.012	0.017
Dry gas	1.02	1.13
LPG	16.28	13.40
Gasoline	33.14	29.60
Heavy naphtha	12.97	12.50
LCO	81.43	19.37
TCO	31.40	31.87
370 + °C	15.80	20.89
Coke	2.36	3.04

Note: Simulated MAT data.
Source: Hydrocarbon Technology, 15th May, 1994.

Feedstock changes generally impact on the aromaticity of the gasoline, with aromatic feed producing high octane number gasoline [Habib, 1989].

Density

With increasing feed density, there is gradual increase in aromatic content. This increases the coking tendency and also reduces feed crackability. This results in more bottom product and less conversion [Rao et al., 1994].

Conradson carbon in feed

With increasing CCR, the regenerator temperature is increased significantly and cat/oil ratio drops which result in lower conversion. Effect of conradson carbon in feed is given in Table 7.10.

Nitrogen

Nitrogen is a dominating contaminant which is usually present in the heavier portion of the crude [Rao et al., 1994]. It has been observed that with increasing nitrogen, LPG and gasoline yields drop significantly with corresponding increase

Table 7.10 Effect of conradson carbon

Run No.	1	Base	2	3
Conradson carbon %wt FF	0.00	0.14	0.40	0.65
Yields				
Dry gas	2.6	2.78	2.62	2.59
LPG	9.9	10.40	10.11	9.77
Gasoline	26.62	27.34	26.89	26.34
Heavy naphtha	13.50	13.67	13.52	13.33
Light cycle oil	30.93	30.13	30.46	30.84
Clarified oil	12.03	11.05	11.71	12.54
Coke	4.52	4.57	4.59	4.55
Reactor				
Conversion	57.14	58.77	57.73	56.58
CSC (%wt)	1.23	1.26	1.29	1.3
Cat/oil ratio	5.26	4.92	4.67	4.38
Regenerator				
Dense temperature (°C)	642.6	654.0	661.6	671.4
CCR T/hour	1017.2	950.6	902.8	848.0
CRC (%wt)	0.53	0.5	0.49	0.46
Air rate (Nm^3/hour)	63903	64645	64830	64227
Regen. Velo. (m/s)	0.59	0.6	0.61	0.61

Source: Hydrocarbon Technology, 15th May, 1994.

in bottom yield [Rao et al., 1994]. High nitrogen feedstocks require higher target activity because nitrogen poison the catalyst. The feed injection system is by far the most critical breakthrough of modern FCC reactor design [www.touchbriefings.com].

Sulfur

FCC gasoline is a major source of sulfur. It has been observed that increasing feed sulfur results in increased SO_x and affects catalyst performance and results in poor product quality LCO in feed.

7.1.4 Salt Fouling in FCC Fractionator

Ammonia salt deposition is a major problem in FCC fractionator due to combination of low temperature and relatively high concentration of ammonia and hydrogen chloride. Impact of salt deposition is plugged trays, product draws, and increased corrosion, loss of fractionation and loss of compressor capacity [Martin and Allen, 2001]. Some of the major methods for reducing salt deposition are desalting, tower water washing and salt dispersant additives. Desalting improves FCC feedstock by reducing feed sodium content, eliminating entrained water or slugs of water and reducing particulates and contaminants in both the water and hydrocarbon [Harris, 1996].

7.1.5 FCC Feed Treatment

FCC gasoline is the major source of sulfur in gasoline pool. Typical gasoline pool composition and their impact on sulfur in gasoline pool is given in Table 7.11 [Shorey et al., 1999]. Fuel specifications are becoming more and more stringent and require low sulfur level. Lower sulfur in the fuel needs either treating FCC feed or treating its product. Feed pretreatment has advantage over products treatment as treating the FCC products requires treatment of a large number of product streams which include—naphtha, light cycle oil and gas stream. However, feed treatment results in increased yield with lower sulfur content in the product stream and lower SO_x emissions. Untreated FCC feedstock properties are given in Table 7.12. Impact of desulfurization on FCC feed properties for four levels of hydrotreating is given in Table 7.13. Impact of hydrotreating on FCC unit performance is given in Table 7.14. Hydrotreating feed directly improves FCC unit performance through increased conversion to naphtha and lighter product, raised gasoline yield, decreased LCO and CSO yield and reduced coke yield [Shorey et al., 1999]. New high performance catalysts and technologies are available for upgrading very refractory heavy feeds into high value products. The process uses dual catalyst system using a staged dual catalyst system consisting of a high activity nickel-molybdenum catalyst over a high activity cobalt-molybdenum catalyst [Dahlberg et al., 2007].

Table 7.11 Typical gasoline pool composition

Gasoline blend stocks	Percent of pool volume	Percent of pool sulfur
Alkylate	12	–
Coker naphtha	1	1
Hydrocracked naphtha	2	–
FCC naphtha	36	98
Isomerate	5	–
Light straight run naphtha	3	1
Butanes	5	–
MTBE	2	–
Reformate	34	–
Total	**100**	**100**

Source: Hydrocarbon Processing [Shorey et al., 1999].

Table 7.12 Untreated FCC feedstock properties

	VGO	CGO	Blend
Feed capacity (bpsd)	31,250	3,750	35,000
Blend contribution (%wt)	89.1	10.9	100.0
Feed properties			
Gravity (°API)	20.9	17.5	20.5
Sulfur (%wt)	2.5	3.4	3.60
Nitrogen (wppm)	700	2,350	880
Carbon residue (%wt)	0.4	0.2	0.4
Bromine number	3	25	7.5
Metal (Ni + V) (wppm)	0.2	1	0.36
Distillation (°F)			
IBP	664	662	664
50%	830	788	817
90%	980	898	965
FBP	1,050	950	1,050

Source: Hydrocarbon Processing [Shorey et al., 1999].

Table 7.13 Impact of desulfurization on FCC feed properties

	Feed desulfurization			
	Untreated	90% HDS	98% HDS	99% HDS
Operating pressure (psig)	–	900	1,000	1,000
Feed properties				
Gravity (°API)	20.5	23.5	24.8	26.0
Sulfur (%wt)	2.6	0.25	0.06	0.02
Nitrogen (wppm)	880	500	450	400
Carbon residue (%wt)	0.4	.25	0.1	0.1
Metal (Ni + V) (wppm)	1	< 1	< 1	< 1
Hydrogen addition to feed (%wt)	0	0.51	0.74	0.94

Source: Hydrocarbon Processing [Shorey et al., 1999].

Table 7.14 Impact of hydrotreating on FCC unit performance

	Feed desulfurization			
	Untreated	90% HDS	98% HDS	99% HDS
Yields (%wt)				
H_2S	1.1	0.1	0.0	0.0
C_2–3.3	3.5	3.2	2.8	
$C_3 + C_4$	16.3	17.6	18.7	19.9
Full range				
Naphtha	48.3	51.5	52.5	53.6
LCO	16.7	15.7	15.0	14.0
CSO	9.0	6.6	5.9	5.2
Coke	5.4	5.0	4.7	4.4
Total	100.0	100.0	100.0	100.0
Conversion (%vol)	74.3	77.7	79.1	80.8
Key product properties				
Naphtha RON	93.2	93.0	92.9	92.7
Naphtha MON	80.5	80.8	81.1	81.0
LCO cetane index	25.7	25.7	26.4	26.5
Product sulfur (wppm)				
H_2S	10,066	753	188	94
Naphtha	36,00,225	55	18	
LCO	29,700	3,400	900	300
CSO	57,800	11,000	3,000	1,100
Coke	30,300	5,700	1,554	516
SO_x (vppm)	2,030	410	120	42

Source: Hydrocarbon Processing [Shorey et al., 1999].

Developments for Maximizing Diesel Production in FCC

Traditionally, the focus has been on the maximization of gasoline. However, there has been shift from gasoline to diesel in recent years. Depending on the region of the world, refiners are looking to meet the growing diesel demand. Diesel blend stock quality depends on the process and for its production. The amount of light cycle oil (LCO) for blending to diesel from FCC unit can be increased by adopting following operation strategies [Yen Yung, 2010].

- Sharper fractionation of the FCC feed to minimize the amount of material boiling below 370 °C
- Minimize the initial boiling point of the LCO as much as possible
- Maximize the end point of LCO
- Lowering the activity of the equilibrium catalyst and use of diesel selective catalysts
- Decreasing operating severity by reducing temperature or increasing feed temperature
- Reduced zeolite-to-matrix ratio to minimize the reaction of LCO cracking to lighter component
- Transforming olefins, LPG and light cracked naphtha produced by FCC into diesel by implementing an oligomerization unit up to 10%
- Operation of FCC in diesel maximization mode which includes adjustment of reactor inlet temperature, reduction in conversion
- Changing the FCC pretreater into a mild hydrocracking unit.

7.1.6 Salient Features of UOP FCC Processes

UOP FCC technology includes optimix feed distributors, vortex separation technology, AFTM spent catalyst stripper technology, catalyst cooler, Rx Cat technology, selective recycle, the combustor regenerator or a two-stage regenerator, power recovery and good catalyst [www.uop.com] recovery.

7.1.7 Optimix Feed Distribution System

Optimix feed distribution system in UOP FCC acceleration zone below the optimix distributors produce a moderate catalyst density to achieve good penetration and mixing of the atomized feed spray. Other benefits are reduced dry gas and delta coke and increased gasoline yield [www.uop.com].

7.1.8 Vortex Separation System (VSS) and Vortex Disengager Stripper[SM] (VDS)

Both systems have critical prestripping features and offer the highest postriser hydrocarbon contaminant. These systems capture the vapor/catalyst mixture at the outlet of the riser and efficiently separate catalyst without letting the vapor enter the reactor vessel. For final clean up, the vapor stream is fed into cyclones.

7.1.9 Advanced Fluidization Spent Catalyst Stripper

The design achieves superior contacting and stripping efficiency [www.uop.com] by incorporating a prestripping zone. The combination of prestripping and primary zone stripping provides the best possible catalyst stripping.

7.1.10 R$_x$ Cat Technology

Carbonized catalyst is recycled back from the reactor to the feed contacting zone resulting in higher cat/oil ratio, very low dry gas and improved overall yield selectivity [www.uop.com].

7.1.11 Catalyst Recirculation

Good catalyst recirculation is the key factor of UOP FCC units. Risers and standpipes are made straight wherever possible.

7.1.12 UOP Millisecond Catalytic Cracking^{SM} (MSCC)

MSCC reactor design is a novel departure from riser departure system. In this process, feed is injected perpendicular to a down-flowing curtain of catalyst. Reaction products move across the reaction zone and are quickly separated in a single stage of external cyclones. The reactor vapor is then carried over the catalyst in primary separation device. Following the primary separation device, the remaining catalyst is further separated in a single stage of external cyclones. The various reactors are then over the main column of the MSSC complex [www.uop.com] [UOP2699D-97 0302RILf].

7.1.13 Residue Fluid Catalytic Cracking (RFCC)

Processing heavier feedstock, containing more metal and asphaltene contaminants, increases the production of coke and gas, and accelerates catalyst deactivations as in case of heavier feedstock a larger fraction of feed does not vaporize under conventional cracking conditions, the feed is more contaminated with heavier metals (V, Ni) and contain higher concentrations of basic and polar molecules, as well as asphaltenes [O'Conner et al., 1991]. Characteristics of some long residues are given in Table 7.15. Impact of heavy feed on FCC performance is given in Table 7.16. Regenerator temperature, catalyst circulation rate, conversion and combustion air flow rate.

Table 7.15 Characteristics of some long residue

	Bombay High	North Gujarat Mix	Dubai	Assam Crude Mix
Range (°C)	370+	370+	370+	370+
Gravity (API)	21.24	18.3	13.97	17.0
Pour point (°C)	+51	+39	+33	+48
Total sulfur (%wt)	0.38	0.09	3.29	4.85
Con carbon (%wt)	4.33	6.18	9.06	0.46
Asphaltenes (%wt)	1.22	0.05	4.2	0.5
V (ppm)	0.8	1.6	82	< 0.1
Ni (ppm)	2.8	10s	33	0.8
Cu (ppm)	< 0.1	0.1	–	0.2
Fe (ppm)	0.35	0.1	–	0.2
Source of data	Ref. 8	Ref. 10	Ref. 11	Ref. 12

Source: Improve Resid Processing, Hydrocarbon Processing, Nov 1991.

Table 7.16 Impact of heavy feed on FCC performance

Parameters	Good quality feed	Heavy feed
Feed gravity (gm/co)	0.89	0.92
Feed CCR (%wt)	0.15	0.70
Feed nitrogen (ppm)	150	450
Yields (%wt FF)		
DG	2.79	3.02
LPG	12.47	7.24
Gasoline	25.27	19.13
HCN	11.15	9.65
LCO	33.59	33.55
CLO	9.46	22.41
Coke	5.27	5.0
Regen. temperature (°C)	650	674
Cat Cir. Rate (temperature/hour)	1053	878
CRC	0.32	0.51
CSC	1.12	1.44
Air rate (Nm^3/hour)	63,738	62,110

Source: Hydrocarbon Processing, 15th May, 1994.

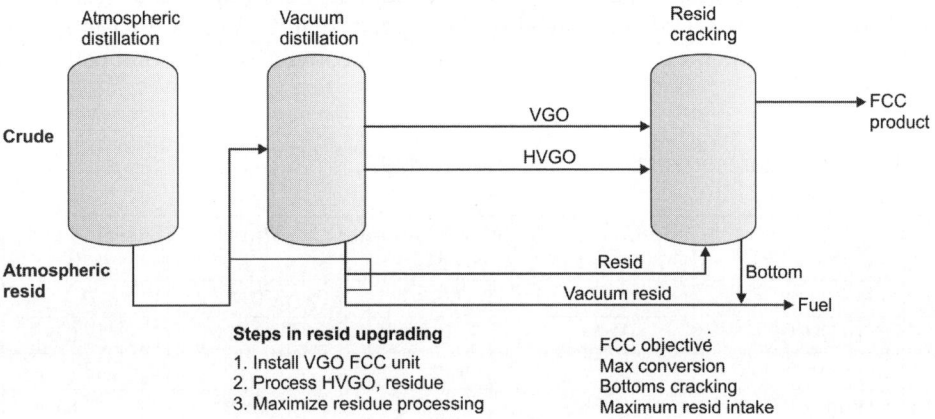

Fig. 7.4 The Role of FCC in Resid Upgrading

RFCC is a fast growing technique which offers better yield structure and flexibility at relatively low capital investment for processing heavier feedstocks. Residue is the hydrocarbon which has minimum 5% boiling above 540 °C. Stone and Webster (S & W) in association with Instut Francais du Petrole (IFP) is the licenser of residual fluid catalytic cracking process and today 26 full technology S & W-IFP processes have been licensed worldwide. Enhanced accessibility of

Resid FCC catalyst's internal active sites for larger incoming resid molecules can significantly improve conversion and bottoms cracking, as well as the overall performance of FCC process [O'Conner et al., 1991]. Role of FCC in resid upgrading is shown in Fig. 7.4.

India's first RFCC has been commissioned in Panipat Refinery with capacity of 0.7 MTPA licensed by Stone and Webster. The unit is designed to process a blend of 85.70% hydrocracker bottoms and 14.3 %wt vacuum residue as well as 100% hydrocracker bottoms. Typical design feed characteristics and yield pattern of RFCC of Panipat is given in Tables 7.17and 7.18.

The advances in resid FCC catalyst development until now have been primarily in the field of metal tolerance (metal traps), with some additional progress in the area of cracking the larger hydrocarbon molecules with an improved active site accessibility [O'Conner et al., 1991].

Table 7.17 Typical feed characteristics of RFCC feed

Parameter	Case 1	Case 2
Feed composition	85.7 %wt OHCU bottom, 14.3 %wt vacuum residue	100% hydrocracker bottoms
API	31.66	35.96
Conradson carbon residue, (wt, ppm)	3.20	0.10
Metals (ppm) Nickel/vanadium	927	10
Sulfur (%wt)	0.76	0.001
Total nitrogen (wt, ppm)	537	10

Source: IOC, Panipat Refinery.

Table 7.18 Design yield pattern for RFCC

Stream	Case I (%wt)	Case II (%wt)
Gas	2.8	2.4
LPG	13.7	16.5
Gasoline	21.5	26.6
TCO (HSD)	49.4	47.7
DCO	4.9	2.0
Coke + loss	7.7	4.8

Source: IOC, Panipat Refinery.

7.1.14 R2R Resid Cracking [HC 1987]

R2R resid cracking where R2R denotes use of one stage of reaction and two stages catalyst regeneration. The process can convert feeds that are difficult to process, having high metal contents (up to 50 ppm) or high conradson carbon (8–10 %wt) special features of R2R resid cracking include [HC, 1987]. R2R resid cracking containing one reactor and two risers has distinct advantage over conventional FCC as it processed heavier feed. Some of the benefits are higher gasoline and gas oil yield, improved regeneration, reduced catalyst consumption [Prezil, 1987].

Flow diagram of R2R technology is shown in Fig. 7.5.

- Feed sprayed into the unit to achieve very fine droplets for easy feed catalyst contacting
- An improved catalyst riser
- Heat balance without antiregenerator heat eliminating devices.

Some of the advantages of R2R resid technology are:
- Gasoline and gas oil yields as high as for conventional cracker feeds
- Improved regeneration to avoid catalyst deactivation
- Reduced catalyst consumption.

Typical yields from R2R cracking are given in Table 7.19 [HC, 1987].

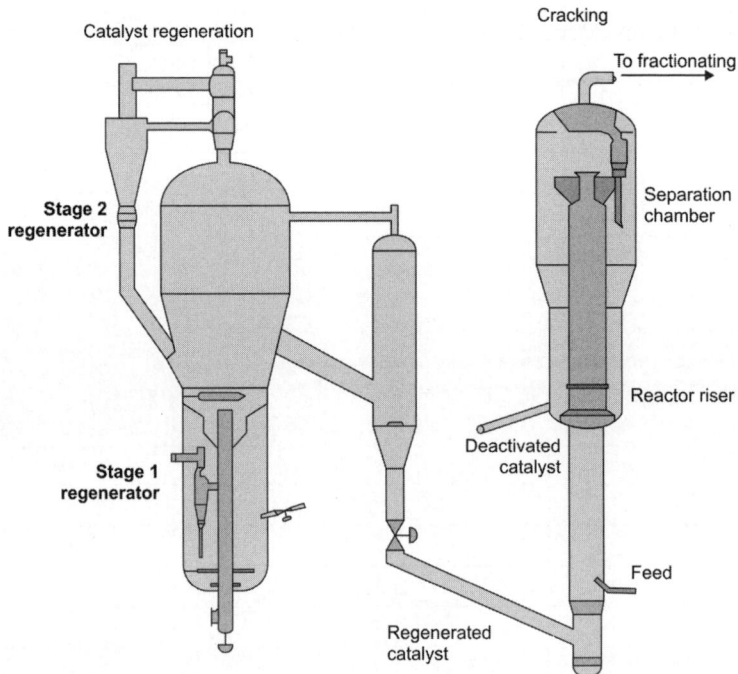

Fig. 7.5 Flow diagram of depicting R2R technology

Source: "New FCC design now commercial" Hydrocarbon Processing, September 1987.

Table 7.19 Typical yields from R2R cracking

Example	1	2	3	4	5
Product choice	Gasoline	Gasoline	Gas oil	Gasoline	Gasoline
Feed properties	A	B	B	C	D
Specific gravity	0.885	0.930	0.930	0.896	0.918
Conradson carbon (%wt)	0.2	4.8	4.8	6.0	5.5
Nickel + Vanadium (ppm)	1	20	20	22	34
Product yield (%vol)					
Gasoline	63.4	60.4	40.6	60.9	62.6
Gas	13.4	17.7	39.7	12.1	17.8

Courtesy: Hydrocarbon Processing, 1987.

7.1.15 Technology for Reduction of FCC Sulfur Content

FCC naphtha typically contributes 90% or more of the sulfur and olefin content to overall gasoline pool. FCC contributes 85–95% of the sulfur to gasoline blends [PTQ spring, 2001]. Extensive research efforts have generated new breed of technologies for reduction in sulfur content. Novel proprietary catalyst and additives are being design to reduce the concentration of sulfur and olefins in the FCC gasoline [reduce the clean gasoline profitability hit with Davison Clean Fuels Techno [Lesemann et al., 2003].

Desulfurization of atmospheric residuals for catalytic cracker feed [Rush and Steed, 1984]:

- Permits increased flexibility in crude selection
- Increases gasoline yield
- Reduces catalytic cracker coke yields
- Decreases fuel oil injection.

FCC sulfur-reduction (SR) additives can be used for primary sulfur management. The new breed of sulfur reduction catalysts can achieve sulfur reductions of 20–30% and 25–35% for full range and cut gasoline respectively [Lesemann and Schult, 2003].

Membrane technology can achieve 90% sulfur removal. Technology can be used as a sulfur removal technology. Another development in the FCC catalyst has been the development of catalytic olefins reduction technology, which has resulted in reduction up to 40% l at constant propylene, butylene and RON. The S content is determined by FCC feed, catalyst properties and usage of other additives as ZSM-5 [Humpries et al., 2003]. ZSM-5 additive has been used by some of the Indian refineries which have resulted in increase in LPG, increase in gasoline yield, overall reduction in heavy cycle oil [Rajaraman et al., 1992].

Another method for control sulfur levels in FCC products to produce ultra low sulfur gasoline is sorbent based process in which sulfur containing molecules react with a novel sorbent in presence of hydrogen. The sorbent retains the sulfur from the molecule while hydrocarbon portion is returned to product stream. The ability to achieve a high degree of sulfur removal while minimizing olefin hydrogenation and subsequent octane loss presents a significant advantage over [www.coptechnologysolutions.com] conventional hydrotreating technologies [Vander Laan et al., 2006]. Summary of the reduction technology is mentioned in Table 7.20. Reduction in cost of treatment technology is shown in Fig. 7.6.

Table 7.20 Summary of sulfur reduction technology

Technology class	Sulfur reduction	Sulfur reduction catalyst	Capital avoiding catalytic sulfur reduction (CACSR)	Membrane process for sulfur reduction
Sulfur-reduction range	Up to 35% GSR Upto 20% FRSR	20–35%	50%+	> 90%
Target gasoline cut	Light and intermediate gasoline	Full-range/ cut gasoline	Full-range/ cut gasoline	Light and intermediate cut naphtha
Capital required	None	None	None	$ 100–500 per bpd capacity
Commercial experiences	25 refiners	11 refiners	3 refiners	Demonstration plat startup, February 2003

Source: Lessemann and Schult, 2003.

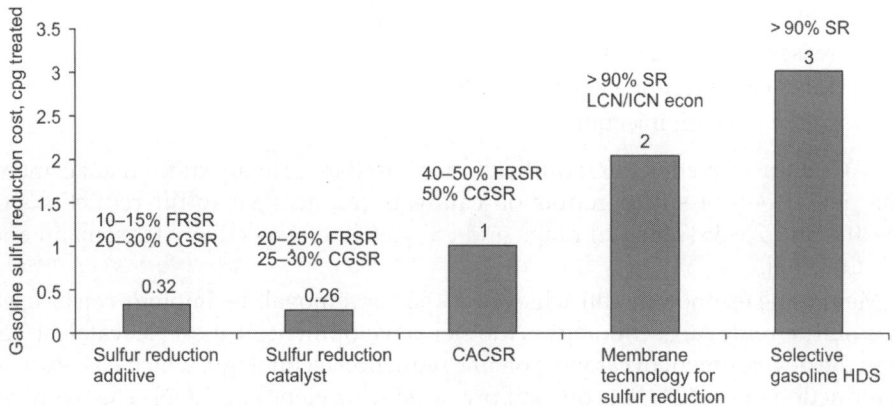

Fig. 7.6 Cost reduction in gasoline sulfur treatment

Courtesy: Lesemann and Schult, 2003; Hydrocarbon Processing, Feb 2003.

7.1.16 Coke Formation in FCC

The mechanism of coke formation is complex and depends on oil feed composition, catalyst properties and operating conditions [Garcia-Dopico, 2006]. Variables that can affect the coke yield for a given amount of feed are primarily the heat of combustion of coke, temperature of the feed and air and heat of reaction. Normal coke yields on gas oil FCCUs are around 5% on fresh feed at high conversion [Letzsch, 2005]. Coke can be divided into four categories [Garcia-Dopico et al., 2006]. Heat of combustion of the coke deposited on the catalyst depends on its hydrogen content and on the ratio CO/CO_2 in the flue gas [Rasseev, 2003].

- *Catalytic coke:* Catalytic coke is produced when a hydrocarbon is cracked on an acid catalyst

  $C_{cat} = k_{cat}\, t^{0.5}$

- *Contaminant coke:* Produce as result of the presence of dehydrogenating pollutants like Ni, Cu, V or Fe

  $C_{cont} = k_{cont}\, [Ni + (Cu/1.23) + (V/4.8) + (Fe/7.1)]$

  where the concentration of metals (Ni, Cu,V, Fe) is stated in ppm.

- *Additive coke:* Additive coke is produced by those feedstock fractions that is not volatile under riser working conditions. Additive coke is related primarily to conradson carbon.

  $C_{ad} = k_{ad}\, (CCR = 0.0015N/(C/O)$

  where concentration of conradson carbon is stated in % weight of oil feed and the nitrogen concentration is stated in ppm.

- *Cat-to-oil coke:* Cat-to-oil coke is thee fraction of oil feed that is trapped or occluded in the catalyst. It is not really coke, as it has high hydrogen content

  $C_{al} = k_{al}\, P_{reac}$

  In above correlations only the kinetic constant k_{cat} depends on temperature, whereas temperature has no effect on other kinetic constants k_{cont}, k_{ad}, k_a [Garcia-Dopico, 2006]. Modeling of coke formation has been presented by Garcia-Dopico et al., 2006.

  In order to reduce coke yield the operator must run to raise the feed preheat temperature as much as possible and run with complete combustion of CO [Letzsch, 2005].

- *Delta coke:* Delta coke is an important factor affecting FCC reactor–regenerator system. Delta coke is difference between coke on spent catalyst and coke on regenerated catalyst. It is related to catalyst type feed type and process conditions [Lieby, 1992].

  Delta coke = Coke yield divided by the catalyst-to-oil ratio in kg/kg [Leiby, 1992].

  Severe hydrotreating reduces the coke precursors which decrease delta coke significantly.

  %wt coke = Delta coke × Cat oil ratio

  ΔT (regenerator–reactor) α-delta catalyst

  Change of 0.1% delta coke changes the regenerator temperature (fixed reactor temperature) by 20–23 °C.

Changing the catalyst is a very effective way to manipulate the delta coke formation. To increase delta coke, the microactivity can be raised or the matrix surface area increased [Letzsch, 2005]. To significantly increase the surface area and raise the delta coke, the catalyst, the catalyst should contain some micropores in the 30–70 Angstrom diameter range. Severe hydrotreating reduces the coke precursors and the delta coke decreases significantly [Letzsch, 2005]. Some of the methods for lowering coke yield are [Letzsch, 2005]:

- Increase the air temperature
- Increase the feed temperature
- Lower the heat of reaction
- Reduce the amount of air to burn the coke.

Restoration of lost activity includes:

- Maintenance of desired selectivity
- Maintenance of physical properties
- Minimisation of liabilities
- Minimization of downtimes.

7.1.17 Technological Development in Fluidized Bed Catalytic Cracking

Catalytic cracking was developed by Houdry who used fixed bed catalytic reactor. Earlier catalytic cracking was prone to coking requiring coke deposits to burn off continuously. First commercial fluidized bed reactor started in 1942 by Exxon in Baton R, Louisianaouge. Since the introduction of FCC technology during early 1940s many major landmarks have been crossed in the appearance and design of FCC. These changes have been brought about by modifications in almost every element involved including feedstock, catalysts, equipment and product in response to shorter reaction time, higher regenerator temperature and heavier feedstock [Murphy, 1992; Chen, 2006]. Major landmarks in the history of FCC are [Ghosh, 2002]:

- Introduction of zeolite catalyst during 1960 which has resulted in lower residence time
- Introduction of ultrastable Y-zeolite in mid 60s
- Switch over from bed cracking to riser cracking
- Introduction of large number of additives for boosting of gasoline octane/yield of light naphtha
- SO_x control
- Nickel and vanadium passivation.

Major development in the FCC hardware includes [Ghosh, 2002]:

- Two-stage regeneration
- Catalyst coolers
- Improved feed distribution
- Quick separation of catalyst from the products at the top of the riser
- Improved stripper design
- Cold wall design, etc.

Reactor system: Earlier most widely used FCC reactor configuration was dense bed reactor which comprises a pressure balanced system using spray dried microspherical aluminosilicate catalyst. In the dense phase reactor, the catalyst is completely backmixed. Thus the feed is contacting with a mixture of fresh and spent catalyst laden with coke and metals. Vapor moves through the reactor dense bed in the form of developing bubbles along the length and as entrained vapor.

Riser: FCC risers are designed with the optimum residence time for maximizing desirable products, Almost all FCC reactors consist of a riser reactor, in which there is short contact time (less than 5 seconds) of catalyst and feed and a catalyst separator [Gupta et al., 2005]. Critical process consideration in design of riser reactor is: uniform catalyst distribution, to avoid unavoidable catalyst backmixing, uniform temperature and uniform gas distribution [Chen Y, 2006]. For achieving both rapid separation and efficient stripping, the new system comprises integrated separation and stripping compartment [Ross et al., 2005]. The riser reactor has the advantage of having good contact between feed and the catalyst with minimum backmixing. The newer flexible dual riser FCC (FD-FCC) can minimize gasoline olefins, increase gasoline octane and has added benefit of a propylene yield [Wang et al., 2003]. FD-FCC inhibits the undesired secondary reaction, promotes the desired secondary reactions [Wang et al., 2003]. Riser hydrodynamics can be improved by having wall baffles which results in both gas and solid velocity profiles [Chen, 2006].

Multiple feed introduction: Another development has been the replacement of single feed injection nozzle with multiple feed nozzles which resulted in improved temperature and density profiles across the riser leading to improved gasoline, dry gas, and coke selectivities due to better mixing at the bottom of riser [Chuang et al., 1992]. New feed nozzle system use increased pressure drop to shear the oil into smaller particles to have faster vaporization and effective mixing with catalyst. [Wrench and Glasgow, 1992].

Riser inlet section and feed introduction: Feed atomization, feed distribution and mixing with catalyst are the three key aspects of feed injection nozzles.

Technological development in FCC feed nozzles are [Chen, 2006]:

1980s	Design based on generating a high velocity liquid jet to impinge on a target, shattering the liquid jet into droplets upon hitting the target with the steam to the nozzle exit.
1990s	Twin fluid atomizers based on internally mixing of steam and liquid feed upstream of the nozzle exit and conveying the two phase mixture to the nozzle exit.
Newest generation of nozzles	Twin-fluid atomizers using two-phase (steam mixed with feed just before the final outlet of nozzle) choking as the atomization mechanism preventing the stratification of two phase flow [www.nt.ntnu.no]

The recent improvements in the feed introduction system are based on two different concepts. In one approach the catalyst is first contacted with lift gas (fuel gas) in the bottom of the riser while in other concept, catalyst is made to move at close to the minimum fluidization velocity at the riser bottom. Advantage of lift gas are: Reduction of possible hydrothermal steam and passivating effect of lift gas on the catalyst metals [Wrench and Glasgow, 1992]. The newest generation of FCC feed nozzles have more uniform feed distribution as a result of better control of homogeneity of two-phase flow at the nozzle exit [www.nt.ntnu.no] [Chen, 2006].

7.1.18 Rise Disengaging Devices

The separation of the catalyst from hydrocarbons after the cracking reactions in the riser is very important in order to avoid further catalytic reaction. Also the residence time of vapor in the high temperature region should be minimum to reduce thermal cracking reaction. For separation of vapor and catalyst, three types of separators: Inertial separators, ballistic separators and cyclone separators have been used. Older FCC units have the reactor mixture discharged from the riser into an open reactor vessel. The modern FCC riser design includes an improved rise termination device allowing quick separation of catalyst and product gas to minimize post riser cracking [Chen, 2006]. For earliest possible disengagement of product gas and catalyst, now modern FCC has a pre stripping cyclone.

FCC cyclones: There has been continuous improvement in cyclone reliability over the past decade. This has resulted in capacity increase with reduction in catalyst losses to stack and to clarified oil. Cyclones are designed with improved geometry that minimizes erosion of walls and build up of coke, compact design [McAuley and Dries, 2001].

Stand pipe flow and third stage separator: In order to improve catalyst circulation rate a new stand pipe inlet designed called catalyst circulation enhancement technology has been incorporated. In order to capture, the small particles and achieve high separation efficiency a third stage separator has been added in the form of swirl tube [Chen, 2006].

7.1.19 FCC Catalyst and Additives

Catalysts and additives play an extremely significant role in the profitability of FCC units. The catalysts are the heart of the FCC process and they provide: (i) low coke yield and effective carbon rejections, (ii) high selectivity to derived products (light olefins and gasoline), (iii) enhancement of desired properties such as octane and (iv) *in situ* control of emissions such as SO_x [Avidan, 1992]. There has been continuous development in catalyst. Advances in FCC catalysts allow refiners to cope with high levels of metals and coke precursors in resid feeds while maintaining conversion levels and product selectivity [Leiby, 1992]. FCC is the largest consumer of the catalyst amounting about 700 tons/day. FCC/RFCC catalyst consumption in India is about 11% of world demand [Ghosh, 2006]. In upgraded manufacturing process have produced FCC catalyst with increased gasoline yield and high hydrothermal stability at high severity operation [Keweshan et al., 2010]. FCC catalyst is an excellent tool for improving overall profitability of refinery. By switching to a rare earth free FCC catalyst output of the premium grade gasoline and operating margin can be increased [Chavdarov et al., 2014].

The optimum cracking catalyst to be used in the FCC process requires that bottoms-cracking and coke selectivity be optimized to match feedstock characteristics and unit constraints, and these catalyst properties becomes key features with significant economic impact on the FCC operation [Lesemann et al., 2005].

Although first FCC catalyst was amorphous silica alumina, real breakthrough in the FCC catalyst was with the introduction of synthetic zeolite in 1962 by Mobil. FCC catalyst is actually composites containing zeolite dispersed in amorphous matrix. The zeolite component comprises 10–50 %wt of the catalyst and provides activity, stability, and selectivity [Leiby, 1992]. The matrix comprises 50–90% of the catalyst and provides desirable physical properties, as well as some catalytic activity [Leiby, 1992]. The matrix consists of binder and filler material further increase in.

Although rare earth have played an important role in Refining industry since 1970s as component of FCC catalyst, however incredibly unpredictable and volatile market of rare earth, zero and low rare earth will plat important role in FCC industry [Srikantharajah et al., 2012].

7.1.20 FCC Matrix

The term Matrix refers to the nonzeolitic component of the catalyst. The proper ratio of zeolite to matrix act synergistically with the zeolite. Although zeolite provides most of the cracking activity, matrix fulfils both physical and catalytic functions that contribute to catalyst performance [Scherser, 1989]. FCC matrix technology innovations have led to improved gasoline yields from FCC units, improved catalyst attraction resistance, improved zeolite stability, enhanced bottoms cracking and improved metals tolerance properties of FCC catalysts [Rajgopalan and Habib, 1992]. Although the initial FCC matrics used commercially in the 1960s were gel based matrix, however, now semi synthetic matrics have replaced the gel based matrics. Some matrix technology developments is given in Table 7.21 [Scherzer,1989]. The proper ratio of zeolite to matrix activity depends on application and feedstock, balancing the multiple objectives of favorable pore space, coke selectivity, bottom upgrading capability, metal resistance etc. [Lieby, 1992]. Another development in FCC catalyst is use of newer high silica (dealuminated) Y zeolites, usually the ultra stable (USY) type which has resulted in substantial increase in research octane number improvements with lower delta coke [Lieby,1992]. Another development in FCC catalyst is the use of ZSM-5 a high silica zeolite having Si/Al ratio from 10 to 500. ZSM selectively cracks low octane straight chain paraffin and olefin components at the higher end of the gasoline boiling range to mainly C_3 and C_4. ZSM-5 does not affect aromatics or naphthenes [Lieby, 1992].

Table 7.21 Properties of FCC matrices

1.	Steamed 6 hours at 1, 400 °F, 100% steam, 5 psig.
2.	MAT conditions: 980 °F, WHSV = 30, C/O = 4, sour imported heavy gas oil feed.
3.	Apparent bulk density (ABD).

Source: Understand FCC Matrix Technology, Hydrocarbon Processing, Sep 1992.

7.1.21 FCC Additives

Additives play major role in enhancing the operating flexibility by reducing some of the major unit limitations [Mandal et al., 2005]. FCC and RFCC additives have grown dramatically during recent the last 20 years. Although the early additive use was limited to CO promoter and ZSM-5, however, overtime the range of additives available for use on FCC and RFCC has grown dramatically due to technology developments, product demand/opportunities at refineries, environmental regulation. Wide range of FCC/RFCC additives used today [Intercat, 2005].

Some of the commonly used additives are given in Table 7.22 [Mandal et al., 2005].

Some of the major development has been use of combustion promoters, ZSM-5 additive, sulfur reduction catalyst, olefin reduction catalyst, desulfurization additives, vanadium trap.

Development incase of residue catalyst is lowest UCS zeolite, high zeolite catalyst, low rare earth level, controlled ratio of matrix-to-zeolite activity, controlled matrix pore size distribution, metal tolerance, traps, and passivation [Avidan,1992]. Some of the new RFCC catalyst technologies are IMPAC^{TM}, PINNACE .

There has been a steady increase in zeolite content of FCC catalysts from 10% in the 1960s to over 35% today [Avidan, 1992]. Some of the today's FCC catalysts contain up to 50% zeolite [Avidan,1992].

ZSM-5, which is a crystalline aluminosilicate chemically, has been widely used as additive to increase octane number and light olefins yield. Raja Raman (1992) has reported increase in the gasoline and LPG yield by use of ZSM-5 as additive in Madras Refinery and they have reported drastic increase in olefins content of cracked LPG. Now most of the refineries in India have also shifted for earlier amorphous catalyst to zeolite catalyst, which has resulted in increase in LPG yield, increase in propylene content in LPG from 5.6 %wt to 15–19 %wt in octane number and reduction in coke. Overview of solid FCC additives is mentioned in Table 7.23. Comparison of processes using ZSM-5 additives is given in Table 7.24. Overview of liquid and gaseous FCC additives is given in Table 7.25. The effect of FCC additives containing SAPO-37: MAT is given in Table 7.26. The effect of zeolite beta as an additive to FCC catalyst is given in Table 7.27.

Table 7.22 FCC additives

Additives	Purpose
ZSM-5 based additive	For boosting gasoline octane number and yield of light olefins
Alumina microspheres with dispersed platinum	For enhancing of CO burning in regenerator dense bed
A specially formulated alumina microsphere	For upgradation of the heavier fractions of the feed
SO$_x$ additive	For reduction of SO$_x$ emission in flue gas
Antimony or bismuth based liquid compound	For passivation of the detrimental effects of nickel
Vanadium trap	For vanadium passivation [www.niscair.res.in]
Sulfur reduction additive	For reduction of gasoline sulfur
ZSM-5 (LPG boosting additive)	For boosting the LPG yield
SO$_x$ additive	For control of SO$_x$ emission.
CO promoter	To achieve complete combustion of CO

Source: Mandal et al., 2005.

Table 7.23 Overview of solid FCC additives

Types of additive	Year introduced	Vendor/developer	Commercial status
Combustion promoter	1975	Akzo Ambur Chevron Davison Engelhard Intercat Katalistiks	In use in > 60% of FCC units worldwide
SO_x scavenger	1977	Akzo Amoco Arco Chevron Davison Engelhard Intercat Katalistiks	In use in > 20% of units worldwide
Octane enhancer	1981	Akzo Chevron Davison Engelhard Intercat Mobil	In use in > 30% of units worldwide
Vanadium trap	1980s	Akzo Chevron Davison Engelhard Intercat Katalistiks	A few commercial tests or under development
Bottoms cracking additives	1980s	Akzo Engelhard Intercat	A few commercial tests or under development
Fluidization aid	1980s	Akzo Engelhard	Not know
Regenerator temperature control additive	1987	UOP	Not know

Source: Hydrocarbon Processing, Nov 1991.

Table 7.24 Comparison of processes using ZSM-5 additives

Pilot plant yields relative to base

Catalyst	Conventional ZSM-5	Improved ZSM-5
C_5–265 cut		
C_5s	1.05	1.01
C_6s	0.98	0.99
C_7s	0.93	1.01
C_8s	0.90	0.93
265–430 cut		
C_8s	0.95	0.97
C_9s	0.96	0.97
C_{10}^+	0.93	0.95

Source: Hydrocarbon Processing, Nov 1991.

Table 7.25 Overview of liquid and gaseous FCC additives

Types of additive	Year introduced	Vendor/developer	Commercial status
Nickel passivator antimony	1977	Betz Chemlink Intercat Nalco Phillips	In use in several units worldwide
Bismuth	1988	Chevron Intercat Nalco	In use in several units worldwide
Proprietary material	1988	Betz	In use in several units worldwide
Vanadium passivator: Tin	1982	Chevron-Nalco Betz	In use in several units worldwide
Lift gas	1983	UOP	In use in several units worldwide

Source: Hydrocarbon Processing, Nov 1991.

Table 7.26 The effect of FCC additives containing SAPO-37: MAT

Catalyst	A	B
Zeolite	15% LZ-210	15% LZ-210 10% SAPO-11
Matrix	Alumina + Kaolin	Alumina + Kaolin
Conversion (%wt)	61.9	62.9
Gasoline (%wt)	46.5	47.6
Gasoline + alkylate (%wt)	55.3	56.7
Coke (%wt)	4.1	4.0
Gasoline analysis: Iso/normal ratio		
C_4	4.8	5.7
C_6	13.6	17.1
C_7	15.6	16.8
C_8	7.1	8.1
Gas product analysis: Olefins/paraffin ratio		
C_3	2.8	3.8
C_4	0.55	0.66

Source: Hydrocarbon Processing, Nov 1991.

Table 7.27 The effect of zeolite beta as an additive to FCC catalyst

Catalyst zeolite	H REY alone	I Beta/REY50/50
Temperature (°C)	505	505
LHSV	0.64	2.62
Conversion—216 °C (%wt)	47.5	46.0
Gasoline + alkylate (%wt)	52.9	54.5
Distillate (216–343 °C) (%wt)	31.0	29.4
Bottoms (343 °C+) (%wt)	20.0	23.8
Coke (%wt)	1.5	0.7
C_5+ gasoline RON	84.0	87.4

Source: Hydrocarbon Processing, Nov 1991.

7.1.22 LPG Additive Families [Intercat, 2005]

- *Standard ZSM-5:* Different activity grades available C_3/C_4 selectivity is same for all activity grades.
- *High Si:Al ratio ZSM-5:* Designed to increase octane with less LPG. Useful for compressor constrained FCC units.
- *Modified selectivity ZSM-5 (XP-ZC1):* Standard ZSM-5 C_3/C_4 selectivity enhanced bottoms/LCO upgrading capability.

- *ZMX additives:* Increased C_4 selectivity with highest bottoms/LCO upgrading capability.

i-Max a ZSM based FCC catalyst additive for boosting LPG yield with increase gasoline yield has been also developed by India Oil Technologies, India. INDMAX employs a proprietary catalyst, which has three different functional components. The bottom cracking component provides highly acidic sites for catalytic cracking of heavy feed molecules [www.iocltech.com]. The component responsible for upgrading naphtha range hydrocarbons is shape elective in nature. This allows cracking to light olefins without increasing coke and dry gas make. The third component contains conventional ultrastable Y-zeolite, which shows synergistic effect with the other two components of the catalyst [www.touchoilandgas.com].

7.1.23 Catalyst Aging

Catalytic cracking catalyst are subjected to an aging phenomenon (hydrothermal aging). Aging of the catalyst is due to repeated contact with steam at high temperature and deposition of heavy metals, especially Ni and V on the surface of catalyst [Rraseev, 2003].

7.1.24 Catalyst Deactivation

Catalyst deactivation has major impact on the catalyst life and activity. Catalyst deactivation can take place because of [Mihacea et al., 1993]:
- Poisoning due to chemisorption of some impurity
- Erosion and breakage
- Hydrothermal ageing, that is loss of surface area because of exposure to high temperatures and steam.

Garcia-Dopico et al., [2006] have presented modeling of coke formation and deactivation in FCCU.

Major impurities responsible to catalytic deactivation are [Krishnamoorthy et al., 2005]:
- Ni and V deposited as FCC catalyst from high boiling metal laden organics in feed
- Iron from organic Fe particulates from corrosion of equipment
- Na and Ca exacerbates the effect of V and Fe
- Oxidized vanadium is mobile within and between particles.

Ni is a strong dehydrogenation catalyst and its presence as contaminate increases hydrogen and coke.

$$2C_xH_y \ (Y-X/2) \ H2 + C_{2x} \ Hx \ (coke)$$

Ni is relatively immobile and does not transfer from particle to particle [Habib, 1992]. Sb effectively passivates Ni through alloy formation. Ni can be passivated by reaction with alumina to form bulk nickel aluminate which has low dehydrogenation activity [Krishna Moorthy et al., 2005].

$$NiO + XAl_2O_3 \ (\text{special type}) \longrightarrow NiAl_2O_4.(X-1)Al_2O_3$$

Vanadium is oxidized and hydrolyzed in the regenerator [Krishna Moorthy et al., 2005].

$$2VC_xH_y + (2.5 + 2X + Y/2) \ O_2 \longrightarrow V_2O_5 + 2XCO_2 + yH_2O$$

Oxidized V (+5 oxidation state) has a higher mobility due to higher vapor pressure and [HABIB, ET, X ZHAO, G YALURIS, WC CHENG, LT BOOCK and ...] is more destructive toward zeolite.

$$\text{Faujasite} \xrightarrow{\ H_3VO_4\ } Al_2O_3.SiO_2 \text{ (amorphous)}$$

Presence of sodium enhances the destructive ability of vanadium. Oxidized vanadium also catalyses coke and hydrogen formation through oxidative hydrogenation. Presence of iron results in dramatic loss of unit activity, loss of bottom cracking [Krishna Moorthy et al., 2005].

In combination with Na and Ca, Fe leads to the formation of low melting temperature Fe rich phases on the exterior surface the catalyst; this restricts the diffusion of large hydrocarbon molecules. However, this effect is less severe on high alumina and high porosity catalysts [Krishna Moorthy et al., 2005].

By use of high alumina, high porosity catalyst, minimising regenerator temperature, reducing Na and Ca content of the feed and reducing acid content of the feed, the de activation capacity can be reduced. Pore structure also has significant impact on cracking. Small pores (< 100 Å) result in high bottom cracking. Medium pore (100–1000 Å) captures unvaporized feed liquid with good bottom cracking and less coke. Large pores (> 1000 Å) result in high catalyst attrition.

7.1.25 INDMAX Technology [Indian Oil Tech/T/2003/001]

INDMAX, a technology for residue upgradation and LPG/light olefins maximization has been developed by Indian Oil Technologies Ltd, India. This is similar to FCC technology except different catalyst system and operating conditions. Comparison of INDMAX and FCC technology is mentioned in Table 7.28. INDMAX catalyst employs a proprietary catalyst with three different functional components. The bottom cracking component provides highly acidic sites for catalytic [www.iocltech.com] cracking heavy feed molecules, which otherwise lead to coke and dry gas formation. The component responsible for upgrading naphtha range hydrocarbons is shape selective in nature, which allows selective cracking to light olefins without increasing the coke and dry gas make. The third component contains conventional ultrastable Y-zeolite which shows synergistic effect with the other two components of the catalyst [www.touchoilandgas.com].

Some of the special features of this INDMAX Technology are: High LPG yield (40–65 %wt of feed), high propylene yield (17–25 %wt of fresh feed), high butylene yield (20–28 %wt of fresh feed), feed CCR ranging from 0.35 %wt to 5.0 %wt, type VGO, RCO and SR, improved coke selectivity which permits a very high cat/oil ratio (15–25), as compared to the other state-of-the-art process ,significantly higher catalyst to oil ratio, lower delta coke results in lower regenerator temperature. The process employs higher riser temperature (560 °C) and a relatively high dilution steam rate (15–20 %wt of feed) to get high catalytic conversion. The high steam rate in the riser allows to minimize the hydrocarbon partial pressure which, among other things, also helps in: (a) Lowering the rate of coke formation, (b) increasing the olefinicity of products by minimizing the hydrogen transfer reactions and (c) increasing the heat demand in the riser and products a lower delta coke on catalyst [www.iocltech.com].

Table 7.28 INDMAX comparison with FCC

Process	FCC	INDMAX
Feed properties		
Density (gm/cc)	0.89	0.872
RCR (%wt)	0.30	0.253
Product yield [%wt (fresh feed)]		
Dry gas	2.9	6.1
LPG	12.1	50.4
Gasoline	31.2	19.4
Diesel	40.0	12.1
Bottom	8.3	5.5
Coke	5.5	6.4
Light olefin yield [%wt (fresh feed)]		
Propylene	3.6	21.2
Isobutylene	1.2	6.1
Total butylene		24.7
Typical operating conditions		
Reactor top temperature (°C)	490–510	550–570
Cat/oil ratio	650–730	670–700
Reactor pressure (kg/cm², abs)	4–7	15–25
	3.0–3.5	2.4–2.6

Maximization of LPG yield in FCC unit [Indian Oil Tech/S/2003/013]: Indian Oil's research development has developed expertise in enhancing the LPG yield by process optimization, catalyst selection and revamp of FCC units. Significant improvement in yield of LPG and propylene production has been achieved by adding ZSM-5 additive in the system even at low concentration [www.iocltech.com] (0.5%). Scoping studies at Indian FCC units indicated an increase in the yield of LPG from 10–20% to 25% of feed and propylene from 4.5 to 12 %wt.

7.1.26 FCC Gases as Petrochemical Feedstock

Although FCC units all over the world continue to be dominate conversion process in the petroleum refinery, however, FCC gases have now become important petrochemical feedstock for production of LPG which can be converted to aromatics and C_3, C_4 and C_5 hydrocarbons, i.e. propylene, butene, isobutene, pentene, etc. Conventional FCC units typically produce 3–6 %wt propylene depending on feed type, operating conditions and type of catalyst. However, addition of ZSM-5 additive can increase the propylene yield to about 8% [Fujiyama et al., 2005]. Significant improvements in FCC design, hardware, operation severity, catalysts, and additives have contributed to higher propylene [www.redorbit.com] yield. A propylene production source is mentioned in Table 7.29. Emerging FCC based propylene technologies are given in Table 7.30 [Aitani, 2004]. The traditional focus of the FCC unit is shifting from that of transportation fuels to that of a supplier of key petrochemical feedstock. Specialty additives and catalysts that contain ZSM-5 zeolite

Table 7.29 Propylene production source

	1998	2002	2010
Steam crackers	71	66	60
Refineries FCC (%)	27	32	35
Others	2	2	5

Source: Ross, 2005.

Table 7.30 Emerging FCC based propylene technologies

Process	Licensor	Propylene yield (%)	Remarks
Deep catalytic cracking	Stone and Webster Sinopec Research Institute of petroleum processing	14–23	Commercial operators at slow catalyst: Oil 7–132
Catalytic pyrolysis process	Stone and Webster Sinopec Research Institute of petroleum processing	18–24	VGO and heavy feeds commercial trials in China
High severity	King	17–25	Down flow; high severity operation
Indmax	Indian Oil Co.	17–25	Upgrades heavy cuts at high cat:oil ratios of 15–255
Maxofin	KBR, Exxon Mobil Corp	15–25	Variations with Superflex to increase propylene yield
Petro FCC	UOP LLC	20–25	Additional reaction severity and the RxCat design
Select component cracking	ABB Lummus Global	24	High severity operation

Source: Hydrocarbon Processing [Fujiyama et al., 2005].

are being used to optimise C_3 and C_4 fractions. The majority of FCC units yield 4–6% propylene whereas ultrahigh ZSM-5 based catalyst additives has the potential to produce upto 15% propylene [hcasia.safan.com] [Hydrocarbon Asia, 2003].

FCC off gases are of considerable importance as feedstock for petrochemicals. Some of the important building blocks, which can be derived from FCC, are: dilute ethylene stream, LPG, propylene, C_4 streams containing butane/butylenes. Typical composition of LPG from RFCC is given in Table 7.31. Typical steam cracker mixed C_4 stream composition is given in Table 7.32 [Morgan, 1998]. Variations in ethylene concentration are affected by the FCC feed composition and cracking severity [Netzer, 1997].

Table 7.31 Typical composition of LPG from RFCCU

Component	Percent	Component	Percent
Methane	0.043	Isobutylene	14.80
Ethane	0.41	*Trans*-butene	16.56
Propane	8.72	1, 3-*cis*-butene	14.53
Propylene	17.82	1, 2-butadiene	0.19
Isobutane	16.38	Isopentane	0.16
n-butane	8.36	n-pentane	0.02
1-butene	2.02		

Table 7.32 Typical steam cracker mixed C_4 stream composition

Feedstock	Naphtha			Gas oil	Propane	Ethane
Severity	ISV	SSV	HSV	HSV	HSV	HSV
Propylene/ethylene ratio	0.54	0.52	0.48	0.60	0.35	0.02
Mixed C_4 yield (% on feed)	10.8	10.2	9.2	9.0	4.5	2.5
Mixed C_4 composition, %						
C_3	0.5	0.5	0.5	0.5	0.5	0.5
n-Butane	4	3	2	2	3	3
Isobutane	2	1	1	1	2	2
Isobutene	23	23	22	23	4	3
Butene-1	14	14	14	17	12	9
Butene-2	10	11	11	13	9	7
1, 3-butadiene	46	47	49	43	69	75
Heavies	0.5	0.5	0.5	0.5	0.5	0.5
Total	100	100	100	100	100	100

Source: Chemistry and Industry, 2 Feb 1998, p90.

Steam crackers accounted for more than 70% of the global propylene production in 1998 which is expected to go down to 60% by 2010 while share of refinery FCC propylene which was 27% in 1998 is likely to rise to 35% by 2010 [Ross, 2005].

The composition and quality of such off gases produced from the catalysts crackers depend upon their design and the various operating conditions including the severity of cracking. It also depends upon the type of the crude processed and the fractions utilised for the cracking unit [Kothary et al., 1988]. Typical composition of C_4 streams from FCCs with available quantities for 1 MTPA and 0.6 MTPA capacities is given in Table 7.33 [Kothary et al., 1988]. There are two basic approaches for ethylene recovery: (i) Recovery of ethylene as dilute gas stream which serves as raw material for production of ethyl benzene and subsequently styrene and (ii) recovery of ethylene as pure polymeric grade liquid [Netzer, 1997].

Table 7.33 Typical composition of C_4 streams from FCCs with available quantities

Components	BP (°C)	Wt. Range	Avg. wt. percentage	Quantity	
				1 MTPA	0.6 MTPA
Isobutane	−11.4	35–36	35.5	26,625	15,975
Isobutene	−6.9	15–17	16.0	12,000	7,200
1-butene	−6.3	10–13	11.5	8,025	5,175
n-butane	0.5	11–14	13.0	9,750	5,850
Cis-butene-2	3.7	9–16	9.5	7,125	4,275
Trans-butene-2	0.9	13-16	14.5	10,875	6,525
Total			100	74,400	45,000

The LPG as well as off gases for the FCC unit are rich in olefins. The former contain up to 20–30% propylene while the latter could have 5–10% ethylene. Typical composition of LPG and straight run LPG is given in Table 7.34. However, use of LPG as petrochemical feedstock would depend on excess availability after meeting the demands for domestic fuel.

Indalin Process: Indalin is a versatile indigenous technology adding value to upstream and downstream oil industries. Indalin is a catalytic cracking process for upgradation of low value naphtha to very high yield of LPG, containing high olefins such as propylene, ethylene, butylenes, etc. Surplus kerosene and gas oil range fraction can also be processed along with naphtha. Indalin can integrate a refinery with petrochemicals complex, and therefore, offers a tremendous opportunity for value addition through upgradation of low value streams to petrochemical feedstock [Bhatacharya, 2011].

Table 7.34 Typical composition of cracked LPG and straight run LPG

Components (percent)	C_2	C_3	C_3+	i C_4	n C_4	Butene	i C_4+	trans-C_4+	Cis-C_4+
Cracked LPG	–	13.7	27.89	15.48	7.36	8.61	10.84	9.77	6.34
St. run LPG	1.5	30.0	0.44	21.8	46.0	0.12	0.14	–	–

REFERENCES

1. Latest Developments in FCC/RFCC Additives INTERCAT Confidential Information, Lovraj Kumar Memorial Trust Workshop on New Delhi, Nov 25–26, 2005.

2. Aitani A. Advances in Propylene Production Routes. Oil Gas European Magazine, Vol 20 (2004) No.1, p36.

3. Avidan AA. FCC is Far from being Mature Technology. Oil and Gas Journal, May 18, 1992, p25.

4. Badoni RP, Kumar Y, Umashanker and Prasada Rao TSR. Emerging Technologies for Light Olefins Production. Chemical Engineering World, Vol 31, No. 12, p105.

5. Bonifay R and Marcilly C. Catalytic Cracking, p169. In Petroleum refining, Vol 3, Conversion Processes, edited by Leprince P. Editions Technip Paris, 2001.

6. Campagna RJ, Kowalcyzk PC and Wilcox JR. Effect of Feed Properties on FCC Unit Performance. Petroleum Technology, Winter, 2001–02, p87.

7. Campagna RJ, Kowalczyk DC and Wilox JR. Effect of Feed Properties on FCC Unit Performance. Petroleum Technology, Winter, 2001–02, p87.

8. Chavdarov I, Stratiev D, Shishkova I and Dinkov R. Role of FCC Catalyst in Refinery Profitability. Petroleum Technology, Quarterly, Q1, 2014, p87.

9. Chen Ye-Mon. Recent Advances in FCC Technology. Powder Technology, 163 (2006), p208.

10. Chuang KC, Young GW and Benslay RM. AIChE Symposium Series, 1992, No. 291, Vol 88, 1992, p1.

11. Dahlberg A, Mukherjee U and Olsen CW. Consider Using Integrated Hydro-processing Methods for Processing Clean Fuels. Hydrocarbon Processing, Sep 2007, p111.

12. Dries H, Muller F, Willbourne P and Williams CP. Consider Using New Technology to Improve FCC Unit Reliability. Hydrocarbon Processing, Feb 2005, p70.

13. Erikson P, Tomlins A and Dash SK. FCCU Advanced Control System Achieves 2 Month Payout. Oil and Gas Journal, Mar, 23, 1993.

14. Fujiyama Y, Okuhara T, Saeed MR, Aitani AM and Dean CF. High Severity FCC: A New Process to Maximize Refinery Propylene. Hydrocarbon Asia, May/June 2006, p20.

15. Fujiyama Y, Redhwi HH, Aitani AM, Saeed MR and Dean CF. Demonstration of Plant for New FCC Technology Yields Increased Propylene. Oil and Gas Journal, Sep 26, 2005, p25.

16. Garcia-Dopico M, Garcia A and Santos Garcia A. Modeling Coke Formation and Deactivation in a FCCU. Applied Catalysis, 303 (2006), p245–250.

17. Ghosh S. Recent Advances in Fluid Catalytic Cracking. Proceeding of Petrotech Summer School, Jul 06.

18. Habib Jr ET. The Effect of Catalyst Feedstock and Operating Conditions on the Composition and Octane Number of FCC Gasoline. Proceedings of the Symposium of The Division of Petroleum Chemistry, Ic. American Chemical Society, Miamai, Sep 10–15, 1989.

19. HC, 1987. New FCC Design Now Commercial. Hydrocarbon Processing, 1987.

20. Harris JR. Use Desalting for FCC Feedstocks. Hydrocarbon Processing, Aug 1996, p63.

21. HC. New FCC Design Now Commercial. Hydrocarbon Processing, Sep 1987, p67.

22. Hemler CL and Smith LF. UOP Fluid Catalytic Cracking Process in Handbook of Petroleum Refining Processes, edited by Meyers RA, 3rd edition, The McGraw Hill Publication Data, 2004.

23. http:www.touchoilandgas.com/articles.cfm?artilce_id=712&level=2.

24. Humpries A, Kuehler C and Reid T. Consider New FCC Technologies to Produce Low Sulfur Gasoline. Hydrocarbon Processing, Sep 2003, p51.

25. Indian Oil Tech/T/2003/001, Indian Oil Technologies Ltd, Faridabad, India.

26. Indian Oil Tech/T/2003/001, Indian Oil Technologies Ltd, Faridabad, India.

27. INDMAX Technology [Indian Oil Tech/T/2003/001].

28. Irion WW and Neuwirth OS. In Oil Refining in Ullman's Encyclopedia of Industrial Chemistry, Vol 24, 2003, p205.

29. Keweshan C, Neuman D, Sexton J, Skurka M and Simon S. Advances in FCC Catlyst Performance. Petroleum Technology, Quarterly, Q4, 2010, p129.

30. Kothary NC, Mulchandani HK, Jain SK, Gomkale AV and Khilnani S. Techno Economics of Utilization of Refiner C-4 Streams. Chemical Engineering World, Vol 23, Dec 1988, p39.

31. Krisnamoorthy SM, Cheng Wu-Cheng, Roberie TG and Nee James RD. Advances in Fluid Catalytic Cracking (FCC) Catalyst Technologies, Lovraj Kumar Memorial Trust Workshop in New Delhi, Nov 25–26, 2005.

32. Long J, Xu Y, Zhang J, Dharia D, Batachari A, Yuan E, Gim S and Xu S. Consider New Processes for Clean Gasoline and Olefin Production. Hydrocarbon Processing, Sep 2011, p85.

33. Leiby S. FCC Catalyst Technologies Expand Limits of Process Capability. Oil and Gas Journal, March 23, 1992.

34. Lesemann M, Nee J, Petti N and Yaluris G. A Complete FCC Catalyst Portfolio Geared Towards Optimizing Refinery Profitability. Hydrocarbon Asia, May/June 2005, p24.

35. Lesemann M, Nee J, Petti N and Yaluris G. Hydrocarbon Asia, May/June 2005, 24.

36. Lesemann M and Schult C. Noncapital Intensive Technologies Reduce FCC Sulfur Content, Part 1. Hydrocarbon Processing, Feb 2003, p69.

37. Letzsch W. Improve Catalytic Cracking to Produce Clean Fuels. Hydrocarbon Processing, Feb 2005, p80.

38. Leuenberger EL. Optimum FCC Conditions Give Maximum Gasoline and Octane. Oil and Gas Journal, May 21, 1988, p45.

39. Mandal S, Das AK, Krishnan V and Makhija S. Improving FCC Margins Through Catalyst/Additive System, Lovraj Kumar Memorial Trust Workshop in New Delhi, Nov 25–26, 2005.

40. Marion P and Benazzi E . A Hydrocracking Strategy for Competitive Market. Petroleum Technology, Quarterly, Summer 2001, p23.

41. Martin DO and Allen RO. Preventing Salt Fouling in FCC Main Fractionators. Petroleum Technology, Spring, 2001, p41.

42. McAuley R and Dries H. FCC Cyclones—A Vital Element in Profitability. Petroleum Technology, Spring, 2001, p21.

43. Miller RB and Macris A. Treating Options to Meet Clean Fuel Challenges. Petroleum Technology, Quarterly, Spring, 2001.

44. Miller RB and Macris A. Treating Options to Meet Clean Fuel Challenges. Petroleum Technology, Quarterly, Spring, 2001.

45. Mihalcea E, Pop G, Bozga G and Mouean O. Roamanian Catal. Soc, 2 (2), 1993, p33.

46. Murphy JR. Evolutionary Design Changes Mark FCC Process. Oil and Gas Journal, May 18, 1992.

47. Nee JRD, Harding RH, Yaluris G, Cheng WC, Zhao X, Dougan TJ and Riley JR. Fluid Catalytic Cracking, Catalysts, Additives. Kirk Othmer Encyclopedia of Chemical Technology, Vol 11, 2006, p678.

48. Netzer D. Economically Recover Olefins from FCC Off Gases. Hydrocarbon Processing, April 1997, p83.

49. Niccum PK, Gilbert MF, Tallman MJ and Santner CR. Future Refinery—FCC Role in Refinery/Petrochemical Integration, 2001 NPRA Meeting, March 18, 2001.

50. O'conner P, Gerritsen LA, Pearce JR, Desai PH, Yanik S and Humphries A. Improve Resid Processing. Hydrocarbon Processing, Nov 1991, p76.

51. Petroleum Refining, Vol 3. Conversion Processes, edited by Leprince P. Editions Technip, Paris, 2001.

52. Bhatacharya D, Brijesh Kumar and Rajgopal S. Indalin: A Versatile Indigenous Process Technology. Jl of Petrotech, July–Sep 2011, p56.

53. Prezelj M. New FCC Design. Now Commercial. Hydrocarbon Processing, Sep 1987, p67.

54. Rajaraman P. Experience of Usage of ZSM-5 Additive in FCC Unit. Hydrocarbon Technology, May 15, 1992, p46.

55. Rajgopalan K and Habib Jr ET. Understanding FCC Matrix Technology. Hydrocarbon Processing, Sep 1992, p43.

56. Rao MR, Bhattacharya D, Mandal S, Das AK and Ghosh S. Resid Processing in Indian FCC Units: Prospects and Problems. Hydrocarbon Technology, 15th May, 1994, p3.

57. Rassev S. Industrial Implementation of Thermal Processes. In Thermal and Catalytic Processes in Petroleum Refining, Marcel Dekker, 2003, p137.

58. Refinig Process, 1998. Hydrocarbon Processing, Nov 1998.

59. Ross J, Roux R, Gauthier T and Anderson LR. Fine Tune FCC Operations for Challenging Fuels Market. Hydrocarbon Processing, Sep 2005, p65.

60. Rush JB and Steed PV. Hydroprocessing Resid for FCC Feed. Hydrocarbon Processing, May 1984, p50.

61. Salbilla DL. Refiners have Several Options to Overcome FCCU Opacity Limitations. Oil and Gas Journal, Jan 11, 1999, p195.

62. Scherser J. Octane Enhancing, Zeolite FCC Catalysts: Scientific and Technical Aspects. Cat Rev Sci. Eng, Vol 31, No. 3, 1989, p125.

63. Shorey SW, Lomas DA and Keesom WH. Use FCC Feed Pretreating Methods to Remove Sulfur. Hydrocarbon Processing, Nov 1999, p46.

64. Srikantharajah S, Baillie C, Zahnbrecher B and Wache W. Evaluation of a Low Rare Earth Resid FCC Catalyst. Ctalysis, 2012, p17.

65. Thota C, Gupta S, Gokak DT, Voolapalli RK, Rao PC and Swaminathan VP. Troubleshooting a FCC Unit Catalysis. PTQ, 2012, p59.

66. Vander Laan J, Sughrue EL, Dodwell G and Meier PF. Control Sulfur Levels in FCC Products. Hydrocarbon Processing, Feb 2006, p49.

67. Wang, Longyan, Yang B, Wang G, Tang H, Li Z and Wei J. New FCC Process Minimizes Gasoline Olefin Increases Propylene. Oil and Gas Journal, Feb 10, 2003.

68. Walther W Irion. Oil Refining. Ullmann's Encyclopedia of Industrial Chemistry 2005, Wiley-VCH, Weinheim.

69. Wilson JW. Troubleshooting FCC Stand Pipe Operations. Petroleum Technology, Quarterly, Autumn, 2000, p71.

70. Wrench RE and Glasgow PE. FCC Hardware Options for The Modern Cat Cracker. Chuang KC, Young GW and Benslay RM. Vol editors Advanced Fluid Catalytic Cracking Technology. AICHE Symposium Series, No. 291, Vol 88, p1.

71. Yen Yung. Petroelum Technology, Quarterly, T, 2010, p27.

8

Hydrocracking

The development of upgrading technology for heavier stocks having high sulfur, nitrogen and heavy metals (Ni, V) are becoming important. Hydrocracking is one of the most versatile and major secondary processes for conversion of low quality feedstocks into high value added products like gasoline, naphtha, kerosene and diesel, and hydrowax which can be used as petrochemical feedstock. Its importance is growing more as refiners search for low investment option for producing clean fuel [Huve et al., 2005]. New environmental legislations require increasing and expensive efforts to meet stringent product quality demands with low sulfur content. Because of the flexibility of hydrocracking chemistry, hydrocracking processes wide variety of feedstocks like naphtha, atmospheric gas oil, vacuum gas oils, coke oils, catalytically cracked light and heavy cycle oils, cracked residue, deasphalted oils and produces different high quality products with low sulfur content. Hydrocracking process involves simultaneous cracking and hydrogenation of hydrocarbons in presence of hydrogen using supported metal bifunctional catalyst [Verma and Narsimhan,1994].

The history of the hydrocracking process goes back to later 1920 when hydrocracking technology for coal conversion was developed in Germany [Adrian Gruia, 2006]. During World War II two-stage hydrocracking was applied in Germany, USA and Britain [http:www.cheresources.com/refining.html]. However, real breakthrough in hydrocracking process was with the development of improved catalyst due to which processing done at lower pressure. Hydrocracking can process wide variety of feedstocks producing wide range of products [Scherzer and Gruia, 1996]. In processing of opportunity crude slurry phase, hydrocracking is a possible solution to improve refining margin as opportunity crudes require more hydrogen addition [Motaghi et al., 2011].

Various feeds for hydrocracker and products are given below:

- *Feed:* Straight run gas oil, vacuum gas oils, cycle oils, coker gas oils, thermally cracked stocks, solvent deashphalted residual oils, straight run naphtha and cracked naphtha.
- *Product:* Liquefied petroleum gas (LPG), motor gasoline, reformer feeds, aviation turbine fuel, diesel fuels, heating oils, solvent and thinners, lube oil and FCC feed.

Various hydrocracking technologies have been in use. Different types of hydrocrackers are being used. Some of the commonly used hydrocrackers are fixed bed hydrocracker, ebullated bed hydrocracker and slurry phase hydrocracker [Motaghi et al., 2011]. Some of the hydrocracking technologies are described in Table 8.1.

Table 8.1 Hydrocracking technologies

Process technology licensers	Description	Remarks
ABB Lummus Global LC-Fining process	Desulfurization, demetalization, CCR reduction and hydrocracking of atmospheric and vacuum residues using LC-Fining process. Typical operating conditions: Temperature: 725–840 °F Pressure: 1400–3500 psi LHSV: 0.1–0.6 h^{-1} Conversion: 40–97% Typical yield for Arabian heavy/ Arabian light blends is given in Table 8.2.	Product: Full range of high quality distillates. Residual product can be used feedstock for RFCC, cokes, visbreaker or solvent deasphalts.
Chevron Research and Technology Co (Chevron isocracking process)	For hydrocracking of naphtha, AGO, VGO and cracked oils from FCCs, cokers, hydroprocessing plant, SDA plants using both amorphous/zeolite and zeolite catalysts. Plant consists of a staged reactor system consisting of one reactor, one HP separator and fractionator. Typical yield from various feedstocks in Chevron iso-cracking process is given in Table 8.3. Some of the developments in Cheveron isocraking technology are optimized partial conversion isocracking technology, selective staging hydro-carcking isocracking technology.	Product: LPG, gasoline, catalytic refiner feed, jet fuel, kerosene, diesel, lube oil and feeds for FCC or ethylene plants.
IFP process (Axen)	Process is based on a new dual catalyst system (using hydrotreatment catalyst —prototype hydrocracking—zeolite catalyst for upgrading of vacuum gas oil alone or blended with light cycle oil, deashphalted oil, visbreaker or coker gas oil. The process uses a refining catalyst usually followed by an amorphous and/or zeolite type hydrocracking catalyst. High tolerance towards feedstock nitrogen, high selectivity toward middle distillate and high activity of the zeolite.	Product: Middle distillates, very low sulfur fuel oil, extra quality FCC feed or high VI lube basestock. Eleven units have been licensed.

Contd.

Table 8.1 Hydrocracking technologies *(Contd.)*

Process technology licensers	Description	Remarks
IFP H-Oil Plants	The process uses an ebullated bed reactor for catalytic hydrocracking and desulfurization of residual and heavy oils. For producing full range distillates and upgraded residuals, transportation fuels, FCC or coker feed and low-sulfur fuel oil. Typical Operating Condition: Temperature 770–840 °F, pressure 1000–2500 psi, LHSV 0.1–0.9 h^{-1}, conversion 40–95%.	Product: Full range of distillates, upgraded residues, transportation fuel, FCC and low sulfur fuel oil
Kellog Brown and Root MAK-Fining HDC process	This process is based on MAK-Fining, HDC process and converts a wide variety of feedstocks including vacuum gas oil, coker gas oil and FCC cycle oils into high quality, low sulfur fuels using the MAK-Fining HDC process. The process uses multiple catalyst system in multi bed reactors that incorporate proprietary advanced quench and redistribution internals [Speight, 2006], typical yield in Kellog Brown and Root MAK-Fining HDC process for Arabian light (AL)/ Arabian heavy (AH) is given in below Table 8.4.	Product: A wide range of high quality, low sulfur distillate feeds and blending stocks including LPG, high octane gasoline, reformer naphtha, jet fuels, kerosene, diesel oil.
Shell International Oil products Hydrocracking process	The process converts heavy VGO and other low cost cracked and extracted feedstocks to high value, high quality products. Two-stage, series flow and single unit design are available. Typical yield of jet fuel and diesel is mentioned in Table 8.5.	Low sulfur diesel and jet fuel with excellent combustion properties, high octane gasoline, and high quality reformer, cat cracker or lube oil feedstocks.
Verba GEL Technologies and Automatisierung GmbH	This processes heavy and extra heavy crude as well as residuals upgrade typical feed characteristics, tield and product qualities of VEBA GEL and automatisierung GmbH Hydrocracking process. Typical feed characteristics, yield and product qualities is given in Table 8.6.	Full range high quality product

Contd.

Table 8.1   Hydrocracking technologies *(Contd.)*

Process technology licensers	Description	Remarks
UOP unicracking process	The process converts wide variety of feedstocks (atmospheric gas oil, vacuum gas oil, FCC/RCC cycle oil, coker gas oil, deasphalted oil and naphtha into lower molecular weight products. Unicracking systems typically use a single stage configuration and operate at a more severe condition because of inherent lower activity of amorphous catalysts. Unicracking catalysts offer the ability to adjust product slate by modest changes in process and distillation conditions. Hydrotreating schemes: Once through, two-stage, separate hydrotreat, hycycle	Product; gasoline, jet fuel, diesel fuel, lube stocks, high quality FCC feed stocks, LPG
Paragan process	This is a two-stage process used for cracking non-saleable VGGO using zeolite catalyst.	Transportation fuels
Petro-Canada HT severe hydrocracking	In this process, the elimination of aromatics and polar compounds is achieved by chemically reacting the feedstock with hydrogen, in presence of a catalyst, at high temperatures and pressures [www.francais.petro-canada.ca].	Transportation fuels
HCat residue hydrocarcking technology	This technology is based on the *in situ* formation of a molecularly dispersed catalyst which is intimately mixed throughout the heavy oil feedstock. The process uses an ebullated bed reactor.	Transportation fuels
LC-Fining	LC-Fining resid hydroconversion process is developed to specifically target most difficult heavy lower value hydrocarbon streams like petroleum residua heavy oil from tar sands, shale oils and solvent refined coal extracts with conversion as high as 90%. The process uses Ebaluating bed hydrocracking.	Full range of high quality distillates

Sources: Refining Processes, 98; Hydrocarbon Processing, Nov 1998, p53; Virdhi et al., 2010, http://www.boucherandjones.com/hydrocracking.htm, Towler GP, Hoehn, RK. Hydrocracking Industrial Processes, Encyclopedia of catalysis vol 3, Willey Interscience Horvath, IT, Light et al., 1981; Kunnas, 2011; Refining Processes, 2008; Akelson, 2004; Bridge and Mukherji, 2004.

Table 8.2 Typical yield for Arabian heavy (AH)/Arabian light (AL) blends in ABB Lummus Global LC-fining process

Feed	Atm. resid		Vac. resid	
Gravity (°API)	12.4	4.73	4.73	4.73
Sulfur (%wt)	3.90	4.97	4.97	4.97
Ni/V (ppmw)	18/65	39/142	39/142	39/142
Conversion %vol (1022 °F)	45	60	75	95
Product (%vol)				
C_4	1.11	2.35	3.57	5.53
C_5–350 °F	6.89	12.6	18.25	23.86
350–700 °F (650 °F)	15.24	30.62	42.65	64.81
700 (650 °F)–1022 °F	55.27	21.46	19.32	11.92
1022 °F	25.33	40.0	25.0	5.0
C_5–API/%wt S	23.7/0.54	22.5/0.71	26.6/0.66	33.3/0.33

Source: Hydrocarboné Processing, Nov 1998, p82.

Table 8.3 Typical yield from various feedstocks in Chevron isocracking process

Feed	Naphtha	LCCO	VGO	VGO
Catalyst stages	1	2	2	2
Gravity (°API)	72.5	24.6	25.8	21.6
ASTM 10%/EP (°F)	154/290	478/632	740/1050	740/1100
Sulfur (%wt)	0.005	0.6	1.0	2.5
Nitrogen (ppm)	0.1	500	1000	900
Yield (%vol)				
Propane	55	3.4	–	–
Isobutane	29	9.1	3.0	2.5
n-butane	19	4.5	3.0	2.5
Light naphtha	23	30.0	11.9	7.0
Heavy naphtha	–	78.7	14.2	7.0
Kerosene	–	–	86.8	48.0
Diesel	–	–	–	50.0
Product quality				
Kerosene smoke point (pt) (mm)	–	–	28	28
Diesel cetane index	–	–	–	58
Kerosene freeze point (°F)	–	–	–65	–75
Diesel pour Point	–	–	–	–10

Courtesy: CPCL.

Table 8.4 Typical yield in Kellog Brown and Root MAK-Fining HDC process for Arabian light (AL)/Arabian heavy (AH) for single pass, moderate pressure operation on middle east VGO and FCC LCO

Feed		AL/AH VGO	LCO
Gravity (°API)		20.2	19.0
ASTM FBP (°F)		1,050	620
Sulfur (%wt)		2.9	1.0
Nitrogen (ppmw)		900	600
Centane index		–	28
Conversion	50%	70%	50%
Naphtha (%vol)	12.9	22.6	54.0
Kerosene (%vol)	14.1	24.5	–
Diesel (%vol)	31.8	32.5	54.3
LSGO (%vol)	50.0	30.0	–
H_2 consumed (scf/bbl)	1,080	1300	1730
Product quality			
Naphtha, RON	64	63	92
Diesel cetane index	53	55	39
Diesel sulfur (ppmw)	< 50	< 50	< 50

Source: Hydrocarbon Processing, Nov 1998, p86.

Table 8.5 Typical yield of jet fuel and diesel from HVGO feed (50/50 Arabian light/heavy)

Quality	Feed HVGO	Jet fuel	Diesel
Sp. gravity	0.932	0.800	0.826
ASTM cut point	405–565	156–525	225–354
Sulfur (ppm)	31,700	< 10	< 20
Nitrogen (ppm)	853	< 5	< 5
Metals (ppm)	< 2	–	–
Cetane index	–	–	62
Flash point (°C)	–	> –40	125
Smoke point (mm, EOR)		26–28	–
Aromatics (%vol, EOR)	–	< 10	13
Viscosity (38 °C, cSt)	110	–	5.3

Courtesy: CPCL.

Table 8.6 Typical feed characteristics, yield and product qualities of VEBA GEL and automatisingerung Gmbh hydrocracking process

Feed	
Gravity (°API)	–3–14
Sulfur (%wt)	0.7–7
Metal (Ni,V) (ppm)	Up to 2,180
Asphaltenes (%wt)	2 to 80
Yields	
Naphtha < 1,800 C (%wt)	15–30
Middle distillate (%wt)	35–40
Vac. gas oil > 3,500 C (%wt)	15–30
Product qualities	
Naphtha	Sulfur < 5 ppm, nitrogen < 5 ppm
Kerosene	Smoke point > 20 mm, cloud point < –50 °C
Diesel	Sulfur < 50 ppm, cetane no. > 45
Vac. gas oil	Sulfur < 150 ppm, CCR < 0.1 %wt, metals < 1 ppm

Source: Hydrocarbon Processing, Nov 1998, p67.

8.1 RECENT DEVELOPMENT IN HYDROCRACKING

There has been continuous development in the hydrocracking technology both in process and catalyst. Some of the important development in hydrocracking has been mild hydrocracking and resid hydrocracking. Mild hydrocracking is characterized by relatively low conversion (20–40%) as compared to convention hydrocracking which give 70–100% conversion of heavy distillate at high pressure. Mild hydrocracking (MHC) route produces low sulfur (10 ppm sulfur as desired by future diesel specification) diesel [Sarranzin et al., 2005]. New mild hydrocracking route produces 10 ppm sulfur diesel which is produced by hydrocracking under mild pressure. MHC allows increasing diesel production through VGO hydro-conversion. The yield of middle distillates obtained from hydrocracker is much more than that obtained from other processes. Also, hydrocracker does not yield coke or pitches as by-product. The increased demand for environmentally acceptable products forced the refiners to accept stringent specifications for gasoline and diesel necessitating the use of hydrocracking technology to limit sulfur and aromatic in petroleum products. No posttreatment is required for the hydrocracker products. Integrating residue hydrocracking operations with advance FCC optimizes upgrading of heavy crude oils. ebaluated bed resdue hydrocracking has been reported highly effective hydrogen addition process that upgrades heavy residue feeds to good quality diesel and FCC feed [Rama Rao et al., 2011].

8.2 HYDROCRACKER IN INDIAN REFINERIES

Hydrocracking is finding increasing use in the refineries due to use of increasing use of heavier feedstocks. Hydrocracking units in some of the Indian refineries are given in Table 8.7 [Verma, 1999].

Table 8.7 Hydrocracking units in Indian refineries

S. No.	Name of the refinery	Capacity MMTPA	Configuration	Name of the licensor
1.	IOCL Gujarat refinery	1.2	Recycle	M/S Cheveron
2.	IOCL Panipat refinery	1.6	Once through	M/S UOP
3.	IOCL Mathura refinery	1.2	Once through	M/S Chevron
4.	MRPL Mangalore refinery	Unit I: 1.05	Recycle	M/S UOP
		Unit II: 1.15	Recycle	M/S UOP
5.	CPCL Chennai refinery	3.16	Single stage with recycle	M/S Chevron
6.	KRL Kochi refinery	3.0	Once through	M/S Chevron
7.	Numaligarh refinery	1.1	Recycle	M/S Chevron
8.	IOCL Paradip refinery	3.0	Once through	M/S Chevron

IOCL: Indian Oil Corporation Ltd, MRPL: Mangalore Refineries and Petrochemicals Ltd, KRL: Kochi Refineries Ltd.

8.3 HYDROTREATMENT AND HYDROCRACKING CATALYST

Hydrocracking processes involved two types of catalysts: Hydrotreatment catalyst and hydrocracking catalyst. Criterion catalyst and technologies and Zeolite international (Zeolyst) have recently new series of Hydrocracking catalyst (pretreat and cracking) which can help refiners extract more value from hydrocracker by increasing liquid yields, lowering gas make and producing products meet future clean fuels specification.

8.3.1 Hydrotreating (Pretreat) Catalyst

The main objective of pretreat catalyst is to remove organic nitrogen from the hydrocracker feed allowing: (i) Better performance of second stage hydroccracking catalyst and (ii) the initiation of the sequence of hydrocracking reactions by saturation of aromatic compounds. Pretreat catalyst must have adequate activity to achieve above objectives within the operating limits of the hydrogen partial pressure, temperature and LHSV. Criterion offers an impressive line of widely used and proven catalyst triobe 424, 411, DN-120 and the newly-developed century DN-190 and Centinel DN-3100. The trilobe shape of the Criterion pretreat catalysts significantly reduces diffusional limitations and provide reduced pressure drop compared with the same size cylindrical catalysts [www.shell.com] [Crirorian Zeolyst].

8.3.2 Hydrocracking Catalyst

A hydrocracking catalyst has four key performances [Rashid, 2007]:

- *Initial activity:* The temperature required to obtain the desired product at the start of run.
- *Stability:* Rate of temperature increases that is required to maintain conversion.
- *Product selectivity:* The ability of a catalyst to produce desired product slate.

- *Product quality:* The ability of a catalyst to make products with desired specifications such as pour point, smoke point or octane number.

Hydrocracking catalyst is a bifunctional catalyst and has a cracking function and hydrogenation–dehydrogenation function. The former is provided by an acidic support whereas the latter is imparted by metals [Fahim, 2010]. Acid sites (crystalline zeolites, amorphous silica alumina, mixture of crystalline zeolite and amorphous oxides) provide cracking activity whereas metals [noble metal (Pd, Pt) or non-noble metal sulfides (Mo, Wo or Co, Ni) provide hydrogenation–dehydrogenation activity. These metals catalyze the hydrogenation of feedstocks making them more reactive for cracking and hetero-atom removal as well as reducing the coke rate [Adrian Gruia, 2006] [Narshiman et al., 2005]. Zeolite based hydrocracking catalysts have following advantages of greater acidity resulting in greater cracking activity; better thermal/hydrothermal stability; better naphtha selectivity; better resistance to nitrogen and sulfur compounds [www.epa.gov]; low coke forming tendency, and easy regenerability [Narshiman et al., 2005]. Hydrocracking catalyst type for desired reaction is given in Table 8.8. List of vendors supplying hydrocracker catalysts is given in Table 8.9.

Table 8.8   Hydrocracking catalyst types for desired reactions

Desired reaction	Catalyst characterization			
	Acidity	Hydrogenation activity	Surface area	Porosity
Naphtha to LPG	Strong	Moderate	High	Low to moderate
Gas oil to gasoline	Strong	Moderate	High	Low to moderate
Gas oil to jet fuel and middle distillates	Moderate	Strong	High	Low to moderate
Gas oil to high lube oils	Moderate	Strong	High	Moderate to high
Solvent deasphalted oils and residues to lighter products	Moderate	Strong	High	Low to moderate
Hydroconversion of non-hydrocarbon S in gas oil	Weak	Strong	Moderate	High
Sulfur and metals removal in residue	Weak	Strong	Moderate	High

Source: IIP, Dehradun, Hydrocarbon Technology, 18 (August 1991), [Narshiman et al., 2005].

Table 8.9   Various hydrocracker catalyst suppliers

Sl. No.	Name of corporation
1.	ABB Lummus Crest Inc., Bloomfield, N.J.
2.	Akzo Nobel (Albamarle), Netherlands and Houston
3.	Catalyst and Chemicals Industries Co Ltd, Tokyo
4.	Chevron Research and Technology Co Ltd, San Francisco

Contd.

Table 8.9 Various hydrocracker catalyst suppliers *(Contd.)*

Sl. No.	Name of corporation
5.	Criterion Catalyst Co Ltd, Houston
6.	Cros Field Catalysts, Warrington, UK and Chicago
7.	Grace Division, Balimore
8.	Kataleuma GmbH, Leuna, Germany
9.	UOP, Desplaines, IL
10.	Zeolyst International, Houston
11.	Haldor topes A/S, Lygby, Denmark
12.	Exxon Research and Engineering Co, Florhm Park, NJ
13.	Acreon, Catalysts/Procatalyse, Houston
14.	Orient Catalyst Co Ltd.,
15.	Sud-Chemie, Louisville, KY

Source: Narshiman and Bask, 2000.

Criterion/Zeolyst developed a new type Y-zeolite with higher cracking activity that provides substantial performance benefits in the production of naphtha (Z-853), jet (Z-3723), diesel/kero (Z-613,Z-623), diesel (Z-513) and high-quality base oils (LH-22) Criterion/Zeolyst. Zeolyst hydrocracking catalysts can result in significant increases in run length due to superior start-of-run activity and stability characteristics of these performance catalysts [www.shell.com].

Aluminum based catalyst has a critical problem of catalyst deactivation caused by coking and deposition of heavy metals even under relatively high hydrogen pressure [Diez et al., 1990; Kore et al., 1995]. Iron/active carbon has effectively cracked heavy residual oil restricting formation of asphaltene suppressing coke formation at higher conversions under low hydrogen pressure [www.eptq.com]. [Terai and Fukuyama, 2000].

8.4 HYDROCRACKING TECHNOLOGY

Hydrocracking process is a versatile process for catalytically hydrocracking heavy petroleum fractions into lighter, more valuable products. Typical feeds to hydrocracker are heavy atmospheric and vacuum gas oils and catalytically or thermally cracked gas oils. These feedstocks are converted into lower molecular weight products, usually maximizing naphtha or middle distillates. With hydro-cracking process sulfur, nitrogen and oxygen are almost completely removed and olefins are saturated, thereby giving products which are a mixture of essentially pure paraffins, naphthenes and aromatics. Typical yield of light and heavy Arabian blends is given in Table 8.10.

Table 8.10 Yield: For Arabian heavy/Arabian light blends

Feed	Atm. resid		Vac.resid	
Gravity (°API)	12.4	4.73	4.73	4.73
Sulfur (%wt)	3.90	4.97	4.97	4.97
Ni/V (ppmw)	18/65	39/142	39/142	39/142
Conversion %vol (1022 °F)	45	60	75	95
Product (%vol)				
C_4 1.11	2.35	3.57	5.53	
C_5–350 °F	6.89	12.6	18.25	23.86
350–700 °F (650 °F)	15.24	30.62	42.65	64.81
700 (650 °F)–1022 °F	55.27	21.46	19.32	11.92
1022 °F	25.33	40.0	25.0	5.0
C_5–API/%wt S	23.7/0.54	22.5/0.71	26.6/0.66	33.3/0.33

Source: Hydrocarbon Processing, Nov 1998, p82.

Process

Depending upon the feed quality, product mix desired and the capacity of unit, different hydrocracking technologies are available which may be mild medium and high pressure processes operating in single or two stages, once through or recycle. Different types of reactor configurations are available which include fixed bed both down flow and up flow reactors, ebullating bed up flow reactor, moving bed reactor and slurry phase reactor. The flow scheme of any hydrocracking process consists of compression of the hydrocarbon feed and hydrogen makeup gas to reaction pressure; heating of the feed and recycle gas to reaction temperature, reaction under suitable condition of temperature and pressure; cooling of the product stream; separation of recycle unconverted hydrogen, and finally stabilization and fractionation of the liquid product streams into the light and heavier product streams. Products obtained are:

- LPG
- Stabilized light naphtha
- Heavy naphtha withdrawn at 146 °C
- Superior kerosene/aviation turbine fuel (TF) at 188 °C
- Diesel at 286 °C
- Some of the by-products from hydrocracker units are filter backflush, unconverted oil, sour water, off gases containing hydrogen, sponge oil absorber off gas.

8.4.1 Single Stage Hydrocracking Process

The unit consists of the following sections:

- Furnace
- First stage reactor section
- Second stage reactor section
- High pressure separator
- Fractionation section
- Light ends recovery section.

In single stage process, both treating and cracking steps are combined in a single reactor. In this process the feed along with recycle unconverted residue from the fractionator is first hydrotreated in a reactor and then the combined stream gases are fed to second reactor where cracking takes place in presence of hydrocracking catalyst. In the single stage process the catalysts work under high H_2S and NH_3 partial pressure. Figure 8.1 illustrates the single stage process.

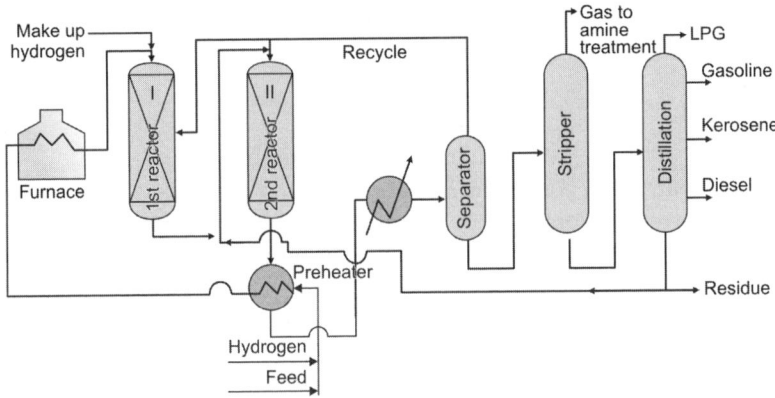

Fig. 8.1 Single stage hydrocracking

8.4.2 Two-stage Hydrocracking Process

The unit consists of the following sections:

- Furnace
- First stage reactor section
- Second stage reactor section
- Third stage reactor
- Fractionation section
- Light ends recovery section.

Preheated feed is first hydrotreated in a reactor for desulfurization and denitrogenation in presence of pretreat catalyst followed by hydrocracking in second reactor in presence of strongly acidic catalyst with a relatively low hydrogenation activity [Speight, 2006]. In the first stage reactor, the sulfur and nitrogen compounds are converted into hydrogen sulfide and ammonia with limited hydrocracking. The two-stage process employs interstage product separation that removes H_2S and NH_3 [www.shell.com]. In case of two-stage process, hydrocracking catalyst works under low H_2S and NH_3. Figure 8.2 illustrates the two-stage process.

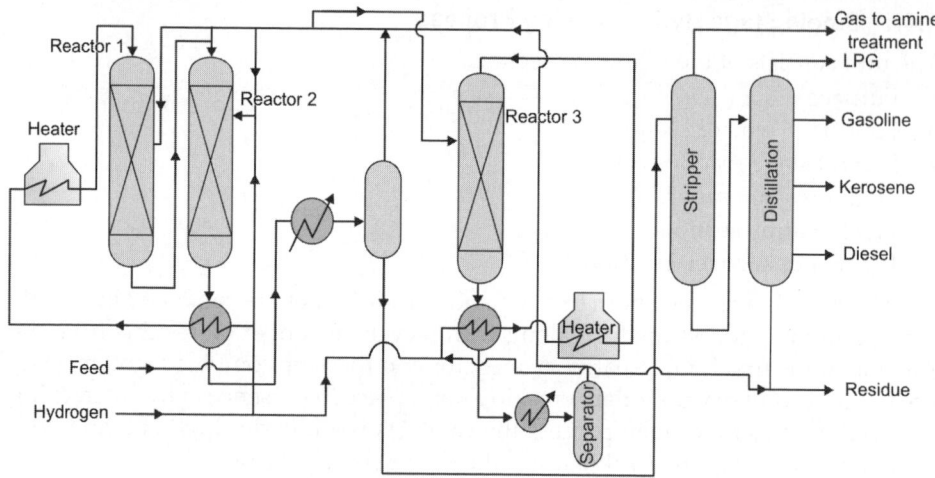

Fig. 8.2 Two-stage hydrocracking process

Isocraking unit

The isocracking unit consists of five principal sections:

1. **Feed and reaction section:** The reaction section contains one reaction stage in a single high pressure loop. The hydrotreating and hydrocracking reactions taking place in the reaction stage occur at high temperature and pressure.

2. **Recycle gas loop:** The recycle gas loop section contains additional equipment for separation of hydrogen-rich gas from the reactor effluents which are compressed and recycled back.

3. **Fractionation section:** The purpose of the fractionation section is to separate reaction section products into sour gas, unstabilized naphtha, kerosene, diesel, and UCO as fractionator's bottoms.

4. **Light ends recovery section:** The light end section is designed to take the sour gas and unstabilized naphtha from the product stripper and fractionator overhead, and produce sweet fuel gas, LPG, and light and heavy naphtha as products.

5. **Makeup hydrogen compression section:** The makeup hydrogen compression section consists of three identical parallel compressor trains, each with three stages of compression.

The primary products from the unit are:

- Cold low pressure separator off-gas to off-plot hydrogen PSA unit/fuel gas
- Sponge oil absorber sweet off-gas to fuel gas
  - o LPG
  - o Stabilized light naphtha
  - o Heavy naphtha
  - o Aviation turbine fuel/superior kerosene
  - o Diesel
  - o Unconverted oil (FCC feed)

- The by-products are:
  - Filter backflush to fuel oil/off-plot or FCC feed
  - Sour water to sour water stripper
  - Spent caustic solution to spent caustic system
  - Brine solution
  - Blow down from steam generators to storm water sewer

Typical feed specifications for an isocracking plant is given in Table 8.11. Make-up hydrogen quality is given in Table 8.12. Typical injection water quality specification is given in Table 8.13. Estimated product yields—start-of-run (SOR) and End-of-run (EOR), product yields: SOR and EOR and product quality: SOR are given in Tables 8.14 and 8.15 [Courtesy: Chennai Petroleum Corporation Ltd (CPCL, Chennai)].

8.4.3 Mild Hydrocracking (MHC) Process

Increasing tightened specifications have forced challenge to go for innovative technology for upgrading heavy cuts into high quality diesel. MHC allows increasing diesel production through VGO hydroconversion [Sarrazin, 2005]. MHC produces diesel with higher cetane and can reduce diesel end point. It has very high flexibility with respect to products' slate [Badhe et al., 2005]. Simultaneously, the hydrotreating reactions yield significant quality improvement for the residual VGO. It becomes a better FCC feed and increases gasoline yield, higher octane retention and low sulfur

Table 8.11 Typical feed specifications for an isocracking plant

Feed specifications	SR VGO	VB VGO
Feedstock sources	CDU/VDU BH/PG Crude	VBU
Distillation (°C), ASTM-D-1160		
Start	320	300
10%	380	324
30%	415	378
50%	440–460	402
70%	490	435
90%	540	473
End point (max.)	585	500
API gravity	20.8–34.8	20.7
Specific gravity	0.825–0.929	0.93
Asphaltenes (%wt, max.)	0.010	0.020
Iron (ppm)	0.0	5
Nitrogen (ppm wt.)	1,000	2,000
Sulfur (%wt)	2.8	5.0

Contd.

Table 8.11 Typical feed specifications for an isocracking plant *(Contd.)*

Feed specifications	SR VGO	VB VGO
Kinematic viscosity at 50 °C (cSt)		46.3
Kinematic viscosity at 100 °C (cSt)		9.6
Kinematic viscosity at 200 °C (cSt)	0.6–1.65	
Kinematic viscosity at 250 °C (cSt)	0.4–1.1	
CCR (%wt)	0.6	1.0
Nickel + Vanadium (ppm)	1.0	9

Courtesy: CPCL.

Table 8.12 Make-up hydrogen quality

Hydrogen purity (mole%)	99.5 (min.)
$CO + CO_2$ (mole)	20 PPM (max.)
Nitrogen (mole)	50 PPM (max.)
Water (mole)	50 PPM (max.)
Chlorine + Chlorides (mole)	1 PPM (max.)
Methane (mole%)	Balance

Courtesy: CPCL.

Table 8.13 Injection water quality

H_2S (ppm)	1,000 (max.)
NH_3 (ppm)	1,000 (max.)
Oxygen content [ppb (wt)]	15 (max.)
Chloride content [ppm (wt)]	50 (max.)
Acidity (pH)	7.5 (min.)
Non-volatile residue [ppm (wt)]	2 (max.)

Courtesy: CPCL.

Table 8.14 Estimated product yields: SOR and EOR

Feed: 90% SR VGO, 10% VB VGO

Total feed rate	=	1.85 MMTPA (112% of design) 249.2 m^3/hour
Total feed rate	=	37,400 BPSD
Chemical hydrogen consumption	=	256 Nm^3/m^3 of feed for SOR, 262 Nm^3/m^3 of feed for EOR
Chemical hydrogen consumption	=	1,520 SCFB for SOR, 1,550 SCFB for EOR

Table 8.14 Product yields: Start-of-run and end-of- run

Products	Start-of-run		End-of-run	
	%wt	LV %	%wt	LV %
H_2S	3.21		3.21	
NH_3	0.13		0.13	
C_1	0.25		0.45	
C_2	0.39		0.66	
C_3	0.67	1.23	1.00	1.83
iC_4	0.73	1.20	1.21	2.0
nC_4	0.73	1.16	1.10	1.75
Light naphtha	3.40	4.7	3.60	4.97
Heavy naphtha	5.10	6.21	5.10	6.20
Kerosene	23.69	27.07	23.60	26.89
Diesel	21.75	23.99	20.02	22.03
Bottoms	42.43	46.33	42.42	46.24
C_5^+	96.37	108.3	94.75	106.32
Total	**102.48**	**111.89**	**102.51**	**111.90**

Note: Yields are expressed as percentages of fresh feed.
Courtesy: CPCL.

Table 8.15 Product quality: SOR

Cut	Light naphtha	Heavy naphtha	Kerosene/ ATF	Diesel	UCO
Cut points (°C)	95	95–130	130–260	260–370	370+
API gravity	79.0	54.0	42.5	36.5	34.8
Density (15 °C)	670–675	763–768	812–817	820–870	850–855
ASTM (°C)	D 86	D 86	D 86	D 86	D 1160
ST	23	102	142	229	337
10	36	107	161	268	394
50	49	114	193	302	473
90	79	128	2238	348	548
EP	91	147	266	376	562
95				370	
Sulfur (ppm)	< 5	< 5	10 (max.)	10 (max.)	< 50 (max.)
Mercaptan sulfur (ppm)	10 (max.)	10 (max.)			
Nitrogen (ppm)	< 1	< 1			

Contd.

Table 8.15 Product quality: SOR *(Contd.)*

Cut	Light naphtha	Heavy naphtha	Kerosene/ ATF	Diesel	UCO
Paraffins (LV %) (estimated)	66	30			
Naphthenes (LV %) (estimated)	32	62			
Aromatics (LV %) (estimated)	2	8	22		
Flash point (°C)			38 (min.)	35 (min.)	
Smoke point (mm)			21		
Cetane index (D 613)				55 (min.)	
Freeze point (°C)			(–)51		
Pour point (°C)				3 (max.)	
Viscosity @ 40 °C (cSt)					
Viscosity @ 100 °C (cSt)				2–5	4.2
Viscosity @ –20 °C (cSt)			8		
Water (ppm)				500 (max.)	
Metals (ppm)					< 0.1
Octane number, estimated RONC/MONC	75/76	64/62			
RVP (kg/cm^2)	0.8 (max.)	0.8 (max.)			

Courtesy: CPCL.

Product properties: Product specifications of a typical isocracking unit are given below:

LPG	
H$_2$S, mole (ppm)	Nil
Copper strip corrosion (1 hour @ 38 °C)	No. 1 (max.)
Mercaptan sulfur (wt, ppm)	5 (max.)
nC$_5$/iC$_5$ (mole%)	1 (max.)
Free water	Nil
Vapor pressure @ 65 °C, KSC (g)	16.8 (min.)
Vaporization @ 2 °C and atm. pressure (%)	95 (min.)
Ethane and lighter (%wt)	2 (max.)
Unsaturated HC	1 %wt (max.)

Courtesy: CPCL.

Light naphtha [ASTM distillation (D 86) °C]

Start	23
5%	29
10%	36
30%	42
50%	49
70%	66
90%	79
95%	84
End point (max.)	91
Density @ 15 °C (kg/m³)	670–375
TBP cut point (°C)	95
PONA	66/0/32/2
Sulfur content (ppm)	< 5
Mercaptan sulfur (ppm)	10 (max.)
Nitrogen content (ppm)	< 1
RONC/MONC	75/76
Metals content	0
RVP, KSC (a)	0.8 (max.)

Heavy naphtha [ASTM distillation (D 86) °C]

Start	102
5%	106
10%	107
30%	111
50%	114
70%	119
90%	128
95%	133
End point (max.)	147
Density @ 15 °C (kg/m³)	763–768
TBP cut point (°C)	95–130
PONA	30/0/62/8
Sulfur content (ppm)	< 5
Mercaptan sulfur (ppm)	10 (max.)
Nitrogen content (ppm)	< 1
RONC/MONC	64/62
Metals content	0
RVP, KSC (a)	0.8 (max.)

Product test	Aviation turbine fuel	Superior kerosene
Density @ 15 °C (kg/m^3)	775–830	812–817
TBP cut point (°C)	130–260	130–260
Start	142	142
5%	154	154
10%	161	161
30%	176	176
50%	193	193
70%	212	212
90%	238	238
95%	249	249
End point (max.)	266	266
Flash point (°C, Abel)	38 (min.)	35 (min.)
Sulfur content (ppm)	10 (max.)	10 (max.)
Cu strip corrosion (2 hour @ 100 °C)	No. 1	No. 1
Freezing point (°C)	–51	
Silver strip corrosion	–	
Aromatic content (%vol)	22	
Naphthalene content (%vol)	3	
Mercaptan sulfur (ppm)	3	
Water tolerance (MI)	1	
Viscosity @ –20 °C (cSt)	8	
Smoke point (mm)	21	21 (min.)
Color saybolt		10 (min.)

High speed diesel

Density @ 15 °C (kg/m^3)	820–870
TBP cut point (°C)	–
Start	229
5%	257
10%	268
30%	287
50%	302
70%	320
90%	348
95%	360
End point (max.)	376

Contd.

High speed diesel *(Contd.)*

Cetane number (ASTM D 613)	55 (min.) EOR
Flash point (°C, Abel)	35 (min.)
Sulfur content (ppm)	< 10 (max.)
Pour point (°C)	3 (max.)
Kinematic viscosity (@ 40 °C)	2–5
Water content, (vppm)	500 (max.)

products [Sarrazin, 2005]. MHC can be applied in ULSD units to process heavier feeds and/or improve diesel product qualities [Catalysis PTQ, 2013, p6].

In MHC process fresh VGO feed is mixed with recycle hydrogen and sent to MHC reactor. After hydrocracking the reactor effluent is cooled, stripped and fractionated in distillation section. The hydrotreated VGO cut is sent to either an FCC unit or to storage while the diesel cut is mixed with fresh hydrogen and sent once through polishing reactor. The high hydrogen partial pressure in the polishing reactor ensures that the converted diesel which is refractory to hydrodesulfurization and hydrogenation, undergoes maximum possible hydrofining [Sarrazin, 2005].

8.4.4 Mobil-Akzo Nobel Kellog (MAK) Hydrocracking

MAK moderate pressure hydrocracking [MPHC] is a single pass hydrocracking process for the partial conversion of heavy gas oils to low sulfur distillates and unconverted oil which is highly upgraded relative to the raw feed. MAK-MPHC can process VGO feeds and can also be integrated with FCC and delayed coking or solvent deasphalting to convert heavy residue to fuels [Patel et al., Feb 1997; Terr and Tracy, 1986; Hunter et al., 1994].

8.4.5 UOP Unicracking Process

UOP's unicracking hydrocraking process is an important conversion technology for producing high value naphtha or distillate products from a wide range of refinery [virtual-e-hosting.com] feedstocks. UOP unicracking processes include single stage and two-stage design. Some of the newer developments of UOP hydrocracking process are once through design, hycycle unicracking design, advance partial conversion unicracking (APCU).

UOP single stage unicracking: In single stage design, fresh feed and recycle oil are converted in the same reaction stage which improves the overall design of the unit by reducing the quantity of equipment in high pressure service, keeping high pressure equipment in a single train [demirel.turkbiz.net].

8.5 HYDROCRACKING CHEMISTRY

Hydrocracking process is a catalytic cracking process which takes place in presence of an elevated partial pressure of hydrogen and is facilitated by bifunctional catalyst having acidic sites and metallic sites. Hydrocracking processes characterized by two types of reactions [Rashid, 2007].

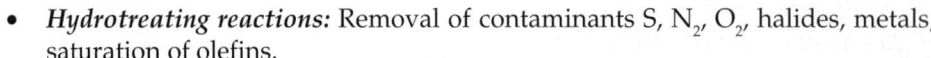
- *Hydrotreating reactions:* Removal of contaminants S, N_2, O_2, halides, metals, saturation of olefins.
- *Hydrocracking reactions:* For yielding valuable products various hydrotreating reactions are hydrodesulfurization, denitrogenation, hydrodeoxygenation, hydrometallization, olefin hydrogenation and partial aromatics saturation. Various hydrocracking reactions are splitting of C–C bond and/or C–C rearrangement reaction (hydroisomerization process). Hydrogenation and dehydrogenation catalysts.

A typical hydrocracking reaction is as follows:

$$C_{22}H_{46} + H_2 \longrightarrow C_{16}H_{34} + C_6H_{14}$$

Hydrocracking catalyst

Hydrocracking is essentially the acid catalyzed carbonium ion reactions of catalytic cracking coupled with hydrogenation reactions. Hydrocracking catalyst is bifunctional having acidic sites and metallic sites. The acidic sites are responsible for cracking reaction whereas hydrogenation/dehydrogenation reaction takes place on metallic sites. Rate of hydrocracking increases with the molecular weight of paraffin.

Strength of hydrogenation and cracking functions is bifunctional hydro-processing catalysts.

Hydrogenation function

$$\xrightarrow{\text{Co/Mo, Ni/Mo} < \text{Ni/W, Pt (Pd)}}{\text{Increasing hydrogenation activity (in low S environment)}}$$

Cracking function

$$\xrightarrow{Al_2O_3 < AlO_3\text{-halogen} < SiO_2\text{-}Al_2O_3, \text{Zeolite}}{\text{Increasing cracking activity (activity)}}$$

Source: Verma, 2002

Hydrocracking feedstocks typically contain sulfur, nitrogen and metals, etc., are removed in the hydrotreating reactor. Some of the first reactions to occur in the unicracking unit are the hydrotreating reactions. These hydrotreating reactions are catalyzed by the metal sites on the catalyst and, in general, are more rapid than the cracking reactions.

8.5.1 Hydrotreating Reactions

The primary hydrotreating reactions are sulfur and nitrogen removal as well as olefin saturation. The products of these reactions are the corresponding contaminant-free hydrocarbon, along with H_2S and NH_3.

- *Sulfur Removal:* The lighter compounds such as mercaptans and disulfides are easily converted to H_2S.
  - a. (Mercaptans)
    $$C\text{–}C\text{–}C\text{–}C\text{–}SH + H_2 \longrightarrow C\text{–}C\text{–}C\text{–}C + H_2S$$
  - b. (Sulfide)
    $$C\text{–}C\text{–}S\text{–}C\text{–}C + 2H_2 \longrightarrow 2C\text{–}C + H_2S$$
  - c. (Disulfide)
    $$C\text{–}C\text{–}S\text{–}S\text{–}C\text{–}C + 3H_2 \longrightarrow 2C\text{–}C + 2H_2S$$

- *Nitrogen removal:*

Pyridine   + 5H$_2$ → + NH$_3$ and + NH$_3$

- *Oxygen removal:* Organically combined oxygen is removed by hydrogenation of the carbon hydroxyl bond forming water and the corresponding hydrocarbon.

Phenol + H$_2$ ⟶ + H$_2$O

- *Olefin saturation:* Olefin saturation reactions proceed very rapidly and have a high heat of reaction.
  a. (Linear olefins)
  $$C–C = C–C–C–C + H_2 \longrightarrow C–C–C–C–C–C \text{ (isomers)}$$
  b. (Cyclic olefins)

  + H$_2$ ⟶

  o *Metals removal:* Removal of metals from the feed normally occurs in plug flow fashion with respect to the catalyst bed. Typical organic metals native to most crude oils are nickel and vanadium. Iron can be found concentrated at the top of catalyst beds as iron sulfides which are corrosion products. Sodium, calcium and magnesium are due to contact of the feed with salt water or additives.
  o *Aromatic saturation:* Aromatic saturation reactions are the most difficult. The reactions are influenced by process conditions and are often equilibrium limited. Unit design parameters would consider the desired degree of saturation for each specific unit. The saturation reaction is very exothermic.
  o *Halides removal:* Organic halides, such as chlorides and bromides, are decomposed in the reactor. The inorganic ammonium halide salts which are produced when the reactants are cooled are then dissolved by injecting water into the reactor effluent, as shown below.

The approximate relative reaction rates and heats of reaction per unit of hydrogen consumption of some of the hydrotreating reactions are given in Table 8.16 [Rashid, 2007].

8.5.2 Hydrocracking Reactions

Hydrocracking catalysts are dual functional, which means that they have both acidcracking sites and metal hydrogenation site. Major hydrocracking reactions are hydrocracking reaction, hydroisomerization reaction, partial or total hydrogenation of aromatic structures, hydrogenolysis of naphthenic rings [Raseev, 2003].

Table 8.16 Relative reaction rates and heat of reaction per unit of hydrogen consumption

Hydrotreating reactions	Reaction rate	Heat of reaction per unit of hydrogen consumption
Olefin saturation	100	1
Desulfurization	80	2
Denitrification	20	1

Reduction in MW by C–C bond cracking:

1. $R_1-CH_2-CH_2-R_2 + H_2 \qquad R_1-CH_3 + CH_3-R_2$
2. $C_6H_5-CH_2-R + H_2 \; C_6H_6 + R-CH_3$
3. Ring opening reaction

Hydroisomerization reaction (improves quality of some cuts):

1. n-paraffins $\longrightarrow$ isoparaffins
2. Rearrangement of naphthenes

Hydrocracking reaction pathways for paraffins, naphthenes and multiring aromatic compounds [Rashid, 2007].

8.6   EFFECT OF OPERATING PARAMETERS

Major controlling parameters in the hydrocracking process are [Rashid, 2007]:
- Reaction temperature
- Hydrogen partial pressure
- Hourly feed velocity of the feed
- Hydrogen recycle ratio.

8.6.1 Reaction Temperature

Increase in temperature accelerates cracking reaction on acid sites and displaces the equilibrium of hydrogenation reactions towards dehydrogenation. Too high temperature limits the hydrocracking of aromatic [Raseev, 2003].

8.6.2 Hydrogen Partial Pressure

The pressure influences significantly the equilibrium of dehydrogenation—hydrogenation reactions that take place on the metallic sites. The increase in pressure for a given molar ratio H_2/feed corresponds to increase in the partial pressure of hydrogen, will produce an increase in the conversion of the aromatic structures to saturated products which will improve the quality of jet fuel, diesel fuel and oil with very high viscosity index [Raseev, 2003].

8.6.3 Effect of Feedstock

A higher aromatic content requires higher pressure and higher hydrogen/feed ratio, the lowest possible temperature and a higher hydrogen consumption of hydrogen and the severity of the process [Raseev, 2003].

8.6.4 Recycle Hydrogen/Hydrocarbon (H_2/HC) Ratio

Maintaining physical contact of hydrogen with catalyst and hydrocarbons is important. Continuous circulation of the recycle gas through the reactor circuit ensures that hydrogen available at the catalytic sites where reaction is taking place [Adrian, 2007]. Recycle gas also helps in maintaining good flow distribution through the catalyst.

8.6.5 Make-up Hydrogen Purity

Hydrogen purity is very important. The makeup gas to the hydrocracker influences the hydrogen partial pressure and recycle H_2/HC ratio [Adrian Gruia, 2007].
Combined feed ratio (CFR):

CFR= (Fresh feed rate + Liquid recycle rate)/Fresh feed rate

Conversion per pass = (Fresh feed rate – Fraction bottoms rate to sorage) × 100/(Fresh feed rate + Liquid recycle rate)

As the CFR increases, the conversion per pass decreases. Therefore, the catalyst temperature requirement is reduced as CFR is increased at constant conversion [Adrian, 2007].

8.6.6 Effects of Feed Impurities

Hydrogen sulfide, nitrogen compounds and aromatic molecules present in the feed affect the hydrocracking reactions. Increase in nitrogen results in lower conversion. Ammonia inhibits the hydrocracking catalyst activity, requiring higher operating temperatures [www.shell.com]. Polymeric compounds have substantial inhibiting and poisoning effects. Polynuclear aromatics present in small amount in the residue, deactivate the catalyst.

8.6.7 Catalyst Life

Effect of various parameters on catalyst life is given below:

Variable	Change	Effect on catalyst life
Feed rate	Increase	Decrease
Conversion	Increase	Decrease
Hydrogen partial pressure	Increase	Increase
Reactor pressure	Increase	Increase
Recycle gas rate	Increase	Increase
Recycle gas purity	Increase	Increase

REFERENCES

1. Adrian Gruia. Distillate Hydrocracking. Handbook of Petroleum Processing, Springer, 2006, Chapter 7, p287.
2. Akelson D. UOP Unicracking Process for Hydrocracking. Handbook of Petroleum Refining Processes, 3rd edition, edited by Meyers RA, 2004.
3. Badhe RM, Basak K, Butley GV, Manna U and Verma RP. Mild Hydrocracking Process—A Flexible Option for Diesel Quality Improvement and Integration with FCC. Proceedings of Petrotech, 2001, New Delhi, 2001.

4. Billion A and Bigeard PH. Hydrocracking in IFP Petroleum Refining. Conversion Processes, Ed Leprince P. Editions Technip, 2001, p334.

5. Bridge AG and Mukherjee UK. Isocracking-Hydrocracking for Superior Fuels and Lubes. Handbook of Petroleum Refining Processes, 3rd edition, edited by Meyers RA, 2004.

6. Catalysis Petroleum Technology, Quarterly, Q, 2013, p6.

7. http:www.cheresources.com/refining5.shtml.

8. Huve LG, Creyghton EJ, Onwehand C, Van Veen JAR and Hanna A. New Catalyst Technologies Expand Hydrocracker's Flexibility and Contributions in Catalyst in Hydrocarbon Processing and Fertilizer Industry—A Compendium. Ed. Siddiqui MA, Verma RP, Lovraj Memorial Trust 8th National Workshop Seminar on Catalyst in Hydrocarbon Processing and Fertilizer Industry, Lovraj Kumar Memorial Trust, New Delhi, Nov 25–26, 2005, criteriancatalysts.com.

9. Kore et al., Ind Engg Chem Res., 1995.

10. Kunnas J. Improving Residue Hydrocarcking Performance. Petroleum Technology, Quarterly, Q3, 2011, p49.

11. Light SD, Berttram RV and Ward JW. Hydrocrack Heavier Feeds. Hydrocarbon Processing, May, 1981, p93.

12. Motaghi M, Ulrich B and Subramanian A. Slurry—Phase Hydrocracking Possible Solution to Refining Margins. Hydrocarbon Processing, Feb 2011, p37.

13. Narshiman CSL, Basak K, Sau M, Manna U, Santra M and Verma RP. Hydrocracking Catalysts in Catalyst in Hydrocarbon Processing and Fertilizer Industry—A compendium. Ed. Siddiqui MA,Verma RP, Lovraj Memorial Trust 8th National Workshop Seminar on Catalyst in Hydrocarbon Processing and Fertilizer Industry, Lovraj Kumar Memorial Trust, New Delhi, Nov 25–26, 2005.

14. Narshiman CSL and Basak K. In Hydroprocessing in Petroleum Refining Industry—A Compendium, Verma RP and Bhatnagar AK. ed. Indian Oil Corporation Ltd, 2005.

15. Patel V, Gentry AR, Hunter MG, Tracy WJ and Groeneveld LR. Upgrade FCC Feed and Product with MAK Hydrocracking. Hydrocarbon Technology, Feb 1997, p52.

16. Rama Rao M, Soni D, Sieli GM and Bhattacharya D. Convert Bottom of the Barrel into Diesel and Light Olefins. Hydrocarbon Processing, Feb 2011, p2011.

17. Raseev S. Hydrocracking in Thermal and Catalytic Processes in Petroleum Refining, Marcel Dekker Inc, 2003, p681.

18. Rashid K. Optimize Your Hydrocracking Operations. Hydrocarbon Processing, Feb 2007, p56.

19. Refining Processes, 98; Hydrocarbon Processing, Nov 1998.

20. Refining Processes, 2008; Hydrocarbon Processing, Sep 2008, p73.

21. Refining Processes, 98; Hydrocarbon Processing, Nov 1998, p53.

22. Sarranzin P, Bonnardot J, Wambergue S and Morel F. New Mild Hydrocracking Route Produces 10 ppm Sulfur Diesel. Hydrocarbon Processing, Feb 2005, p57.

23. Scherzer J and Gruia AJ. Hydrocracking Science and Technology, Marcel Dekker Inc, 1996.

24. Terai S and Fukuyama H. Petrolum Technology, Quarterly, 2000, p31.

25. Towler GP and Hoehn RK. Hydrocracking Industrial Processes. Encyclopedia of Catalysis, Vol 3, Willey Interscience Horvath, IT.

26. Verma RP and Narshimhan CSL. Modeling of Hydrocracking Reactors. Hydro-carbon Technology, May 15, 1994, 82.

27. Verma RP. Hydroprocessing: Indian Scenario. Indian Chemical Engineer Special Issue, Vol 1, 2002.

28. Virdi H, Sieli G and Torchia D. Strategies for Hydrocraking Refractory Feeds. Compendium of 16th Refinery Technology Meet, organized by Center for High Technology, Indian Oil Corporation, Kolkata, Feb 17–19, 2011.

9

Catalytic Reforming

Catalytic reforming is one of the important conversion processes in petroleum and petrochemical industries. Catalytic reforming is a major conversion process that is being used in refinery for conversion of low octane straight run naphtha and cracked naphtha feedstock to high octane reformate for using as a gasoline blending component to make high octane lead free petrol. In petrochemical industry, catalytic reforming is an important conversion process for production of aromatics—benzene, toluene and xylenes which are important feedstocks in synthetic fiber, explosive, dyes intermediates, pharmaceuticals, etc. Valuable by-products from catalytic reforming process are hydrogen gas (about 90% purity) and LPG. In the current world market there is strong demand for new reforming capacity which is driven by [Stine, Mararet; 2006]:

- Strong demand for gasoline
- Increasing demand for gasoline in India and China
- Strong demand for petrochemicals worldwide.

9.1 HIGH OCTANE GASOLINE

Requirement of the high octane gasoline is increasing with phasing out of TEL and recent restriction with MTBE. Catalytic reforming process plays an important role in modern refinery to meet the requirement of high octane gasoline. Octane number of important hydrocarbon is given in Table 9.1. The octane number of straight run gasoline is around 66 whereas requirement of octane number is around 86–93. Thus catalytic reforming process which produces reformate rich in aromatics having octane number above 90 provides high octane motor gasoline blending stock.

9.2 AROMATICS

Aromatics are important petrochemical feedstocks. There has been continuous increase in the demand of aromatics during last three decades. Benzene is a versatile petrochemical building blocks and is used for manufacture of important petrochemicals like ethyl benzene, linear alkyl benzene, cumene, cyclohexane, chlorobenzene and aniline and large number of other benzene derivatives. Increasing demand of polyester has further increased the demand of p-xylene by requiring more and more catalytic reforming units. O-xylene obtained during p-xylene manufacture as by-product served as feedstock for phthalic anhydride. Major portion of the toluene from reformate fraction is converted to benzene and xylene. Uses of m-xylene is also increasing for the manufacture of isophthalic acid which along with terephthalic acid is being used for manufacture of PET resins blend. Some of the major processes for producing aromatics from reformate are sulfonane extraction, parex, isomar and tatoray.

Table 9.1 Octane number of important hydrocarbons

Hydrocarbons	Research octane number (RON)	Motor octane number (MON)
Butane	95	92
Isopentane	92	89
Cyclohexane	83	77.2
Methylcyclohexane	74.8	71.1
1, 3-dimethylcyclohexane	71.7	71.0
Butane	94	
n-heptanes	0	
2-methyl pentane	74	
3-methyl pentane	75	
Benzene	114.8	> 100
Toluene	120	103.5
m-xylene	117.5	115.0
MTBE	115	99
ETBE	114	98

9.3 CATALYTIC REFORMING PROCESSES

Although catalytic reforming process was developed in 1940 by using activated alumina containing molybdenum oxide, the process called hydroforming; however, real breakthrough in catalytic reforming process was with the commercialization of UOP Platforming process in 1949 using platinum catalyst at Old Dutch refinery in Muskegon. Subsequently, other catalytic reforming processes were developed by other companies. First UOP platforming was semiregenerative or fixed bed employing monometallic catalyst. Since then UOP has made innovations and advances in process variables optimization, catalyst formulation, equipment design, and maximization of liquid and hydrogen yields, controlling coke deposition and catalyst deactivations through an advanced bimetallic catalyst formulation [Lapinski, 2004].

The various catalytic reforming processes which have been commercialized can be characterized in four major categories depending upon the frequency of catalyst regeneration—semiregenerative (SR), continuous catalyst regenerative (CCR), cyclic and hybrid.

There are three main types of catalytic reformers:

1. *Semiregenerative fixed bed reactors:* In this type of reformers the catalyst generally has a life of one or more years between regeneration. The time between two regenerations is called a cycle. The catalyst retains its usefulness over multiple regeneration.

2. *Cyclic fixed bed reformers:* Cyclic reformers run under more severe operating conditions for improved octane and yields. Individual reactors are taken offline by a special valving and manifold system and regenerated while the other reformer unit continues to operate.

3. *Continuous reformers:* In these reformers the catalyst is in moving bed and regenerated frequently. This allows operation at much lower pressure with a resulting higher product octane. Summary of performance and yield of CC reforming unit is given in Table 9.3.

Various types of reforming process technologies available from major process licensers are given in Table 9.2 [Little, 1985; Prasada Rao et al., 1992]. Yield advantage and operating condition of CC reformer over semiregenerative fixed bed reactor is mentioned in Tables 9.4 and 9.5 respectively.

Table 9.2 Catalytic reforming processes

Process name	Licenser	Reformer type
Catalytic reforming	Institut francais du petrole (IFP) IIP-IFP (India)	Semiregenerative, continuous catalyst regenerative, hybrid semiregenerative, continuous catalyst regenerative
Magnaforming	Engelhard Industries, Div. of Englehard ARCO: Atlantic Richfield Co.	Semiregenerative, semicyclic
Platforming	UOP process	Semiregenerative, continuous, catalyst regenerative hybrid
Cyclex TM	UOP process	Fixed bed reforming
Powerforming	Exxon Research and Engg. Co.	Cyclic, semiregenerative, semicyclic
Rheniforming	Chevron Research Co.	Semiregenerative
Ultraforming	Standard Oil Co. (Indiana) Anoco R & D Dept.	Cyclic, semiregenerative, semicyclic
Houdriforming	Houdry Process Corporation	
Octanizing	Axen	CCR

Comparison of conventional and octanizing process is given in Table 9.4.

Table 9.3 Typical performance and yield summary of continuous catalytic reforming unit

	Average	Normalized yield
RONC	98.41	
Reformate	95.10	94.2
H$_2$ rich gas	5.82	5.80
Hydrogen	2.48	2.40

Source: IOC, Panipat Refinery.

Table 9.4 Comparison of conventional and octanizing process

Yields: Typical for a 90 °C to 170 °C (176 °F to 338 °F) cut form light Arabian feedstock

	Conventional	Octanizing
Operating pressure (kg/cm^2)	10–15	< 5
Yields (%wt on feed)		
Hydrogen	2.8	3.8
C$_5$$^+$	83	88
RONC	100	102
MONC	89	90.5

Installation: Over 100 units licensed, 53 units are designed with continuous regeneration technology capability.
Licensor: IFP.
Source: Refining Processes, 1998 [Hydrocarbon Processing, Nov 1998, p61].

Table 9.5 Typical condition and yield pattern in semiregenerative and continuous catalytic reforming

Yields

Operating mode	Semiregenerative	Continuous
On stream availability (days/year)	330	360
Feedstock	Middle East	Middle East
IBP/EP (°F)	200/380	200/380
Operating conditions		
Reactor pressure (psig)	200	50
C$_5$$^+$ octane, RONC	100	100
Catalyst	R-72 staged loading	R-134
Yields information		
Hydrogen (scfd)	1,290	1,640
C$_5$$^+$ (lv%)	77.5	82.8

Licensor: UOP LLC.
Source: Hydrocarbon Processing, Nov 1998, p61.

9.4   CATALYTIC REFORMER IN INDIA

Due to increasing demand of high octane and aromatics, catalytic reforming has become integral part of Indian refineries and petrochemical complexes. Some of the refineries which were not having catalytic reforming earlier have added reformers. List of the catalytic reforming units in India is given in below:

Number of catalytic reformers = 15,   Capacity = 5.57 MMT
Semiregenerative reformers = 9,   Capacity = 1.7 MMT
Continuous catalytic reformers = 6,   Capacity = 3.8 MMT

9.5   PROCESS DESCRIPTION

A typical catalytic reforming process includes following six sections:
1. Naphtha splitting unit
2. Naphtha hydrotreating
3. Catalytic reforming
4. Catalyst circulation and regeneration
5. Nitrogen generation unit
6. Ammonia refrigeration unit.

9.5.1 Naphtha Splitting

The naphtha coming from is fractionated to have naphtha of desired cut for producing suitable fraction having lower benzene and higher octane number naphtha for further processing in catalytic reforming.

9.5.2 Naphtha Hydrotreatment

Naphtha hydrotreatment of the reformer feed is one of the important steps in catalytic reforming process for removal of the various catalyst poisons. The naphtha hydrotreatment process eliminates the impurities such as sulfur, nitrogen, halogens, oxygen, water, olefins, diolefins, arsenic and other metals present in the naphtha feedstock to have longer life catalyst. Figure 9.1 illustrates hydrodesulfurization process.

Sulfur in the naphtha is present in the form of mercaptans, disulphide and thiophenes and poison of the platinum catalyst. The sulfur content may be as high

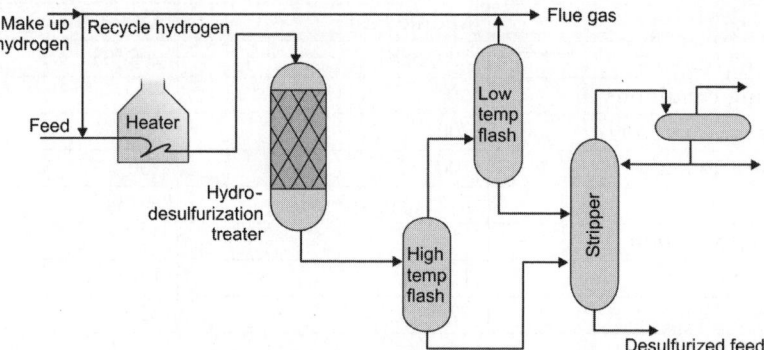

Fig. 9.1 Hydrodesulfurization

as 500 ppm. Maximum allowable sulfur content naphtha is 0.5 ppm or less and water content reduces to less than 4 ppm.

Hydrotreatment of naphtha is a catalytic process using a nickel-molybdenum catalyst using fixed bed reactor where both hydrodesulfurization reactions and hydrodenitrification reactions take place. The catalyst is continuously regenerated. Liquid product from the reactor is then stripped to remove water and light hydrocarbons.

Various Sections in Naphtha Hydrotreatment Unit

- *Charge heater:* This is used to heat the preheated reactor feedstock to reaction temperature of 340 °C. Charge heater has four passes and four gas burners. Heater tubes are made up of SS 321.
- *Reaction section:* The reactor consists of two catalyst beds where hydrorefining and hydrogenation reaction takes place.
- *Stripping section:* Stripping section uses air for stripping the light ends mainly hydrogen sulfide from reactor product, stripper pressure 14 kg/cm^2 and temperature 172 °C
- *Stripper reboiler:* Stripper reboiler supplies heat required for stripper.

Operating Variables

Various operating variables are reactor temperature, space velocity, hydrogen partial pressure, H$_2$/HC ratio, feed quality and stripper bottom temperature.

Reactions

- *Desulfurization:* Mercaptides, sulfides and disulfides reaction leading to the formation of corresponding saturated or aromatic compounds.

$$R–SH + H_2 \longrightarrow RH + H_2S$$
$$R–S–R' + 2H_2 \longrightarrow RH + R'H + H_2S$$
$$R–S–R' + 3H_2 \longrightarrow RH == R'H + H_2S$$

- *Denitrification:* Nitrogen compounds in the naphtha are eliminated by producing ammonia.

$$\text{Methyl pyrrol} + 4H_2 \longrightarrow C_5H_{12} + NH_3$$
$$\text{Quinoline} + 5H_2 \longrightarrow C_6H_6 + C_3H_8 + NH_3$$

- *Hydrodeoxygenation:* Oxygen compounds are eliminated by production of water.

$$C_6H_5OH + H_2 \longrightarrow C_6H_6 + H_2O$$

- *Hydrogenation:* Olefinic compounds provoke coke deposit and are eliminated by formation of saturated compounds.
- *Arsenic and metal compounds removal:* Arsenic, lead, mercury, sodium and others deteriorate reforming catalyst and are removed by adsorption on catalyst.

9.5.3 Various Catalytic Reforming Processes

9.5.3.1 Semiregenerative Catalytic Reforming Process

In typical semiregenerative catalytic reforming process, the naphtha hydrotreated feed is sent to series of three reactors having preheater furnace to maintain temperature of about 500 °C and pressure of about 300 kg/cm^2. As the main catalytic reaction naphthene dehydrogenation is endothermic there is drop in temperature requiring inter stage heater. The catalyst is regenerated *in situ* during catalyst regeneration. It reformed naphtha then passes to a separator where hydrogen and other gases are separated and the separated reformate goes to a stabilizer section for separation of LPG and reformate. In the modified semiregenerative reformer called cyclic reformer, an additional reactor called swing reactor is provided which is brought into operation when a reactor is bypassed for regeneration.

9.5.3.2   *UOP Continuous Catalytic Reforming Process (CCR Platforming Process)*

In the CCR reforming process the naphtha passes over a slow moving bimetallic catalyst bed in adiabatic reactors, in the presence of hydrogen at relatively high temperatures and low pressure. As high temperature (in the range of 500 °C) is required to promote the chemical reactions which improved octane number, the feed is heated up before entering the reactor. The catalyst is withdrawn from the reaction section at a fixed rate, regenerated in the continuous catalytic regeneration unit (CCR) and returned fresh to the reaction section.

The rate of catalyst withdrawal and regeneration ensures a consistently high active catalyst with a low carbon content and controlled chloride/water content. This maximizes yields of both reformate- and H_2-rich gas.

UOP basic CCR platforming has a single-stacked reactor. Stacked reactors offer many advantages which include lower capital, fewer pieces of equipment, lower catalyst attrition and easier maintenance. UOP has developed four types of CCR^{TM}: (1) catalyst regenerators; (2) atmospheric CCR regenerator; (3) pressurized CCR regenerator, cycle Max^{TM} CCR, and (4) the recent sequential CCR regenerator [Stine, Margaret; 2006]. UOP reactors use a proprietary up flow center pipe. In order to reduce both chloride consumption and treating costs in CCR platforming units, UOP has commercialized chlosorb system which recovers chlorides (HCl and Cl_2) from vent gases and recycles the chlorides back into the process resulting in reduced total chloride consumption [Stine, Margaret; 2006]. Flow diagram of CCR is shown in Fig. 9.2. Operating condition and yield pattern of CCR are given in Tables 9.9 and 9.10 respectively.

UOP Cylex ^{TM} is a fixed bed reforming unit with increased hydrogen production and is characterized with low capital and operating cost, increased hydrogen and reformate yields and increased fixed bed catalyst cycle length [uop.com Internet Source] [Peters T, UOP].

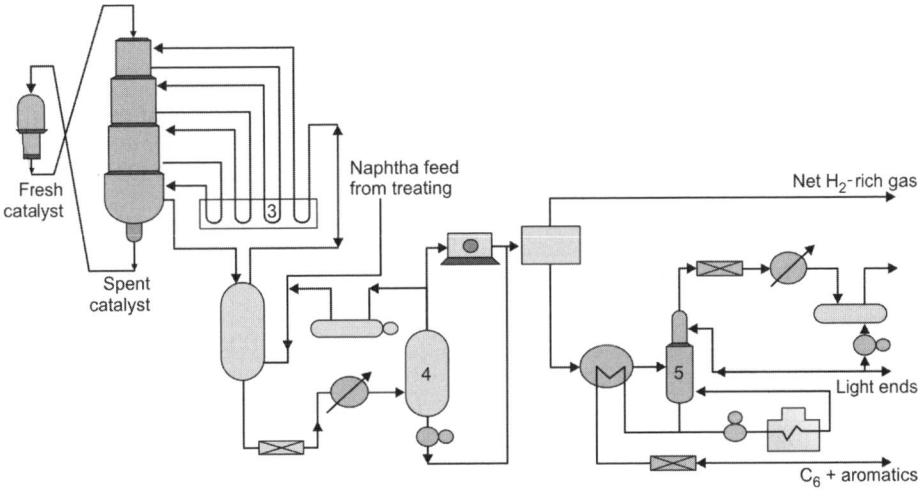

1. Stacked radial flow reactors 2. Catalyst regenerator section, 3. Interheaters,
4. Separators, 5. Stabilizer, 6. Gas recovery system

Fig. 9.2 Continuous catalytic reforming process

Source: Reproduced from Hydrocarbon Processing Copyright, Gulf Publishing House and Reproduced by Permission: Petrochemical Processes, 2003.

Heat generated when reaction in reformer takes place is mentioned in Table 9.6. Operating condition for different catalytic reformers is given in Table 9.7. Typical operating condition, yield pattern of a catalytic reformer is given in Tables 9.8 to 9.16.

Table 9.6 Reforming reactions, heat of reaction, relative rate of reaction

Reactions	Heat of reaction (1) Kcal/mole
Naphthenes dehydrogenation	–50
Paraffin dehydrocyclization	–60
Isomerization: Paraffins	+2
Naphthenes	+4
Cracking	+10

Table 9.7 Catalytic reforming operating conditions for different reformers

	Typical semi-regenerative process with R-62 catalyst	Full conversion to CCR platforming with R-134 catalyst	State-of-the-art CCR platforming with R-134 catalyst
Feed rate (m^3/hour)	99.40 (15,000)	139 (21,000)	139 (21,000)
Reactor pressure [kg/cm^2 (psig)]	21.1 (300)	7.0 (100)	3.5 (50)
C$_5$ + RONC	98	102	102
Catalyst cycle length (months)	12	Continuous operation	Continuous operation
C$_5$ + yield (LV%)	74.2	78.8	80.0
H$_2$ yield [(nm^3/m^3) (SCFB)]	141 (835)	262 (1,552)	289 (1,709)

Source: Sharma and Halgeri, 2005.

Table 9.8 Typical operating conditions of continuous catalytic reforming unit

Operating parameters	Naphtha hydrotreater	Returner
WAIT (°C)	302	503
DELTA temperature	+2	(−)256
WHSV (h^{-1})	–	
LHSV (h^{-1})	5.02	–
H$_2$/HC (mol/mol)	–	1.58
H$_2$/HC (m^3/m^3)	186.4	–
Separator		
Pressure (kg/cm^2)	15.9	2.23

Source: IOC, Panipat Refinery.

Table 9.9 Typical yield pattern of continuous catalytic reforming unit of IOC Panipat refinery

	Quantity in '000' tons	%wt
Feed: Hydrotreated naphtha	62.50	100.00
Product		
• Reformat	58.8	
• LPG	0.6	
• Component	2.8	
• Hydrogen	0.3	
• Rich gases		
Gas and loss	62.5	
Total		**100.00**

Source: IOC, Panipat Refinery.

Table 9.10 Typical characteristics of naphtha hydrotreater unit feed, reformer feed and reformate

	NHU* feed	Reformer feed	Reformate
Sp. gravity	0.7613	0.7615	0.8109
IBP	84	87	43
5%	95.3	96.6	80.4
10%	97.6	98.6	89.8
40%	107	106.7	110.2
50%	110.7	110.4	115.8
70%	119.7	119	128.2
90%	133	132.2	147.8
95%	138.7	138.4	162
FBP	148.7	149.6	177.4
Sulfur (ppm)	210	0.9175	
RONC			98.41
RVP			0.462
PONA:			
P		48.9	25.75
O		0.2	0.0
N		26.02	11.91
A		24.84	62.34

*NHU: Naphtha Hydrotreater Unit
Courtesy: IOCL, Panipat Refinery.

Table 9.11 Characteristics of naphtha feed to catalytic reforming before and after hydrotreatment

Metals distribution is as follows:

Arsenic (wppb)	10
Lead (wppb)	20
Copper (wppb)	20
Nickel (wppb)	20
Heavy metals (wppb)	50
Total	**120**

Courtesy: CPCL.

Table 9.12 Typical specification of hydrotreated naphtha, which is fed to CRU, is given below.

Sulfur	< 0.1 ppmw
Nitrogen	< 0.5 ppmw
Halogens (Cl, F)	< 1.0 ppmw
Arsenic	< 5.0 ppmw
Heavy metals	< 5.0 ppmw
Water and oxygenated	< 5.0 ppmw
Olefins and diolefins	0.0

Courtesy: CPCL.

Reactor Operating Conditions

In order to maintain the expected performance, the hydrotreater reactors are designed to operate under specific conditions, for catalyst SOR and EOR operations.

Table 9.13 Typical condition in hydrotreater reactor

First hydrotreater reactor	Case I	
	SOR	EOR
Reactor outlet pressure (kg/cm²g) Reactor temperature (°C)	27.5	27.5
Inlet	160	180
Outlet	165	185
Maxi-space velocity LHSV (h⁻¹)	8	8
Second hydrotreater reactor	**Case I**	
	SOR	EOR
Reactor outlet pressure (kg/cm²g) Reactor temperature	22.1	22.1
Inlet	270	300
Outlet	304	10
Hydrogen partial pressure [kg/cm²g (min.)]	10	333
Maxi-space velocity LHSV (h⁻¹)	20/4	20/4

Courtesy: CPCL.

Table 9.14 Conditions maintained in catalytic reforming unit

	Case I	
	SOR	EOR
Reactor inlet/outlet temperature		
1st reactor 206-R3	499/432	519/463
2nd reactor 206-R4	499/466	519/491
3rd reactor 206-R5	499/484	519/506
WHSV h^{-1}	1.20	1.20
Recycle ratio	7.0	7.0
Reactor average operating pressure (kg/cm^{2}g)	17	17

Table 9.15 Typical feedstock characteristics of CCR pretreated naphtha feed

Hydrotreater feed	Straight run naphtha			Hydrocracker naphtha	Visbreaker naphtha
	Arab mix.	Arab mix.	Bombay high		
Steam origin	83–125		70–125	95–130	C$_5$–165
TBP cut points (°C)	0.724	0.726	0.762	0.761	0.722
Sp. gr.	73.1	73.8	48.2	C$_7$/P:2 C$_8$/P:28	35
Paraffin (%vol)	–	0.5	–	–	40

Courtesy: CPCL.

Table 9.16 Characteristics of catalytic reforming pretreated naphtha

Hydrotreater feed	Straight run naphtha			Hydrocracker naphtha	Visbreaker naphtha
	Arab mix.	Arab mix.	Bombay high		
Unsaturated (%vol)	–	0.5	–	–	40
Naphthenes (%vol)	19.7	16.8	25.2	C$_7$/N:16.6 C$_8$/N: 46	20
Aromatics (%vol)	6.7	8.9	25.2	C$_7$/A:6 C$_8$/A:1.4	5
Sulfur (wppm)	500	500	500	1.0	12,500
Nitrogen (wppm)	3.0	3.0	3.0	< 1.0	50
Mercaptans (wppm)	200	200	200	Nil	4000
Chloride (wppm)	2.0	2.0	2.0	< 1.0	1.0

Contd.

Table 9.16 Characteristics of catalytic reforming pretreated naphtha *(Contd.)*

Hydrotreater feed	Straight run naphtha			Hydrocracker naphtha	Visbreaker naphtha
	Arab mix.	Arab mix.	Bombay high		
Bromine, No			–		80
MAV (mg/g)	–	–	–	–	20
Total metals (wppb)	120 max.	120 max.	120 max.	120 max.	120 max.
Others	Nil	Nil	Nil	Nil	Nil

Courtesy: CPCL.

9.5.4 Catalyst Regeneration

During catalytic reforming process the catalyst gets deactivated due to coking and sintering. The deactivation becomes faster under higher severity conditions, thus requiring frequent regenerations. Regeneration is done by burning the coke in inert atmosphere containing small amount of oxygen. Regeneration may be *in situ* or continuous. Bimetallic catalyst is easy to regenerate by burning coke under an air diluted atmosphere. Steps involved in regeneration are burning, oxychlorination, calcinations and reduction.

9.5.5 Nitrogen Generation and Ammonia Refrigeration

In order to provide inert gas nitrogeneration unit is required. Ammonia refrigeration unit is required for increasing purity of hydrogen in recycle gas by eliminating condensable hydrocarbons.

9.6 CATALYST IN CATALYTIC REFORMING

Catalytic reforming catalyst has undergone continuous development since starting of the first catalyst reforming process with catalyst followed by UOP monometallic catalyst based on platinum in order to meet the better reformate yield, control of coke formation and deactivation of catalyst to have longer catalyst life and cycle length. New reforming catalysts are bimetallic or multimetallic. The catalysts are now bifunctional in nature with metallic function and acid function. Dehydrogenation reaction and hydrogenation reactions are catalyzed by metal activity while the acidic function catalyzes the structural rearrangement of the molecules involving reorganization of the carbon bonds. Compared to bimetallic catalyst, the multimetallic catalyst has advantage of (a) higher C_5 reformate RONC at lower operating temperatures, (b) low benzene content in the reformate and (c) operation at lower temperature, pressure and H_2/HC ratios, etc. [Sharma and Halgeri, 2005]. Some of the broad composition of commercial reforming catalysts are given below:

Monometallic : Pt
Bimetallic : Pt and Re; Pt and Sn
Multimetallic : Pt, Re, Ge; Pt, Re, Ir; Pt, Ir, Cu, Se

Platinum content is approximately 0.3 %wt while chlorine content is 1 %wt. Most of the catalysts particularly those containing rhenium must be sulfide prior to reaction (sulfur, 0.1–0.5 %wt) in order to decrease the high hydrogenolytic activity [Little, 1985]. Addition of iridium, rhenium, tin or germanium has positive impact and results in better yield and coking resistance [Martino, 2001]. Catalyst supports in naphtha reforming catalyst is chlorinated Al_2O_3.

9.6.1 Effect of Acid Function

Increase or decrease in acid function affects the catalytic reforming process. Increase in catalyst may be due to overchlorinated catalyst or high water in the recycle. Increase in acid function results in slight decrease in octane, decrease in liquid product, increase in LPG production, decreased C_1 production, decreased recycle gas H_2 purity, decrease in MT in last reactor.

Decrease in acidic function may be due to elution of chloride due to high water content in the recycle gas and nitrogen in the feed. Decrease in acid function results in decreased octane, decreased LPG production, increased C_1 production, increased recycle gas H_2 purity, increased liquid product yield.

9.6.2 Effect of Metal Function

Decrease in metal function may be due to change in activity of the catalyst due to temporally reversible poisoning by sulfur or permanent poisoning by metals. Decrease in metal function results in large decrease in octane, decrease in MT in first reactor, decreased $C_1 + C_2$ production, increase in liquid production, large decrease in H_2 production and decreased recycle gas H_2 purity.

9.6.3 UOP CCR Platforming Catalyst [Stine, Margaret 2006, Poparad et al., 2012]

UOP has a long history and commitment to catalyst innovation. Innovation for alumina base includes maximizing the acid strength and distribution while maintaining or increasing the surface area stability, chloride retention and strength. For the metal function, the innovation include continued investigations of metallic interactions, metal levels, and advanced manufacturing methods. Some of the recent developments of catalyst are given below:

1990s: R-130™ with improved surface area stability allowing larger number of cycles and longer catalyst lives and inhibit high chloride retention

1996: High yield catalyst

1998–1999: High density R-160™

2000: R-234

2001–2002: R-230™ and R-270™ series of CCR platforming catalysts having increased catalyst stability and maximum product yields at significantly lower coke production than R-130 series catalyst.

2004: R-264 high activity, high selectivity

2002: R-274 high selectivity

2010: R-254 high selectivity, good activity

2010: R-284 highest selectivity

- *New R-264^{TM} catalyst:* As compared to R-130 series catalyst, R-264 catalyst has higher activity, produces less coke (8–10%) and allows higher pinning margin. R-264 catalyst also offers operating flexibility by its ability to be run in a high yield mode or high-activity mode. Commercial experience of R-264^{TM} catalyst shows increased of C_5^+ (1–2%) increased hydrogen production, decreased chloride injection by 70%, reduced coke and can operate in high activity or high yield mode.

- *R-500 fixed bed platforming catalyst* was introduced by UOP in 2010. It has a very high catalyst activity and lower start-of-run reactor temperatures [Poparad, 2012]

9.6.4 Criterion Catalytic Reforming Catalysts [Criterion Catalyst Bulletin]

Criterion catalyst is one of the major catalyst manufacturers has developed full range of reforming catalysts from monometallics used in the front end of cyclic reformer, to new fixed bed catalysts that combine high stability with a significant boost to the yields of hydrogen and reformat. These catalysts add value by increasing the reformer's flexibility to increase throughput, octane or upgrading lower quality naphtha.

9.6.5 Platinum Only Catalysts

- *Low platinum loaded catalysts:* Recommended when feedstocks contain less than 2 ppm. Usually used in the lead reactor(s) position of fixed bed semi-regenerative or fixed bed cyclic reformer.

- *High platinum loaded catalyst:* Recommended when feedstocks contain sulfur more than 2 ppm. Usually used in lead reactors(s) cyclic reformer units with relatively high severity.

- *Platinum/rhenium catalysts:* Recommended for general purpose reforming. Feedstocks sulfur should be less than 1 ppm with a target 0.5 ppm. This gives longer cycle length.

Modified platinum/rhenium catalysts or modified Pt/Re catalysts have been adjusted to increase hydrogen, C_5^- and aromatics yield compared to standard platinum/rhenium catalysts or platinum/tin catalysts. In low pressure operations these catalysts offer higher hydrogen and C_5^- yields than platinum/rhenium promoted or modified platinum/rhenium promoted catalysts. These catalysts are used in CCR units and also suited for using in CCR units and are also suited for using in fixed bed cyclic designs units. The platinum/tin system preserves the ring compounds in the feed to obtain higher yields of aromatics and hydrogen. A C_5^+ yield improvement of 2 %wt greater than base (platinum/rhenium) has been observed.

9.7   OPTIONS FOR IMPROVING CATALYTIC REFORMING

In order to meet the increasing demand of high octane reformate and high value aromatics, hydrogen products in large number of reforming units are being operated with changing feedstocks and product specification. Design and practice in catalytic reforming is evolving to meet challenges, including lower gasoline pool benzene and increased demand for hydrogen [Zhou and Baars, 2010]. The important options for improving reforming performance are use of higher performance catalyst with improved activity, contaminant tolerance resistance, high selectivity, improved stability, sound regeneration practices, reduced fines make with low catalyst attrition

[Poparad et al., 2012]. Pretreatment of naphtha feedstock to remove catalyst poisoning impurities is an important step in catalytic reforming.

9.8 STATUS OF THE REFORMING CATALYSTS/TECHNOLOGY IN INDIA

Stages in historical development of reforming catalyst in Indian scene are given in Table 9.17.

Table 9.17 Stages in historical development of reforming catalyst in Indian scene

- Development of low Pt monometallic catalyst IRC-1002 by IPCL for BT production.
- Commercialization of IRC-1001 catalyst in the first reactor of IPCL's three-reactor system for xylenes production, 1987.
- Scale up and manufacture of bimetallic catalyst IPR-2001 at IPCL's catalyst division.
- Commercialization of bimetallic catalyst at MRL for gasoline production, 1990.
- Commercialization of bimetallic catalyst IRC-1002 by IPCL for BT production.
- Commercialization of monometallic catalyst at IRC-1002 in BPCL reformer for BT production, 1990.
- Development of improved versions of reforming catalysts:
  High rhenium catalyst—recipe ready for scale up.
  Multimetallic catalyst—recipe ready for commercial trial.
  Spheroidal catalyst—CCR operations recipe in advanced stage.

Source: Sharma and Halgeri, 2005.

9.9 REACTIONS IN CATALYTIC REFORMING

Major reactions involved in catalytic reforming process are: Dehydrogenation of naphthenes, isomerization of naphthenes and paraffins, dehydrocyclization of paraffins and hydrocracking of paraffins [Liittle, 1985; Raseev, 2003; Mohan Lal, 2011]. Effect of catalyst function on various reactions is given in Table 9.18.

Table 9.18 Various reactions and effect of catalyst function

Reaction	Catalyst function	Promoted by	
Naphthene dehydrogenation	Metal	Temperature	Pressure
Naphthene isomerization	Acid	High	Low
Paraffin isomerization	Acid	Low	–
Paraffin dehydrocylization	Metal/acid	Low	–
Hydrocracking	Acid	High	Low
Aromatic dealkylation	Metal/acid	High	High
Demethylation	Metal	High	High

9.9.1 Naphthenes Dehydrogenation

Naphthenes dehydrogenation is the most important reaction in catalytic reforming and is highly endothermic and is favored by high temperature and low pressure.

Dehydrogenation of naphthenes increases the octane number of gasoline. In the dehydrogenation, naphthenic compounds, i.e. cyclohexane, methylcyclohexane, dimethyl cyclohexane up to C_{10} naphthenes are dehydrogenated respectively into benzene, toluene, xylenes, C_9 and C_{10} aromatics with the production of 3 moles of hydrogen per mole of naphthenes. Naphthenes are obviously the most desirable feed components because in addition to being easy to promote they produce by-products, hydrogen as well as the aromatic hydrocarbon.

The cyclohexane reaction, for instance, proceeds as follows:

Cyclohexane   Benzene

9.9.2 Paraffins Dehydrocyclization

Paraffins dehydrocyclization is a multiple step process involves in dehydrogenation of normal paraffins or isoparaffins with a release of one hydrogen mole followed by a molecular rearrangement to form a naphthene and the subsequent dehydrogenation of the naphthene. Dehydrocyclization is promoted by both catalytic metallic and acidic functions and is favored at low pressure and high temperature.

C_7H_{16}

C_7H_{14}

$3H_2 +$

Toluene   Methylcyclohexane

9.9.3 Isomerization of Paraffins

Paraffins isomerization: Isomerization of paraffin occurs readily in reforming reactions. This reaction leads to an increase in octane when rearranging to the corresponding branched isomer.

C_7H_{16} C_7H_{16}

These reactions are fast, slightly exothermic and do not affect the number of carbon atoms. The thermodynamic equilibrium of isoparaffins to paraffins depends mainly on the temperature. The pressure has no effect.

9.9.4 Naphthenes Isomerization

The isomerization of an alkylcyclopentane into an alkylcyclohexane involves a ring rearrangement and is desirable because of the subsequent dehydrogenation of the alkylcyclohexane into an aromatic. The reaction is slightly exothermic and higher carbon number is desirable.

CH₂CH₃

This type of reaction is also easier for higher carbon number.
Reaction is as follows:

C_7H_{16} C_7H_{16}

These reactions are fast, slightly exothermic and do not affect the number of carbon atoms. The thermodynamic equilibrium of isoparaffins to paraffins depends mainly on the temperature. The pressure has no effect.

9.9.5 Hydrocracking Reaction

Hydrocracking reactions of paraffins are the undesirable reaction in catalytic reforming and is favored by high temperature and pressure resulting in lower gasoline and hydrogen and increase in LPG production.

N-decane ⟶ 3-methyl pentane

9.9.6 Hydrodealkylation and Hydrogenolysis

Hydrodealkylation and hydrogenolysis are two other side reactions which result in lower gasoline yield. Hydrodealkylation results at high temperature and results in dealkylation of toluene to benzene.

9.10 FEED AND PROCESS VARIABLES [LIITTLE, 1985; RASEEV, 2003; MOHAN LAL, 2011]

9.10.1 Feedstock for Catalytic Reforming

Naphthenic feedstock is desirable for higher cetane and aromatic production. Preferred naphtha for naphtha reforming for gasoline production is in the range of 80–185 °C. Depending on the objective of catalytic reforming, naphtha of particular range is taken. A high end boiling point of the feed means greater amount of polyaromatics and then higher coking tendency.

9.10.2 Effect of Major Variables in Catalytic Reforming

Some of the major variables in catalytic reforming are feedstock, reaction temperature, space velocity, hydrogen to hydrocarbon ratio, catalyst type and catalyst poisons Fukuyama. Aromatic yield is increased by high temperature, low pressure and low hydrogen to hydrocarbon ratio.

9.10.3 Feed Quality

The feedstock properties greatly affect the reformate yield and aromatic content of reformate. Naphthenes dehydrogenate very fast and thus high naphthenic feed gives the highest reformate yield where as feedstock having paraffins require more severe conditions and give lower reformate yields. Cyclization is faster for C_8 paraffin than for C_7, and for C_7 than for C_6. Most suitable fraction for catalytic reforming process is the C_7–C_{10} fraction. General recommended feed ranges for production of benzene, toluene, xylene and octane blending stock are 60–90 °C, 90–110 °C, 110–140 °C and 90–140 °C fractions respectively.

N + 2A: N + 2A (where N is naphthenes %, A is aromatic %) are commonly used to judge the reformer yield and with higher N + 2A aromatic yield will be higher. Characterisation factor is related to N + 2A by

$$\text{Characterization factor} = 12.6 - \frac{N + 2A}{100}$$

Higher N + 2A results in higher reformat yield and octane number.

Although N + 2A is commonly used as measure of feed quality IFP now uses 0.85 N + A, which is found to be more representative. Higher is this index, the lower is the severity to meet the same product specification. Lower is this index, the higher the severity of operation to meet the same product specifications as the dehyrocyclisation of paraffins becomes important [Saxena, 2006].

9.10.4 Reactor Temperature

Reaction temperatures affect the catalytic reforming reactions as catalyst activity is directly related to temperature. High temperature increases the undesirable reaction resulting in lower octane and yield. An increase in temperature favours the kinetics of dehydrogenation, isomerization and dehydrocyclization, however, it also favors undesirable reactions and results in higher coke formation. A controlled temperature is required for high octane and longer catalyst life. Normally reforming reactor inlet temperature is maintained around 470–550 °C. Temperature below 420 °C or above 550 °C results in lower reformates yield.

9.10.5 Pressure

Pressure affects the catalytic reforming process as all the desirable reaction are favoured by lower pressure. Although decrease in pressure increases dehydrogenation of naphthenes and dehydrocyclization of paraffins which results in higher production of aromatic and hydrogen, however, it increases the rate of coking of catalyst resulting in deactivation of catalyst and short cycle length. Higher pressure causes higher rates of hydrocracking reducing reformate yield, but decreases coking of catalyst and results in longer life cycle and poor reformate yield. An optimum pressure is to be maintained keeping balance of aromatic yield and coke formation.

9.10.6 Hydrogen and Hydrocarbon Ratio (H_2/HC Ratio)

Molar ratio of hydrogen and hydrocarbon has direct impact on formation of coke. H_2 partial pressure is linked to the H_2/HC ratio and total pressure. Increase in the partial pressure of hydrogen results in increase in the hydrogenation of the coke precursors prior to their transformation into aromatic polycyclic hydrocarbons [Raseev, 2003]. Higher the H_2/HC ratio results in higher cycle length. Recycle hydrogen is necessary in the operation for purpose of catalyst stability. An increase in H_2/HC ratio will move the hydrocarbon at a faster rate in the reactor and supply a greater heat sink-increased stability.

$$\frac{H_2}{HC} = \frac{\text{Pure hydrogen (mole/hour) in recyle}}{\text{Naphtha flow rate (mole/hour)}}$$

Effect of temperature, pressure and H_2/HC ratio in the process is mentioned in Table 9.19.

Table 9.19 Effect of temperature, pressure and H_2/HC ratio

Increase of	Effect on dehydrogenation due to thermodynamics to kinetics		Effect on dehydrocyclization due to thermodynamics to kinetics	
Temperature	Increases	Increases	Decreases	Decreases
Pressure	Decreases	Unaffected	Increases	Increases
H_2/HC ratio	Slightly decreases	Slightly decreases	Slightly decreases	Slightly decreases

Increased	RONC	Reformate yield	H_2 yield	Coke deposit
Pressure	Decrease	Decrease	Decrease	Decrease
Temperature	Increase	Decrease	Increase	Increase
Space velocity	Decrease	Increase	Decrease	Decrease
0.85 N + A	Increase	Increase	Increase	Decrease

9.10.7 Space Velocity

Space velocity directly affects the kinetics of reforming reactions and has direct impact on the octane rating. Lower space velocity results increase in octane number. Highly naphthenic feedstock requires high space velocity whereas more paraffinic stocks require lower space velocities. A decrease in space velocity means an increased residence time, hence higher severity, resulting in increased octane, lower reformate yield and higher coke deposit.

Weight hourly space velocity (WHSV)

$$\text{WHSV} = \frac{\text{Weight of feed per hour}}{\text{Weight of catalyst}}$$

Liquid hourly space velocity (LHSV)

$$\text{LHSV} = \frac{\text{Hourly feed flow rate in } m^3/ \text{ hour at } 15\,°C}{\text{Volume of catalyst in } m^3}$$

9.10.8 Influence of Chloriding on Reforming Reactions

Careful balance has to be maintained between the hydrogenation/dehydrogenation function of Pt and the acid function of alumina in bifunctional catalyst [Menon and Paal, 1997]. Moisture in the naphtha and chlorine in the catalyst are two main parameters which control the acidity of the catalyst [Menon and Paal, 1997]. Water–

chlorine balance should be continuously monitored and water content in the feed should be about 3–6 ppm. Water formed by combustion strip chloride from the catalyst so chlorine addition is used to maintain chlorine level as well as keep platinum dispersed [Alkabbani, 1999].

9.10.9 Catalyst Poisoning

Two types of catalyst poisoning materials (temporary poisons and permanent poison) are present in the reformer feed which deteriorate the performance of reformer catalyst by reversible and irreversible deactivation [Querini, 2003].

1. *Temporary poisons:* Temporary poisons are those which can be removed from the catalyst without a shut down. The activity and selectivity of the catalyst is restored once the poison disappears. Some of the temporary poisons are sulfur, oxygenates organics, halogens, organic nitrogen and water.

2. *Permanent poisons:* Permanent poisons are those which have irreversible damage to the catalyst, are arsenic, lead, copper, mercury, iron, silicon, nickel, and chromium. Sources and permissible limit for various catalyst poisons are given in Table 9.20. Sources and permissible limits of catalyst poisons is given in Table 9.21.

Table 9.20 Maximum allowable impurities (wt) in the feed

Component	Max. allowable (wt)
Sulfur (as sulfur)	0.5 ppm max
Organic nitrogen (as nitrogen)	0.5 ppm max
Water or oxygenated products	5.0 ppm max
Chlorine as chloride	0.5 ppm max
Fluorine as fluoride	0.5 ppm max
Arsenic	1 ppb max
Lead	5 ppb max
Copper	5 ppb max
Mercury	5 ppb max
Iron	< detection limit
Silicon	< detection limit
Nickel	< detection limit
Chromium	< detection limit
Sodium	< detection limit
Cadmium	< detection limit
Potassium	< detection limit

Table 9.21 Sources and permissible limits of catalyst poisons

Poisons	Max. level, ppm (wt.)	Source
Arsenic	1	SR or cracked naphtha
Lead	5	Recycle
Copper	5	Corrosion
Mercury	5	Naphtha condensate
Iron	5	Corrosion
Silicon	5	Corrosion
Nickel	5	Food additives
Chromium	5	Corrosion

Source: Stine, Mararet; 2006.

- *Sulfur:* Sulfur compounds present in the feedstock are converted to hydrogen sulphide which poisons the catalyst. Sulfur content in the feed to reformer should be less than 0.5 ppm by weight. Sulfur poisoning is caused by the H_2S which is either contained in the feedstock or formed as decomposition product. H_2S reacts with platinum and rhenium and consequently reduces the activity of the catalyst as the metallic contact area reduces. Sulfur poison inhibits the metal function of the catalyst and results in decrease of H_2 yield, recycle purity.

  $$Pt + H_2S \longrightarrow Pt\,S + H_2$$

  Sulfur contamination inhibits the metal function of the catalyst. This is indicated by:
  - A decrease in hydrogen yield
  - A decrease in recycle gas purity
  - An increase in hydrocracking (LPG yield increase)
  - An increase coking rate.
- *Nitrogen:* It is present relatively in less quantity than sulfur, but it also affects the reformer performance appreciably and reduces the acid function and results in decrease in octane, a slight increase in H_2 production. Organic nitrogen is converted to ammonia under the reforming condition and may poison the catalyst and depress the activity of isomerization, hydrocracking and dehydrocyclization. Nitrogen content should be less than 0.5 ppm by weight.
- *Water:* Although little water is necessary to activate the acidic function of the catalyst, however, excess of water leads to decrease in catalyst activity and decreases the dehydrocyclization reactions. Maximum allowable concentration of water in reformer feed is 5 ppm by weight.

9.11 BENZENE REDUCTION IN NAPHTHA CATALYTIC REFORMING

The source of benzene in the gasoline pool vary for each refinery [Chitnis et al., 2010]. Of the various refinery streams which are blended into gasoline 70–85% of the benzene is contributed by reformate from catalytic reforming, 10–25% by FCC gasoline. Some of the other sources are alkylate, isomerate, light hydrocrakate, light coker naphtha [Palmer et al., 2008]. Naphtha reformate is one of the major gasoline blending components. New stringent gasoline specifications regarding benzene reduction in gasoline require careful operation to minimize the benzene precursors in the reformer feed to achieve benzene specification of 1.0 max.

In reformer benzene is formed by:
- Dehydrogenation of cyclohexane, then dehydrogenation to benzene
- Isomerization of methyl cyclopentane to cyclohexane followed by dehydrogenation
- Hydrocraking and dealkylation of heavier aromatics to benzene

Some of the basic approaches for achieving significant reduction in benzene content [Larraz, 2001; Palmer et al., 2008]:
- Removing benzene precursors from the reformate feed, or fractionate out a light reformate for subsequent conversion
- Saturating benzene contained in light straight and/or light hydrocrakate
- Operating the catalytic reforming at low severity in order to minimize benzene formation
- Removal of benzene from reformate via solvent extraction.

Although prefractionation reduces the benzene content in the pool, but hydro-dealkylation processes contribute to increase in the benzene content. Post-fractionation is the most efficient option for reducing benzene in the gasoline.

9.12 MODELLING, SIMULATION AND OPTIMIZATION OF NAPHTHA CATALYTIC REFORMING PROCESS

Modelling, simulation and optimization of naphtha catalytic reforming process has become very important in order to optimize the catalytic reforming process for designing new processes, monitoring process performance, troubleshooting, diagnosing faults, optimizing process and process control [Hu et al., 2003]. Naphtha reforming model has been developed by Larraz, 2001; who simulates how feed composition and operating conditions affect product compositions and yield. Modelling of reforming reactions can be accomplished using correlation models or kinetic models. Reaction kinetics has a paramount role in catalytic reforming model performance. Kinetic based model has been widely reported. Kinetic models can easily handle interdependent and feed forward/reverse reactions and are extent tools for predicting individual component yields as well as yield patterns such as C_5^+ yield in a reformer [Singh and Chopra]. A kinetics based mathematical model for commercial catalytic reforming process has been reported by Hu et al., 2003; for predicting reaction temperature and concentration profiles of each reactor, heater duties, catalyst deactivation, recycle gas composition and octane number resulting from different feedstocks or operating conditions. Jorgo and Eduardo, 2000; has proposed a lumped kinetic model for naphtha catalytic reforming process. Saxena et al., 1994; have reported a simulation (Simpack) and optimization package (Refopt) useful for design and optimal operation of semiregenerative type catalytic reformers.

9.13 AROMATIC COMPLEX

Aromatic production involves catalytic reforming of naphtha, separation of non-aromatics, separation of aromatics and transalkylation and isomerization process.

By best choice of feedstocks, appropriate identification of reformer feed cut points and proper selection of reformer severity, the benzene and para-xylene production can be controlled [Kim et al., 2001]. Figure 9.3 shows aromatic production process. A typical aromatic complex consists some of the following steps:

- Catalytic reforming of naphtha to produce reformate containing aromatics
- Separation of non-aromatics from aromatics using liquid extraction or extractive distillation
- Separation of benzene, toluene and xylene from reformate by fractional distillation
- Dispropornation and transalkylation of toluene to benzene and toluene
- Separation of o-xylene from mixed xylenes by fractional distillation
- Separation of p-xylene from m-xylene by adsorption or crystallization
- Isomerization of m-xylene to produce p-xylene.

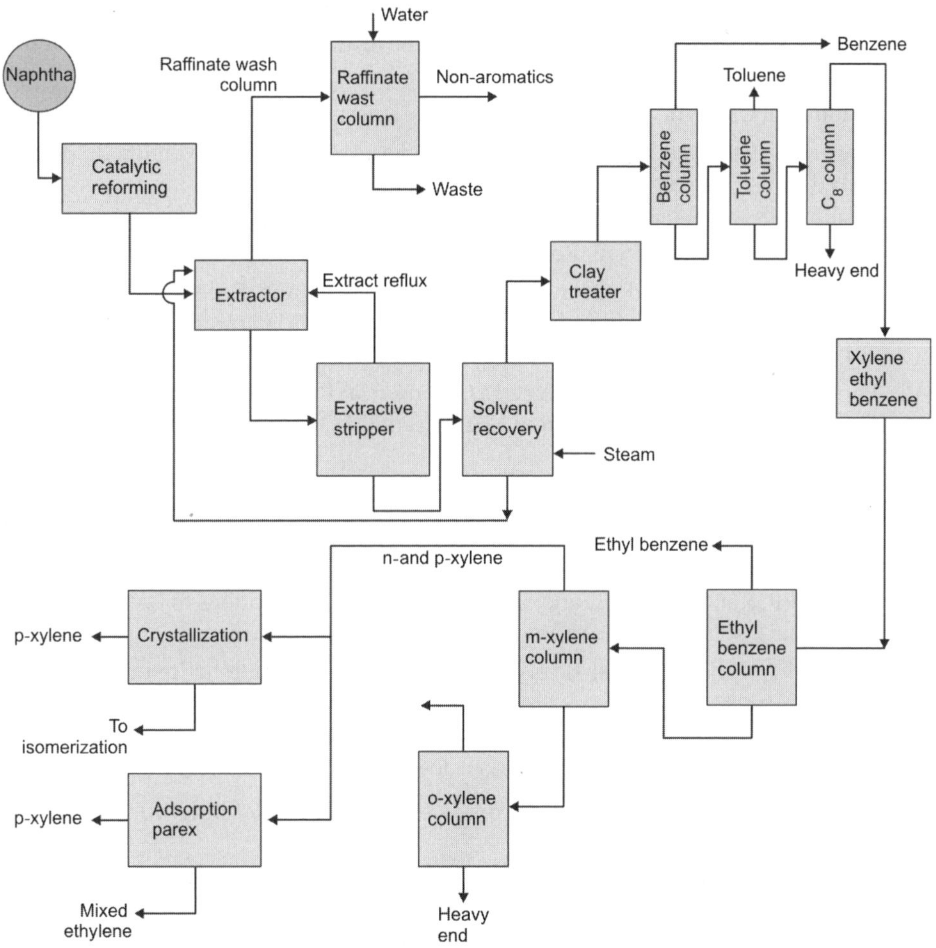

Fig. 9.3 Aromatic production

Source: Macmillan India, Mall, 2006.

REFERENCES

1. Chitins G, Meekki EI and Prasad A. Gasoline Benzene Reduction Through Exxon Mobil Research and Engineering Company's Reformate Alkylating Cataltic Technology: BenzOUT ™, Compendium of 16th Refinery Technology Meet, organized by Center for High Technology, Indian Oil Corporation, Kolkata, Feb 17–19, 2011.

2. http://en.wikipedia.org/wiki/catalytic_reforming.

3. Hu Y, Su H and Chu J. Modeling, Simulation and Optimization of Naphtha Catalytic Reforming Process. Proceedings of the 42nd IEEE Conference on Decision and Control, Maui, Hawali, USA, Dec 2003 FrE04-1.

4. Jorge AJ and Eduardo VM. Kinetic Modeling of Naphtha Catalytic Reforming Reactions. Energy Fuels, Vol 14, No. 5, 2000, p1032.

5. Kim HC, Yoon HS, Turpin L, Lakshman and Park NS. A Reformer Model for A Petrochemical Complex. Petroleum Technology, Quarterly, Jl, Spring, 2001, p61.

6. Lapinski ML, Baird L and James. In Handbook Petroleum Refining Ed. Meyers RA, The McGraw Hill Companies, R. 4.32004.

7. Larraz R. Benzene Reduction in Naphtha Catalytic Reforming. Petroleum Technology, Quarterly, Jl, Autumn, 2001, p61.

8. Liittle DM. Catalytic Reforming Penn Well, Tulsa, Okla, 1985, p2.

9. Martino G. Catalytic Reforming in IFP Petroleum Refining, Conversion Processes, 3rd editions, Technip, 2001, p101.

10. Menon PG and Paal. Some Aspects of the Mechanisms of Catalytic Reforming Reactions. Ind. Eng. Chem. Res., Vol. 36, 1997, p3282.

11. Mohan Lal. Catalytic Reforming Process, Catalysts and Reactors. 6th Summer School on Petroleum Refining and Petrochemicals, IIPM Gurgaon, June 6–10, June 2011, organized by IOCL and Petrotech.

12. Parera JM and Figo INS. Reactions in Commercial Reformer in Catalytic Naphtha Reforming, (Eds. Antos GJ, Aitani MA, Parera JM) Marcel Dekker Inc, New York, 1995, p45.

13. Petrochemical Processes, 2003, Hydrocarbon Processing.

14. Prasada TSR. Catalytic Reforming and Recent Indian Achievements. Oncology, Vol 6, No. 12, May 1992, p1.

15. Palmer ER, Kao SH, Tung C and Shipman DR. Consider Options to Lower Benzene Levels in Gasoline. Hydrocarbon Processing, June 2008, p55.

16. Poparad A, Ellis B, Glover B and Metro S. Improved Catalytic Reforming. PTQ, Q1, 2012, p111.

17. Querini CA. Reforming—Heterogeneous in Encyclopedia of Catalysis, edited by Horvath, IT Vol 6, Willey-Interscience, 2003, p1.

18. Raseev S. Thermal and Catalytic Processes in Petroleum Refining, Marcel Dekker Inc, New York, 2003, p749.

19. Saxena AK. Catalytic Reforming Advances in Petroleum Refining. Petrotech Summer School, IIPM Gurgaon, 5th July 2006.

20. Saxena AK, Das G, Goyal HB, Kapoor VK, Singh H and Garg MO. Simulation and Optimization Package for Semiregenerative Catalytic Reformer. Hydrocarbon Technology, Aug 15, 1994, p71.

21. Sharma N and Halgeri AB. Naphtha Reforming Catalysts: New Vistas Stine, Catalyst in Hydrocarbon Processing and Fertilizer Industry—A Compendium. Ed. Siddiqui MA, Verma RP, Lovraj Memorial Trust 8th National Workshop Seminar on Catalyst in Hydrocarbon Processing and Fertilizer Industry, Lovraj Kumar Memorial Trust, New Delhi, Nov 25–26, 2005.

22. Stine M. CCR Platforming ^{TM} Process—Advancements and Consideration of Super-sized Units. Hydrocarbon, Asia, May/Jun 2006, p12.

23. Zhou T and Baars F. Catalytic Reforming Options and Practices. Petroleum Technology, Quarterly, Q2, 2010, p21.

10

Alkylation, Isomerization and Polymerization

Alkylation, isomerization and polymerization have received considerable interest and can be used across wide spectrum applications in refining and petrochemical production for production of high octane gasoline, feedstock for alkylation, petrochemical feedstocks. Alkylation of isobutene with C_3–C_5 olefins, C_4 and C_5 olefins, isomerization of n-butane and light naphtha, polymerization of lower octane light olefins to high octane hydrocarbons, petrochemicals are some of the major applications given in Table 10.1. Phasing out of MTBE has led to the requirement of more alkylates for blended gasoline for which these processes are the most attractive and feasible option both economically and environmentally. Already large number of plants are working for production of alkylate and isomerate for gasoline pool.

Table 10.1 Alkylation, isomerization and polymerization

	Feedstock	Source	Process	Typical product	Application
Alkylation	Petroleum gas Olefins Isobutane	Distillation, cracking Catalytic or hydro-cracking Isomerization	Unification	High octane gasoline	Gasoline pool
Isomerization	n-butane n-pentane n-hexane	From various refinery processes	Rearrange-ment	Isobutene Isopentane Isohexane	Alkylation Gasoline pool Gasoline pool
Polymerization	Light olefins (ethylene, propylene, butylenes)	Various refinery processes	Unification	High octane naphtha Petrochem-ical feed-stocks LPG	Gasoline pool

10.1 ALKYLATION

Alkylation process which was commercialized in 1938, since then there has been tremendous growth in the process. Alkylation processes are becoming important due to growing demand for high octane gasoline and requirement of low RVP, low sulfur, and low toxics. Alkylate is an ideal blendstock to meet the requirement of low RVP, low sulfur and low toxics because it has no olefins or aromatic compounds [albemarle.com, internet source], allow sulfur content, a limited heavy end, a low vapor pressure and octane [D'Aquino and Mavridis, 2007]. Summary of conventional and improved catalyst for gasoline production is given in Table 10.2.

Table 10.2 Summary of conventional and improved catalysts for gasoline production

Process	Objective	Conventional/ old generation catalysts	Improved new generation catalysts	Benefits of improved catalysts
Alkylation	Production of branched alkanes for gasoline octane enhancement	H_2SO_4 HF	• Supported BF_3 • Modified SbF_3 • Supported liquid-acid catalysts	• Less corrosive • Safe handling • Fewer environmental problems • Continuous operation
Isomerization	Conversion of C_5/C_6 alkane streams into high-octane branched isomers	Pt/Al_2O_3 Pt/SiO_2-Al_2O_3 Modernito (zeolite)	• Solid super acid catalysts (e.g. suffocated zirconia)	• Low-temperature (thermodynamically favorable) operation • Increased conversion, less cracking

Source: Hydrocarbon Processing, 1997 [Abasi-Halabi, Stanislaus A, Qabazard H. Hydrocarbon Processing, Feb 1997, p45].

The process of alkylation of different isoparaffins using olefins were developed during thirties using aluminum chloride box catalyst, this catalyst was later replaced by HF and sulfuric acid. Although butylenes alkylation is one of the most commonly used processes, however, alkylation of amylenes obtained from C_5 fraction of FCC can be another route to increase the availability of alkylate. Alkylation of C_5 cut from FCC can significantly reduce RVP of finished gasoline pool [Jezak, 1994].

The process of HF alkylation produces high octane blendstock for isoparaffins (mainly isobutane) and olefins (propylene, butylenes and amylenes) in the process of HF catalyst to meet all the criteria of reformulated gasolines. In the US and Europe, alkylate is about 11–12% and 6% in gasoline pool respectively. Replacing high risk toxic liquid acids such as hydrofluoric acid (HF) and sulfuric acid with a solid acid catalyst is a challenging goal [Mukherjee et al., 2006].

The efficiency of alkylation unit can be evaluated by using two parameter RONAL and MONAL which are defined as:

$$RONAL = RON \times \frac{\text{Quantity of alkylate IP} - 190\ ^\circ\text{C (IP} - 374\ ^\circ\text{F)}}{\text{Total quantity of alkylate}}$$

$$MONAL = MON \times \frac{\text{Quantity of alkylate IP} - 190\ ^\circ\text{C (IP} - 374\ ^\circ\text{F)}}{\text{Total quantity of alkylate}}$$

10.1.1 Reactions in Alkylation

The reaction involved in alkylation of isobutane with olefin is quite complex due to large number of side reactions. Alkylation involves a series of consecutive and simultaneous reactions occurring through carbocation intermediates and resulting in formation of complex mixture of isoparaffins (alkylate) and polymers soluble in the acids called red oils [Mukherjee and Nehlsen, 2006]. Alkylation reactions are catalyzed by strong acids is shown in Fig. 10.1.

$$\underset{\text{isobutane}}{\text{i-}C_4H_{10}} + \underset{\text{1-butene}}{C_4H_8} \longrightarrow \underset{\text{iso-octane}}{\text{i-}C_8H_{18}}$$

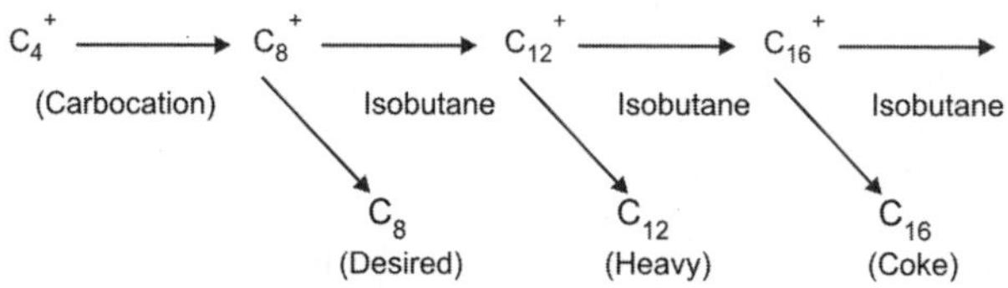

Fig. 10.1 Isoparaffin alkylation mechanism

Source: Hydrocarbon Processing (Mukherjee and Nehlsen, 2006).

The side reaction results in increased isobutane consumption, increased acid consumption, increased acid soluble red oil formation, coke formation, equipment handling and for corrosion problem [Funk and Fieldman, 1983].

Some of the other side reactions are the formation of paraffin which boils above and below the desired product. Impurities in the feed of acid and normal operating practices all can contribute to additional side reactions.

The key factors to be controlled in alkylation process are [Jezak, 1994]:

- Maintaining proper composition of reaction mixture which includes isobutene olefins and the HF acid.
- Maintaining the proper reaction environment which includes adequate contacting, controlled temperature, and freedom from surges.
- Making a proper separation of the reactor effluent into its various components.

10.1.2 Sulfuric Acid Alkylation

Sulfuric acid alkylation is the oldest alkylation process and consists of following steps:

- A reaction zone where an emulsion of acid and hydrocarbons is formed and the alkylation reaction takes place.
- A settling zone for separation of acid and hydrocarbons with acid recycled to the reactor and hydrocarbons routed to the separation zone.
- An unreacted isobutane/n-butane and alkylate separation zone consisting of stabilization column.

Process details

The process of sulfuric acid alkylation involves alkylation of feedstocks (propylene, butylenes, amylene, and fresh isobutene) in stirred autorefrigerated cascade type sulfuric acid alkylation reactor which is divided into zones. Concentrated sulfuric acid (85–95%) which is used as catalyst. Use of concentrated sulfuric acid minimizes corrosion problem which is more severe in dilute condition. Description of alkylation process is given in Table 10.3.

Table 10.3 Common alkylation processes

Process	Description
Exxon alkylation	The process involves alkylation of propylene, butylenes and perylene with isobutene in the presence of sulfuric acid catalyst using autorefrigeration method. Autorefrigeration alkylation results in low maintenance, low operating cost and high service factor with trouble free operation. Products: A low sensitivity, highly iso, low RVP, high octane gasoline blendstock paraffinic. Alkylate quality 96 RON/94 MON alkylate yield 1.78 bbl C_5^+/bbl butylenes feed. Isobutane required: 1.17 bbl/bbl butylenes feed.
Stratco process	The process involves reaction of propylene, butylenes and amylenes with isobutene in the presence of strong sulfuric acid using the effluent refrigerated alkylation process to make high branched chain hydrocarbons The stracto reactor allows minimum contact time to reduce the side reactions that may lead to formation of red oils. Product: Total debutanized alkylate RON: 92–96, MON: 90–94.
ConocoPhillips process (ReVAP process)	Alkylation of propylene, butylenes, pentenes and isobutane to high quality motor fuel using HF catalyst. Product: Motor or aviation fuel blending stock. Typical feed and alkylate specification is given in Table 10.6.
UOP HF alkylation process	Alkylation of isobutane with light olefins (propylene, butylenes and amylenes to produce branched chain parafinic fuel) using hydrofluoric acid catalyst. More than 100 commercial processes.
UOP alkylene™	UOP alkylene process is based on solid catalyst (HAL-100) for alkylation of light olefins and isobutane to form a complex mixture of isoalkanes which are highly branched trimethyl-pentanes (TMP) that have high octane blend values of approximately 100.

Source: Refining Processes, 1998; Refining Processes, 2008; Graves, 2004; Roeseler, 2004; Himes et al., 2003; Gravely, 2004.

The reaction is exothermic and reaction heat is removed by use of auto-refrigeration system. The reactor effluent is separated into hydrocarbon and acid phases in settler. The settled sulfuric acid is recycled to the reactor. The hydrocarbon phase from the alkylator are washed with caustic and send to deisobutanizers for separation of alkylate and isobutene which is recycled. Figure 10.2 illustrates conventional sulfuric alkylation process.

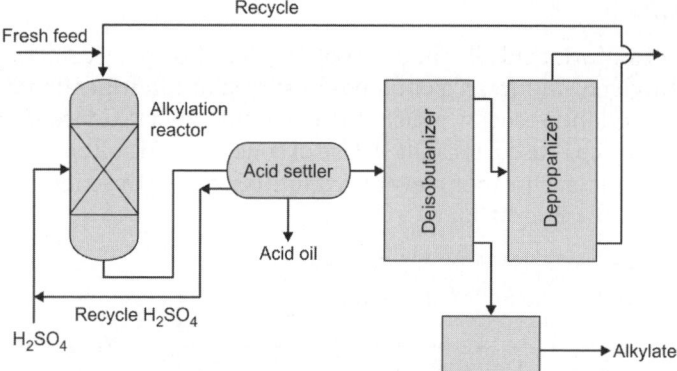

Fig. 10.2 Conventional sulfuric alkylation process

Some of the major operating variables are catalyst composition, reaction temperature, isobutene/olefin molar ratio and stirring power of H_2SO_4 units. Isobutane/olefin ratio is one of the most important parameters in alkylation which controls the catalyst consumption and yield of the product. Reaction temperature range is around 0–10 °C. Although temperature below 0 °C results in higher octane, however, acids viscosity becomes viscous and poses operational problem.

Acid consumption is dependent on olefin feedstock and impurities in the feedstock. However, significant improvement has been made in the sulfuric acid alkylation process to reduce acid consumption and improve alkylate quality.

10.1.3 HF Alkylation

HF alkylation involves alkylation of light olefins (propylene, butylenes and amylenes) with isobutene in presence of HF as catalyst resulting in formation of branched-chain paraffinic fuel having high octane blending stock.

HF alkylation process: The olefin and isobutane feed along with recycle isobutene is charged to alkylator. HF alkylation units do not have mechanical stirring system due to low viscosity of HF. The reaction product from the alkylator containing hydrocarbon phase and HF phase are separated in the acid settler separated to hydrocarbon phase containing alkylate and other hydrocarbon and acid phase. Hydrocarbon phase is fractionated and isobutene is recycled to the reactor and the bottom product containing alkylate is washed with KOH and send to gasoline pool. The light end from the isostripper is separated from HF in HF stripper. When the feedstock contains propane also then an additional dispropanizer is used to separate propane from isobutene which is then alumina treated to remove traces of HF. Typical HF process flow diagram is shown in Fig. 10.3. Typical yield of alkylate from a butylenes and propylene-butylene mix is given in Table 10.4. Hydrotreatment of an HF alkylation feed significantly improves the performance of alkylation by hydrogenation of polyunsaturated, conversion of butadiene into butenes, conversion of 1-butene to 2-butene using alumina supported catalyst using nickel and palladium [Chaput et.al., 1992].

Table 10.4 Typical stabilized alkylate yield from butylene and propylene-butylene mix

Feed type	Butylene	Propylene-butylene mix
Propylene	0.8	24.6
Propane	1.5	12.5
Butylene	47.0	30.3
i-butane	33.8	21.8
n-butane	14.7	9.5
i-pentane	2.2	1.3
Alkylate, stabilized		
Gravity (°API)	70.1	71.1
RVP (psi)	6–7	6–7
ASTM 10% (°F)	185	170
ASTM 90% (°F)	236	253
RONC	96.0	93.5
Per bbl olefin converted		
i-butane consumed (bbl)	1.139	1.175
Alkylate produced (bbl)	1.780	1.755
Installation: 107 units licensed worldwide		

Source: Hydrocarbon Processing, Nov 1998, p55.

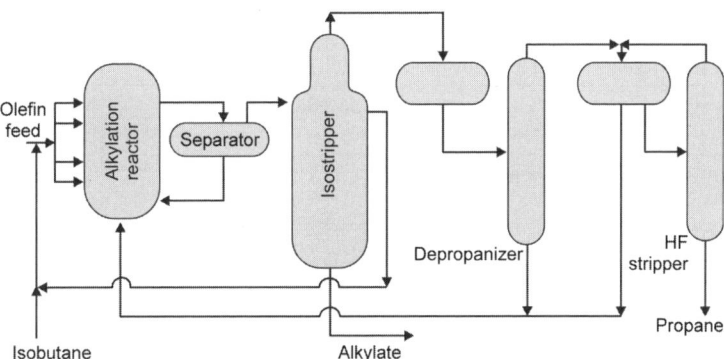

Fig. 10.3 UOP HF alkylation process (Feed: butene)

The alkylate produced from UOP alkylation process is a high quality motor fuel with excellent octane qualities with octane in the range 94+ to 97+ MON [HC, 1998]. Isobutene alkylate has higher octane values for HF alkylation compared to H_2SO_4 alkylation [Chaput et al., 1992] (Table 10.5 and 10.6).

10.1.4 Quality Factor

A quality factor (F) is a term frequently used to indicate relative quality of alkylate measured by research octane number, aviation blending octane number or superior blending index number [Funk and Smith, 1990]

F = IC4/OLE) (IC4)/(SV)

IC4/OLE = Isobutane-olefin volume ratio

IC4 = Volume fraction isobutene in reactor effluent

SV = Olefin space velocity, volume/reactor volume/hour

Alkylation of MTBE raffinate:

Table 10.5 Comparison of alkylene technology with modern sulfuric acid and hydrofluoric acid technologies

Parameter	Modern sulfuric acid technology	Modern hydrofluoric acid technology	Alkyclean
Base condition	C_4 = feedstock	C_4= feedstock	C_4= feedstock
Product RON	95	95	95
Product MON	Base	Base or better	Base or better
Alkylate yield	Base	Base	90% of base
Total installed cost	Base	855 of base	50% of base
Total installed cost, including OSBL (regeneration, facilities, and safety installations)	Base	Less	None
ASO yield	Base	Less	None
Equipment maintenance	High	High	Much lower
Corrosion problems	Yes	Yes	Higher
Reliability and on stream factor	Base	Base	Match FCC or better/shorter
Turn around frequency/ dilution	Varies/longer	Varies/longer	Match FCC or better/shorter
Safety	Unit specific safety precautions as well as transport precautions and unit specific precautions	C safety precautions required that extend throughout refinery very specific	No special precautions other than those for any refinery process unit
Catalyst	H_2SO_4	HF	Zeolite
Environmental	Significant waste streams generated	Significant waste streams generated	No emissions to air, water or ground

Source: Vn Rooijen, Edwin, Hydrocarbon Processing.

Table 10.6 RON and MON of alkylates from various olefins in HF and H_2SO_4 alkylations

	1-butene		2-butene		Isobutene		Propene		Pentene	
	HF	H_2SO_4	HF	H_2SO_4	HF	H_2SO_4	HF	H_2SO_4	HF	H_2SO_4
RON	94–95	98–99.6	97–98	98–99	94–95	90–91	91–93	89–92	91–92	92–93
MON	91–92	94–95	93–94	94–95	90–91	88–89	89–91	88–90	90	91

10.1.5 C₅ Alkylate

Amylene alkylation has two-fold advantages: It increases the volume of alkylate available while decreasing reid vapor pressure and olefinic content of gasoline blendstocks [Jezak, 1994]. Alkylating amylene fraction of FCC gasoline results in substantial reduction in RVP is given in Table 10.7. The alkylate consists of 8–9% isopentane—typically for full range FCC mixed amylene feed [Jezak, 1994].

Table 10.7 C₅ vapor pressures

C₅ vapor pressures	RVP, psi
1-pentene	19.1
2-pentene	15.3
2-methyl-1-butene	18.4
2-methyl-2-butene	14.3
3-methyl-1-butene	26.4
Cyclopentadiene	12.5
C₅ alkylate	3.5

Source: Hydrocarbon Processing [Jezak, 1994].

10.1.6 Alkylation Catalyst Development

Although hydrofluoric acid and sulfuric acid are proven liquid alkylation catalysts, but their corrosively requires special materials of construction, and their recovery generates copious amounts of waste water during their use [D'Aquno and Mavridis, 2007]. Replacing high risk liquid acids, such as hydrofluoric acid or sulfuric acid with benign solid acid catalysts is a challenging goal for isoparaffin alkylation technology due to rapid catalyst deactivation [Mukherjee and Nehlsen, 2006].

10.1.7 Solid Acid Catalyst (SAC) Alkylation

Both H_2SO_4 and HF alkylation are dangerous and lead to corrosion induced accidents. The hazards and costs associated with using and regenerating corrosive liquid acids are eliminated with SAC alkylation. SAC can produce high quality alkylate without drawbacks of existing H_2SO_4 and HF acid liquid technologies [Mukherjee and Nehlsen, 2006, 2007; D'amico et al., 2006]. Key to the economic viability of SAC is an adequate catalyst stability as SAC deactivated on order of minutes to hours [Mukherjee and Nehlsen, 2007]. Operating parameter of SAC process is given in Table 10.8 [Mukherjee and Nehlesen, 2007]. Catalyst regeneration requires 6–18 hours. The catalyst service life is the most critical design parameter. Service life is controlled by: Isoparaffin to olefin ratios and olefin hourly space velocity. Figure 10.4 shows the key variables that influence the design and operation of an alkylation process velocity [Mukherjee and Nehlsen, 2006]. Cost of alkylation process is mentioned in Table 10.9. Alkyclean technology and the associated solid acid catalyst were jointly developed by Albemarle's, Lummus Technology and Neste Oil. Alkyclean process significantly improves the safety of refinery alkylation over conventional liquid based processes. It reduces potential hazards associated with transportation of corrosive sulfuric acid and hydrofluoric acid [Catalysis, 2012]. Typical solid acid catalyst (SAC) process is shown in Fig. 10.5 [Refining Processes, 2008].

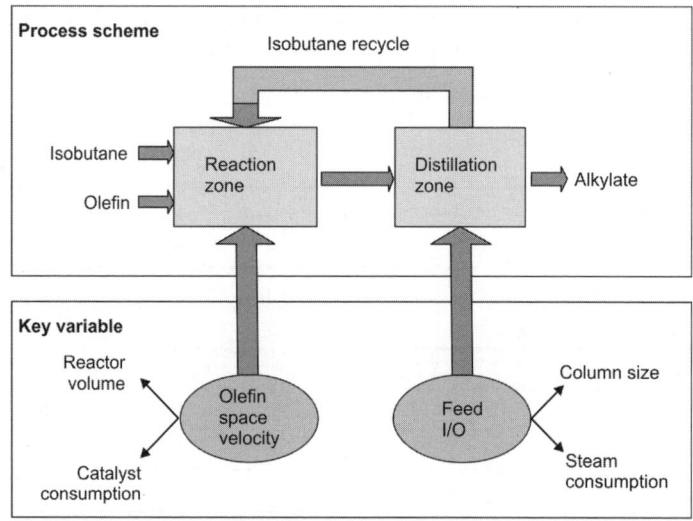

Fig. 10.4 Key variables that influence the design and operation of an alkylation

Courtesy: Hydrocarbon Processing, 2006; Mukherjee and Nehlsen, 2006.

Solid acid catalyst (SAC) process: Process flow diagram of SAC process is given in Fig. 10.5. The SAC process consist of:

Feed treatment: For removal of contaminants

Multistage fixed bed reactor: For alkylation of isobutane with olefin (isobutene)

Debutanizer: For separation of alkylate from unreacted feedstream

Deisobutanizer: For separation of isobutene and n-butane

Catalyst regeneration: For regeneration of catalyst

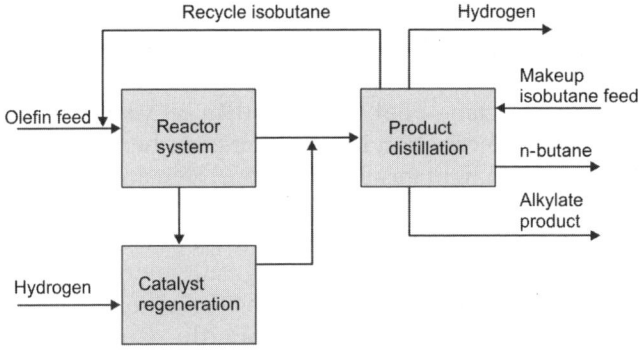

Fig. 10.5 Typical solid acid catalyst (SAC) process

Table 10.8 Typical operating parameters for the SAC process

Reaction temperature (°C)	50–100
Reaction pressure (bar)	20
Feed isobutane/olefin ratio (mol/mol)	10–15
Olefin space velocity	0.2–0.5

Source: Mukherjee and Nehlsen, 2007.
Courtesy: Hydrocarbon Processing.

Table 10.9 Economics of conventional and new alkylation process

Parameter	Sulfuric acid process	SAC process
Capital investment ($/bpsd)	36,000	2,600
Yield vol/vol olefin Alkylate yield Isobutane consumption	1.78 1.17	1.83 1.21
Utilities (per bbl of alkylate) Steam 60 psig (lb) Power (kWh) Water cooling Sulfuric acid (lb) NaOH, 100% (lb) Hydrogen (lb)	200 10.5 2.2 20 0.1	237 3.5 0.227 – – 0.15

Source: Mukherjee and Nehlsen, 2006.
Courtesy: Hydrocarbon Processing.

10.2 ISOMERIZATION

In recent years, the isomerization process has become the strategic gasoline process that ensures octane chracteristics of the overall gasoline pool and is playing a significant role to maintain gasoline pool octane and meeting the stringent gasoline pool requirement [Mirimanyan and Rudin, 2007]. Recent pricing trends show isomerization could be the significant contributor to octane pool which will offset the loss from gasoline desulfurization and aromatic reduction. Isomerization is also a widely used petrochemical process. Light paraffin isomerization has significantly contributed to world's octane pool for more than 40 years [Rosin, 2004]. Light naphtha and paraffin isomerization is recognized another emerging technologies in order to boost octane in light gasoline fractions. Demand for high performance naphtha isomerization has increased because its ability to reduce benzene concentration in the gasoline pool while maintaining or increasing the pool octane. Today total worldwide capacity of isomerization units amounts to about 50 million tons/year. During the last 20 years in Europe, iosmerization capacity has increased fourfold, and now the region constitutes one-third of worldwide capacity [Mirimayan and Rudin, 2007].

Petroleum fraction contains significant amounts of n-alkanes. The isomerization of alkanes into corresponding branched isomers is one of the important processes in petroleum refining [Matsuda et al., 2003]. The highly branched paraffins with 7–10 carbon atoms would be the best to fulfill the recent requirements of the reformulated gasoline [Rossini S, Catalysis today, 77, 2003, p467]. Therefore, the production of paraffin bases high octane gasoline blendstock such as isomers from isomerization of light- and mid-cut naphtha might be a key technology for gasoline supply to cope with future gasoline regulation [Visanandham and Garg, 2010]. The crackability of the paraffins increases with chain length and hence from n-heptane onwards the cracking of paraffins increases over isomerization in the homogeneous series of alkanes, leading to the production of undesired lower paraffins rather than high octane isomerate [Blomasma et al., 1997].

Isomerization converts the n-butane, n-pentane and n-hexane into their respective isoparaffins having higher octane number [www.orosha.org, Internet Source]. Isomerate is used in gasoline blends to increase its octane [uop.com, Internet Source] number and reduce aromatic, benzene and olefin content. Isomerate as percentage of gasoline used is USA 8%, Western Europe 16%. In India, isomerization unit based on UOP, Penex™ Process Technology has been commissioned Indian Oil Corporation, Mathura refinery in U.P.

Two distinct types of isomerization processes are in commercial use: Pentane/hexane (C_5/C_6) isomerization which produce high octane isomerate and butane C_4 isomerization which served as feedstock for alkylation. Some of the commercial isomerization processes and economics are given in Tables 10.10 and 10.11.

Two types of isomerization catalyst, zeolite and chlorinated alumina, have been used. Zeolite catalyst requires higher temperatures and provide lower octane boost while chlorinated alumina results in highest octane, however, it has higher sensitivity to feedstock impurities requiring strict feed pretreatment to eliminate oxygen, water, sulfur and nitrogen containing compounds [Domergue and Matthews, 2001]. Zeolite iosmerization catalyst is a platinum-carrying zeolite (mordenite) and does not require use of any halogens as an activator or promoter. Chlorinated catalysts have the highest iosmerization activity for C_5 and C_6 hydrocarbons, controlled by continuous supply of organic chlorine [Mirimanyan and Rudin, 2007]. The current families of isomerization catalyst offer relative advantages and disadvantages: Chlorinated alumina have the highest activity and yield, but they require an organic chloride feed and they are water-sensitive and requires expensive driers. Zeolite catalysts are regenerable and relatively contaminant tolerants, but they have the lowest activity. Sulfated zirconia catalysts are also regenerable and contaminant tolerants, with activities significantly higher than zeloite catalysts, but they are still less active than chlorided alumina catalysts [Rosin, 2004]. Two or three higher octane number is obtained. Zircon based catalysts are suitable in operating isomerization units designed for zeolite catalysts.

UOP uses dual functional catalysts in the Penex^{TM} and Butamer^{TM} processes which are platinum on chlorided alumina support. The process has been described in Table 10.12. These catalysts offer the highest activity to take advantage of higher thermodynamic equilibrium iso-to normal ratios achievable at lower temperatures [www.uop.com internet source]. Some of the recent developments in the Penex and Butamer catalysts are given in Table 10.13 [Andele and Metro, 2004]. Feed quality and makeup hydrogen purity are very important in isomerization process. Allowable contaminant levels in feed and hydrogen makeup are given in Tables 10.14 and 10.15 respectively.

Table 10.10 Commercial isomerization processes

Isomerization process	Description
UOP TIP and Once-through zeolite isomerization processes (UOPO-T zeolite isomerization process)	The process involves vapor phase isomerization of low octane normal pentane or normal hexane or both to high octane isoparaffins in presence of hydrogen. Reactor: Temperature and pressure.
UOP isomer process	The process involves catalytic isomerization of pentanes, hexanes, and mixtures in a fixed bed of catalyst operating at moderate temperature and pressure.
UOP Penex process	The process is specially designed for catalytic isomerization of pentanes, hexanes and mixture of two at operating conditions that promote isomerization and minimize hydrocracking.
ABB Lummus Global	The process upgrades the octane number of C_5/C_6 refinery streams by conversion of normal paraffins to their higher octane isomers using Akzo Nobel's high activity chlorided alumina catalyst. The technology also achieves saturation with subsequent octane upgrading of resulting cyclohexane.
Lyondell petrochemical's isomerization technology (ISOMPLUS process) CDTECH and Lyondell petrochemical	In this process, isobutylene is produced by the catalytic skeletal isomerization of normal butenes available in the hydrocarbon stream such as MTBE raffinate. The preheated feed is isomerized in fixed bed catalytic reactor to isobutylene. The reactor effluent is compressed and fractionated to separate C_5^+ stream from butane isomerate.

Contd.

Table 10.10 Commercial isomerization processes *(Contd.)*

Isomerization process	Description
IFP process	IFP process involves isomerization of C_5/C_6 paraffin-rich hydrocarbon streams to produce high RON and MON products. The process involves isomerization of the feed in affixed bed reactor containing zeolite or chlorinated alumina catalyst. Ipsorb isom process: Hexorb iso process:
ISOFIN process Kellog Brown and Root	The process produces iso-olefins from normal olefins via olefin skeletal isomerization. The iso-olefins (isobutylene and isoamylene) are typically used for production of oxygenates such as MTBE, ETBE, TAME and TAEE. Typical reactor condition: Temperature 300–400 °C, psig 5–50.
Hydrisom Process Phillips Petroleum Co.	The process involves liquid phase selective isomerization of butene-1 to butene-2 and 3-ethyl-butene-1 to 2-methyl-butene-1 and 2-methyl-butene-2 in fixed bed catalytic reactor in presence of hydrogen. Installation of Hydrisom unit of an etherification and /or alkylation unit can result in improved etherification, increased ether production, increased alkylate yield, reduced chemical and HF acids costs, reduced alkylation unit cost.

Source: Refinig Processes, 1998, 2008; Kunchal et al., 1993; Petroleum Technology, Quarterly, Winter 1997. HC Tech, 1991; Sullivan, 2004; Cusher, 2004a; Cusher, 2004b; Cusher, 2004c; Cusher, 2004d .

Table 10.11 Processing economics of various isomerization processes

Parameters	Once through process	Deisohexanizer	Ipsorb	Hexorb
Isomerate RON	83	87.5	89	91.5
Isomerate MON	80.5	85.5	86.5	89
Operating cost (million $/year) (utilities-catalyst-adsorbants)	0.8	3.3	3.4	4.7
Product revenue (million $/year) (Octane-barrel feed/isomerate)	11.1	16.2	17.9	20.5
Investment cost	5.9	14.2	15.2	20.8

Economics for four processing schemes 8000 bpd 40:60 C_5/C_6 feed chlorinated alumina catalyst.
Source: Domergue, 2001.

Table 10.12 UOP light paraffin isomerization technology

Process	Description
Penex^{TM}	UOP Penex process upgrades light naphtha components to high octane motor gasoline blendstocks and is being used by more than 120 refineries all over the world. Advantage of Penex process: Maximum octane bbls, high octane, best long-term profitability higher investment cost. It can handle undesired feedstocks including feed and process high benzene content feeds. It has wide range of operations. Penex once through, Penex plus^{TM} for extra high benzene levels DIH, DIP/ DIH, MDEX TMHigher octanes, higher product yields more than 120 licensed units.
Butamer process	The process used for isomerization of n-butane to isobutene which may be used as feedstock for aliphatic alkylation as source of isobutylene by dehydrogenation. The catalyst used is Pt/chlorinated Al_2O_3.
Par-isom^{TM}	Par-isom process with PI-242 catalyst: Best LPG production, good octane, rapid payback, low investment cost. UOP introduced Par-isom^{TM} in 1996 using sulfate of zirconium catalyst, lower equipment cost, multiple catalyst approach, zeolite and chlorided sulfated zirconium.

Source: Rajaram Panchapakesan. Latest development in UOP isomerization catalysts. National workshops on catalysts in hydrocarbon processing, Nov 25–26, 2005, organized by Memorial Trust, New Delhi 110001.

Table 10.13 Recent development in UOP Penex and Butamer catalysts current

Process	Penex		Butamer	
Catalyst	1–82 TM	1–8 Plus TM	1–120 TM	1–12 TM
Performance	Premium	Standard	Premium	Standard
Support	Trilobe	Cylinder	Trilobe	Cylinder
Pt, %wt	0.24	0.24	0.12	0.12

New

Process	Penex		Butamer	
Catalyst	1–82	1–84 TM	1–122 TM	1–124 TM
Performance	Premium	Premium	Premium	Premium
Support	Trilobe	Trilobe	Trilobe	Trilobe
Pt, %wt	0.24	0.14	0.12	0.08

Source: Andele and Metro, 2004.

Table 10.14 Allowable impurity levels in the feed

Total sulfur	0.5 wt ppm max.
Total nitrogen	0.1 wt ppm max.
Water + Oxygenates	0.1 wt ppm max.
Metals	5 wt ppb max.

Table 10.15 Maximum allowable contaminant levels in hydrogen makeup

Total sulfur	1.0 wt ppm max.
Total organic nitrogen	1.0 wt ppm max.
Water	0.5 wt ppm max.
$CO + CO_2$	10 wt ppm max.
Olefins	10 mol ppm max.

10.2.1 Advance Isomerization Technologies

For full conversion of all normal paraffins, recycling normal paraffins to extinction is required to convert them entirely to branched isomers [www.axens.net, Internet Source] [Domergue and Mathews, 2001]. Recycling of normal paraffins achieved by separating from isomerate using molecular sieve separation. The desorbed normal paraffins are again completely isomerized. Two patented processes Ipsorb and Hexorb are available [Domergue and Mathews, 2001]. Historical development of light paraffin isomerization catalyst is given in below Table 10.16.

Table 10.16 Historical development of light paraffin's isomerization catalyst

Isomerization catalyst	
1st generation	Fredel and Crafts $AlCl_3$ catalysts, exhibit very high activity at low temp 98–100 °C.
2nd generation	Metal/support bifunctional catalyst essentially Pt/alumina sensitivity to poisons are less acute, however, require higher temperature (550 °C).
3rd generation	Metal/support bifunctional catalysts with increased acidity by halogenations of the alumina support. Sensitive to poisons and need pretreatment, corrosion problem. High activity at low temperature (120–160 °C).
4th generation	Bifunctional zeolite catalysts, very resistant to catalyst poison and feed does not need pretreatment.

Source: Travers, 2001.

10.2.1.1 UOP Penex Process

UOP Penex process is one of the major refinery isomerization processes being used all over the world and more than 120 refineries are operating with this process. The process involves isomerization of low octane light straight run naphtha (60–70 RON). The process uses a highly active, low temperature hydroisomerization catalyst (Pt on chlorinated alumina). In this process light naphtha is dried in molecular sieve drier, the preheated hydrogen and the dry naphtha feed is charged to two reactors. The reactor effluent is stabilized and the bottom isomerate product is sent to gasoline pool. Light end gases are scrubbed with caustic and used as fuel (Fig. 10.6).

10.2.1.2 UOP Conventional Once through (O-T) Zeolite Isomerization of Light Naphtha (Pentane and Hexane)

The process involves fixed bed vapor phase isomerization of light naphtha containing n-pentane or n-hexane using zeolite catalyst in presence of hydrogen. This is carried out at 245–270 °C and pressure of 20–35 kg/cm^2 in presence of hydrogen. The reaction products are cooled subsequently in the product separator. The hydrogen separated is compressed and recycled and the isomerate is stabilized and separated from the light end gases.

10.2.1.3 UOP Par-Isom Process

The Par-Isom process is a light naphtha isomerization process using high performance sulfated metal oxide catalyst (LPI-100) with activity approaching to chlorided alumina [Cusher, 2004]. In this process, the preheated feedstock combined with recycled hydrogen is charged to the reactor and the isomerized product is sent to product separator where the isomerate is separated from the other light end product and sent to gasoline blending. Alternatively, the stabilizer bottoms can be fractionated to separate the n-hexane and low octane methyl pentanes from the high octane isomerate. Figure 10.7 illustrates the process.

10.2.1.4 Isomerization of Butane (Butamer Process)

Butane isomerization process involves isomerization of n-butane to i-butane which may be the feedstock for alkylation process. The process is a fixed bed vapor phase process promoted by the injection of the trace amounts of organic chloride. The reaction is carried out in presence of hydrogen. Major process variables are reaction temperature, liquid hourly velocity, hydrogen to hydrocarbon ratio and pressure. The process is shown in Fig. 10.8.

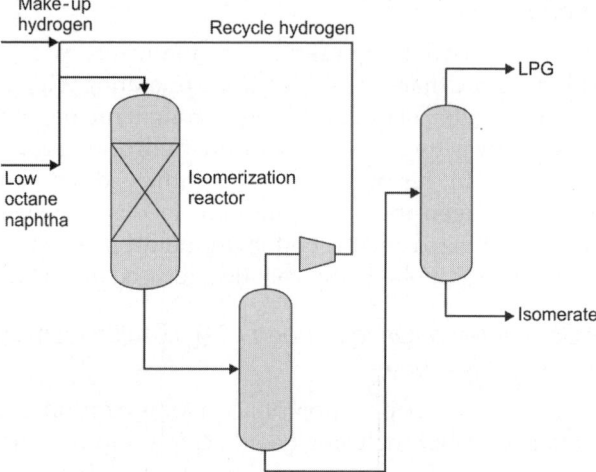

Fig. 10.6 Once through isomerization process

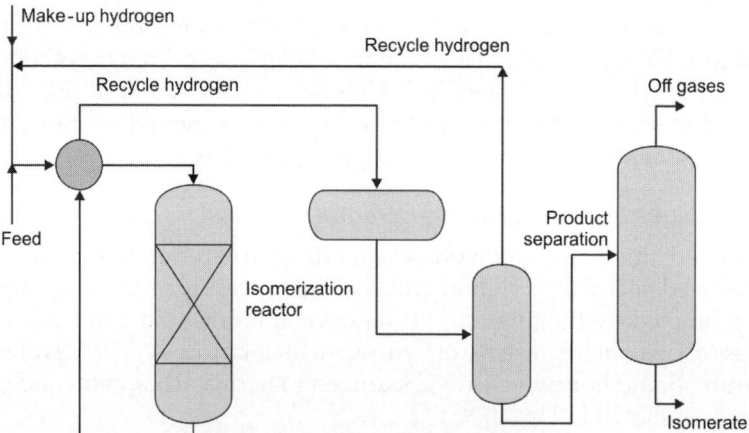

Fig. 10.7 UOP Par-Isom process

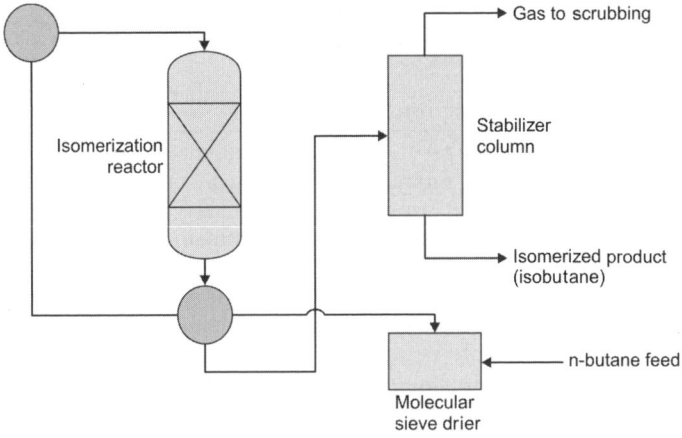

Fig. 10.8 UOP Butamer process

10.2.1.5 IFP Isomerization Process [Domergue and Mathews, 2001; Travers, 2001]

IFP isomerization process uses following three types of catalysts: Zeolite catalyst made of platinum deposited on mordenite, Pt on alumina catalyst which is chlorinated in the isomerization unit and Pt deposited on chlorinated alumina. In order to lower the n-paraffin content in the isomerate and increase the paraffin content in the feed IFP has developed two processes: Ipsorb and Hexorb which are based on molecular sieve separation. In the Hexorb process the molecular sieve is integrated with a downstream deisohexanizer that separates the raw isomerate from the molecular sieve into a isomerate overhead product rich in isopentane and dimethyl butanes and low octane methyl pentanes [Domergue and Mathews, 2001].

10.3 POLYMERIZATION

Polymerization processes have received considerable interest in petroleum refining because of the higher requirement of reformulated gasoline and phasing of MTBE. The process may be attractive in two main areas [Leprince, 1998].

- Upgrading of C_2 and C temperature, 150–200 °C; pressure, 30–50 bar; space velocity, 0.3–0.5 $m^3/hour/m^3$ cuts from catalytic cracking for oligomerization ethylene and propylene to olefinic gasoline.
- Producing high quality and middle quality

Some of the commercial polymerization processes are given in Table 10.17.

Table 10.17 Commercial polymerization processes

Process	Description
UOP catpoly process	The process is based on polymerization of propylene, butane or a mixture of both using phosphoric acid catalyst laid on a solid support of kieselguhr type. The process consists of polymerization of C_3/C_4 stream in multibed catalytic reactor. Propane is used as coolant to adjust the temperature from the reactor the polymerate is sent to separator for removal of entrained acid and stabilization section for separation of polymerate and light end products. Process condition: temperature, 180–200 °C; pressure, 30–40 bar; space velocity 0.3–0.5 m^3/hour/m^3.
IFP dimersol process	The process involves dimerization of olefin in a liquid phase reactor using soluble catalyst consisting of an organic nickel salt. The process units consist of demerization reactor, catalyst recovery, and stabilizer column. Process condition: Temperature, 40–50 °C; pressure, 10–30 bar; residence time 1–5 hours.
IFP polynaphtha process	The process converts C_3, C_4 and C_5 olefins into a gasoline cut in a fixed bed reactor using silicoalumina catalyst. The process stream consists of preheated, reactor and stabilization section for separation of polymerate from the light gases. Process shown in Fig. 10.9. Process condition: Temperature, 150–200 °C; pressure, 30–50 bar; space velocity, 0.3–0.5 m^3/hour/m^3. Product: Gasoline/gas oil.
Mobil olefin to gasoline and distillate (MOGD)	The process is based on oligomerization of light olefins to iso-olefins using ZSM-5 catalyst in four fixed bed adiabatic reactors. Three remains in reaction phase while fourth under regeneration. Process condition: Temperature, 150–200 °C; pressure, 7–12 bar; space velocity, 1–2 m^3/hour/m^3 of catalyst.
NExOCTANE^{TM}	The process involves dimerization of isobutylene to isooctane which can be subsequently hydrogenated to produce iso-octane.
Octol process	Normal butene containing isobutene depleted C_4 hydrocarbons are oligomerized using the octol process in liquid phase on a heterogeneous catalyst system to yield mainly C_8 and C_{12} olefins. The product will have high octane as well as petrochemical feedstock for plasticizer or detergent.

Source: Leprince, 1998; Irion and Neuwirth, 2003; Birkhoff and Nurminen, 2004; Nierlich, 1991.

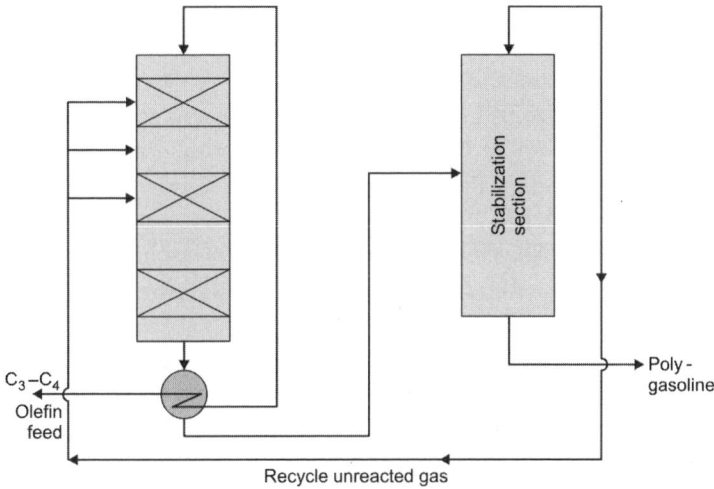

Fig. 10.9 Polymerization process for polygasoline

REFERENCES

1. Abasi-Halabi, Stanislaus A and Qabazard H. Hydrocarbon Processing, Feb 1997, p45.

2. Andele C and Metro S. UOP Commercializes New Light Naphtha Isomerization Catalysts. Technology and More UOP, Spring, 2004.

3. Birkhoff R and Nurminen M. NExOCTANE™ Technology for Isooctane Production, Chapter 1.1, p1.3 in McGraw Hill Handbook of Petroleum Refining Process, 2nd edition, edited by Meyers RA, McGraw Hill Companies, 2004, p1.33.

4. Blomasma E, Martens JA and Jacobs PA. Studies in Surface Science and Catalysis, 105, 1997, p909.

5. Catlysis, 2012. Catalysis Petroleum Technology, Quarterly, 2012, p5.

6. Chaupt G, Laurent J, Boltiaux, Cosyns and Sarrazin. Pretreat Alkylation Feed. Hydrocarbon Processing, Sep 1992, p51.

7. Cusher NA. UOP Par-Isom Process in McGraw Hill Handbook of Petroleum Refining Process, 2nd edition, edited by Meyers RA, McGraw Hill Companies, 2004d, p9.41.

8. Cusher NA. UOP Penex Process in McGraw Hill Handbook of Petroleum Refining Process, 2nd edition, edited by Meyers RA, McGraw Hill Companies, 2004b, p9.15.

9. Cusher NA. UOP TIP and Once-through Zeolite Isomerization Process in Handbook of Petroleum Refining Process, 2nd edition, edited by Meyers RA, McGraw Hill Companies, 2004c, p9.29.

10. Cusher NA. UOP Butamer Process in Handbook of Petroleum Refining Process, 2nd edition, edited by Meyers RA, McGraw Hill Companies, 2004a, p9.7.

11. D'Amico V, Gieseman J, Van Broekhoven E, Van Rooijen E and Nousiainen H. Consider New Alkylate Methods to Debottleneck Clean Alkylate Production. Hydrocarbon Processing, Feb 2006, p65.

12. D'Aquino R and Mavridis L. Solid Acid Catalysts Shape for Alkylation. CEP, Jan 2007, p8.

13. Detrick KA, Himes JF, Meister JM and Nowak FM. UOP HF Alkylation Technology Handbook of Petroleum Refining Process, 2nd edition, edited by Meyers RA, McGraw Hill Companies, 2004, p1.33.

14. Domergue B and Matthews R. Advanced Recycle Paraffin Isomerization Technology. Petroleum Technology, Quarterly, Spring, 2001, p29.

15. Dunham D. Upgrade Alkylation for Refining Environment Hydrocarbon Processing, Sep 2005, p93.

16. Edwin van Rooijen. Alkyclean—A True Solid Acid Gasoline Alkylation Process.

17. Funk GL and Feldman JA. Better Alky Control Pays. Hydrocarbon Processing, Sep 1983, p92.

18. Funk GL and Smith DF. Emerging Economics Incentive for a Computer Control System: An Application Approach. IEE Transaction on Industry Applications, Vol 57, No. 2, 4 (Jul/Aug) 1979, p394.

19. Gravely ML. ConocoPhillips Reduced Volatility Alkylation Process (ReVAP) in McGraw Hill Handbook of Petroleum Refining Process, 2nd edition, edited by Meyers RA, McGraw Hill Companies, 2004, p1.79.

20. Graves DC. Stratco Effluent Refrigerated H_2SO_4 Alkylation Process in McGraw Hill Handbook of Petroleum Refining Process, 2nd edition, edited by Meyers RA, McGraw Hill Companies, 2004, p1.3.

21. Himes JF, Meister JM, Nowak and Franz-Marcus. UOP HF Alkylation Technology in McGraw Hill Handbook of Petroleum Refining Process, 2nd edition, edited by Meyers RA, McGraw Hill Companies, 2004, p1.33.

22. Hydrocarbon Technology International, 1991, p73.

23. Irion WW and Neuwirth OS. In Oil Refining in Ullman's Encyclopedia of Industrial Chemistry, Vol 24, p205.

24. Jezak A. C_5 Alkylate: A Superior Blending Component. Hydrocarbon Processing, Feb 1994, p47.

25. Kunchal SK, Dwyer FG and PAmbler CP. ISOFIN—A New Process for Olefin Isomerization. Paper AM-93–45, NPRA Annual Meeting, San Antonio, Texas, March 21–23, 1993.

26. Matsuda T, Sakagami H and Takahashi. Catalysis Today, 81, 2003, p31–42.

27. Leprince P. Oligomerization, Chapter 9, p321 in IFP Petroleum Refining, Part 3. Conversion Processes, edited by Leprince P. Editions Technip, 1998.

28. Mirimanyan AA, Vikhman AG and Rudin MG. FSU Refiners to Build More Isom Capacity. Oil and Gas Journal, Feb 12, 2007, p56.

29. Mukherjee M and Nehlsen J. Consider Catalyst Developments for Alkylation Production. Hydrocarbon Processing, Sep 2006, p85.

30. Mukherjee M and Nehlsen J. Reduce Alkylate Costs with Solid Acid Catalysts. Hydrocarbon Processing, Oct 2007, p110.

31. Nierlich F. Oligomerize for Better Gasoline. Hydrocarbon Processing, Feb 1992, p45.

32. PTQ, 97. Expand Revamp Opportunities for Isomerization Units with Par-Isom Process. Petroleum Technology Quarterly, Winter, 1997.

33. Rajaram Panchapakesan. Latest Development in UOP Isomerization Catalysts. National Workshops on Catalysts in Hydrocarbon Processing, Nov 25–26, 2005, organized by Memorial Trust, New Delhi 110001.

34. Refining Processes, 1998; Hydrocarbon Processing's Refining Processes, p93, 98.

35. Refining Processes, 2008; Hydrocarbon Processing's Refining Processes, p93, 98.

36. Roeseler. UOP Alkylene™ Process for Motor Fuel Alkylation in McGraw Hill Handbook of Petroleum Refining Process, 2nd edition, edited by Meyers RA, McGraw Hill Companies, 2004.

37. Rosin R. New Solutions to Light Paraffin Isomerization Technology and More UOP, Spring, 2004.

38. Travers C. Isomerization of Light Paraffins in IFP Petroleum Refining, Part 3. Conversion Processes, edited by Leprince P. Editions Technip, 1998.

39. Travers C. Isomerization of Light Paraffins, p231, Chapter 6, IFP Petroleum Refining Part 3. Conversion Processes, edited by Leprince P. Editions Technip, 1998.

40. Visanandham N and Garg MO. A Novel Catalyst for Improved C^+ Isomerization of Naphtha. Chemical Industry Digest, May 2010, 65.

11
Lubricating Oil and Waxes

Lubricating oils and wax are the important products of refinery which makes use of the residual low value distillate to value aided product. Lubricating oils may be based on animal and vegetable oil or mineral oil. Vegetable oil have excellent lubrication property. Some of the limitations of vegetable oil are very poor oxidation stability, high pour point and rapid thickening. Vegetable oils like castor oil are commonly used. At present most of the liquid lubricants used all over the world are mineral oils, which are petroleum based oils [iocl.com]. Mineral based lubricating oils are having more importance because of availability from heavy residue from crude oil distillation at much cheaper rate than vegetable based lubricating oil. Normally heavier crude yield more lubricating oil fraction. Characteristics of lubricating oil feedstock is given in Table 11.1.

Waxes are very important product from lubricating oil manufacturing unit and find large scale application in packaging, adhesives, candles, cosmetics, inks, chewing gum, plastics, rubber, polish. Wax coated paper find application in packaging industry an insulation.

Global demand for lubricants is forecast to increase 2.2% per year to 39.3 million tons in 2008 value at 436 billion. Among specific products, the best opportunities are of high quality lubricants formulated using synthetic or highly refined petroleum base stocks [Hydrocarbon Processing, April 2005, p21]. The largest lubricant market is the US, followed by China, Japan, Russia, Germany, Canada and India—each of which consumes over 1 million tons per year [Hydrocarbon Processing, April 2005, p21].

11.1 LUBRICATING OIL BASE OIL

Some of the key properties of lube base oils are: Viscosity, viscosity index, pour point, color, flash point, volatility, conardson carbon ash content, water content and oxidative and thermal stability, acidity. Undesirable characteristics of lube base oils are high pour points, large viscosity changes with temp, low viscosity index, poor oxygen stability, poor color, high cloud points, high organic acidity, high carbon and sludge forming tendencies. Multigrade oil has gradually replaced monograde oils in passenger car service [Hournace, 1981].

Base of the lubricating oil is an important factor affecting quality of lubricating oil. Good lubes need good base oils. Although naphthenic crude gives appropriate freezing point without going further particular treatment, however, gives lower viscosity index lube oil. Paraffinic crude provides higher viscosity index lube oil which is desirable [Hournace,1981]. Some of the requirements of lubricating base oil are [Bose, 2010]:

- Appropriate viscosity
- High viscosity index
- Low pour point
- Low volatility and high flash
- Point
- Good thermal and oxidation
- Stability
- Non-corrosive, oiliness and wettability.

Characteristics of lubricating oil feedstock is given in Table 11.1.

Table 11.1 Lubricating oil feedstock

Crude	Properties 370–509 °C cut			
	KV@100 °C (cst)	Wax content (%wt)	Dewaxed oil properties	
			VI/pour point (°C)	Density (15 °C g/cm³)
Upper zakum	7.10	11.0	62/−12	0.930
Arab light	6.74	10.5	64/−15	0.9275
Basra light	7.06	11.3	68/−18	0.9275
Kuwait	7.49	9.1	62/−12	0.9325
Bombay High	8.49	49.5	34/−12	0.9395
Labuan (Malaysia)	8.65	23.5	−165/0	0.9740
Bonny light (Nigeria)	11.13	18.5	−17/−15	0.9475

Source: Ghosh, 2006.

Properties of lube oil components are given in Table 11.2.

Table 11.2 Lube base oil component properties

Property	n-paraffins	Isoparaffins	Naphthenes	Aromatics
Viscosity index	High	High	Medium	Low
Pour point	High	Low	Low	Low
Oxidation stability	Good	Good	Fair	Poor
Thermal stability	Good	Good	Fair	Poor

Source: Ghosh, 2006.

11.2 LUBE PLANT

Lube refinery processes heavy crude oil produces various grades of lubricating oils which include light, medium and heavy lubestocks and transformer oil basestocks. There are three lube refineries in India: Hindustan Petroleum Corporation limited, Madras Refinery Limited and Haldia Refinery. These units produce lubricating crudes like Arabian mix crude.

A typical lube oil plant consists of following units:

- Vacuum distillation column
- Deasphalting
- Solvent refining
- Dewaxing
- Hydrogen finishing
- Base oil product storage.

Block diagram of lube oil plant is shown in Fig. 11.1.

Process

Lube refinery unit consists of vacuum distillation unit which produces lube oil of different viscosities. Product quality is controlled by viscosity of the reduced crude oil and the type of crude oil processed. Typical condition in the vacuum distillation unit is 400 °C (flash zone) temperature and 130 mm Hg absolute vacuum. Various streams of different viscosities—motor oil distillates and the transformer oil distillates are produced.

Major products from a vacuum distillation unit are:

Vacuum gas oil	Obtained as overhead of vacuum column and goes to high speed direct pool.
Excess vacuum gas oil	After solvent extraction and propane deasphalting, it is used in high speed diesel pool.
First side stream **Second side stream** **Third side stream**	Used for producing different grades of lubricating oils after solvent extraction and propane dewaxing.
Vacuum tower bottom	Used for manufacture of bitumen.

The raw distillate side streams from the vacuum column are further treated in process to meet the specifications of color and color stability.

11.2.1 Solvent Extraction of Lubricating Oils

Solvent extraction has been used from very beginning of the petroleum refinery for removing aromatics and other undesirable constituents to improve viscosity index, color, oxidative stability, thermal stability and inhibition response of base oils [Sequeria et al., 1979]. Viscosity index is one of the important characteristics of lubricating oil as viscosity plays a vital role in the lubrication of moving part of machine. Low viscosity index is due to presence of aromatics and can be improved by removal of aromatics.

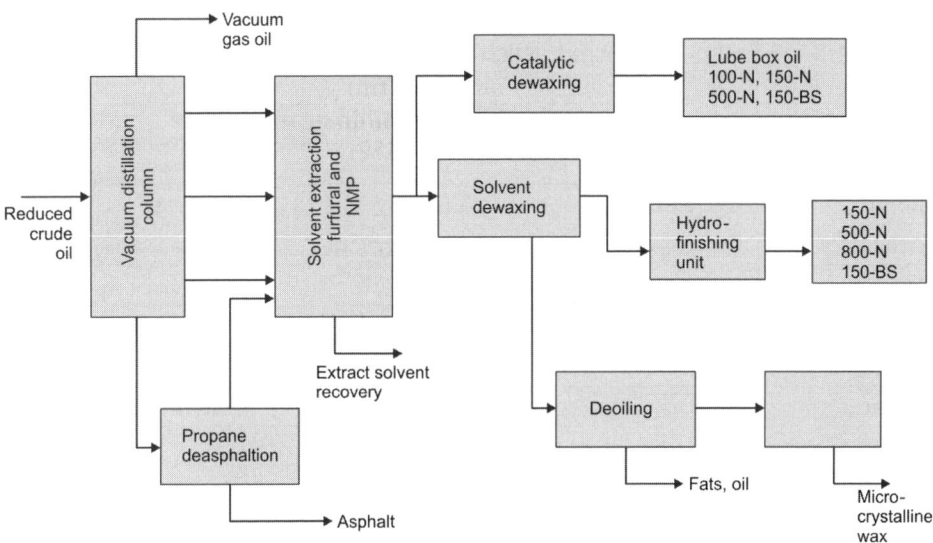

Fig. 11.1 Block diagram of a typical lube plant

Although liquid SO_2 which was initially used for treatment of kerosene for removing aromatics it was also used for lube oil treatment, however, its toxicity and environmental problem restricted its application. Initially, furfural and phenol were the most commonly used solvents, however, due to its nontoxic nature, high solvent power, good selectivity, and adaptability to the extraction of both paraffinic and naphthenic feedstocks, n-methyl-2-pyrrolidine (NMP) is attractive, alternative to furfural. NMP exhibits better solvent power, reduced solvent recirculation, better chemical and thermal stabilities and lower toxicity, reduced specific utilities consumption, reduced solvent loss, reduce equipment fouling over either furfural or phenol [Seqeira et al., 1979; Bertagnolio, 1983]. Comparison of furfural, NMP and phenol in solvent extraction process is given in Table 11.3 [Ghosh, 2005; Seqeira et al., 1979].

Criteria of ideal solvent for extraction of lubricating oil are [Sequeria et al., 1979]:

- High selectivity—for undesirable components
- Good solvent power—low solvent feed
- High extraction temperature—for good mass transfer
- Easy recovery—by simple flash distillation
- Low vapor pressure—avoid use of pressure equipment
- High density—rapid separation of oil and solvent phases
- No emulsification—rapid separation of oil and solvent phases
- Stability—no chemical or thermal degradation
- Adaptable—to a wide range of feedstocks
- Available—at reasonable cost
- Non-corrosive—to conventional metals of construction
- Nontoxic—environmentally safe.

Table 11.3   Effect of solvent extraction on lube oil

Effect on composition		Effect on properties	
Asphaltene	Decrease	Density	Decrease
Resin	Decrease	Viscosity	Decrease
Aromatics	Decrease	Viscosity index	Increase
Naphthenes	Increase	Pour point	Increase
Paraffins	Increase	Color	Improve
Nitrogen	Decrease	Stability	Improve
Sulfur	Decrease	Additive response	Improve

Source: Ghosh, 2005.

11.2.2 Furfural Extraction

Typical furfural extraction units consist of deaeration, furfural extraction, raffinate separation, extract separation, solvent recovery, neutralization with sodium carbonate. The process involves liquid-liquid extraction where the extract phase is aromatic oil plus major solvent while the raffinate phase is paraffinic oil plus solvent. Solvent from the extract and raffinate is recovered by heating, flushing and stripping. Furfural is still used as a solvent because of lower toxicity, cost, availability, adaptability to extraction of gas oils and better selectivity in the lubestocks which exhibit excellent refining response at low solvent to oil ratios [Sequeria et al., 1979].

11.2.3 N-Methylpyrrolidone (NMP) Extraction

The various steps involved in NMP extraction process are same as furfural extraction process except that instead of aerator, in NMP process absorber is there where traces of solvent from vent gases are absorbed in the feed oil to extractor. However, NMP process has certain advantages over other solvents as less solvent ratio is required in comparison to furfural. Some of the advantages of NMP extraction process are [Beragnolio, 1983]:

- Higher solvent power
- Reduce specific utilities consumption
- Reduced solvent loss
- Reduced equipment fouling because of lower solvent degradation
- Reduce ecological problem because of lower toxicity.

11.2.4 Phenol Extraction

Although phenol cost is lower than other solvent, however, phenol extraction, it has poor selectivity compared to furfural and NMP. Phenol is suitable for extraction of paraffinic stocks which require high extraction temperatures and is less suitable for extraction of naphthenic lubestocks [Kosters, 1977]. Comparison of solvent is given in Table 11.4.

Table 11.4 Comparison of furfural, NMP and phenol in solvent extraction process

	Furfural	NMP	Phenol
Formula structure			
Specific gravity	1.162	1.04	1.08
Boiling point	162	202	182
Vaporization enthalpy at 60 mm Hg kj/kg	451	493	479
Biodegradability	Good	Good	Good
Toxicity	Moderate	Low	Severe
Solvent power	Good	Excellent	Very good
Selectivity	Excellent	Very good	Good
Stability	Good	Excellent	Very good
Adaptability	Excellent	Very good	Good
Emulsibility	Low	Moderate	High
Solvent-to-oil ratio	Moderate	Very low	Low
Extraction temperature	Moderate	Low	Intermediate
Refined oil yield	Excellent	Very good	Good
Product color	Very good	Excellent	Good
Corrosiveness	Intermediate	Low	Good
Energy cost	Moderate	Low	Intermediate
Relative cost	1.0	1.5	0.36
Investment cost	Intermediate	Low	Moderate
Operating cost	Intermediate	Low	Moderate
Maintenance cost	Intermediate	Low	Moderate

Source: Ghosh, 2005; Seqeira et al., 1979.

11.2.5 Lube Hydroprocessing

Shell lube hydroprocessing is a two-stage hybrid process which is the combination of solvent extraction and one-stage hydroprocessing for processing wider range of crudes. The process gives higher yield of base oil crude having lighter crude [Refining Process, 1998].

11.2.6 Hy-Raff Lube Hydrotreating Process

Hy-Raff process is a new process to hydrotreat praffinates from an extraction unit of a solvent-based lube oil plant. The integration of this process into an existing base oil plant allows cost effectively upgrade base oil [Refining Processes, 2008].

11.3 DEWAXING

Dewaxing process is used to remove the paraffin wax from the lubricating oil which lowers the pour point and improves the flow at lower temperature. Various dewaxing processes are: Ketone dewaxing, propane dewaxing and catalytic dewaxing.

11.3.1 Propane Dewaxing

Propane which is available in abundance from refineries has received considerable interest as a solvent for dewaxing. Hot feed containing dewaxing aid is mixed with propane, warmed and send to series of chillers where the solution is cooled to desired filtration temperature. Wax is separated from the solvent using rotary drum filter with simultaneous fresh solvents.

11.3.2 Ketone Dewaxing

Ketone dewaxing uses methyl-ethyl-ketone in conjunction with methyl-isobutyl-ketone or toluene. Ketone dewaxing process consists of mixing with solvent, crystallization, refrigeration, filtration, vacuum blow gas, solvent recovery and solvent drying. Ketone dewaxing uses methyl-ethyl-ketone in conjunction with methyl-isobutyl-ketone or toluene.

Ketone dewaxing process consists of mixing with solvent, crystallization, refrigeration, filtration, vacuum blow gas, solvent recovery and solvent drying.

11.3.3 Catalytic Dewaxing Process

The catalytic dewaxing process uses selective shape catalyst which selectively cracks n-paraffins in base oils resulting in low pour point [Meenakshi Sundram, 2004].

11.3.3.1 Mobil Selective Catalytic Dewaxing (MSDW)

The process steps involved are hydrotreatment of raffinate feedstock involving desulfurization, denitrogenation, olefin saturation and partial aromatic saturation, selective catalytic dewaxing by wax isomerization, hydrofinishing followed by product purification by distillation.

Catalyst: **Hydrotreatment:** EM-7420 NI-Mo hydrogenation catalyst

           Catalytic dewaxing: Pt/Pd hydrogenation catalyst for HDS, HDN and aromatic saturation and EM7300 Pt isomerization catalyst

11.3.3.2 Chevron Lummus Global's Isodewaxing

Isodewaxing technology revolutionized catalytic dewaxing since its commercialization in 1993. Isodewaxing technology instead of removing the wax, it catalytically isomerizes the molecular structure of the wax into $C_{20}{}^+$ isoparaffins which have high VI, low pour points and excellent resistance to oxidation [www.chevron.com] [ABS Lummus]. The process involves dewaxing by isomerizing paraffins followed by hydrofinishing with high noble metal catalyst. The process involves isomerization of normal paraffins wax to desirable isoparaffins lube molecules, resulting in high viscosity index, low pour point base oils, while producing small quantities of high quality middle distillate [Bridge and Mukherjee, 2004]. The process offers improved product quality and also increased yield of lube oil basestock typically; the isodewaxing process uses noble metal catalyst with mildly acidic shape selective sieves as support and, hence has a stringent requirement on sulfur content of the feed [Meenakshisundram, 2004].

11.4 DEASPHALTING

Deasphalting is a carbon rejection process and involves separation of residual into oil and asphalt fraction. It removes asphaltenes and resins with relatively low H/C ratios. Removal of asphalt helps in preventing coke build up in further processing of deasphalted oil. Solvent deasphalting is commonly used. Liquefied saturated light hydrocarbons $(C_3–C_4)$ are used as solvents which promote precipitation. The choice of the solvent is vital to the flexibility and performance of the deasphalting unit [Speight and Ozum, 2003]. Propane is commonly used for lube oil basestock while butane/pentane is used for cracker feedstock production from gas oil.

11.4.1 Propane Deasphalting

Propane deasphalting is the most commonly used process for deasphlating as it gives both better yields and high grade refined product [Hournac, 1981]. Propane deasphalting unit produces deasphalted oil (DAO) of better quality by removing asphalt from vacuum residue obtained from vacuum distillation column. A characteristic of vacuum residue feed to the unit is mentioned in Table 11.5. Typical characteristic of deasphalted oil and asphalt is given in Tables 11.6 and 11.7. Process flow diagram for propane deasphalting is given in Fig. 11.2.

Table 11.5   Typical characteristics of vacuum residue feed to deasphalting unit

Characteristics	
Degree API	5.24
Specific gravity	1.023
Mol wt.	878
TBP cut point (°C)	574
CCR (%w/w)	23.01
Kin. viscosity (CST)	At 98.9 °C, 4788
Pour point (°C)	60
Flash point (°C)	277.5
Penetration number	80
Wax content (%wt)	1.8
Asphaltenes	10.96

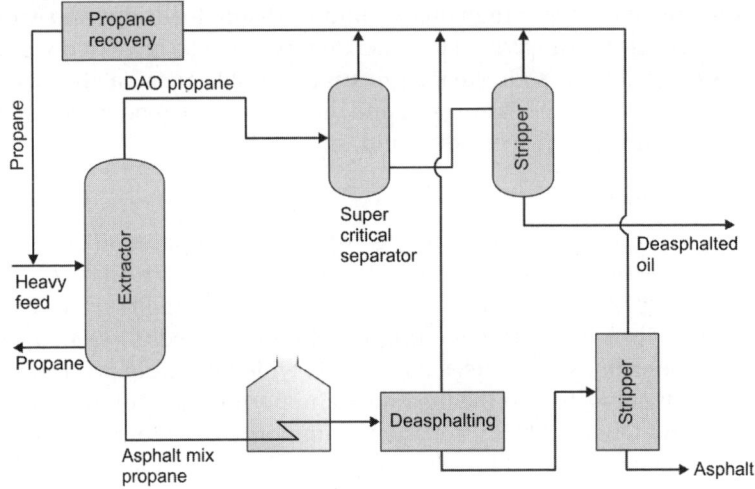

Fig. 11.2 Propane deasphalting

Table 11.6   Typical characteristics of deasphalted oil

Degree API	20.6
Specific gravity	0.93–0.04
Mol wt.	607
CCR (%w/w)	23.01
Kin. viscosity (CST)	At 98.9 °C, 42–48
Pour point (°C)	39
Flash point (°C)	304

Table 11.7   Typical characteristics of asphalt

Degree API	5.24
Specific gravity	1.023
Mol wt.	878
Kin. viscosity (CST)	At 98.9 °C, 42581

11.4.2 UOP/FWUSA Solvent Deasphalting [UOP 4223–29]

The UOP/FWUSA solvent deasphalting process produces a low contaminant deasphalted oil (DAO) which is rich in paraffinic type molecules. In this process the feed is mixed with a light paraffin solvent typically butane, where the deasphalted oil is solubilized in the solvent and the insoluble pith precipitate out of the mixed solution. In the extractor, separation of the deasphalted oil phase (solvent—DAO mixture) and the pith phase occurs. DAO phase after separation is then heated to conditions where the solvent becomes supercritical. Under these conditions the separation of the solvent and DAO is relatively easier and the separation occurs in the DAO separator [uop.com]. The solvent from the pith is recovered and the recycled solvent from the DAO phase is tripped and recycled along with solvent from pith phase.

11.4.3 Low Energy Deasphalting (LED) Process

The LED process is used to extract high quality lubricating oil brightstock or prepare catalytic cracking and hydrocarcking feeds and asphalt. The process uses low boiling hydrocarbon solvent. Residue is extracted in a rotating disc extractor (RDC) with solvent. RDC provides more extraction stage. Deasphalted oil separator recovers solvent at supercritical conditions and asphalt flash recovers solvent [Refining Processes, 1998; Speight and Ozum, 2003].

11.4.4 Major Operating Variables in Deasphalting [UOP 4223–29]

- *Yield versus product quality:* Increasing DAO increases contaminants in the DAO Solvent selection as the solvent gets heavier, the yield of DAO increases, but quality declines.
- *Temperature:* A lower temperature of extraction, increases the DAO yield and decreases the quality at a constant solvent composition and pressure. Increase in the extraction temperature reduces the solubility of the heavier feedstock components. The critical temperature of the solvent limits the maximum extraction temperature, since at critical temperature, no portion of the feedstocks would be soluble in the solvent and no separation would occur [uop.com].
- *Contact time:* As contact time increases, the asphaltene yield increases until maximum separation is achieved.
- *Solvent to feed ratio (s/f):* Increase in s/f ratio increases selectivity, main variable for increasing DAO quality. Typical condition of s/f = 12. Addition cost is associated with the increased solvent recirculation and solvent requirements.

11.5 WAXES

The origin of the wax in the crude oil is, as for the crude oil itself. The discovery of wax in wood took place about 1830 and by 1833 wax was separated from crude oil from crude seepage in Burma [Jowett F. Petroleum Waxes]. In India, Assam Oil Refinery Company (now IOC Digboi) at Digboi was the first commercial plant which started manufacturing different grades of waxes.

Waxes are fully saturated hydrocarbon with formula C_nH_{n+2} and are paraffinic (C_{18}–C_{40}), intermediate (C_{25}–C_{60}) and naphthenic (C_{25}–C_{85}). Paraffinic wax is linear while intermediate waxes have increased branching. Microcrystalline waxes are complex, branched structure with little or no linear hydrocarbons. Melting point of the wax has direct or indirect significance in the utilization of wax [Speight, 1998]. Parafin wax is an essential component of the candle making industry. Microcrystalline wax possesses a fine grained crystal structure rendering them more flexible.

- *Paraffinic wax:* Low melting, white, hard, brittle, translucent, crystalline, glossy
- *Microcrystalline wax:* Higher melting, colored, soft, malleable, opaque, amorphous, adhesive

Hardness, rigidity, brittleness and low viscosity of petroleum wax are associated primarily with its degree of crystallinity [Jowett, 1984]. Microcrystalline wax consists of high n-paraffin with predominance of branched and cyclic paraffins which result in the formation of crystal needles.

Microcrystalline waxes are of laminating grades, coating grades and hardening grade. Laminating grade waxes are use in packing, adhesive, cosmetics, rubber, candles. Coating grade waxes are used in adhesive, packaging, chewing gum, inks, plastics, rubber. Hardening grade waxes are used in adhesives, inks, chewing gum, candles and speciality products. Classification of various types of microcrystalline waxes is given in Table 11.8. Paraffin waxes are used for waterproofing and waxed paper and in textile industry.

Table 11.8   Classification of microcrystalline waxes

Microcrystalline wax	Melt point (°C)	Needle penetration, (dmm)	Properties
Laminating grade	54.4–76.7	20–24	Flexible, tacky
Coating grade	76.7–85	15–25	Harder, low tackiness
Hardening grade	85–93.3	5–12	Very hard, higher viscosity

Source: IGI, 2007.

Typical characteristics of various grades of wax manufactured by IOC are given in Tables 11.9 to 11.11.

Table 11.9   Characteristics of paraffin wax of superior grade used for food packaging, cosmetics and pharmaceuticals

Properties	Ensure
n-paraffin (%wt > 95)	Very good feedstock for preparing derivatives
Melting points (60.0–62.5 °C)	High quality film formation, higher gloss and blocking points Structural stability, well-defined crystalline structure No chances of contaminating food products
Oil content (< 0.5 %wt)	No deterioration of color during storage
UV absorption test: Passes	Nil toxicity
Color < 1.0Y	Dazzling whiteness

Source: [File://h;bitumen%20&%20wax/Indian%20Oil%20Corporation%2].

Table 11.10 Characteristics of best quality wax for paper and textiles sizing, manufacturing candles, polishes, tires and metal casings

Properties	Ensure
n-paraffin (> 95 %wt)	High quality film formation, higher gloss and blocking points
Melting points (60.0–62.5 °C) Oil content (< 0.5 %wt)	Structural stability, well-defined crystalline structure
Hardness	No oozing of oil during transportation and storage
Needle pentration: 12–14 (1/10 mm, 5 sec, 100 g at 25 °C)	No deterioration of color during storage Hardness, no bending candles

Source: [File://h;bitumen%20&%20wax/Indian%20Oil%20Corporation%2].

Table 11.11 Characteristics of ideal wax for matchsticks, tarpaulins and wax emulsions

Properties	Ensure
Melting points (60.0–62.5 °C)	Structural stability, well-defined crystalline structure
Oil content (< 3.5 %wt)	No oozing of oil from applied surface
ASTM color L2.0	Bright appearance

Source: [File://h;bitumen%20&%20wax/Indian%20Oil%20Corporation%2].

11.5.1 Manufacture of Wax

Waxes are obtained as by-products during manufacture of lubricating oil. The paraffin waxes from solvent dewaxing are further processed for separation of wax. The steps of the process are deoiling which involves sweating or solvent deoiling, followed by bauxite treatment or hydrotreatment, steam stripping for odor removal. The steps involved in the manufacture of wax is given in Fig. 11.3. Depending upon the various grades of light, medium and residual lube distillate from the vacuum column, different grades of wax such as paraffin wax, intermediate wax and microcrystalline wax respectively can be produced.

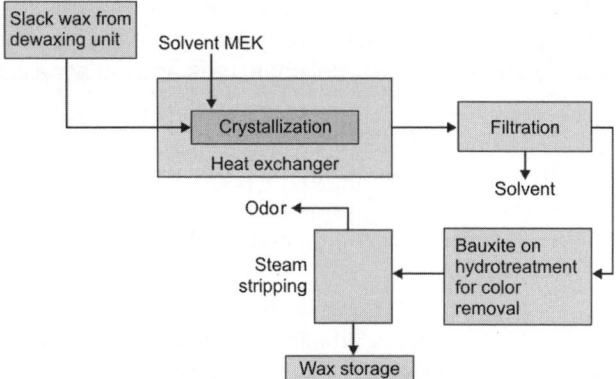

Fig. 11.3 Wax manufacture

11.5.1.1 Sweating Deoiling

Wax sweating process was used earlier and has been replaced with solvent deoiling in modern refinery. The process involves slow heating of the slack wax feed to a temperature where lower melting wax separates by sweating (dripping) and higher melting wax is left which is further processed for color and odor removal.

11.5.1.2 Solvent Deoiling

Recent trend in the deoiling is the use of solvent for separation of wax. Methyl-isobutyl-ketone (MIBK) is commonly used. The process takes advantage of different solubilities of wax in MIBK. The feed is mixed with solvent in suitable ratio and dissolved by heating. The wax is separated by cooling to predetermined temperature. The wax gets crystallized after cooling and filtered and sent for color and odor removal.

11.5.2 Chlorinated Paraffin Wax (CPW)

Chlorinated paraffin wax varies from pale yellow viscous liquid to soft or brittle resinous solid depending upon the chlorine contents. Chlorinated paraffin wax is used as plasticizer for PVC, in textile and tent canvas, electrical wire insulation, as additive, in wool carding, in cleaning fluids, flameproof coating on textile wax and for making carbon paper, surface coating. Chlorinated paraffin wax is physically homogeneous, viscous liquid or low melting solid depending on the chlorine content.

Process

Wax is melted in jacketed vessel by steam and charged to glass lined chlorinator jacketed for steam heating. Chlorination of the melted wax is done by chlorine gas at 100 °C for 18–20 hours. Chlorinated wax is mixed with additives, filtered and cooled. Hydrochloric acid which is a by-product and is removed by scrubbing with water.

REFERENCES

1. Bertagnolio M. Modernizing a Lube oil Plant. Hydrocarbon Processing, March 1993, p103.
2. Bose S. Lube Oil Processing. 5th Summer School on Petroleum Refining and Petrochemicals, IIPM Gurgaon, June 7–11, 2010, organized by IOCL and Petrotech.
3. Bridge AG and Mukherjee UK. Isocracking-hydrocracking for Superiors Fuels and Lubes Production. McGraw Hill Handbook of Petroleum Refining Process, 2nd edition, edited by Meyers RA, McGraw Hill Companies, 2004.
4. Ghosh R. Lube Oil Production. Petrotech Society's Summer School Program on Advances in Petroleum Refining Industry, July 3–8, 2006.
5. Hournac R. Good Lubes Need Good Base Oils. Hydrocarbon Processing, Jan 1981, p207.
6. Hournace R. Good Lubes Need Good Base Oils. Hydrocarbon Processing, Jan 1981, p207.
7. Hydrocarbon Processing, April 2005, p21.
8. IGI (The International group). Use of Microcrystalline Waxes in Candles. Presented to ALAFAVE and the NCA, June 28, 2007.
9. Jowett F. Petroleum Waxes in Modern Petroleum Technology, Part 2, edited by Hobson GD, 1984, p1021.
10. Kosters WCG. The Role of Extractives in Lube manufacture. Chemistry and Industry, Nov 2, 1977, p67.
11. Meenakshi Sundram A. Hydroprocessing for Fuels and Lubes Production. Bulletin of Catalysis Society of India, 3 (2004), p1–9.
12. Richter W. Wax in Modern Petroleum Technology, Vol 2, edited by Lucas AG, 2001, p431.
13. Refining Processes, 1998. Hydrocarbon Processing, Nov 1998, p104.
14. Refining Processes, 2008. Hydrocarbon Processing, Nov 2008, p78.
15. Sequeria A, Sherman PB, Douceiere JU and Mcbride EO. MP Refining of Lubes. Hydrocarbon processing, Sep 1979, p155.
16. Speight JM. The Chemistry and Technology of Petroleum, 3rd edition, Marcel Dekker, 1998, p809.
17. Speight JG and Ozum B. Petroleum Refining Processes, Mracel Dekker, 2002.

12
Bitumens

Bitumen is a very highly viscous liquid or solid, non-crystalline solid obtained as a residue from the vacuum distillation of asphaltic crude or by precipitation from residual fractions by propane deasphalting. Bitumen is extremely complex in nature and can be liquefied and reduced to low viscosity by heating. Bitumen may be regarded as colloidal system of highly condensed aromatic particles (asphaltenes) suspended in a continuous oil phase [Chersources, 2002. http://www.cheresources.com/ refining9.shtml]. Bitumen is normally produced from non-waxy crude. Most of the Middle East crude is excellent for bitumen production. Arab heavy crudes include asphalt, lubricants, waxes and fuel oils. Bitumen content of Arab heavy crude may be in the range of 27% while light Arabian crude has approximately 12% of bitumen [Hollern, 2002. Manufacturing Bitumen, VSS Technology Library, http://www.slurry.com/techapers_mfrbit.shtml].

Various varieties of bitumens are produced with varying amount of volatile material in the product. Higher the volatile matter, smaller the amount of volatiles, the harder the residual bitumen [www.r-t-o-l.com]. Some of the varieties of bitumens produced are crumb rubber modified bitumen and paving grade bitumen.

Because of its unique polymer like properties, relative chemical inertness, relative ease of application and low cost, bitumen finds wide application. Bitumen is used as adhesive, coating, sealant and preservative [van Gooswilligen, 2001]. Major application of bitumen is in paving application, such as a binder in asphalt mixes for load sealing, for surface treatments, for bridge decks and as joint sealant. Some of the other major applications of bitumen in building and construction are waterproofing, pipe coating, flooring, briquetting, paints, carpet backing, sound insulation and slip layers. Waterproofing is the largest application of bitumen in domestic and industrial sectors. Roofing felt is made by using penetration grade bitumen. The insulating properties of bitumen enable their applications for cable jointing compounds and heavy duty cable wrapping.

12.1 BITUMEN PROPERTIES AND TESTING

Bitumen possesses strong adhesives and excellent waterproofing properties and have high resistance to oxidation and weathering processes. Bitumens are soluble in carbon disulfide, benzene and chlorinated hydrocarbons. Bitumen possesses very good electrical properties, having low electrical conductivity [Chipperfield, 1984]. Some of the common tests used for characterization of bitumen are penetration, softening point, flow, viscosity, stiffness modulus, dynamic shear, ductility, frass breaking point, cold bending test, thermal fracture, contraction, stain index, oliensis compatibility, oliensis spot test, density, thermal stability and flash point [van Gooswilligen, 2001].

The various grades of bitumen are paving grade bitumens and, roofing grade bitumens, polymer modified bitumens, emulsions, cut back bitumens. Characteristics of various grades of bitumen are given in Table 12.1.

Table 12.1   Various grades of paving bitumen conforming to IS: 73–1961 (revised)

Sl. No.	Characteristics	Requirement for grades					Method	GR typical value	
		S.35 30/40	S.45 40/50	S.65 60/70	S.90 80/100	S.200 180/200	Oftest	S.65 60/70	S.90 80/100
1.	Specific gravity at 27 °C (g/mL, min.)	0.99	0.99	0.99	0.98	0.97	IS: 1202–1958	1.1014	1.0147
2.	Water (%wt, max.)	0.2	0.2	0.20	0.2	0.2	IS: 1211–1958	0.1	0.1
3.	Flash point (PMCC) (°C, min.)	175	175	175	175	175	IS: 1209–1958 (Method A)	205	200
4.	Softening point (R & B) (°C)	50–65	45–60	40–55	35–50	30–45	IS: 1205–1958	45	42
5.	Penetration at 25 °C (100 g)	30–40	40–50	60–70	80–100	175–225	IS: 1203–1958	65	90
6.	Ductility at 27 °C (cm, min.)	50	75	75	75	–	IS: 1208–1958	88	80
7.	a. Loss on heating (%wt, max.)	1	1	1	1	2	IS: 1212–1958	0.5	0.5
	b. Penetration of residue (%wt, min.)	60	60	60	60	60	IS: 1203–1968	62	60
8.	Matter soluble in carbon disulfide, (%wt, min.)	99	99	99	99	99	IS: 1216–1958	99.6	99.2

Additional test requirements can be met according to customer's specifications (expressed as percentage of item 5).

Carbon tetrachloride or trichloroethylene may also be used instead of carbon—disulfide, the method of test being the same. This is as per Amendment No. 2—March, 1983.

12.2 BITUMEN MANUFACTURE

Bitumen can be produced either by air blowing of vacuum residue or blending of asphalt with aromatics extract of lube basestocks. Bitumen blowing process is used in many of the Indian refineries. Bitumen is normally produced from non-waxy crudes. Most of the Middle East crude is excellent for bitumen production.

12.2.1 Bitumen Blowing Process

Feedstock for bitumen plant is high boiling vacuum residue having boiling point around 530 °C. Air blowing of bitumen at high temperature improves the binding properties of bitumen as it increases the contents of gums and asphaltenes. Asphaltenes content of bitumen influenced bitumen solidity and softening point, higher the asphaltene content, the more solid is the bitumen. The quality and the quantity of feedstock and desired grades of bitumen, determine the rate of air to be used for blowing. Air blowing of bitumen is a chemical conversion process involving dehydrenation, oxidation and polymerization.

The process of bitumen consists of air blowing into vacuum residue at about 200–220 °C in a reactor. Oxygen reacts with the asphaltenes and resins and the progressive loss of hydrogen polymerizes and condenses the bitumen and improves its properties. The reaction is exothermic and continuous removal of the heat is achieved by injecting small amount of dimineralized (DM) water. Careful control of temperature and feed rate is necessary. The reactor pressure is maintained around 2 kg/cm². The various units in atypical bitumen are: Cold feed pump and preheating, bitumen furnace, reactor, gas separator, finished bitumen processing, reactor overhead section, oxidation gas separator and incinerator are given in Fig. 12.1.

Cold bitumen is preheated to about 150–180 °C and pumped to reactor through bitumen furnace where it is heated to about 220–230 °C. Bitumen furnace, natural draft furnace having convection and radiation zone having horizontal and vertical feed coils respectively. Oxidation of the residue is carried out in the reactor. Air is passed in the reactor through sparger. The conditions of air blowing are regulated so that blown bitumen of desired properties is produced. The finished air-blown bitumen is pumped out of the reactor at about 240–260 °C and cooled. Cooled

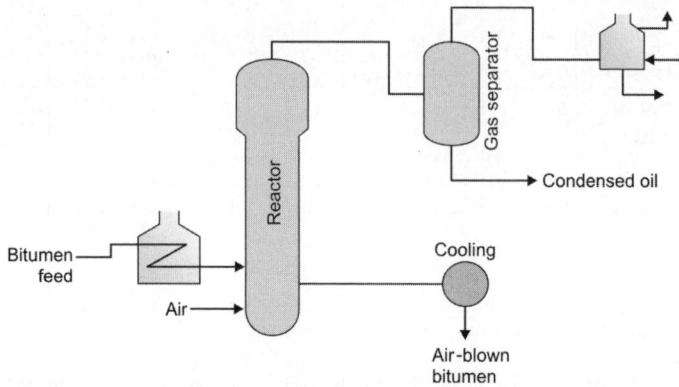

Fig.12.1 Bitumen blowing unit

bitumen is sent to the bitumen storage. Reactor overhead gases containing hydrocarbons, steam and unreacted air go to air coolers and separator. The condensed oil is separated and the uncondensed vapor goes to incinerator.

12.3 USES OF BITUMEN

Bitumen is used in civil engineering works and industries. Various grades which are being manufactured in IOC refineries are bitumen 80/100, bitumen 60/70, bitumen 30/40, bitumen CRMB and bitumen emulsion—Indmul [http://www.iocl.com/products/Bitumen.aspx].

- *Civil engineering works:* Construction of roads, runways and platforms, water-proofing to prevent water seepage, canal lining to prevent eroding, dumpproof courses for masonry, tank foundation, joint filling material for mason [iocl.com].
- *Industrial use:* Electrical cable and junction boxes, battery manufacture as sealing compound, paint industries for manufacturing black paints and anti-corrosive paints, ceramic industries, printing inks, waterproof papers, electrical capacitors, bituminous felts, bituminous grease for lubricating open gears [iocl.com].

REFERENCES

1. Chersources, 2002. http://www.cheresources.com/refining9.shtml.
2. Chipperfield EH and Harper FD. Bitumen in Modern Petroleum Technology, Vol 2, Fifth edition, edited by Hobson GD, John Willey & Sons, 1984.
3. Hollern, 2002. Manufacturing Bitumen. VSS Technology Library, http://www.slurry.com/ techapers_mfrbit.shtml.
4. http://www.iocl.com/products/Bitumen.aspx.
5. Van Gooswilligen G. Bitumen in Modern Petroleum Technology, Vol 2. Downstream, edited by Lucas AG, John Wiley & Sons Ltd, 2001.

Lubricating Grease

Lubricating grease is one of the important petroleum products because proper lubrication is essential for smooth and long life of rolling bearings. Although evidence of archaeological findings in the West Asia suggests that art of making grease was known about 4000 years ago, however, the technology for the current generation of lubricating grease is less than 100 years. A wide variety of greases are available in the market in order to meet the requirement of large production facilities having variety of equipments. A wide variety of greases have been formulated using variety of base oils, thickeners and additives to maximize grease lubrication effectiveness, minimize cost and minimize risk of application induced failure [machinerylubrication.com]. These greases range from slow to high speeds, and from low to high loads in an effort to provide a single product to meet multiple requirements [Johnson, 2002] [machinerylubrication.com]. The worldwide grease market survey of Lubricating Grease Institute, USA, indicates that worldwide grease volume of 2.4 billion lbs with mineral oil-based grease accounting for about 92% followed by synthetic and 7% semisynthetic oil-based grease. Bio-based greases are also available [Kumar, 2014].

Lubricating greases are semi-solid or solid mixture consisting of gelling or thickener agent in a liquid lubricant (base oil) having varying consistency and additives. Other constituent of grease may be additives and fillers. The liquid lubricant is mineral oil, synthetic oil or vegetable oil. As per ASTM, grease is a solid or semi-fluid product of dispersion of a thickening agent in a liquid lubricant. Lubricating oil is the cheapest oil of varying viscosity and is commonly used in the grease. Low viscosity oil-based grease is used in low temperature application. Synthetic oil may be used in case of speciality greases having wide range temperature application, oxidation stability, fire resistance, low volatility, or a high viscosity index [Spicer, 1986]. Greases may be soap-based or nonsoap-based. Normally, the gelling agent soap is formed by reacting a metal oxide with fatty acid. The type of soap used depends on the conditions in which the grease is to be used. Different soaps provide different levels of temperature resistance, water resistance and chemical. Some nonsoap-based greases contain carbon black, organobentonites, silica, clay, sodium octadecyl terephthalamates, phthalocyanine, aryl substituted urea base greases, diamido-dicarbonyl base grease.

Additives are added to the grease to improve their lubricating properties, service life at elevated pressure with higher temperature stability, rust resistance and oxidation resistance, cohesive and adhesive properties. Some of the commonly used additives are glycerol and sorbitan esters, graphite, molybdenum sulfide, amines, zinc dialkyl dittiophosphate, etc.

Various types of lubricating oils are automotive lubricating oils, automotive speciality oils, rail road oils, industrial lubricating oils, industrial speciality oils, metal working oil and agricultural spray oils [www.indianoilcorp.com], marine lubricating oils and defence grade oils. Various types of greases manufactured are automotive grease: Calcium base grease, sodium base grease, lithium base grease; rail road greases, industrial greases. Indian Oil Corporation is one of the leading manufacturers of lubricating oils and greases and produces wide variety of greases.

13.1 VARIOUS TYPES OF GREASES

A wide variety of lubricating greases are available in the market depending on the end application of the grease which may be soap-based or nonsoap-based having petroleum oil or synthetic oil. Some of the commonly used greases are calcium soap grease, sodium soap grease, lithium soap grease, aluminum soap grease, barium soap grease, multipurpose grease, silicon greases, polyurea greases, clay greases, and carbon black greases. Greases are also classified on the basis of their end applications, e.g. ball roller bearing grease, ball bearing grease, automotive grease, laboratory grease, cement plant grease, extreme pressure grease. Various types of greases are given below [Spicer, 1986; Mead, 2001; http://www.iocl.com/products_lubes_8.aspx, http://en.wikipedia.org/wiki/silicone_grease, http://www.iocl.com/products_lubes_10.aspx, http://www.reliabilityweb.com, NLGI 99, Speight, 1999]:

Types of grease	Description
Calcium soap greases	: Calcium soap greases are made by thickening calcium soap with lubricating oil. Because of water evaporation calcium soap greases are not suitable for high temperature application and limited to application to 60 °C in case of normal calcium soap and up to 120 °C in case of high quality calcium grease and it dehydrates at temperature around 79 °C at which its structure collapses, resulting in softening and phase separation. It is suited for general chassis lubrication including suspension and stirring system. Oxidation stability of these greases is poor.
Calcium complex grease	: Calcium complex grease is prepared by adding calcium acetate which provides extreme pressure characteristics [Biolubricants, 2013].
Sodium soap-based grease	: Sodium soap grease is made by mixing sodium soap and lubricating oil. Sodium soap grease gives excellent performance even when subjected

to high shear and can be used over a wide range of temperature has fibrous structure and shows reasonable mechanical stability over long service. Sodium soap-based grease withstands excessive churning effectively and recommended for wheal bearing and other automotive grease applications. Sodium soap-based grease has high melting point.

Lithium soap-based grease : Lithium soap grease are made from lithium soap which is made by reacting lithium hydroxide with hydrogenated castor oil or hydroxystearic acid. Lithium soap grease has high drop point, excellent mechanical stability, and good water resistance and good thermal and structural stability and provides high degree of resistance to oxidation and protection against rusting and corrosion.

Multipurpose lithium grease : Multipurpose lithium grease is compounded with molybdenum, sulfide or other complexing agent like boric acid or azealic acid with improved thermal stability, long service life, water resistance, mechanical stability and drop point. IOC multipurpose lithium grease having molybdenum has extremely well anti-stuffing and provides protection to moving parts under severe shock load conditions. Another IOC is lithium base grease containing specialized extreme pressure additives having excellent shear stability, resistance to water washout and excellent antifriction properties.

Silicon grease : Silicon grease is amorphous fumed silica thickened, polysiloxane based compound, non-oil base grease and have good water resistance and can be used up to 120 °C. It has good resistance to both high temperature and oxidation, if combined with synthetic fluids. It is commonly used for lubricating rubber parts. Silicon greases prevent the rubber from drying out. It has good corrosion resistance.

Aluminum soap greases : Aluminum soap grease has an attractive, transparent, smooth and polished and are very sticky making them perfect for applications requiring surface lubrication appearance. It finds application for harshest applications such as steel mills, brick kilns, etc., they have extreme water and heat resistance. Drop point is low and not recommended above 80 °C. Aluminum soap grease has good water resistance, good adhesive properties and rust resistance.

Aluminum complex grease : Aluminum complex grease has good water and chemical resistance, but have shorter life in high temperature and high speed applications.

Barium soap grease : Approximately 350 °C or higher. Barium soap greases are general purpose grease, a valued for their applicability to work over a wide temperature range.

Polyurea greases : Polyurea is the most important organic non-soap thickener and is made by reacting amines with isocyanates in presence of a polyxyethylene/polypropylene glycol copolymer. It has good temperature properties with longer performance. The polyurea is a unique thickener in that its composition is also an antioxidant where conventional grease thicken acts as oxidation promoters. Polyurea complex grease is produced by incorporating complexing agent like calcium acetate or calcium phosphate in the polymer chain.

Clay greases : Finally divided clay particles of bentonite and hectorite are used as thickeners. Clay grease has excellent temperature and water resistance, but requires additives for oxidation and rust resistance.

Carbon black grease : Carbon black greases are made by using acetylene black and are suitable for high temperature environment and can be used up to 200 °C. Carbon black greases are the most heat resistant and are use in high temperature applications.

Indanthrene grease : Indanthrene grease is made from indanthrene blue and silicon fluid. This grease is used at extreme temperatures for rolling bearings and other high temperature applications. It has good water resistance, good antioxidant properties and good mechanical property.

Phthalocyanines grease : Phthalocyanines grease has high resistance and uses copper phthalocyanines.

Sodium octadecyl terephthalamate grease : Sodium octadecyl terephthalamate grease has very high temperature properties, high water resistance, and good mechanical stability.

Fluoroether grease : Fluoro polymers containing C–O–C (ether) bonds for flexibility are soft, often used as greases due to its inertness to solvents, acids, bases and oxidizers.

| Graphite grease | : | Graphite grease is a calcium base grease containing graphite filler. It is used for general lubrication under comparatively high load and low relative displacement of interacting surfaces. It is recommended for leaf springs, hydraulic rams, plungers, slides, elevator cables, pantograph pans, as protective for steel wire ropes [apar.com]. |
| Cement plant grease | : | Typical cement plant grease is sprayable adhesive lubricant with high quality colloidal graphite. It is suitable for very severe lubrication conditions encountered in very low speed and large open gear. |

13.2 SELECTION OF GREASE

The proper choice and selection of greases is very important for better performance and long life of equipments. First step is to characterize the equipment and plant conditions. Some of the operational characteristics to be considered are size and type, load, speed, atmosphere, temperature, moisture, airborne contaminants [Johnson, 2002]. The most significant factors in selecting the proper grease for bearing lubrication are machine type, bearing type and size, operating temperature, operational load conditions, speed range, operating conditions (such as vibration and horizontal/vertical operation of the shaft), cooling conditions, sealing efficiency and external environment [www.machinerylubrication.com, Snyder, 2005].

13.3 MANUFACTURE OF GREASE

The process of grease manufacture involves: Saponification for making base thickening soap, dehydration and base oil addition, cooling, additive addition, finishing, removal of entrained air and packaging. Lubricating greases are generally made in batch process, however, continuous processes are also being used for large scale production.

Saponification is the first step in the manufacture of grease and the process consists of saponification of animal fat or vegetable oil with a metal hydroxide (calcium hydroxide, caustic soda, lithium hydroxide, barium hydroxide, etc.) and a small portion of base oil in a kettle or in autoclave. Excess water is evaporated off. After saponification the charge is transfer to a mixing kettle where remaining oil is added to the soap oil mixture, and cooling with or without stirring depending on type of grease. The rate of cooling and amount of agitation greatly affect the grease fiber formed while making sodium soap grease. In order to secure desired structure proper cooling is essential. The grease is usually then homogenized or milled to improve dispersion of soap fiber in the oil and smoothness and uniformity of the product. The lubricating value of the grease is chiefly dependent on the quality and viscosity of base oil [Speight, 1999].

13.4 ASTM TESTS FOR GREASE CHARACTERISTICS

Some of the commonly tests for evaluating the grease characteristic and performance are apparent viscosity/pumpability, consistency and shear stability, corrosion and rust resistance, dropping point, evaporation, heat resistance/consistency, leakage, oxidation stability, water resistance, wear resistance, rust/corrosion resistance.

REFERENCES

1. hhttp://en.wikipedia.org/wiki/siliconegrease.
2. http://www.iocl.com/products lubes 10.aspx.
3. http://www.iocl.com/products lubes 10.aspx, http://www.reliabilityweb.com, NLGI 99.
4. http://www.iocl.com/products lubes 8.aspx.
5. Johnson M. Selecting a General Purpose Grease without Compromising Performance. Machinery Lubrication Magzine, Jan 2002.
6. Kumar A. Selecting the Right Lubricating Grease. Chemical Industry Digest, April 2014, p73.
7. Mead HB. Grease in Modern Petroleum Technology, Vol 2. Downstream, edited by Lucas AG, 2001, p397.
8. NLGI, 99. Chapter 5 grease NLGI EM1110-2-1424, Feb 28, 1995, p3.
9. Snyder DR. Proper Lubrication Keeps Bearing Rolling Along. Machinery Lubrication Magazine, May 2005.
10. Speight JG. The Chemistry and Technology of Petroleum, Marcel Dekker Inc, New York, 3rd edition, 1999, p777.
11. Spicer FEH. Lubricating Grease in Modern Petroleum Technology, edited by Hobson GD, 1986, p1009.

Hydroprocessing, Sulfur Removal and Sulfur Recovery

Sulfur compounds (hydrogen sulfide, mercaptans, disulfides, thiophenes) are inherently present in the natural gas and crude oil and their presence even in small quantities is undesirable and must be removed to ppm level in order to avoid catalyst poisoning in various conversion processes, reducing corrosion problem and meet the environmental standards which are becoming more and more stringent during the last few years, the so called bottom of barrel upgrading has resulted in deeper sulfurization of heavier fractions [CEW, Feb 2004]. Consequently, larger amount of hydrogen sulfide has to be processed. Sulfur recovery has been an area of great interest and intense development during recent years. Major sources of sulfur compounds are gas and liquid sweetening and sour water stripping section. The level of sulfur in crude oil has steadily increased during last two decades due to increasd use of heavier crude sulfur content. Future refining configuration and cost of gasoline, diesel and other fuel will be influenced by sulfur content of crude and required sulfur specs as now refineries are preparing for ultralow sulfur gasoline and ultralow sulfur diesel. There are thousands of sulfur recovery units worldwide converting hydrogen sulfide into elemental sulfur [Heisel and Rameshni, 2011].

Sulfur in crude oil varies considerable. Sulfur content in some crude oil is given below [Goel et al., 2008]:

Crude oil	S (%wt)
Mumbai High	0.12
Dubai	1.8
Murban	1.8
Al shaheen	2.2
Arab Mix	2.8

Sulfur compounds present in the fuel result in toxicity, corrosion, inhibition of catalytic conversion processes, inhibit antioxidant performance. The sulfur content in the crude oil varies widely depending upon source and quality of crude being processed. Major sulfur, nitrogen and metals present in the hydrocarbon are given below.

- **Sulfur compounds:**Thiphenol, ethyl mercaptan, diethyl sulfide, biphenyl sulfide, 3-methyl-1-butanethiol, diethylsulfide, dipropylsulfide, di-isomyl sulfide, thiophenes.
- **Nitrogen compounds:**Pyridine, quinoline, isoquinoline, pyrrole, indole, carbazole, alkylamines, aniline.
- **Metals:** Nickel, vanadium.

Sulfur compounds are inherently present in the natural gas and crude oil. Emission of sulfur has been one of the major environmental issues during recent years and petroleum industries have been forced to incur extremely high costs for environmental compliance to meet the sulfur emission standards which is become more and more stringent. Sulfur removal and recovery process in a refinery involves following operations to meet above objectives are hydrotreating, desulfurization, amine absorption of H_2S, merox sweetening and sulfur recovery by Claus process, modified Claus process.

Due to increasing environmental concerns, stringent limits on sulfur levels in fuel are being implemented all over the world to achieve sulfur standards below 100 ppm from 330–350 ppm sulfur. Deep hydrodesulfurization is required which is an additional capital cost as well as an energy intensive step [Ultradeep sulfur diesel by oxidative desulfurization of HDS diesel, Garg, 2010]. Sulfur output in refinery takes place as one of the following [Goel et al., 2008].

- Sulfur content in finished product
- Sulfur emission into atmosphere in the form of SO_2
- Sulfur recovery in sulfur recovery unit.

Typical sulfur distributions in a refinery are [Goel et al., 2008]:
- Sulfur in various products (58%)
- Product sulfur (41%)
- Sulfur emission (1%).

14.1 HYDROTREATMENT PROCESSES

Hydrotreatment of the various streams from refinery and petrochemical industries has become integral part in order to meet the feed standards of various processes in order to avoid catalyst poisoning, improving quality of products and meet the environmental standards. Hydroprocessing has emerged as the pivotal technology in the modern refinery in order to meet the challenges arising because of consistently increasing impurities present in the crude oil due to processing of heavier feedstocks which may have adverse impact on the catalyst and product quality and for meeting the stringent environmental standards of the fuel and lubricants.

Hydroprocessing technologies consist of any one of the following processes:
- Pretreatment (hydrotreatment) of naphtha and gas oil, residue for catalytic reforming, catalytic cracking and hydrocracking in order to remove the impurities like sulfur, nitrogen, heavy metal, etc. Sources and recovery of sulfur are shown in Fig. 14.1.
- Hydrocracking processes
- Hydrotreatment of the fuels and lubricants

Hydrotreatment of naphtha, gas oil and residue for catalytic reforming, catalytic cracking and hydrocracking processes are given below. These processes have been discussed separately in the Chapters catalytic reforming, catalytic cracking and hydrocracking.

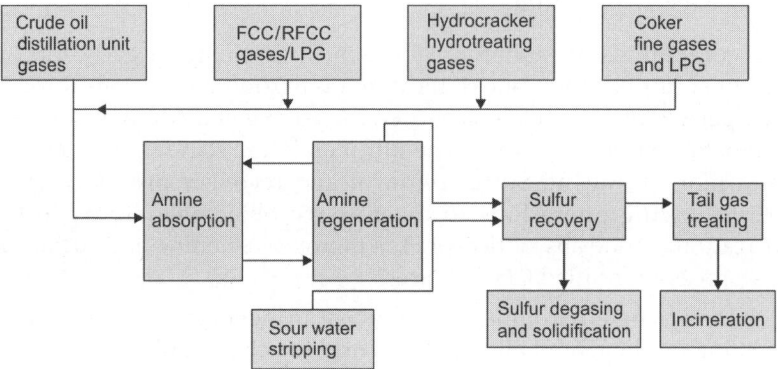

Fig. 14.1 Major sources of sulfur and recovery processes in refinery

Hydrotreating of the fuels and lubricants are used to reduce the impurities and improve the quality of fuel products (FCC gasoline, kerosene, diesel), and lubricating oil. Some of the major hydrotreating processes for improving the quality of fuel and lubricating are given in Table 14.1.

Some of the major hydrotreating processes in refinery are given in Figs. 14.2 and 14.3.

Table 14.1   Hydrotreating processes for improvement of quality of product stream

Hydrotreating process	Improvement
FCC gasoline hydrotreatment	Reduces sulfur content and produces ultralow sulfur FCC gasoline with maximum retention of olefins and octane.
Kerosene hydrotreating	Reduces mercaptans, sulfur, aromatics and improves smoke point
Diesel hydrotreatment	Reduces mercaptans, sulfur, aromatics and improves cetane number, thermal stability
Lube hydrotreatment	Reduces sulfur, conradson carbon and improves color, oxidation stability

Hydrotreating process	Improvement
Hydrotreating of naphtha for catalytic reforming	Removal of sulfur, nitrogen, metals
Hydrotreating of vacuum gas oil for FCC	Removal of sulfur, nitrogen, metals
Hydrotreating of pyrolysis gasoline	Removal of sulfur, nitrogen, metals

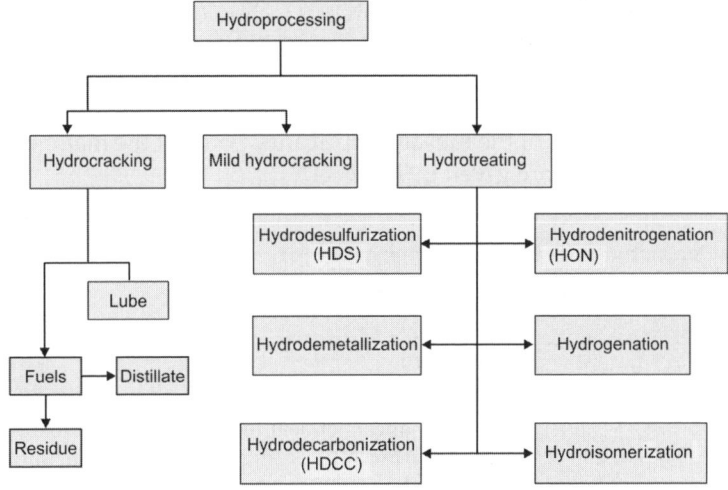

Fig. 14.2 Hydrotreating processes in petroleum refinery

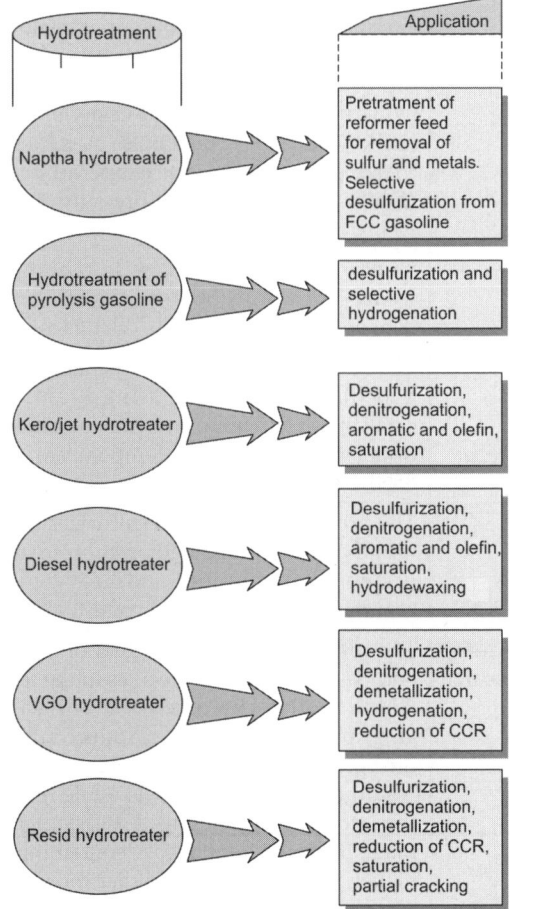

Fig. 14.3 Hydrotreatment processes in refinery

Hydroprocessing has emerged as the pivotal technology in the modern refinery in order to meet the challenges arising because of consistently increasing impurities present in the crude oil due to processing of heavier feedstocks which may have adverse impact on the catalyst and product quality and for meeting the stringent environmental standards of the fuel and lubricants. Some of the major commercial hydrotreating processes are given in Table 14.2.

Table 14.2   Various hydrotreating processes

Process	Description
UOP unionfining	UOP unionfining process involves hydrotreatment of petroleum and chemical feedstocks. The reaction involved hydrodesulfurization, hydrodenitrification, saturation of olefins, and saturation of aromatics. Catalyst CoO or NiO 1–6 %wt, MoO_3 6–25 %wt, Al_2O_3 balance.
UOP RCD unionfining	RCD unionfining reduces the sulfur, nitrogen, conradson carbon, asphaltene, and organometallic contents of heavier residue derived feedstocks to allow them to be used as fuel oils or as feedstocks for downstream processing units such as hydrocracker, fluidized catalytic cracker, resid catalytic cracker and coker. The process uses fixed bed catalytic reactor that operates at moderate temperatures and under moderate to high hydrogen partial pressure. SOR temperature, 330–355 °C.
UOP SRC uniflex process	UOP SRC uniflex process is a high conversion residue hydroprocessing process for the production of gas oil conversion unit feedstock, distillates and naphtha.
UOP unisar process for saturation of aromatics	The process saturates the aromatics in naphtha, kerosene, and diesel feedstocks using highly active noble-metal catalysts on either an amorphous or molecular sieve support under mild conditions (temperature, 205–270 °C; pressure 3500–8275 kPa; space velocity, 1.0–5.0 vol/vol. hour).
CDHydro and CDHDS (CDTECH)	The process selectively desulfurizes the FCC gasoline with minimum octane loss with maximum retention of olefins and octane. The light, mid and heavy cut naphtha is treated separately, under optimal conditions.
RDS/VRDS hydrotreating process (Chevron Research Technology Co)	The process upgrades residual oil by removing impurities and cracking heavy molecules in the feed to produce lighter product oils. The process involves hydrotreatment of atmospheric or vacuum residuum feedstocks to reduce sulfur, nitrogen, metals and asphaltene and carbon residue contents resulting in lighter products while reducing the viscosity of the unconverted bottoms. Catalyst: Small, extruded pellets made from an alumina base containing active metals cobalt, nickel, molybdenum and other more proprietary materials. Size, 0.8–1.3 mm.

Contd.

Table 14.2   Various hydrotreating processes *(Contd.)*

Process	Description
Chevron Lummus Global Ebullated Bed Bottom of the barrel Hydroconversion (LC-fining) process	LC-fining processes heavy oil feeds are hydrogenated and converted to a wide spectrum of lighter, more valuable product such as naphtha, light and middle distillates, and atmospheric and vacuum gas oils.
Exxon Hydrotreating Technology	Exxon hydrotreating technology processes wide variety of feedstocks for removal of sulfur, nitrogen, metals, saturation of olefins, diolefins, aromatics and improvement of product quality. **Hydrofining:** Naphtha and distillate hydrotreating. **DODD:** Diesel oil deep desulfurization **GO-fining:** Gas oil hydrotreating **Resid-fining:** Atmospheric or vacuum reside hydrotreating **Hydrofining** is ideally suited for the production of reformulated gasoline. GO-fining and resid-fining are suited for the treatment of feeds to catalytic cracking for removal of conradson carbon, metals, sulfur and nitrogen and saturation of multiring aromatics. Some of the recent developments to meet all current and possible future premium diesel requirements are ULSD HDS: Ultradeep hydrodesulfurization, HDHC: Heavy distillate mild hydrocracking, MAXSAT: High activity aromatic saturation process, CPI: Diesel cloud point improvement.
Haldor Topsoe hydrotreating process	The process has a wide range of applications, including the purification of naphtha, distillates, and residue, as well as the deep desulfurization and color improvement of diesel fuel and the pretreatment of FCC and hydrocracker feedstocks. Ultralow sulfur diesel (ULSD) is designed to produce diesel 5–50 wppm using special catalyst. Both low pressure reactor and high pressure reactor configuration are available.
Howe-baker Engineers	The process is used for the reduction of sulfur, nitrogen, and metals content of naphtha, kerosene, diesel or gas oil streams.
IFP Hyvahl process	The process upgrades or converts high metal atmospheric or vacuum residues using fixed bed and dual catalyst hydrotreating process.
Axen ultralow sulfur diesel (ULSD)	Produces ultralow sulfur diesel and high cetane and improved color diesel from wide range of distillate.

Source: Refining Processes, 1998 [2008; Brossard, 2004; Kokayrff, 2004; Gowdy, 2004; Gills, 2004; Gupta, 2004].

14.2 HYDROTREATING REACTIONS

Major reactions taking place in the hydrotreating reactions are:

- Desulfurization process where sulfur compounds (mercaptans, sulfide, disulfide, thiophenes, and benzothiophenes, etc.) are converted to hydrogen sulfide using Ni/Co and molybdenum catalyst.
- Denitrification in which organic nitrogen compounds (3-methylpyridine, quinoline, 3-methylisoquinoline, pyrrole, indole, carbazole) are converted to ammonia.
- Conversion of organic oxygen compounds (naphthenic acids) to water
- Conversion of halides to hydrogen halides
- Demetallization removal of organometallic compounds
- Saturation of olefins
- Saturation of aromatics.

Main reaction involved in desulfurization is the removal of sulfur compounds in form of H_2S. Degree of desulfurization varies from feed to feed with nearly complete removal to about 50–70% for heavier residual materials.

Relative desulfurization reactivity is in order of increasing difficulty is given below [Kokayeff, 2003].

Thiphenol > ethyl mercaptan > diethyl sulfide > diphenyl sulfide > 3-methyl-1-butanethiol > diethylsulfide > dipropylsulfide > diisomyl sulfide > thiophene

Sulfur species reactivity is given in Table 14.3.

Table 14.3   Sulfur species reactivity

Sulfur compound	Relative reaction rate	Boiling point (°F)
Thiophene	100	185
Benzothiophene	50	430
Dibenzothiophene	30	590
Methyl dibenzothiophene	5	600–620
Dimethyl dibenzothiophene	1	630–650
Trimethyl dibenzothiophene	1	660–680

Source: Nalamura, 2004.

14.3 NAPHTHA HYDROTREATING

Pretreatment of naphtha in presence of hydrogen is one of the major steps in the catalytic reforming. The naphtha hydrotreatment process eliminates the impurities such as sulfur, nitrogen, halogens, oxygen, water, olefins, diolefins, arsenic and other metals presents in the naphtha feedstock to have longer life (See Chapter Catalytic Reforming).

14.4 DIESEL HYDRODESULFURIZATION (DHDS)

Nowadays, more and more stringent specifications are imposed upon sulfur content of diesel produced by refineries. The DHDS unit is extensively used in petroleum refining to reduce sulfur content in the diesel and produce diesel as per standard. The diesel/gas oil from following unit is treated for removal of sulfur.

1. Straight run gas oil
2. Vacuum diesel
3. Visbreaker gas oil
4. Total cycle oil.

Process: Feedstocks are blended from various sources, straight run or cracked products. Sulfur and nitrogen contents depend upon the crude. Cracked products are characterized by the presence of unsaturated hydrocarbons (olefins, diolefins and aromatics). Following reactions take place during hydrodetreatment:

- Conversion of organic sulfur compounds to hydrogen sulfide
- Conversion of organic nitrogen to ammonia
- Conversion of organic oxygen compounds to water
- Saturation of olefins
- Conversion of halides to hydrogen halides
- Removal of organometallic compounds

The purpose of the section is to reduce 90% sulfur in feed diesel using hydrogen from catalytic reformer or hydrogen generation units. The required level of desulfurization is achieved by hydrotreating over a specially selected catalyst. The presence of olefins or diolefins called for additional bed installed in upstream of the desulfurization bed. The DHDS section consists of:

Reaction section: It consists of fresh feed pretreatment, feed preheat, make-up hydrogen system, recycle hydrogen section, recycle gas scrubbing, reactor containing multiple catalyst bed, quench section, reactor effluent cooling and water wash.

Fractionation section: For separation of sour gas and naphtha from diesel product. Typical flow diagram of ultralow sulfur DHDS is given in Fig. 14.4. Characteristics of feed before and after desulfurization in typical DHDS is given in Table 14.4.

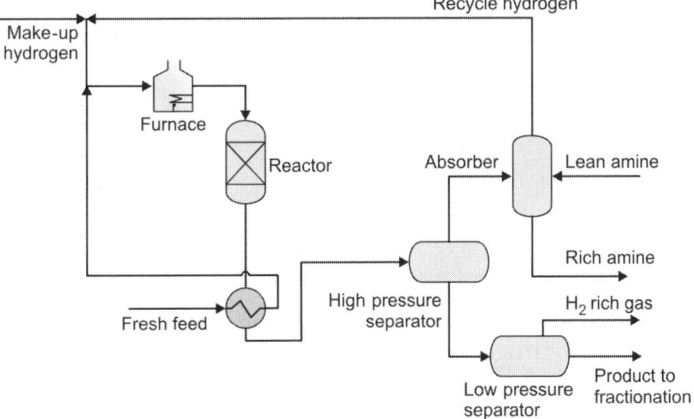

Fig. 14.4 Block diagram of ULSD DHDS unit

Source: Refining Process, 2008; Hydrocarbon Processing, Sep 2008.

Table 14.4   Typical characteristics of feed before and after desulfurization in a refinery

Properties	Feedstock components						
	GO (Plt 1)	GO (Plt 15)	LVGO (Plt 15)	SPO blend	LCO-FCCU	HCO-FCCU	Blended feedstock
Sp. gravity	0.8532	0.8572	0.8929	0.9016	0.8055	0.9205	0.869
Total sulfur (%wt)	1.5	1.5	2.6	2.1	0.4	3.3	1.86
Bromine no. (g/100 g)	< 2	< 2	< 2	< 2	46	22	8
Metals (ppm, wt)							< 1
Cetane index ASTM D976	52.6	52	44.9	45	22.3	31.9	48.8
Flash point (°C)							36*
Aniline point (°C)	75.5	74.5	71	72	28.5*	28.5*	
Nitrogen (ppm, wt)	300*	300*	300*	400*	700*	700*	402

* Values estimated as basis.
Courtesy: CPCL, Chennai.

Properties	Diesel product	
Start-of-run	End-of-run	
Sp. gravity	0.852	0.852
ASTM D86 50% (°C)	306	308
ASTM D86 90% (°C)	378	378
Sulfur (wt, ppm)	2,000 max.	2,000 max.
Water content (wt, ppm)	< 135	< 135
Nitrogen (wt, ppm)	285 max.	285 max.
Flash point (°C)	At least identical to feed	
Cetane index ASTM D976	52.9	52.9
Bromine number (g/100 g)	< 1	< 1
Pour point (°C)	Same as feed	
Color	Same as feed	

Courtesy: CPCL, Chennai.

14.4.1 Amine Treatment and Amine Recovery

Amine treatment is one of the very important sections in the refinery which is designed to remove hydrogen sulfide (H_2S) from gaseous hydrocarbon effluents.

Capacity DEA < MEA < MDEA < DGA

Effectiveness of the various gas treating amines in the removal of COS.

MDEA < DEA < MEA < DGA

Increasing COS removal

Solvent MEA < DGA < DEA < MDEA

MDEA < DEA < MEA < DGA

Decreasing bar strength

MEA < DEA < DGA < MDEA

Increasing solubility of hydrocarbons

H_2S removal from gaseous hydrocarbons effluents is achieved by means of a continuous absorption/regeneration process using a 25 %wt diethanol amine (DEA) for H_2S removal.

14.4.1.1 Amine Absorption

The various gaseous streams containing hydrogen sulfide is collected in header system from where it is processed in amine unit for removal of H_2S. The sour gases after cooling fuel gas amine absorber feed cooler is passed to fuel gas amine absorber knockout drum for removal of any liquid streams remaining in the gaseous stream. The gaseous stream is then passed to amine absorption unit where it is scrubbed with lean amine obtained from amine regeneration unit to remove H_2S. Rich amine from bottom of the amine absorber is sent to amine recovery unit. The sweetened gas from the amine absorber from the top of the absorber goes to knockout drum for removal of any entrained amine. And finally sent to refinery gas stream.

$$2RNH_2 + H_2S \longrightarrow (RNH_3)_2S$$
$$RNH_2 + H_2S \longrightarrow R_2NH_3S$$

14.4.1.2 Amine Recovery Unit

The rich amine obtained from amine absorption unit is processed in amine recovery unit for regeneration of lean amine to reuse in amine absorption column. Reactions involved in regeneration are:

$$(RNH_3)_2S \longrightarrow 2RNH_2 + H_2S$$
$$R_2NH_3S \longrightarrow RNH_2 + H_2S$$

Rich amine from amine absorber is first sent to rich amine flash drum for removal of any entrained gas and hydrocarbon then sent to amine strippers. The rich amine is heated by LP steam in the reboiler. Amine stripper consists of valve trays. Steam strips off the H_2S and CO_2 present in the amine solution. Stripping steam generated in the amine stripper reboiler. The liberated sour gases leave the regenerator from top and enters overhead condenser. The sour gases are sent to sulfur recovery unit for production of sulfur. The lean amine is recycled. Figure 14.5 illustrates amine absorption process.

Operating condition in the regenerator: Top temperature 105 °C and middle temperature 115 °C.

Cost comparison of the installation of membrane and amine/glycol plant is shown in Fig. 14.6 and Table 14.5. Cost basis of plant is given in Table 14.5.

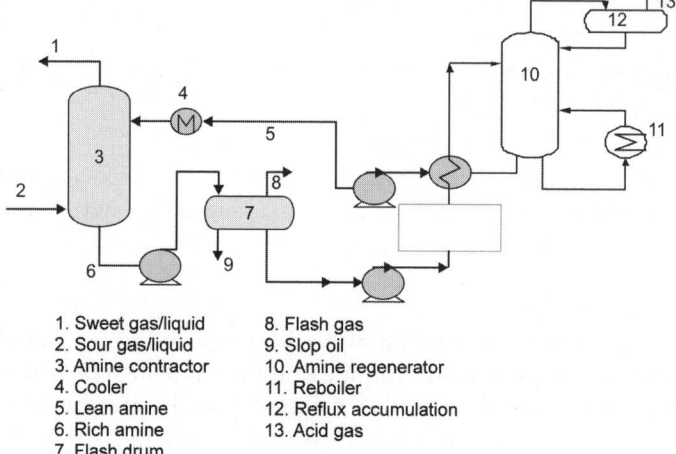

1. Sweet gas/liquid
2. Sour gas/liquid
3. Amine contractor
4. Cooler
5. Lean amine
6. Rich amine
7. Flash drum

8. Flash gas
9. Slop oil
10. Amine regenerator
11. Reboiler
12. Reflux accumulation
13. Acid gas

Fig. 14.5 Amine absorption unit

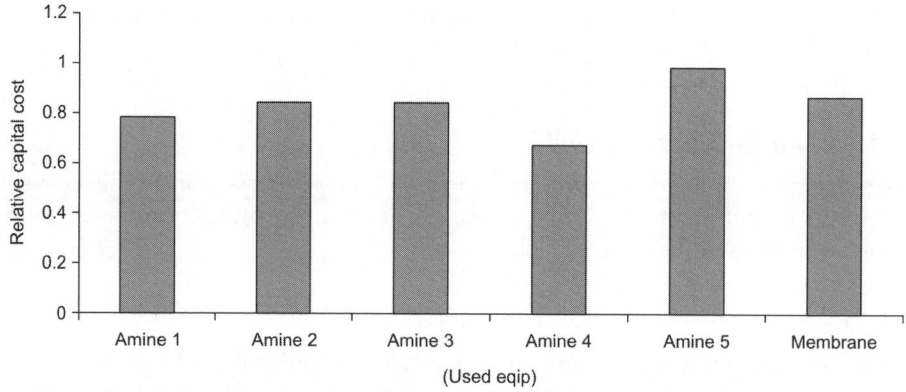

Fig. 14.6 Comparison of relative installed capital cost

Table 14.5   Cost comparison of amine process with membrane process

Basis: Natural gas sweetening unit of 37 MMSCF.

CO_2 content in feed (%)	5	10	15
Amine			
Capital (millions of $)	3.35	4.54	5.45
Expense (millions of $)	1.22	1.81	2.33
Lost product (millions of $)	0.02	0.04	0.07
Capital charge (millions of $)	0.91	1.23	1.48
Processing cost ($ per MSCF feed)	0.17	0.24	0.3

Contd.

Table 14.5   Cost comparison of amine process with membrane process *(Contd.)*

Membrane (multistage)			
Capital (millions of $)	1.86	3.33	3.87
Expense (millions of $)	0.53	0.85	0.97
Lost product (millions of $)	0.43	0.69	0.93
Capital charge (millions of $)	0.51	0.9	1.05
Processing cost ($ per MSCF feed)	0.11	0.19	0.23
	DEA amine	Membrane	Membrane/ DEA
Relative capital cost	1.0	0.26	0.72
Relative operating cost	1.0	1.51	1.14
Relative net present cost @ 15%	1.0	0.76	0.89

Source: Joshi et al., 2010.

The above comparison table shows that the various cost parameters including the gas processing cost is lower for a membrane process is that its capacity can easily be increased or decreased by changing the number of modules in the skid mounted structure.

14.5 SWEETENING PROCESSES

Some of the sweetening processes are given in Table 14.6.

Table 14.6   Sweetening processes

Sweetening processes	Sweetening process involves removal of mercaptans in two steps: Transform mercaptans into mercaptides—caustic and oxidize mercaptides to disulfide.
Caustic wash	Removes H_2S, mercaptans, thiophenes, thiophenols and naphthenic acids. Mercaptan solubility decreases with increase in mol wt.
Copper chloride process	Direct oxidation of mercaptans to disulfide in presence of cupric chloride catalyst. Cu_2Cl_2 + mercaptan $\longrightarrow$ oil-soluble products
Doctor treatment	The process involves treatment of the sour distillates with alkaline sodium plum bite (doctor solution) in presence of small amount of free sulfur.
Chelate sweetening (MOBIL process for gasoline)	The process involves extraction of mercaptans by NaOH followed by oxidation of mercaptan by oxygen supplied by chelate. RSH + NaOH $\longrightarrow$ RSNA + NaOH $RSNa + \dfrac{1}{2}O_2 + H_2O \longrightarrow$ RSSR + NaOH Chelate is a reaction product of N, N'-disalicylidine-1, 2 propane diamine and cobaltous chloride hexahydrate.

Contd.

Table 14.6   Sweetening processes *(Contd.)*

Mercapfining	Meracptans are converted to disulfide in fixed bed reactor. $$\text{Mercaptans} \xrightarrow[\text{catalyst}]{O_2} \text{disulfide}$$
Inhibitor sweetening	The process uses a phenylenediamine type inhibitor, air and caustic to sweeten low mercaptan content gasoline.
Merrichem Technology (Thiolex, Mericat)	The process involves removal of mercaptan by caustic extraction followed by catalytic oxidation in presence of metal phtalocyanines oxidation temperature 30–50 °C.
Sulfuric acid treating	This process uses weak (80%) or concentrated sulfuric acid (> 100%) for removal of sulfur compounds. During the treatment asphaltene is also removed. Sulfuric acid has been also used earlier for treatment of kerosene and lubricating oils.
Nalfining process	The process uses acetic anhydride and a caustic rinse to convert contaminants into less objectionable, but oil-soluble compounds.
INDE treat and IDE sweet H_2S/mercaptan removal and sweetening Process Technologies	The process uses Continuous Film Contractor (CFC). It can remove H_2S from LPG, naphtha and gasoline, mercaptans from LPG, naphtha, gasoline, gasoline and ATF/kerosene, naphthenic acids from diesel, acid gases from natural gas, fuel gas [www.iocltech.com].
Solutizer process	The process involves extraction of mercaptans in alkaline solution (KOH, NaOH) in presence of promoters (phenols, carboxylic acids, etc. extractive solution. Solulized is mixture of KOH and tricresols in water. **Dualayer (Mangolia Petroleum Co):** Two-phase mix of cresols in KOH (43%). **Mercasol (Pure Oil Co):** Mixed cresols in NaOH. Single phase. **Shell solutizer:** Potassium isobutyrate in KOH. **Tannin solutizer (Mobil & Shell):** One percent tannin in solutizer solution **UNISOL (Atlantic Refining Co):** Ethanol in 450Be caustic)
Mercapsol process	Process for extraction of mercaptans using sodium (or potassium) hydroxide, together with cresols, naphthenic acids and phenol
Merox process	The process involves either mercaptan extraction or sweetening
Caustic Free Merox	Mercaptans are catalytically converted to disulfides which remain in the hydrocarbon feed using preimpregnated fixed bed catalysts. Merox no. 21 catalyst for gasoline and Merox-31 catalyst for kerosene. Liquid activator, Merox CF provides an active, selective, and stable sweetening environment in the reactor [www.accessengineeringlibrary.com].

Contd.

Table 14.6   Sweetening processes *(Contd.)*

Minalk process	Minalk process is suitable for light and heavy gasoline from catalytic cracking and characterized by minimum use of alkali (3% wt. dilute NaOH). The process involves mixing of gasoline with dilute caustic which goes to fixed bed reactor containing activated carbon impregnated with merox catalyst reactor. Temperature 40–50 °C and pressure 8–20 bar.
Kerox process	Kerox process is a fixed bed sweetening process suitable for feeds boiling > 140 °C containing high molecular weight and branched mercaptans (e.g. kerosene).
HS process	HS process is for selective removal of H_2S from natural gas. The HS process technology combines three principal elements: A formulated selective chemical solvent, a unique contactor design and a specially designed selective contractor tray. The solvent is MDEA based solvent. The contractor is multistage.
Konox process	The process new approach to iron oxide process and consists of absorption of the hydrogen sulfide with a strong oxidizing agent and regeneration of the reacted agent to produce elemental sulfur.
Thiuolex/regen	The process extracts H_2S, COS and mercaptans from gases and light liquid streams, including gasoline, with caustic using fiber fil contractor technology. It can be also used to hydrolyze COS contained in LPG.
LO-CAT process	Catalytic sweetening process scrubs H_2S out of gas streams and converts it to solid.
IIP-BPCL LPG sweetening process	The process uses cobalt phthalocyanine sulfonamide catalyst for sweetening of LPG.

Source: Basu, 2005; Gatan, 2004; Marty C. White Products Refining Sweetening. IFP, Daziables, 2004; Speight, 1999; Cabodi and Hardison, 1982; Sigmund et al., 1981; Kasai, 1975; Oil and Gas Journal, Aug 12, 1985, p67; Gas Processes, 2006, p67; Heguy and Nagl, 2003; Rathore et al., 2011.

14.6 MEROX PROCESS

Merox process is the world's most widely applied refining technology and today more than 2000 Merox units are in operation. The process efficiently and economically treats petroleum fractions to remove mercaptan sulfur (Merox extraction) or to convert mercaptan sulfur to less objectionable disulfides (Merox sweetening). Merox process can be used to treat wide range of liquids such as LPG, natural gas liquid (NGL), naphthas, gasoline, kerosene, jet fuels and heating oils. It can be also used to treat gases such as natural gas, refinery gas and synthesis gas in conjunction with conventional pretreatment and posttreatment processes [UOPhttp://www.UOP.com/refining/1110.html.].

Mercaptan is a predominant sulfur compound present in many refinery products and has unpleasant smell and affects fuel stability with initiation of gum formation, and corrosive environment to form SO_x during combustion, if not removed.

Mercaptan malodorous sulfur compounds are formed highly corrosive compounds as condensation. Merox process is commonly used in the refinery which either removes the mercaptans or controls it to less harmful disulfides. Merox process involves two stages: Merox extraction and Merox sweetening, these processes can be used singly or in combination. Process diagrams are shown in Figs 14.7 and 14.8.

In general merox treatment can be used in following ways [Ddziabis, 2003]:

- For improving lead susceptibility of light gasolines
- For improving the response of gasoline stocks to oxidation inhibitors added to prevent gum formation during storage
- For improving odor of all stocks
- For reducing the mercaptan content to meet product specifications
- For reducing the sulfur content of LPG and light naphtha products
- To reduce sulfur content of coker FCC olefins to save acid consumption in alkylation.

Sweetening process involves removal of mercaptans in two steps: (1) transform mercaptans into mercaptides—caustic and (2) oxidize mercaptides to disulfide.

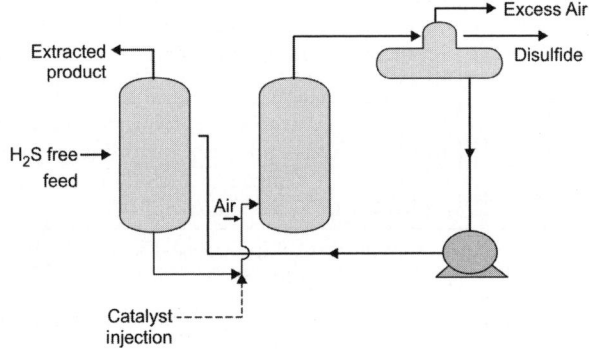

Fig. 14.7 Merox extraction/sweetening

Source: Hydrocarbon Processing, Jan 2000, p69.

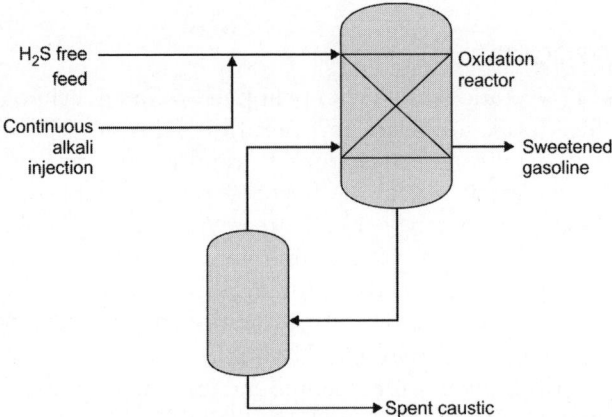

Fig. 14.8 UOP Merox process

14.6.1 Merox Extraction

Merox extraction involves counter current extraction of mercaptan in extraction column using caustic soda.

14.6.2 Merox Sweetening

In Merox sweetening, mercaptans are directly oxidized to disulfides in presence of catalyst and remaining the product, total sulfur remaining same in the product stream.

Caustic extraction reaction:

$$RSH + NaOH \longrightarrow NaSR + H_2O$$

Regeneration: $4NaSR + O_2 + H_2O \longrightarrow 2RSSR + 4NaOH$

$$RSH + 1/2\,O_2 \longrightarrow 1/2\,RSSR + 1/2\,H_2O$$

$$2CH_3SH + 1/2\,O_2 \longrightarrow 1/2\,CH_3SSCH_3 + 1/2\,H_2O$$

Merox process is used for treatment of LPG, gasoline, naphtha, kerosene, diesel. In Merox process, these fuels are treated in merox unit with ethanol amine and caustic where mercaptans are removed and the mercaptans are converted to disulfides in presence of Merox catalyst. Description of Merox Process is given in Table 14.7.

Table 14.7 Merox process and description

Merox processes	Description
Straight run and cracked LPG merox	Straight run LPG directly from atmospheric column and cracked LPG is first treated in amine absorber from which the treated LPG is sent to prewash in caustic column where it is treated with 3% caustic and then the gas enters to an extractor from bottom through spargers where 10% caustic solution is used. The overhead gas goes to a caustic settler where caustic separated and the treated LPG goes to LPG storage vessel. Rich caustic is then sent to oxidizer where sodium mercaptide is oxidized to sodium hydroxide water and disulfide, the regenerated caustic is separated in the separator and recycled to reuse caustic which is separated.
FCC gasoline merox unit	FCC gasoline is first caustic washed to remove thiphenol and followed by oxidation of mercaptans to disulfide in reactor containing charcoal impregnated with merox catalyst. The treated gasoline goes to a collector which allows the separation of caustic aqueous phase which is sent to effluent treatment plant. Antioxidant is added to the treated gasoline and sent to storage.

Contd.

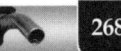

Table 14.7   Merox process and description *(Contd.)*

Merox processes	Description
Naphtha merox	Visbreaker naphtha is first prewashed with caustic and then sent to an extractor where it is treated with caustic for removal of mercaptans. The rich caustic after separation goes to caustic regeneration unit. Treated naphtha is mixed with air and sent to reactor containing charcoal impregnated with Merox catalyst where the mercaptans are oxidized to disulfide. The treated product is passed through sand filter and routed gasoline pool. The aqueous caustic phase separated in a collector is gone to effluent treatment plant.
Kerosene/ATF (kero/ATF) merox unit	**Kero/ATF** is first sent to coalescing media for separation of water droplets. Then the kero/ATF is sent to weak caustic (2–3%) prewash column from where it is sent to Merox reactor along with air. The reactor contains charcoal impregnated with Merox catalyst. The treated feed from the reactor goes to settler where caustic phase is separated from the hydrocarbon stream. The hydrocarbon is washed with demineralized water and then it goes to salt drier for removal traces of water and finally it goes to clay filters. The treated product is sent to storage tanks. During course of time the activated charcoal is regenerated by removing the organic and soap from the pores.

14.6.3 The ADIP Process

The Shell ADIP process alkalomines such as di-isopropaanolamine (DIPA), methyl di-ethanol amine (MDEA) or mixture of two or more are used for acid gas removal. The ADIP Process is suitable for removal of H_2S from natural gas and CO_2, synthesis gas and refinery gas streams, selective removal of H_2S from gases containing H_2S, removal of H_2S and COS from light liquid hydrocarbon streams (LPG, NGL) [JB Rajani treating Technologies of Shell Global Solutions for Natural Gas and Refinery Gas (mk@MSITstore:E:\congress.chem::/UMB1630.htm)].

    ADIP-X process is a regenerable amine process for acid gas removal by utilizing a mixture of two or more alkanol amines, in general a base amine such as MDEA and an accelerator. The process achieves larger loading capacity compared to single amine solvents.

14.7 ADVANCES IN DESULFURIZATION RESEARCH

Merox^{TM}: UPO Merox^{TM} is a sweetening process by extracting and/or converting mercaptan sulfur to less objectionable sulfide and is often used for treating LPG, naphtha, gasoline, kerosene, jet fuel and heating oils.

14.7.1 SCANfining

SCANfining developed and commercialized by Exxon Mobil Research and Engineering lowers sulfur with minimal olefin saturation/octane loss. The process removes sulfur of naphtha to less than 10 ppm with only small octane loss due to olefin saturation. The process uses a proprietary catalyst formulation from Akzo nobel [Sweed NH and Ellis ES. Meeting the Three Low Sulfur Mogas Challenge. Petroleum Technology, Quarterly, Autumn, 2001, 45].

SCANfining ^{TM}

Broad variety of processing options for full range or narrow boiling range feeds and uses conventional fixed bed hydrotreater equipmenmt meets < 100 ppm sulfur targets. First commercial application in an Exxon Mobil Refinery in 1995.

- *SCANfining I:* Single stage unit preferred for low sulfur feed (e.g. from FCC with hydrotrated feed). It is preferred for moderate to high levels of sulfur.
- *SCANfining II:* Two-stage unit with H$_2$S removal between stages, effective for high sulfur. This is preferred when very high HDS is required (with very high sulfur feeds or very low sulfur products) and octane loss must be minimized. It is more cost effective when the sulfur feed is high, the sulfur product is low and olefin feed is low [Sweed and Ellis, 2001].
- *SCANfining I with EXOMER:* SCANfining I followed by extraction of mercaptans, effective for high sulfur [Petrotech 2003, Delhi, India; Japan, 2003].

Octane loss increases as sulfur product decreases (as percent desulfurization increases) because of more severe operating conditions needed to get lower sulfur product are more severe for saturating olefins [Sweed and Ellis, 2001].

14.7.2 OCTGAIN^{SM} Technology

Deep hydrodesulfurization process for < 5 wppm S, complete octane retention with modest C$_5^+$ conversion to LPG. Attractive where C$_3$/C$_4$ has high value, very low olefins product content. Excellent sulfur feed flexibility, proven in high sulfur feed (> 1%).

Octgain hydrotreats the naphtha over an HDS catalyst to remove essentially all sulfur and intentionally saturate the olefins. The process is followed by cracking low octane paraffins to propane and butane, smaller paraffins in second reactor using zeolite based catalyst [Sweed and Ellis, 2001].

Choice of technology for meeting low sulfur mogas depends on:
- Configuration of the existing refinery
- Gasoline species for sulfur and olefins—content, new, future
- Availability of existing hardware
- Refinery octane balance
- Value of octane, LPG, C$_5^+$
- Cost availability of hydrogen, isobutane, olefins and steam

14.7.3 BSR Gasoline Hydrotreating Process

The process has been designed to produce ultralow sulfur gasoline of 30 ppm sulfur. The process is based on selective hydrogenation of FCC gasoline to olefin-rich light-cat gasoline and separation of the sulfur-rich heavy-cat gasoline. The process converts lighter mercaptans and light sulfides to heavier sulfur species and also saturate unstable dienes with no octane loss and minimum hydrogen consumption [Sanghavi and Schmidt, 2011].

14.7.4 INDE Treat and IDE Sweet

This is for H_2S/mercaptan removal and sweetening process technology. This process has been developed by Indian Oil Technologies Limited. The process uses Continuous Film Contractor (CFC). The process can remove H_2S from LPG, naphtha and gasoline, mercaptans from LPG, naphtha, gasoline, gasoline and ATF/Ker, naphthenic acids from diesel, acid gases from natural gas, fuel gas. The process is a low energy consumption process due to mixing at low pressure drop, efficient contacting, efficient separation of hydrocarbons and aqueous stream, higher hydrocarbon to caustic ratio so smaller caustic circulation, less settling time, hence smaller settler. Salient features of CFC are nondispersive contacting, enormous surface area, high mass transfer efficiency, based on caustic amine, efficient removal of contaminants, small size, non-aqueous phase entrainment, low caustic/amine consumption and low cost [www.iocltech.com].

Merox removal options: Removal of mercaptans from feedstocks and/or convert them to another form Merox is economically viable technology and inexpensive.

14.7.5 Super Sour Process

Stringent environmental regulations have necessitated higher recovery of H_2S from sour water stripper unit designs. Super sour process ensures minimum H_2S loss. The process employs additional hot feed flash drum upstream of cold feed surge drum. The H_2S-rich vapors from hot feed flash drum upstream of cold feed surge drum is routed to a small amine scrubber to absorb liberated H_2S. The H_2S lean gas primarily containing hydrocarbons is then routed to incinerator of the sulfur recovery unit. The absorbed H_2S-rich amine is recovered in the amine regenerator and is fed to the sulfur unit for converting it to sulfur [Sharma et al., 2011].

14.7.6 S Zorb Sulfur Removal Technology

S Zorb sulfur removal technology has been commercialized ConocoPhillips Petroleum Company for removal of sulfur from gasoline and diesel [Sughrue, 2004; Sughrue and Parsons, 2005]. The process uses fluidized bed reactor. The process produces ultralow sulfur gasoline and diesel.

14.8 SULFUR RECOVERY PROCESSES

The production of sour natural gas and the refining of higher sulfur content opportunity crude oil are increasing all over the world. The overall long-term trends to indicate an increase in sulfur content of crude oil, coal and natural gas. As the sulfur content in the feedstocks increases, the acid gas stream may contain up to a few percent of mercaptans as opposed to only a trace amount of mercaptans found in processing of low sulfur energy sources. The presence of mercaptans in the acid gas to the classes unit will increase and the air demand and heat duty of the unit [Chou et al., 1991]. Due to increasing environmental awareness, current air quality imposes stringent limitations on the emission of sulfur compounds; recovery of sulfur from hydrogen sulfide have become integral parts in refineries and fertilizer plants. The Claus process that was originally invented for the purpose of recovering sulfur during production of soda ash by Leblanc method is being used for conversion of hydrogen sulfide to elemental sulfur. H_2S is removed from the sour gases by amine treatment process. The Claus process consists of introducing a mixture of H_2S and air over a pretreated catalyst bed at constant temperature. The effluent gas is cooled and liquid sulfur is produced. Due to overall poor recovery of sulfur in the original Claus processes, several modifications of Claus process have been made with very high sulfur recovery up to 99.5%. Typical conventional Claus process is shown in Fig. 14.9. Reaction of Claus process is given below:

Reaction:

$$H_2S + \frac{3}{2}O_2 \longrightarrow SO_2 + H_2O$$

$$2H_2S + SO_2 \longrightarrow \frac{3}{2}S_2 + 2H_2O$$

$$H_2S + \frac{1}{2}O_2 \longrightarrow \frac{1}{2}S_2 + H_2O$$

$$2NH_3 + \frac{3}{2}O_2 \longrightarrow N_2 + 3H_2O$$

$$CH_4 + 2O_2 \longrightarrow CO_2 + 2H_2O$$

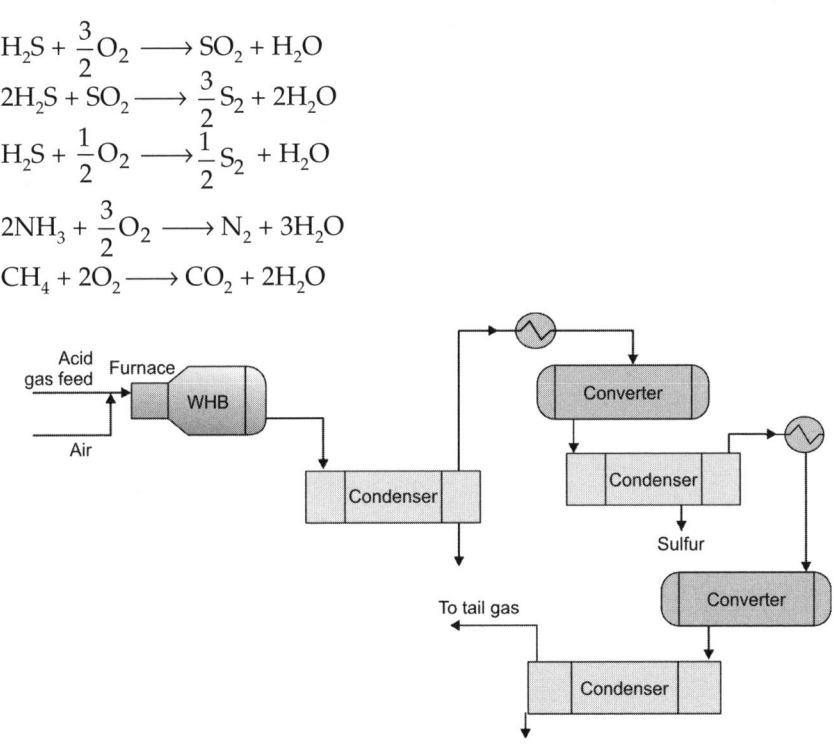

Fig. 14.9 Conventional Claus process

Sulfur recovery in traditional Claus plants varied from 90–96% for a two-stage plant and from 95–98% for a three-stage plant [Lagas et al., 1989]. Super Claus process was developed to overcome the traditional Claus process limitation and can be applied in either existing or new plants to increase cost effective sulfur recovery by adding a new active metal oxide catalyst in the last reactor of conventional clauss unit. Summary of sulfur recovery processes are given in Table 14.8.

Table 14.8　Sulfur recovery processes

Claus process	Description
Super Claus process	The process consists of thermal stage followed by three catalytic reaction stages with sulfur recovered between stages by condenser [www.hofung-technology.com]. Overall 99.5% sulfur recovery. Over 70 commercial plants have been erected.
Euroclaus process	This is an improvement of super Claus process with yields up to 99.7%.
SCOT (Shell Claus off-gas treating) process	The process is used to remove sulfur components from Claus plant tail gas to further reduce SO_2 emissions. *Stabdrad scot process:* Sulfur components in Claus tail gas are catalytically converted into H_2S which is selectively absorbed in amine solvent. *LS-scot process:* The process uses additive which improves the regeneration of the solvent with better H_2S removal efficiency. *Suoer scot process:* The process achieves a H_2S concentration of 10 ppm or a total sulfur content of less than 50 ppmv and is based on two-stage stripping, optionally with a lower lean solvent temperature [www.hofung-technology.com].
Amco direct oxidation process	This process does not use thermal reactor. The process uses a *met chemistry* (redox) technique for reacting H_2S with oxygen in presence of vanadium and other reactive promoters.
Selectox process	This process is a catalytic process and does not contain combustion stage.
Sulfreen process (cold bed sub-dew point processes)	This is also cold bed sub-dew point type Claus process.
Oxygen enrichment process (COPESM)	This process is based on Claus process and uses oxygen resulting in increased H_2S handling capacity with a significant reduction in tail gas volume.
LO-CAT process	This process uses an iron chelate system to remove H_2S directly from gas stream by an oxidation/reduction reaction where H_2S is directly converted to solid sulfur.
LO-CAT process II	LO-CAT process II uses the autocirculation principle to avoid pumping costs, but uses an improved configuration and catalyst resulting

Contd.

Table 14.8   Sulfur recovery processes *(Contd.)*

Claus process	Description
	in smaller liquid volumes, lower air rates and pressure subsequently reduced air compressors. The process preserves inherent high efficiency.
Sulfreen process (Lieammi Lurgi Bamay GmbH Ei)	This process is based on Claus reaction and is based on catalytic purification of tail gas or lean H_2S waste gas. Overall sulfur recovery 99–99.9%. The catalyst consists of impregnated activated alumina. Flow diagram is shown in Fig. 14.12.
Oxy Claus process (Lurgi Oil Gas Cheme GmbH the Pntchard Corp)	The modified Claus reaction is carried out with direct oxygen combination using proprietary thermal reactor burner.
MTE sulfur recovery	MTE sulfur recovery process uses a flowing catalyst in a single reactor/regenerator loop to perform a catalyzed Claus process reaction both above and below the dew point. The unit consists of a furnace, sulfur condenser, reheater, regenerator, water condenser, blower, pipe reactor and reactor vessel.
CASTOL-II (developed by EIL, India and GNFC, India)*	This process uses redox and oxidizes H_2S present in the feed gas to produce elementary sulfur.
Improved oxidation process	The process uses partial oxidation catalysis to improve the operation and reliability of Claus unit. The process uses catalytic combuster instead of conventional Claus burner and thermal reactor. The process uses double catalyst in a short reactor to achieve near equilibrium hydrogen sulfide conversion.
SDP process (SDP1, SDP2, SDP3)	In SDP process the catalytic Claus reaction is continued at lower temperature which shifts the chemical equilibrium to favor more sulfur formation. Temperature 125–130 °C which is chosen between sulfur solidification point and the SDP. Alternate SDP processes are the combination of Claus process and sub-dew point tail gas treatment process such as SDP1 and SDP2

Source: Goar, 1986; Lagas and Wetzels, 1993; Borsboom J et al., 2003; Hardison, 1984; Ayer, 2006; Gas Processes, 1998; Simek, 2001; Kasai, 1975; Hardison and Ramshaw, 1992; Vazquez et al., 2013; Heisel and Rameshni, 2011.

14.8.1 LO-CAT Process

LO-CAT process is a single step liquid phase oxidation process for removing H_2S selectivity from gas stream and producing element sulfur using a dilute aqueous solution of iron, held on solution by organic chealting agents [Haridson LC]. The LO-CAT in hydrogen sulfide oxidation process. LO-CAT process has been illustrated in Fig. 14.10.

The combined gases stream containing H_2S is scrubbed in the high pressure venturi and high pressure absorber. In the absorber the H_2S gas absorbs very rapidly into the catalyst solution where it is oxidized to elemental sulfur. The reduced iron is then recirculated to oxidize unit where the iron is oxidized to and after regeneration the regenerated catalyst solution is recirculated.

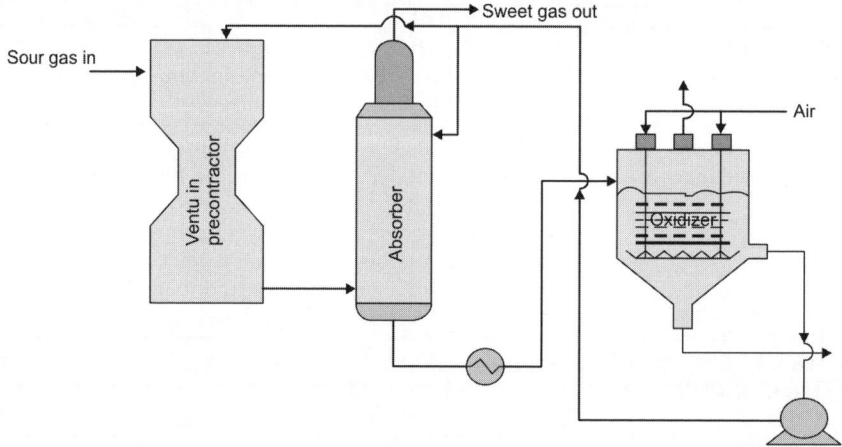

Fig. 14.10 Conventional LO-CAT process

Reaction:

$$H_2S\ (g) + H_2O \longrightarrow H_2S\ (aqueous) + H_2O$$
$$H_2S\ (aqueous) \longrightarrow H^+ + HS^-$$
$$HS^- \longrightarrow H^+ + S^-$$
$$S^{--} + 2Fe^{+++} \longrightarrow 2H^+ + S^0 + 2Fe^{++}$$

$$\frac{1}{2}O_2\ (gas) + H_2O + 2Fe^{++} \longrightarrow 2(OH^-) + 2Fe^{++}$$

$$H_2S + \frac{1}{2}O_2 \longrightarrow H_2O + S^0$$

14.8.2 CATSOL II Process

The process uses redox chemistry. The ferric ions involve oxidation of H_2S. H_2S present in feed gas is first absorbed in alkaline solution of catalyst in the absorption section. The pH of the solution is maintained around 7.2–7.8.

$$H_2S\ (g) \qquad\qquad\qquad H_2S$$
$$H_2S\ (aq.) \qquad\qquad\qquad 2H^+$$

Sulfur compounds inherently present in the natural gas and crude oil. Emission of sulfur has been one of the major environmental issues during recent years and petroleum industries have been forced to incur extremely high costs for environmental compliance to meet the sulfur emission standards which is become more and more stringent. Sulfur removal and recovery process in a refinery involves following operations to meet above objectives are: Hydrotreating, desulfurization, amine absorption of H_2S, merox sweetening and sulfur recovery by Claus process and modified Claus process. Figures 14.11 to 14.13 illustrate the process for acid gas treatment.

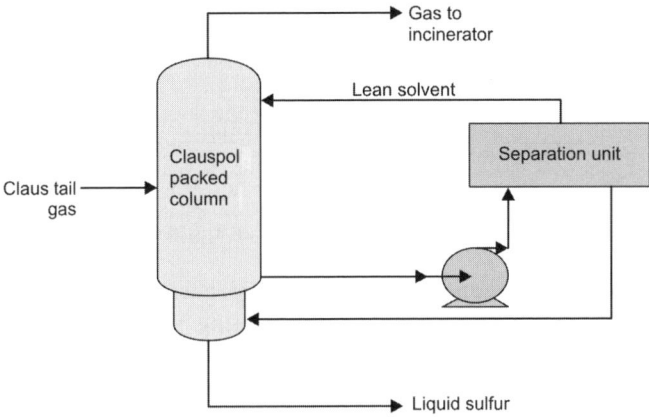

Fig. 14.11 Clauspol process

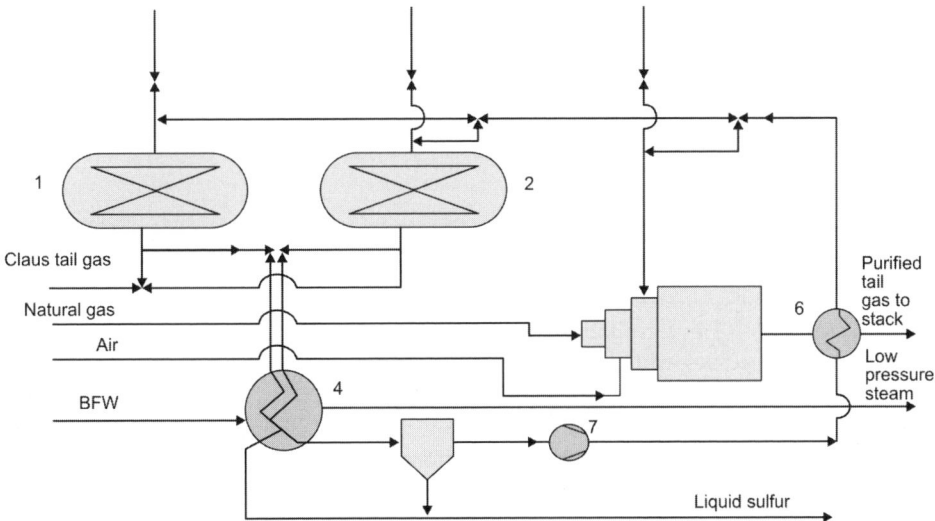

Fig. 14.12 Sulfreen process

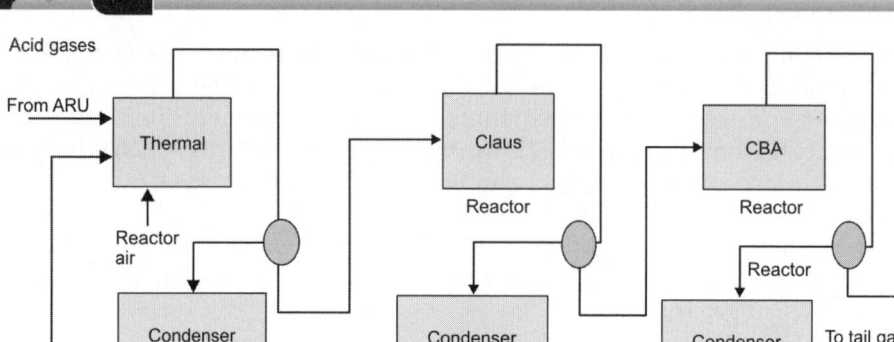

Fig. 14.13 Treatment of acid gases

Lean acid gas enrichment process: Lean acid gas enrichment process can be used to upgrade low quality off gas from treating unit to higher quality Claus plant feed or to smaller volume stream that are suitable for reinjection. Acid gas enrichment technology can be used to selectively absorb H_2S from lean gases that contain H_2S less than 10% to high quality acid gas with an H_2S concentration up to 75%. Double absorption can be used to improve the quality of the acid gas from conventional acid gas removal streams [Zarenezhad, 2011].

Claus tail gas units: In order to recover sulfur recovery above 98% and to meet the emission standards seconadry tail gas treating is required. A typical two-stage Claus achieves 95–97% recovery and a secondary tail gas treating is required for total recoveries from 98% to 99.98%. Three principles of Claus tail gas treatment process are recycle process, cyclic catalytic sub-dew point (SDP) process and selective oxidation process [Heisel and rameshni, 2011].

REFERENCES

1. Ayer J. Desulfurization Technologies. Chhemical Industry Digest, Aug 2006, p58.
2. Basu B. Sweetening Processes in Refining IOC R & D Development. Petrotech Society Training Program, 2005.
3. Brossard DN. Chevron Global RDS/VRDS Hydrotreating. Handbook of Petroleum Refining Processes, edited by Meyers RA, 3rd edition, The McGraw Hill Publication data, 2004.Chapter 8.3, p8.3.
4. Borsboom J, Lagas JA, Wolfer W, Bosch J and Goar BG. Superclaus Process Proves Reliability. Hydrocarbon Technology, International '90/'91, p31–36.
5. Borsboom J and Van Nisselrooij. Unconventional Cooling of Tail Gas in Sulfur Plants by A New Deep Cooler. Paper Presented at the 46th Annual Laurence Reid Gas Conditioning Conference, Norman, Oklahoma, USA, March 3–6, 1996.

6. Borsboom J, Van Grinsven R, Van Nisselrooij P and Butler E. Superclaus/Euroclaus, The Latest Developments, Paper presented at the Brimstone Symposium, Banff Canada, April 2003.

7. Cabodi AJ and Hardison LC. First Commercial Test is Success for Catalytic Hydrogen Sulfide Oxidation Process. Oil and Gas Journal, July 5, 1982, p107.

8. Chou JS, Chen DH, Walker RE and Maddox RN. Mercaptans Affect Claus Units. Hydrocarbon Processing, April 1991, p39.

9. Daziabis. UOP Merox Process. Handbook of Petroleum Refining Processes, edited by Meyers RA, 3rd edition, The McGraw Hill Publication data, 2004, p11.31.

10. Gas Processes 2006. Hydrocarbon Processing, Jan 2006, p67.

11. Garg MO. Ultradeep Sulfur Diesel by Oxidative Desulfurization of HDS Diesel. Chemical Industry Digest, June 2010, p86.

12. Gatan. UOP Green Chemistry Technologies. Presented at Bendictine University, Nov 11, 2004.

13. Goar B. Sulfur Recovery Technology. Energy Progress, Vol 6, No. 2, June 1986, p71.

14. Goel SK, Khanna S and Zutshi DK. Sulfur Management in Refineries. Chemical Industry Digest, May 2008, p81.

15. Gills DB. UOP RCD Uninonfining Process. Handbook of Petroleum Refining Processes, edited by Meyers RA, 3rd edition, The McGraw Hill Publication data, 2004. Chapter 8.4, p8.43.

16. Hardison LC and Ramshaw DE. H_2S to S: Process Improved. Hydrocarbon Processing, Feb 1992, p89.

17. Hegy DL and Naga GJ. Consider Optimized Iron-redox Processes to Remove Sulfur. Hydrocarbon Processing, Jan 2006, p53.

18. Heisel MP and Rameshni M. Minimize Carbon Footprint from Claus Tail Gas Units. Hydrocarbon Processing, Feb 2011, p71.

19. Indian Oil Tech/T/2003/001, Indian Oil Technologies Ltd, Faridabad, India.

20. JB Rajani. Treating Technologies of Shell Global Solutions for Natural Gas and Refinery Gas (mk@MSITstore:E:\congress.chem::/UMB1630.htm).

21. Joshi MK, Goyal GD, Deshpande A and Sarkar DK. Mmembrane Process for Gas Sweetening. Journal of Petrofed, Jan–March 2010, p20.

22. Kasai T. Konox Process Removes H_2S. Hydrocarbon Processing, Feb 1975, p93.

23. Kokayeff P. UOP Unionfining Technology. Chapter 8.3, p8.31, in Handbook of Petroleum Refining Processes, edited by Meyers RA, 3rd edition, The McGraw Hill Publication data 2004. Chapter 11.1, p11.1.

24. Lagas JA, Borsboom J and Berben PH. Selective oxidation Catalyst Improves Claus Process. Oil and Gas Journal, Technology, Oct 10th, 1988, p68–71.

25. Lagas JA, Borsboom J and Van Nisselrooij P. The Superclaus Process in the Light or Future Developments. Paper presented at the Sulfur '94 International Conference, Tempa, Florida, USA, Nov 6–9, 1994.

26. Lagas JA and Wetzels MLJA. Recent Development to the SCOT Process. Sulfur no. 227, July/August 1993, p33–44.

27. Legg D and Giaslason J. The S Zorb Sulfur Removal Technology Applied to Gasoline. Chapter 11.4, p11.43. Handbook of Petroleum Refining Processes, edited by Meyers RA, 3rd edition, The McGraw Hill Publication data, 2004.

28. Marty C. White Products Refining Sweetening. IFP Petroleum Refining, Vol 3 Conversion Processes, edited by Leprince P. Editions Technip, Paris, 2001.

29. Nakamura D. Product Sulfur Specs will Determine Future Refinery Configuration. Oil and Gas Journal, Oct 18, 2004, p48.

30. Oil and Gas Journal, Aug 12, 1985, p67.

31. Quinlan M. KBR Refinery Sulfur Management. Handbook of Petroleum Refining Processes, edited by Meyers RA, 3rd edition, The McGraw Hill Publication data, 2004. Chapter 11. p11.1.

32. Rathore V, Rao PVC, Suresh V, Das G, Kumar S and Garg MO. Sweetening of LPG. Journal of Petrotech, July–Sep, 2011, p60.

33. Refining Process, 1998; Hydrocarbon Processing, Nov 1998.

34. Refining Process, 2008; Hydrocarbon Processing, Nov 2008.

35. Sames JA. Sulfur Recovery Process Fundamentals. Paper presented at the Strok Comprimo Sulfur Workshop, Katwoude, The Netherlands, Nov 1987 and Jan 1988.

36. Sanghavi K and Schmidt J. Achieve Success in Gasoline Hydrotreating. Hydrocarbon Processing, Sep 2011, p59.

37. Sharma MK and Nag A. Super Sour Process. Jl of Petrotech, July–Sep, 2011, p49.

38. Sigmund PW, Butwell KF and Wussler AJ. HS Process Removes H_2S Selectively. Hydrocarbon Processing, May 1991, p118.

39. Simek IO. Sulfur Unit Circulates Catalyst. Hydrocarbon Processing, April 2001, p65.

40. Speight JG. The Chemistry and Technology of Petroleum, 3rd edition, revised and expanded, Marcel Dekker Inc, New York, 1999.

41. Sughrue Ed and Parsons JS. ConocoPhillips S Zorb Diesel Process. Handbook of Petroleum Refining Processes, edited by Meyers RA, 3rd edition, The McGraw Hill Publication data, 2004, Chapter 11.5, p11.51.

42. Sweed NH and Ellis ES. Meeting the Low Sulfur Mogas Challenge. Petroleum Technology, Quarterly, Autumn, 2001, p45.

43. UOP http://www.UOP.com/refining/1110.html.

44. Vazquez RG, Seeger G and Thundyil M. Improved Oxidation Processes Increase Sulfur Recovery in Claus Unit. Hydrocarbon Processing, Sep 2013, p99.

45. Zarenezhad B. Consider Different Alternatives for Enriching Lean Acid Gases. Hydrocarbon Processing, Jan 2011, p37.

15
Natural Gas Processing

Natural gas is one of the important feedstocks for fertilizer and petrochemicals. Natural gas is a mixture of gaseous hydrocarbons with methane as major constituent. Dry natural gas is a gas that does not contain readily condensate hydrocarbon. Wet natural gas is a gas, which contain about more than two gallons condensate per 1,000 cubic feet of gas. Wet natural gas has lower methane content. Sour natural gas contains appreciable quantity of hydrogen sulfide. Natural gas liquids consist of constituents condensed from natural and associated gases and are hydrocarbons ethane through pentanes. Classification of natural gas is given in Table 15.1.

Typical characteristics of natural gas are given in Fig. 15.1 [Mall, 2007]. There are two major sources of natural gas—associated gas found along with crude oil and unassociated gas which is in isolation of heavier fraction and found in gas field underground reservoirs broadly similar to oil reservoirs. The wet natural gas contains relatively higher ethane and is a co-product of crude oil production. The other promising source may be shale gas hydrate and coal bed methane. Shale gas has emerged as an important source of natural gas for energy and as petrochemical feedstock. Schematic geology of natural gas is given in Fig. 15.2. Although shale gas formation was known for long time, however, with development of horizontal drilling combined with hydraulic fracturing has allowed the development of shale gas resources. The US energy Information Administration estimated technically recoverable shale gas resources in India at 96 trillion cubic feet in Cambay, KG, Cauvery and Damodar valley basins [Goel, 2014]. The shale gas revolution has spurred an unprecedented wave of ethane-to-ethylene cracker construction and expansion [Processing Shale Feedstocks, PTQ Supplement, 2014, p11].

15.1 OIL AND NATURAL GAS PROCESSING

The processing and separation of associated gas involves stabilization of wellfluid for separation of water, oil and gas. A block diagram showing various steps involved in processing of gas and oil is given in Fig. 15.3.

Well-fluid with associated gas from the oil field is stabilized in order to facilitate its storage at close atmospheric conditions. Oil, gas and water so separated in crude stabilization unit are routed to different processes for recovery of useful products. The oil goes for removal of water emulsion and salts before it is used in the refinery. The water after separation of oil is recycled. The gas along with condensate goes to separator for separation of gas and condensate. The natural gas liquid is fractionated into their separate components. Typical condensate processing unit involves removal of H_2S and CO_2 and recovery of LPG and natural gas liquid. The condensate processing plant consists of separator for water removal, stripping column for removal of H_2S and LPG recovery column.

Table 15.1   Classification of natural gas

Associated gas	Natural gas present as associated gas along with oil in the porous reservoir.
Non-associated gas	Natural gas present in the reservoir with no oil.
Dry natural gas	Natural gas that does not contain readily condensable hydrocarbons.
Sour natural gas	Natural gas containing appreciable amount of hydrogen sulfide and carbon dioxide.
Natural gas condensate or natural gas liquid	Condensable hydrocarbon present in the natural gas having heavy hydrocarbon. NGL are removed from natural gas by cooling and the uncondensed gas is mainly methane.
Liquefied natural gas (LNG)	LNG consists of mixture of methane, ethane, propane, small amount of higher hydrocarbons and possibly nitrogen.
Compressed natural gas (CNG)	Compressed natural gas is a pressurized natural gas stored at very high pressure of 200 bar.
Natural gas hydrates (NGHs)	Natural gas hydrates are ice-like crystals composed of water and natural gas and are characterized as the natural gas trapped in cage-like structure formed by ice. Natural gas hydrate is a modified ice structure enclosing methane and other hydrocarbons. However, they can melt at temperatures well-above normal ice [www.rediff.com]. There are three types of methane hydrate structures. They all include pentagonal dodecahedra of water molecules enclosing methane.
Shale gas	Shale gas is a natural gas produced from shale and has become important feedstock for petrochemicals and alternative source of energy. With the advent of horizontal drilling and hydrofracturing, shale gas availability has increased and number of shale gas-based petrochemical complexes has been planned in the US.
Coalbed methane	Coalbed methane is a form of natural gas extracted from coal beds and has become important source of energy all over the world. Methane adsorbed into solid matrix of the coal. Unlike much natural from conventional reservoirs, coalbed methane contains very little heavier hydrocarbons such as propane or butane, and no natural gas condensate. Porosity plays an important role in building up methane gas reserves in the coalbed [Gas (CNG): An alternative to LNG].

Source: http://en.wikipedia.org/wiki/coalbed methane 2002, Malhotra et al., 2012; Tulsi, 2011.

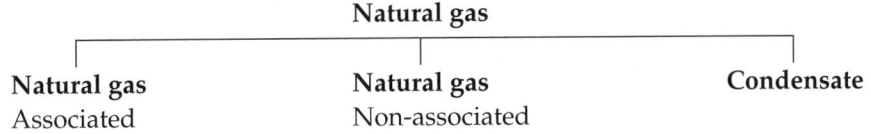

	Associated gas (%vol)	Non-associated gas (%vol)
Methane	80.70	74.6
Ethane	9.90	10.5
Propane	5.20	9.0
Butane	1.60	3.5
Pentane	0.40	1.0
Carbon monoxide	0.70	0.6
Nitrogen	1.50	0.8

Components	Boiling point at atmospheric pressure (°C)	Critical temperature (°C)
Methane	−162	−83
Ethane	−89	32
Propane	−42	97
1-butane	−12	135
n-butane	−1	152
n-pentane	36	197
C_1	Fraction	Fertilizer and petrochemicals
C_2–C_3	Fraction	Petrochemicals
C_3–C_4	Fraction	LPG (domestic, fuel and petrochemicals)
NGL and condensate fraction		Petrochemical

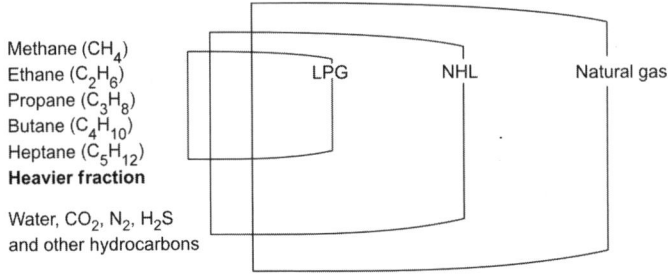

Fig. 15.1 Natural gas and their characteristics

Source: Mall, 2007.

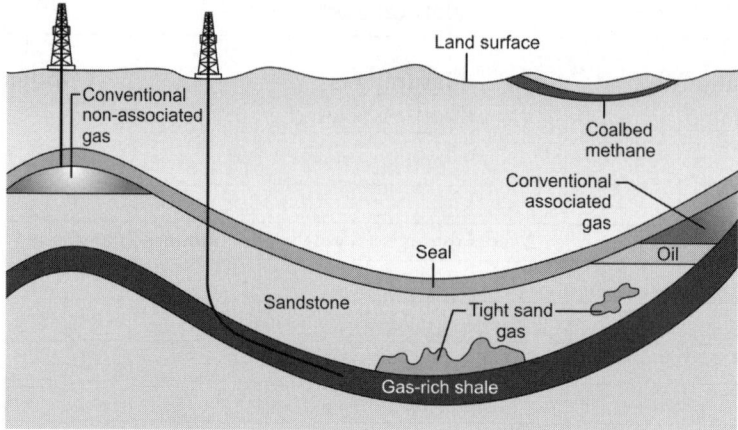

Fig. 15.2 Schematic diagram of geology of natural gas resources

Courtesy: Energy Information Administration.

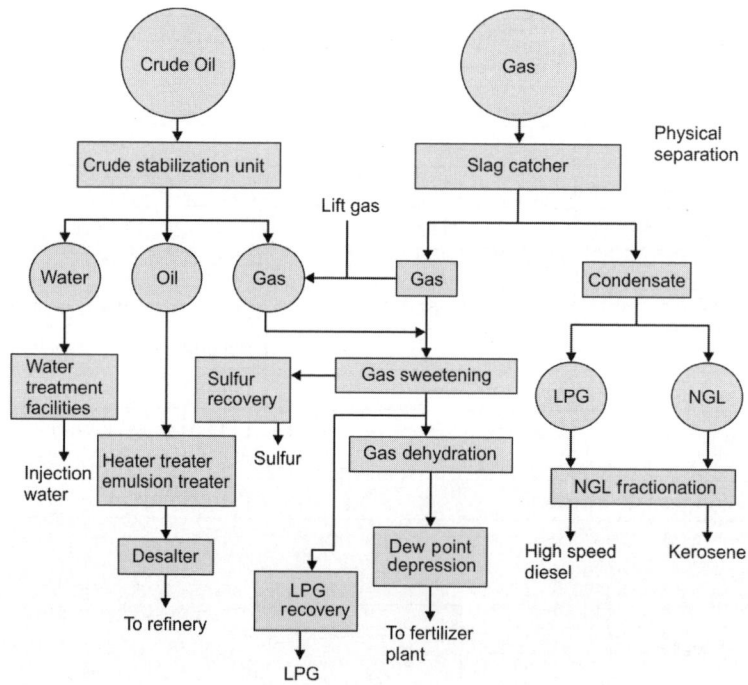

Fig. 15.3 Oil and gas processing

Source: Ravindranath and Habibula, 1992.

15.2 NATURAL GAS PRETREATMENT AND PROCESSING

Natural gas is present as associated and non-associated natural processing. Natural gas contains impurities like sulfur compounds, carbon dioxide, water, heavy metals, etc. and their pretreatment is done before transportation or processing. The various processes of pretreatment and processing of natural gas are:

- Gas sweetening (removing CO_2, H_2S, etc.)
- Gas dehydration
- Removing water (molecular sieves)
- Minor impurities—heavy metals (Hg, V, etc.)
- Separation—by cooling, expansion and distillation.

Gas sweetening

Removal of acid gases CO_2 and H_2S, etc.
- Chemical, solvent process (MEA, DEA and alkaline salts)
- Physical solvent process (Sulfonate and KOH)
- The sulfinol process uses a mixture of sulfolane and alkanolamines
- DGA is used in econamine process.
- Dimethyl ether of polyethylene glycol is used in Selexol process.
- Direct conversion process for dry-bed process.

Gas dehydration

- Sweetened gas is dehydrated to adjust humidity to prevent hydrate formation.
- Liquid desiccants (DEG, TEG)
- Solid desiccants (mol. sieves, silica gel, alumina and calcium chloride)

Separation of natural gas fractions

- Currently, processes based on external refrigeration or turbo expanders are used to recover ethane plus components.
- Processes based on adsorption using regenerable adsorbent and membrane separation processes are under development.

The processing and separation of associated gas involves stabilization of well fluid for separation of water, oil and gas. The oil goes for removal of water emulsion and salts before it is used in the refinery. The water after separation of oil is recycled. The gas along with condensate goes to separator for separation of gas and condensate. The natural gas liquid is fractionated into their separate components. Typical condensate processing unit involves removal of H_2S and CO_2 and recovery of LPG and natural gas liquid. The condensate processing plant consists of separator for water removal, stripping column for removal of H_2S and LPG recovery column.

A typical non-associated natural gas processing consists of gas receipt terminal, gas sweetening, gas dehydration, dew point suppression unit, condensate fractionation unit, LPG recovery unit and kerosene recovery unit. At the gas terminal the sour gas is collected and then the liquid and condensate are filtered and sent for processing.

Wet gas receiving terminal consists of pig receiving trap for removal of foreign matter, residual solids, pressure reducing valves station for maintaining pressure, slug catcher for separation of condensate. The gas from slug catcher is sent for filtration for separation of entrained condensate from gas. The gas is then sent to gas sweetening unit. The H_2S content from the sweetening plant is 4 ppm vol max. A generalized schematic diagram of a typical gas treating plant is given in Fig. 15.4.

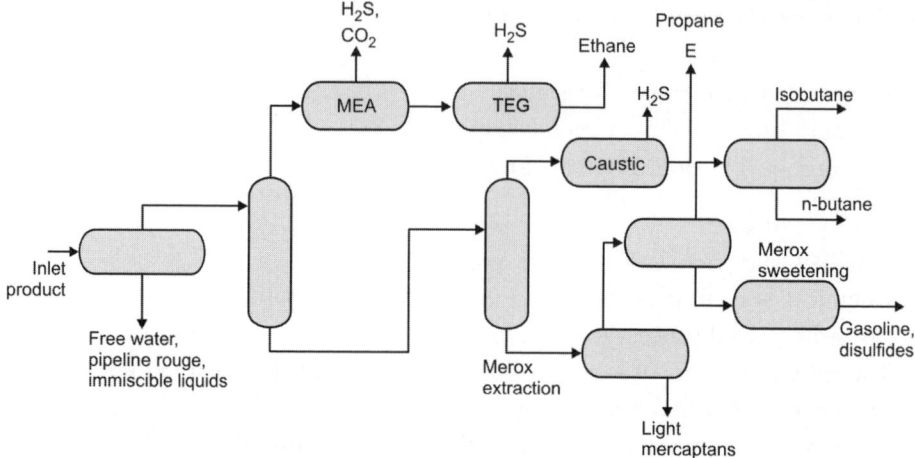

Fig. 15.4 A generalized schematic diagram of a typical gas treating plant

Courtesy: Hydrocarbon Processing, Hatch and Mate, 1997.

15.2.1 Gas Sweetening

Further processing of natural gas depends upon the types of compound present in it. A dry gas low in sulfide content needs little pretreatment or no treatment except to adjust the moisture content while processing of the sour natural gas requires removal of hydrogen sulfide and carbon dioxide. The processing of sour natural gas involves sweetening for removal of H_2S and CO_2 using monoethanol amine and diethanol amine as absorbing media. The sulfide carbonates and steam stripping regenerates bicarbonate, thus formed. Other solvents used for separation of H_2S gas are sulfolane, sulfolane-carbonate, and mixture of sulfolane and alkanol amines (sulfil process, merox process). Molecular sieves are used for removal of moisture, H_2S, mercaptans (RSH) and carbonyl sulfide. Membrane process is an another promising process for treatment of natural gas for removal of sour gases. Process flow diagram for high pressure amine absorption system and amine regeneration is given in Figs. 15.5 and 15.6.

Based on lower operating costs, comparable capital cost and only slightly higher product loss (including fuel), membranes have demonstrated a flexible, cost-effective alternative to amine treating for some natural gas processing applications [Cook and Losin, 1995]. Separation of CO_2 by membrane is based on the principle that some gases pass more readily and others are retained by the membranes. Membrane process showed economic advantage over amines system over a wide range of feed gas compositions and flow rates. Membranes have been found competitive at the lower flow rates and for high carbon dioxide concentrations [Spillman, 1989].

15.2.2 Gas Dehydration

The treated gas leaving the gas sweetening unit enters the dehydrating unit at a pressure of 60 kg/cm^2 and temperature of about 400 °C where dehydration takes place using triethylene glycol.

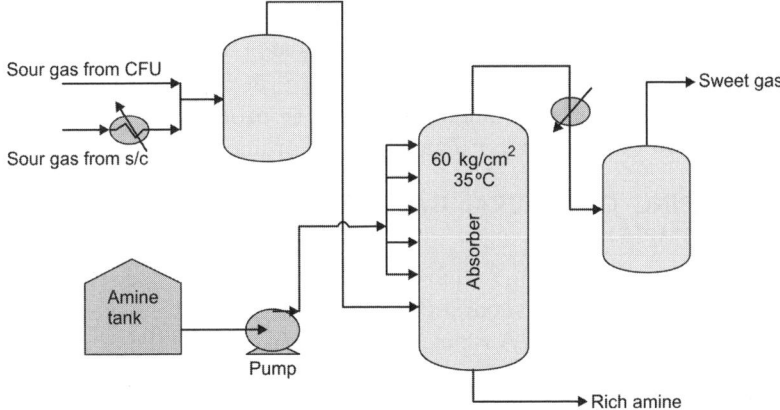

Fig. 15.5 HP Absorption

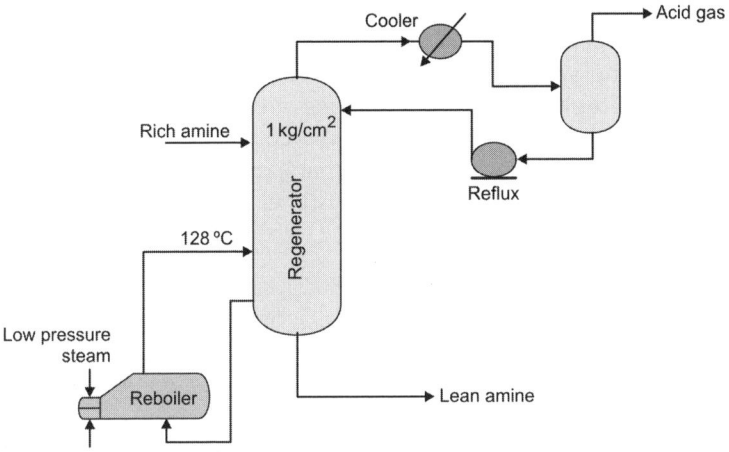

Fig. 15.6 Amine regenerator

- *Dew point depression unit:* The sweet and dehydrated gas contains significant quantities of heavy hydrocarbons which are removed in the dew point depression unit to avoid condensation at lower temperature. The dew point depression unit consists of chill down section and propane refrigeration section. The feed gas is chilled to temperature of −2.5 °C by evaporating propane. The chilled is sent to filter separator to knock out hydrocarbon condensate and traces of water. The separated gas is then sent for transportation.

- *Condensate processing:* Gas condensate from various gas processing units contains H_2S and valuable hydrocarbons which are separated by fractionation. The process involves filtration of condensate followed by fractionation of condensate into various streams like H_2S, LPG and aromatic. The condensate is passed through the cartridge type filters to remove scale/dust/debris.

  Stripper is used to strip from the condensate in the H_2S stripper. The condensate from the stripper contains maximum 4 ppm H_2S. Stripper bottom liquid condensate enters the LPG column where LPGs (propane and butane) are recovered as top products. The LPG is sweetened in a caustic unit to remove the H_2S from the LPG as the remaining 4 ppm H_2S in condensate appears mostly in the LPG.

- **LPG recovery unit:** The LPG recovery plant unit consists of expansion and consequent cooling of sweetened gas/gases to produce condensate which is distilled to give LPG and natural gasoline, i.e. aromatic-rich naphtha.

15.3 PROCESSING OF NATURAL GAS FOR C_2/C_3 EXTRACTION FOR PETRO-CHEMICAL COMPLEX

A typical natural gas processing unit in a petrochemical plant consists of sweetening of natural, dehydration and separation of natural gas. Natural gas as feedstock contains gas-sweetening unit. A typical composition of natural gas is given below:

Percentage Composition

$C_1 = 84.06$, $C_2 = 9.38$, $C_3 = 0.99$, $C_4 = 0.04$, $CO_2 = 5.52$, $H_2S = 4$ ppm, $N_2 = 0.01$

The sweetened gas from gas sweetening plant is fed to the feed gas knock out drum for separation of any entrapped liquid from the feed gas and then it is compressed and cooled to about 17 °C. It is then sent to feed gas moisture separator and feed gas-drying section for removal of traces of moisture. The gases after passing through filter are sent through series of exchangers for cooling to below –60 °C and then send to cold section. In the cold section the gases are sent to demethanizer column for separation of methane from the C_2/C_3. The temperature in the demethanizer is maintained about –101 °C. Typical composition of C_2 and C_3 feed to gas cracker is shown in Table 15.2.

Table 15.2   Typical compositions of C_2 and C_3 feed to gas cracker

C_1	6.04%
C_2	90.45%
C_3	0.81%
C_4	0.62%
C_5	0.03%
Others	Balance

15.4 NATURAL GAS LIQUEFACTION

The LNG value chain consists of liquefaction plant and receiving terminals. Major components for LNG value chain include [Economides and Mokhatab, 2007]:

- NG production
- The liquefaction process where pretreated NG becomes liquefied at a temperature of approximately 256 °F
- Transportation
- Regasification
- Distribution

LNG projects are inherently capital intensive, with liquefier making up around 25–50% of total cost. The balance is for storage, send-out terminals, jetties and ships (base load) or vaporizers (peak shave). Features of a LNG chain is given in Fig. 15.7. [Finn et al., 1999]. LNG requires very large reserves near the facilities to get acceptable return on capital investment [Economides and Mokhatab, 2007]. Break down cost of liquefaction plant is shown in Fig. 15.8.

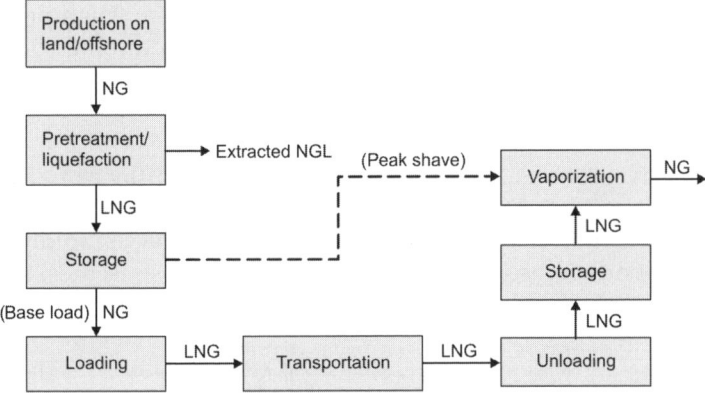

Fig. 15.7 Feature of LNG chain

Courtesy: Hydrocarbon Processing.

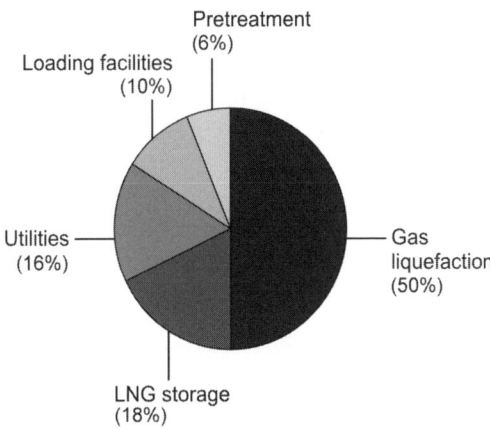

Fig. 15.8 Typical breakdown cost of liquefaction plant

Source: Hydrocarbon Processing, Finn et al., 1999.

Liquefactions

Commonly used refrigeration cycles are the cascade, mixed refrigerant and expander cycle [Figs. 15.9 to 15.11].

15.4.1 Cascade Refrigerant Cycle

In cascade refrigerant cycle natural gas is cooled, condensed and subcooled in heat exchange with propane, ethylene or ethane. Methane finally is cooled in three discrete stages having multistage refrigerant expansion and compression, each typically operating at three evaporation temperature levels [costain-floating-lng.com].

- *Mixed refrigerant cycle:* In this process a single mixed refrigerant is used instead of the multiple pure refrigerants in the cascade cycle. A mixture of nitrogen and hydrocarbons (usually in the range C_1 to C_5 range) is normally used to provide optimal refrigeration [www.timesb2b.com].
- *Expander cycle:* In expander cycle process refrigeration is provided by compression and work expansion of a single gas stream. High pressure cycle gas is cooled in counter current heat exchange with returning cold cycle gas. [costain-floating-lng.com].

15.5 COMPRESSED NATURAL GAS

CNG technology provides an effective way for shorter distance transport while LNG is an effective long distance delivery method, constitutes 25% of global gas movement [Economides and Mokhatab, 2007]. Globally, CNG is rapidly recognized as an NG transportation solution for certain stranded gas reserves, markets and associated gas production [www.energytribune.com].

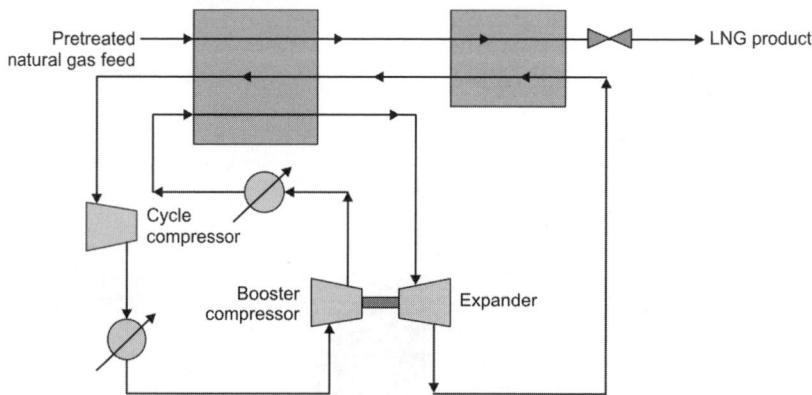

Fig. 15.9 Typical expander cycle (single cycle)

Source: Hydrocarbon Processing; Johnson GL and Tomlinson TR; Development in Natural Gas Liquefaction.

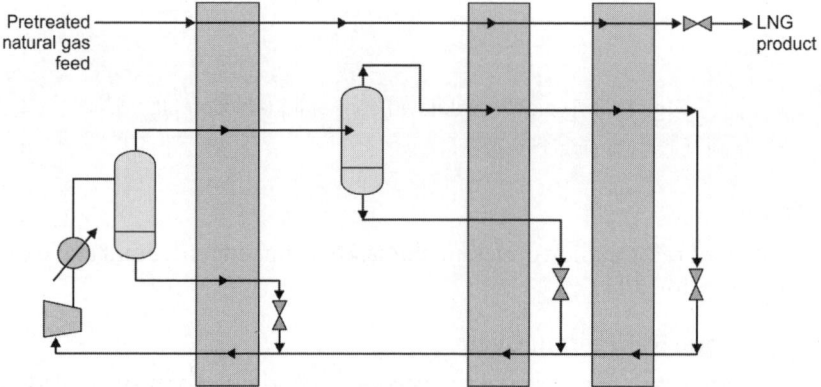

Fig. 15.10 Typical three-stage MRC

Source: Hydrocarbon Processing; Johnson GL and Tomlinson TR; Development in Natural Gas Liquefaction.

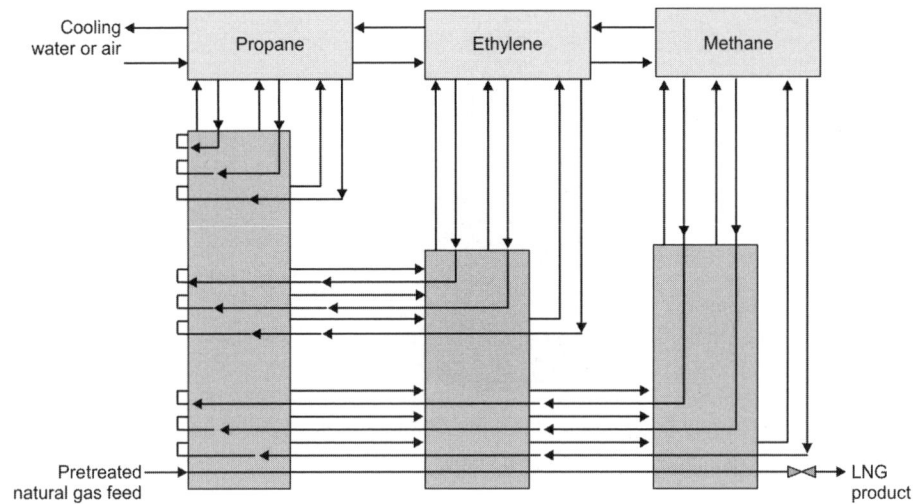

Fig. 15.11 Cascade cycle

Source: Hydrocarbon Processing; Johnson GL and Tomlinson TR; Development in Natural Gas Liquefaction.

15.6 NATURAL GAS TRANSPORTATION

Global natural gas demand is expected to double by 2020 which has resulted in increased activity of natural gas exploration. The efficient and effective movement of natural gas from producing regions to consumption regions requires an extensive and elaborate transportation system. Constructing natural gas pipelines requires a great deal of planning and preparation globally 75% of the natural gas produced is transported through pipelines and remaining 25% is transported as liquefied natural gas (LNG) [Saraf, 2007]. For short distances, pipelines—where feasible—are usually more economical. LNG is most competitive for long distance routes since overall costs are less affected. For large deliveries (around 1Tcf/year), transporting gas by high pressure pipelines appears very competitive. For long distances, LNG is competitive for capacities below 1 Tcf/year [Economides and Mokhatab, 2007].

With increasing demand of natural gas there may be additional load for supply of gas. Pipeline components of natural gas supply system consist of pipelines; compression stations, metering stations, number of valves, control stations, and supervisory control data acquisition, pipeline inspection and safety.

Compressed natural gas (CNG) and natural gas hydrates (NGHs) may be the alternative transportation options. CNG is an economic alternative to monetize stranded gas reserves and creates new market where pipelines and LNG deliveries are not practical [Economides and Mokhatab, 2007]. Figure 15.12 shows product distribution from the das Island associated gas liquefaction.

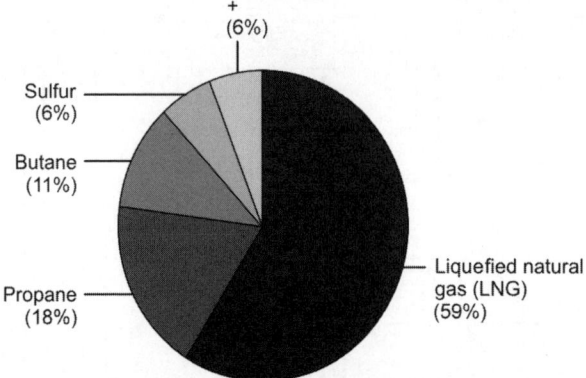

Fig. 15.12 Product distribution from the das Island associated gas liquefaction

Courtesy: Hydrocarbon Processing [Finn et al., 1999].

15.7 USES OF NATURAL GAS

Major industrial use of natural gas is as energy resource, transportation fuel, petrochemical and fertilizer feedstocks for manufacture of hydrogen, ammonia, urea, petrochemicals and industrial and domestic fuel. With commercialization of GTL process natural gas is going to play an important role in meeting the demand of gasoline.

15.7.1 Natural Gas as Energy Resource

With increasing availability of natural gas, it is being increasingly used as major source of electricity generation through cleaner and produces less carbon dioxide than burning petroleum and about the use of gas turbines and steam turbines. Natural gas has advantage over other fuels like oil and coal as burning natural gas produces about 30% less carbon dioxide than burning petroleum and about 45% less than burning coal [Natural Gas, http://en.wikipedia.org/wiki/Natural gas, Natural gas and the Environment; http://www.natural gas.org/environment/natural gas.asp#greenhouse/].

Gas accounts for 10.6% of the total installed capacity is given in Table 15.3 and share of gas and oil in power by 2052 is expected to be 16% against the present share of 11.5%.

Table 15.3   Installed capacity as per energy sources

Energy source	Capacity (MW)	Share (%)
Coal	69,616	54.1
Gas	13,582	10.6
Oil	1,202	0.9
Hydro	34,034	26.5
Nuclear	3,900	3.0
Renewable	6,191	4.8
Total	1,28,432	100.00

Source: Gharpure, 2007.

15.7.2   Natural Gas Transport Fuel

Compressed natural gas is finding increasing uses of transportation fuel because of high octane number which permits high compressions and resultant combustion efficiency and clean burning. Natural gas is considered as the most environment friendly of the fossil fuels because of lowest CO_2 emission per unit of energy. Major bottleneck is the cost of compression and supply system.

15.7.3   Natural Gas as Domestic Fuel

Natural gas as such and LPG recovered from natural gas are the major source of domestic fuel.

15.7.4   Natural Gas as Fuel through GTL Technology

GTL Technology has become one of the promising sources of fuel as replacement of crude oil and several GTL projects are expected to come on stream during the next decade and almost all of the world's major international oil and gas companies have in GTL through demonstration plants or projects at various stages of commercialization [Jamieson et al., 2007]. Some of the main forces for GTL as a gas modernization strategy are [Jamieson et al., 2007; Garg et al., 2003]:

- Natural gas supply at lower cost
- Technology advancement leading to the reduction in GTL cost
- Large liquid markets for products like diesel and naphtha
- High quality products: Diesel, naphtha and lubricant basestocks. Diesel produced from GTL is of high cetane number, and low aromatics content. Naphtha from GTL is highly paraffinic, giving higher ethylene yield than refinery naphthas.
- Higher return in a high oil price environment.

Typical specification of GTL diesel is given in Table 15.4. Diesel produced from GTL has proved cleaner than that from petroleum refining.

Table 15.4   Typical diesel specification

Properties	Conventional ULSD	GTL diesel
Sulfur (ppm)	10	5
Cetane number	48 minimum	75
Specific gravity	0.82–0.86	0.78
Polyaromatics	11	5

Courtesy: Oil Gas Journal, Jamieson et al., 2007.

The process of GTL involves three steps:

- Reforming of natural gas in autothermal reformer
- Fischer-Tropsch (FT) reaction in FT fixed bed reactor using cobalt based catalyst
- Upgrading of the FT product hydroprocessing to produce sulfur-free diesel fuel and naphtha.

Various FT technologies are given in Table 15.5. Process flow diagram of typical GTL process is given in Fig. 15.13. Commercial projects have been mentioned in Table 15.6. The main products from GTL plants are diesel, naphtha and lubricant base oils with smaller volumes of LPG, n-paraffins and waxes [Jamieson et al., 2007].

Table 15.5   Fischer-Tropsch technologies

Company	Syngas	Fischer-Tropsch	Upgrading
BP	BP/Davy compact reformer	BP fixed-bed process. Co catalyst	–
ConocoPhillips	Catalytic partial oxidation (CoPox)	Slurry phase process. Co catalyst	–
Axens	–	Eni/IFP slurry phase process. Co catalyst	Axens HDK process. Eni/IFP catalyst.
Exxon Mobil	Exxon Mobil fluidized combined reforming	Exxon Mobil AGC-21 slurry phase process. Co catalyst	Exxon Mobil
Rentech	–	Slurry phase process. Fe catalyst	–
Sasol	–	Sasol slurry phase process (SSPP) Co catalyst	Chevron
Shell	Shell Partial Oxidation	Shell middle distillate synthesis (SMDS). Fixed-bed process.	Shell
Statoil	–	Slurry phase process, Co catalyst	–
Syntroleum	Syntroleum air-blown autothermal reforming	Slurry phase process. Co catalyst	–

Source: Fedou et al., 2008.

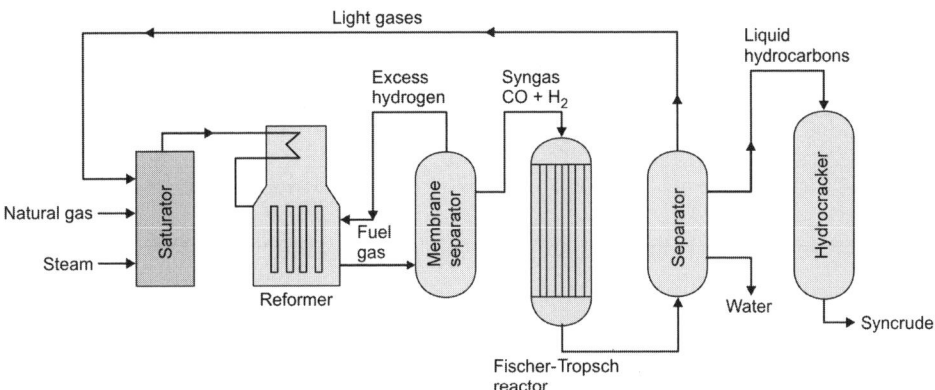

Fig. 15.13 Typical gas to liquid technology

Coutesy: Davy Process Technology [CEP, May 2005].

Table 15.6   GTL commercial projects

Owner	Syngas technology	FT technology	Capacity, kbph	Status	Start-up-date
Qatar petroleum (51%), sasol (49%) Oryx GIL project-Qatar	HTAS ATR	Sasol SBCR	Two phases: 34 + 66	06/06/06 plant inauguration start up completed in Jan 2007. Operating rates lower than planned	1st Q 2007 expansion in 2009
NNPC (25%), Chevron (75%) Escravos GTL Project (EGTL)– Nigeria	HTAS ATR	Sasol SBCR	34	Feed completed in 2004 by Foster Wheeler, EPC phase. Contractor JSK; consortium (JGC, KBR, Snamprogetti)	2009
Shell peari GIL Project– Qatar	Shell POX	Shell fixed-bed reactor	Two phases: 70 + 70	Feed completed, EPC awarded to KBR/JGC, construction phase launched in 27/07/06	2009, expansion in 2011
Qatar Petroleum Sasol Chevron	HTAS ATR	Sasol SBCR	130	Evaluation phase, project delayed	–
Marathon Oil	Syntroleum air-blown ATR	Syntroleum	120	Pre-feed work completed in the end of 2003. Project delayed.	–

Contd.

Table 15.6   GTL commercial projects *(Contd.)*

Owner	Syngas technology	FT technology	Capacity, kbph	Status	Start-up date
Conoco-Phillips	C-POX	SBCR	Two phases: 80 + 80	Statement of inents (SOI) with QP signed in Dec 2003. Pre-feed initiated. Project delayed.	
Exxon Mobil	HRAS ATR	Exxon SBCR	150	Head of agreement (HDA) with QP signed in July 2004. Project dropped in Feb 2007.	–

Source: Fedou et al., 2008.

- **New GTL process:** A new GTL process for converting natural gas and light hydrocarbons into a high value, easily transportable gasoline product has been developed. The process is an integrated conversion process in which the cracking reactor operates by combusting fuel gas with oxygen to generate a very high temperature flame for cracking the preheated feed to produce ethylene and acetylene in the range of 40–80%. First acetylene is converted to ethylene in ethylene reactor and then the ethylene product is further converted to gasoline in a product reactor. The product blendstock primarily consists of C_6–C_8 with some lighter components [Cantrell et al., 2013].

15.7.5   Natural Gas Fertilizer and Petrochemical Feedstocks

Natural gas is an important feedstock for petrochemical and fertilizer plants. Product profile of natural gas as feedstock is given in Figs. 15.14 and 15.15 [Mall, 2007]. Natural gas is used for the production of synthesis gas, olefins and aromatics which are the backbone of chemical industry.

Although LPG is the major source of domestic fuel, however, it is also a promising chemical feedstock for the manufacture of aromatics through cycler process which converts LPG to aromatics [Gosling et al., 1991]. Reaction pathways for aromatics and hydrogen production are given in Fig. 15.16. The process for conversion of LPG to BTX consists of reaction, catalyst regeneration and product recovery sections. The major process variables controlling product yields with a given catalyst formulation are feed composition, space velocity, reaction temperature and reaction pressure.

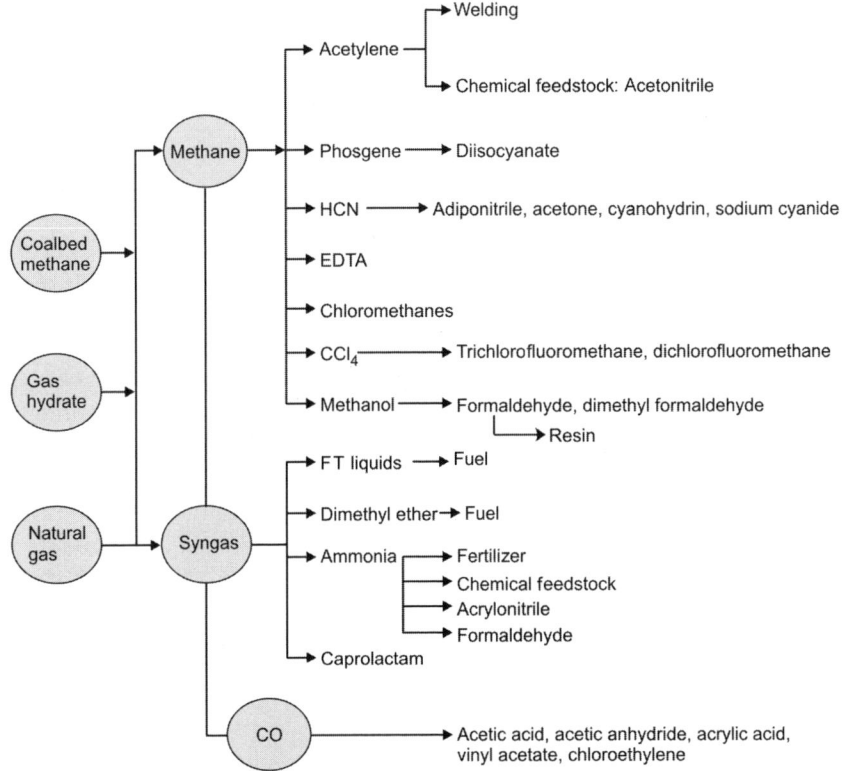

Fig. 15.14 Natural gas, coalbed methane, gas hydrate as petrochemical feedstocks

Source: Mall, 2007.

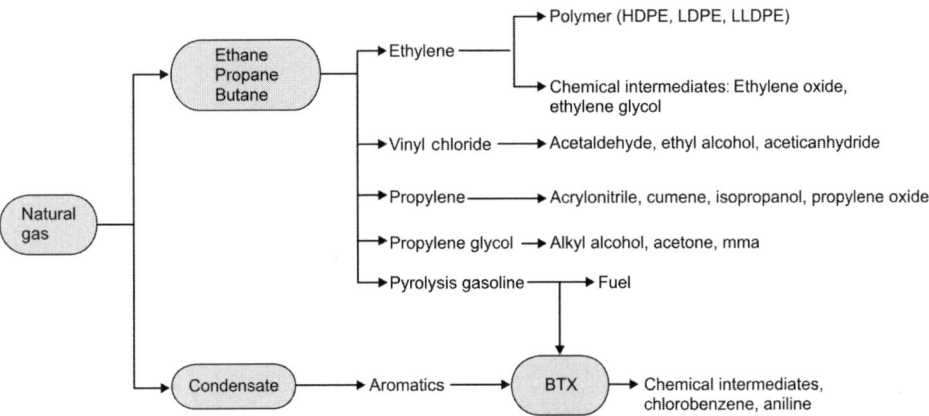

Fig. 15.15 Natural gas as petrochemical feedstock

Source: Mall, 2007.

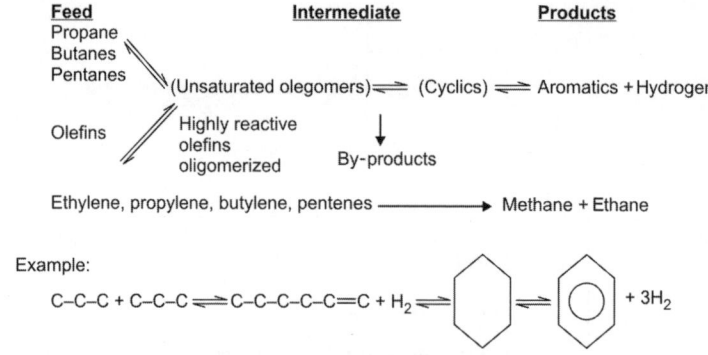

Fig. 15.16 Reactions involved in cyclar process

Courtesy: Hydrocarbon Processing [Gosling et al., 1991].

REFRENCES

1. Cantrell J, Bullin JA, Mcintyre G, Butts C and Cheatham B. Consider GTL as an Economic Alternative for Stranded Natural Gas and Ethane. Hydrocarbon Processing, Oct 2013, p77.

2. Cook PJ and Losin MS. Membranes Provide Cost Effective Natural Gas Processing. Hydrocarbon Processing, April 1995, p79.

3. Economides MJ and Mokhatab S. Compressed Natural Gas—Another Solution to Monetize Stranded Gas. Hydrocarbon Processing, Jan 2007, p59.

4. Fedou S, Douziech D and Axens EC. Synthesis Gas to High Cetane Diesel: The Gasel Process. Journal of the Petrotech Society, Jan 2008, p26.

5. Finn AJ, Johnson GL and Tomlinson TR. Developments in Natural Gas Liquefaction. Hydrocarbon Processing, April 1999, p47.

6. Garg MO and Bhadoni RP. Gas to Liquid Technologies: Challenges and Opportunities. Chemical Digest, May–June, 2003.

7. Gharpure YH. Energy Requirements of India. Chemical, Weekly, June 26, 2007, p213.

8. Gosling CD, Wilcher FP, Sulvian L and Mountford RA. Processes LPG to BTX Products. Hydrocarbon Processing, Dec 1991, p69.

9. Goel A. Shale Gas. Chemical Industry Digest, Annual, Jan 2014, p65.

10. Hatch F and Matar S. Hydrocarbon Processing, May 1977, p192.

11. http://en.wikipedia.org/wiki/coalbed methane 2002.

12. Jamleson A, McManus G and Mackenzie W. GTL Production Will Partially Ease Regional Diesel and Naphtha Imbalances. Oil and Gas Journal, Mar 19, 2007, p49.

13. Mall ID. Petrochemical Process Technology. MacMillan India, 2007.

14. Natural Gas, http://en.wikipedia.org/wiki/Natural gas, Natural Gas and the Environment: http://www.natural gas.org/environment/natural.

15. Processing Shale Feedstocks. PTQ Supplement, 2014, p11.

16. Parkinson G. Gas-to-Liquid Technology Gains Momentum. Chemical Eng. Progress, May 2005, p6.

17. Saraf S. Shipping Natural Gas—New Frontiers. Hydrocarbon Processing, Jan 2007, p5.

18. Spillman RW. Economics of Gas Separation Membranes. Chemical Engineering Progress, Jan 1989, p41.

19. Tuli D. Shale Gas Journal of Petrofed, January–March, 2011, 36.

16

Oxygenates

Oxygenates portion of the gasoline was the fastest growing gasoline components of reformulated and oxygenated gasoline since early 1990s. Fuel oxygenates are less photochemically reactive than most gasoline hydrocarbons. Substitution of more reactive components such as olefinic materials would result in ozone reduction [Brockwell et al., 1991]. Oxygenates reduce CO emissions and unburned hydrocarbons in the automobile exhaust [Piel and Thomas, 1990]. Some of the important classes of oxygenates are ethers and alcohol. Ethers include methyl tertiary butyl ether (MTBE), tertiary amyl methyl ether (TAME) and ethyl tertiary butyl ether (ETBE), tertiary amyl ethyl ether (TAEE), tertiary hexyl methyl ether (THEME) and diisopropyl ethyl ether (DIEE). Important alcohols used as oxygenates are methanol, ethanol, gasoline grade tertiary butyl alcohol (GTBA), ethyl tertiary butyl alcohol (ETBA), isopropyl alcohol (IPA) and tertiary amyl alcohol (TAA). Ethers show both physical and thermodynamical properties that result in low emission gasoline for high ozone areas. Ethers provide a large potential volume of non-aromatic octane even at relatively low oxygen levels in gasoline [Piel and Thomas, 1990]. Although ethanol and biodiesel production are increasing, but greater challenge is to produce biofuels at lower cost than that of petroleum in transportation fuels.

The driving force for biofuels are—reduce greenhouse gas (CHG) and efforts by various governments to reduce dependence on imported oil and to improve supply of transportation fuel. MTBE was fastest growing oxygenates during 1980s because of phasing out of tetraethyl lead and increasing demand of high octane gasoline. Number of MTBE plants came into operation during 1990s. Although MTBE was major additive in reformulated gasoline earlier, however, due to recent concern over the potential water pollution problems caused by leaking underground gasoline storage tanks, and contamination of ground water because MTBE is soluble in water and does not degrade easily in many countries are adopting legislation to either phase out MTBE or restrict the MTBE usage as a gasoline additive. One of the major reasons for fast growth of the MTBE and TAME was due to the availability of large amount of FCC C_4 and C_5 gases containing isobutylene and isoamylene. Gasoline component of FCC case is given in Table 16.1.

The oxygenated MTBE and ETBE are produced by the reaction of methanol/ethanol and isobutylene. TAME is produced by etherification of isoamylene.

Isobutylene and isoamylene can be recovered from C_4 and C_5 stream of steam crackers.

Table 16.1 Gasoline pool components—FCC case

Component	FCC without ethanol (%vol)	FCC with ethanol (%vol)
Heavy	27.4	25.2
Isomerate	23.2	25.7
Alkylate	8.9	7.9
FCC gasoline	32.5	28.2
Butanes	7.1	3.7
Ethanol	0.0	5.0
Heavy naphtha	0.0	3.4
Reformate	0.0	0.9

Source: Hydrocarbon Processing, Feb 2007, p78.

A comparison of gasoline component typically obtained from 1,00,000 MT of either propylene or isobutylenes is mentioned in Table 16.2.

Table 16.2 Typical yields per 1,00,000 tons feed

Process	Mogas component	Volume cu m	Octane (R + M)/2	Pool contribution 10^8 (octane) (cu m)
Propylene feed				
Dimerization	Isohexenes (dimate)	1,47,000	90	13.2
Hydration	Isopropanol (IPA)	1,77,000	106	18.8
	IPA + MeOH (40/60)	4,42,000	106	46.8
Isobutylene feed				
Dimerization	Iso-octenes	1,39,000	101	14.0
Alkylation	Alkylate	2,95,000	95	28.0
Hydration	Tertiary butanol (TBA)	1,58,000	100	15.8
	TBA + MeOH (40/60)	3,95,000	106	41.9
Etherification	MTBE	2,10,000	106	22.3

Source: Hydrocarbon Processing, Sep 1987, p68.

Some of the commonly used alcohols as oxygenates are methanol, ethanol, iso-propanol and tertiary butanol. Pure methanol has excellent antiknock quality and has highest octane. However, it has a number of disadvantages like a demixing tendency at low temperature, a corrosive effect, formation of azeotropes with light components in the motor fuel and self-ignition tendency [Travers, 2001]. Properties of the compounds are given in Table 16.3.

Table 16.3   Properties of oxygenation compounds

	MeOH	EtOH	IPA	TBA	SBA	MTBE	TAME
Atmospheric boiling point (°C)	64.6	78.5	82.4	82.6	99.5	55.4	86.3
Density (25/4 °C)	0.79	0.79	0.78	.78	0.80	0.74	0.74
Molecular mass	32.04	46.07	60.09	74.12	74.12	88.15	102.18
Oxygen content (%wt)	50.0	34.7	26.6	21.6	21.6	18.2	15.7
Combustor, heat (kJ/kg)	22,707	26,945	33,300	35,590	35,690	38,220	39,392
Vaporization heat (kJ/kg)	1,104	839	666	536	562	337	326
Flash point (°C)	6.5	12	13	11	24	−28	−
Ignition point (°C)	464	425	456	470	380	460	−
Water azoetrope (Atm. BP, °C)	(none)	78.2	80.5	79.9	87.5	52.2	73.8
Water in azoetrope (%wt)	−	4.4	12.3	11.8	27.3	3.2	9.0
Research (ON)	122	121	117	106	108	115	108
Motor (ON)	93	97	95	94	91	97	96
Blending (R + M)/2 (clear *)	107	109	106	100	99	106	102
Admix limit in gasoline (%vol **)	3.0	5.0	10.0	7.0	10.0	15.0	15.0

*Typical at 10 %vol in unleaded premium base gasoline.
**EC methanol with compulsory co-solvent 60/40%.
Source: Hydrocarbon Processing, Sep 1987, p68.

In addition to increasing the octane of blending stock and replacing high octane aromatics in the fuel, all oxygenated fuels reduce CO unburned hydrocarbons in the automobile exhaust. Blending characteristics of fuel oxygenates and various oxygenate (ether) manufacturing processes are given in Tables 16.4 and 16.5 [Brockwell et al., 1991].

Table 16.4   Blending characteristics of fuel oxygenates

Oxygenates	Boiling point (°C)	Blending RVP (kg/cm²)	Average octane (R + M)/2	Oxygen content (%wt)	Blending limits (percentage)	
					O_2 (%vol)	O_2 (%wt)
Ethers						
MTBE	131	8	110	18	15	2.7
ETBE	161	4	111	16	16	2.0
TAME	187	1	106	16	17	2.0
DIPE						

Contd.

Table 16.4 Blending characteristics of fuel oxygenates *(Contd.)*

Oxygenates	Boiling point (°C)	Blending RVP (kg/cm²)	Average octane (R + M)/2	Oxygen content (%wt)	Blending limits (percentage)	
					O₂ (%vol)	O₂ (%wt)
Alcohols						
Methanol	148	60	120	120	50	
Ethanol	173	18	115	115	35	3.7
MeOH Blend	145/180	31	108	108	35	3.7
GTBA	181	12	100	100	21	3.7

Courtesy: Hydrocarbon Processing, September 1991, p133.

Table 16.5 Various oxygenate (ether) manufacturing processes

Oxygenates	Processes	Description
MTBE, ETBE, TAME	Huls AG and UOP Ethermax process	The process consists of etherification of FCC C_4 and C_5 streams using methanol for MTBE/TAME and ethanol for ETBE. The process consists of washing section, reaction system containing fixed bed reactor, reactive distillation column containing proprietary packing where simultaneous reaction of remaining isobutylene and distillation occur. The process is reported to be environmentally compatible and does not produce hazardous waste. Process variation includes MTBE, TAME, ETBE, MTBE/TAME coproduction
MTBE, TAME	EOP olefin isomerization (BUTESOM PROCESS) coupled with Ethermax process.	The process consists of isomerization of normal C_5 olefins to reactive isoamylene for conversion to TAME by etherification using Ethermax process. Isomerization section consists of feed heater, fixed bed reactor using proprietary catalyst separator and tripper for separating isobutylene from light gases.
MTBE, TAME	CDTECH	The process produces MTBE from methanol and isobutylene (from C_4 gases). The same process scheme can be used for production of TAME. The process consists of a fixed-bed downflow adiabatic reactor, catalytic distillation where simultaneous reaction and product separation take place, methanol recovery section.
MTBE, ETBE, TAME	IFP process	The process produces oxygenates (MTBE, ETBE, TAME and/or higher molecular weight ethers) as high octane reformulated gasoline components

Contd.

Table 16.5   Various oxygenate (ether) manufacturing processes *(Contd.)*

Oxygenates	Processes	Description
		from C_1 to C_3 alcohols and reactive hydrocarbons present in C_4 to C_6 cuts using acid resins catalyst. The process units consist of alcohol and hydrocarbon purification sections, main adiabatic using an expanded bed technology reactor finishing and fixed-bed reactor, reactive distillation column for separation of ethers, alcohol separation unit consisting of water wash and standard distillation column.
MTBE, ETBE, TAME	Phillips process	The process produces octane enhancing ethers (MTBE, ETBE, TAME, TAEE) by reacting methanol or ethanol with isobutene or iso-amylenes. The process unit consists of two reactors, MTBE fractionator, methanol extractor and methanol fractionator.
MTBE, ETBE, TAME	Snamprogetti-SpA,	The process produces high octane, low vapor pressure oxygenates ethers (MTBE, ETBE, TAME) Process flow consists of two fixed-bed tubular reactor, product fractionation, methanol removal tower (by extraction with water) and finally methanol—water fractionator where methanol is recovered and recycled.
MTBE, TAME and heavier ethers	NEXETHERS ™	The process involves reaction of iso-olefins and methanol in presence of cation exchange resin. The process achieves a significant increase in isoolefin conversion by combining alcohol recovery and circulation with oxygenated removal.
Diisopropyl ether	UOP's oxypro process	It is a low lost, refinery based catalytic process for the production of diisopropyl ether using propylene derived from FCC. The process flow scheme consists of a fixed-bed reactor, fractionator for separation of light ends and propane, product recovery section where IPA and DIPOE are separated.

Source: Ancillotti et al., 1987; Brockwell et al., 1991; Ignatius et al., 1995; Krupa et al., 2004a, 2004b, 2004c; Refining Processes, 1998; Jezak, 1995; Rock, 1992; Tsai et al., 2002; Bader and Guesneux, 2005; Przelj, 1987; Chavel and Lefebvre, 1985.

16.1 METHYL TERTIARY BUTYL ETHER (MTBE)

MTBE was originally used as octane booster to offset the gasoline octane reduction due to phasing out of lead. MTBE fulfilled oxygen requirement while offering good octane blending value [Hunszinger et al., 2003]. MTBE is one of the most important oxygenates used in the production of lead-free gasoline and is produced on a large scale throughout the world. The incorporation of 10–15% MTBE in reformulated gasoline has achieved reductions of common air pollutants as compared to conventional gasoline [Woo, 2007]:

- 20–25% less CO
- 10–15% less unburned hydrocarbons
- About 30% less Pm
- 20–30% less benzene
- 15% less evaporative emissions
- Reduction of ground level ozone.

MTBE is produced by reaction of methanol with isobutylene contained in C_4 streams from thermal crackers in presence of ion exchange resin at 40–90 °C and pressure of 5–10 kg/cm². Catalytic cracking butylenes and field butanes are additional possible source of isobutylene.

Reaction involved during production of MTBE is:

| Isobutylene | Methanol | Methyl tertiary butyl ether |

Side reactions:
- Isobutylene hydration with formation of tertiary butyl alcohol (TBA)
  Isobutylene + Water → TBA
- Isobutene dimerization with formation of diisobutylene
  Isobutylene + Water → TBA
- Methanol dehydration with formation of dimethyl ether, DME
  Methanol + Methanol → DME

Today all etherification process licenses use basically the same type of catalyst, independent of reactor configuration. The catalyst is a macroreticular ion exchange resin based on sulfonate styrene divinyl benzene copolymer. The MTBE synthesis reaction is governed by a thermodynamic equilibrium and is shifted towards the MTBE formation at low temperature. Reaction kinetics on the other hand, is favorable at higher temperature. The reactor design is, therefore, a compromise between thermodynamic and kinetic considerations.

The acidic ion exchange resin catalysts used for MTBE production is susceptible to some impurities present in the feedstock which includes basic compound, cations and nitrites, surface compounds. These poisons neutralize the acid sites of the catalyst causing deactivation. Acetonitrile is the most significant poison in the FCC. These impurities are removed by washing thoroughly with water. After water wash, the feed is mixed with methanol and fed to reactor. The inlet temperature and pressure of the reactor is controlled and sufficient care is taken not to allow to increase in temperature of the reactor. Presence of butadiene in the feedstock also reduces catalyst activity. Butadiene, if present, in relatively high concentration in the feedstock can result in isomerization of the butadiene inside the matrix of catalyst. Until 1981, all MTBE production were by conventional technology using two fixed bed reactors in which first reactor is either a cycle reactor or tubular reactor, which performs most of the isobutylene conversion. The finishing reactor which performs at low temperature and low LHSV is to provide equilibrium conversion. Process flow diagram for MTBE by fixed-bed technology is given in

Fig. 16.1. The MTBE from the reaction mixture is separated in distillation column. In the catalytic distillation process, fixed-bed reactor which provides the bulk of conversion and the finishing reaction is done by installing catalyst in the column in a special packing to provide simultaneous reaction and distillation.

In the catalytic distillation (Fig. 16.2) process after the reaction, the equilibrium converted reaction mixture from reactor is fed to the catalytic distillation reaction column where the reaction is continued and the product MTBE is separated from the unreacted C_4 in the catalytic distillation. The C_4 distillation stream from the catalytic distillation column is fed to the methanol extraction column where water is used to separate methanol and water extract containing methanol is sent to methanol recovery column for separation of methanol. The C_4 stream is recovered from the top of the extract column. MTBE processes have been modified to tolerate butadiene so that the process can be placed upstream of butadiene extraction [Ancillotti et al., 1987].

The combined etherification technology where reaction of isoamylene in C_4 stream with methanol takes place in two reactors using palladium loaded ion

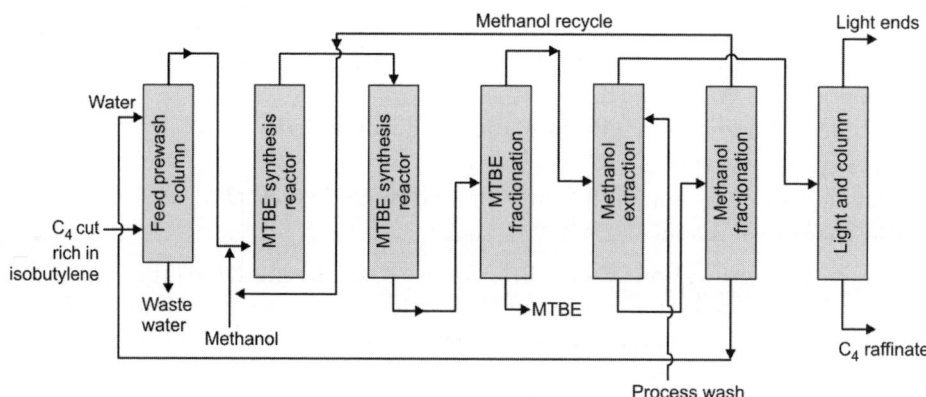

Fig. 16.1 MTBE by conventional process

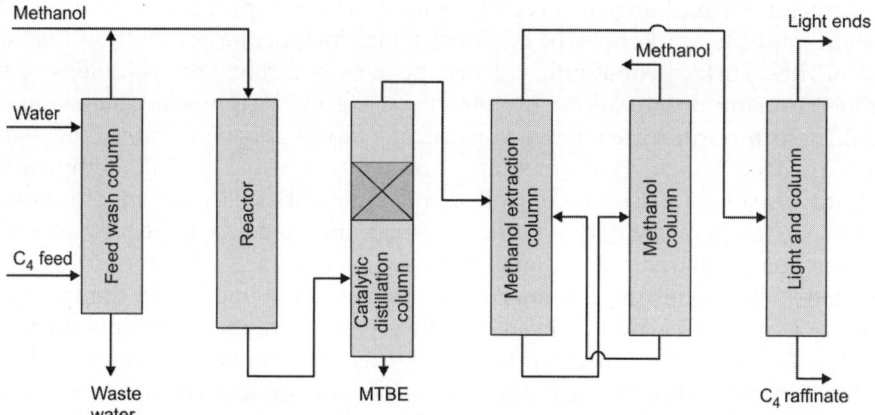

Fig. 16.2 MTBE by catalytic distillation process

exchange resin followed by catalytic distillation in a fractionation column containing patented catalytic packing that allows simultaneous catalytic reaction and distillation.

16.1.1 New Technologies for Replacement of MTBE

Due to restriction in the use or phasing out of MTBE, because of concern over water pollution problem, new process technologies are being worked out as an MTBE replacement. Possible new processing technologies are [Tsai et al., 2002]:

- Convert isobutene to iso-octene and then hydrogenate iso-octene to iso-octane.
- Retrofit an existing MTBE unit to ethyl tertiary butyl ether (ETBE) production.

The iso-octene/iso-octane technology is applicable to all typical feedstocks used in MTBE including pure isobutene, steam cracked C_4 raffinate, FCC butane-butene streams. The main reaction involved is the dimerization of isobutene to 2, 4, 4-trimethyl-1-pentene (DIB).

$$
\begin{array}{ccc}
\underset{\text{Isobutylene}}{H_3C-\overset{CH_2}{\underset{CH_3}{C}}} & + & \underset{\text{Methanol}}{H_3C-\overset{CH_2}{\underset{CH_3}{C}}} \longrightarrow \underset{\text{Di-isobutylene}}{[CH_3-\overset{CH_2}{\underset{}{C}}-CH_3]_2}
\end{array}
$$

2 i C_4 2,4,4-trimethyl-1-pentene (DIB)

In ETBE process, the mixed C_4 feed is combined with make-up ethanol at a controlled ratio to isobutene and the mixture is fed to the liquid phase, fixed-bed reactor containing ion exchange resin catalyst.

The main reaction involved is:

$$
\underset{\text{Isobutylene}}{H_3C-\overset{CH_2}{\underset{CH_3}{C}}} + \underset{\text{Ethyl alchol}}{HO-CH_2-CH_3} \longrightarrow \underset{\text{Ethyl tertiary butyl ether}}{H_3C-CH_2-O-\overset{H_3C \quad CH_3}{\underset{CH_3}{C}}}
$$

2 i C_4 + Ethanol ETBE

16.2 TERTIARY AMYL METHYL ETHER (TAME)

Amylene alkylation gives refiners two-fold advantages. It increases the volume of alkylate available while decreasing the Reid vapor pressure and olefinic content of gasoline blendstocks [Jezak, 1994]. Although TAME has slightly lower octane, it has favorable comparisons for vapor pressure, boiling point, energy density and water miscibility [Rock, 1992].

TAME is an another important oxygenate which is produced by reaction of methanol with Isoamylene, which is present in the FCC gases and steam cracker C_4 streams. Although TAME has slightly lower octane, it has favorable comparisons for vapor pressure, boiling point, energy density and water miscibility [Rock, 1992]. TAME is produced from isoamylene which is obtained as by-product from FCC and steam crackers.

How that a refiner can produce same volume of TAME as for MTBE using FCC olefins [Rock, 1992]? Preliminary estimates TAME is obtained by reaction of methanol with the reactive isoamylene. Out of the three isoamylene, 2-methyl-1-butene and 2-methyl-2-butene are reactive while 3-methyl-1-butene is non-reactive. Reaction involved is given below [Brockwell, 1991]:

CH_3
|
H_3C   C
     \\   // \\
      CH_2   CH_2
2-methyl-1-butene

CH_3
|
H_3C   C
     \\   // \\
      CH   CH_3
2-methyl-2-butene

CH_3
|
H_3C   ⊕ C
     \\   \\
      CH_2   CH_3 + CH_3OH
Tertiary butyl cation

CH_3
|
H_3C   C
     \\   / | \\
      CH_2   CH_3
            OCH_3

Tertiary amyl methyl ether (TAME)

In the conventional two fixed-bed reactors, etherification process is accomplished in two reactors using strong acid macro porous palladium loaded ion exchange resins as catalysts. For TAME applications, the palladium resin selectively hydrogenates dienes in the feed, ensuring a long etherification catalyst life and a water white clear product [www.adcats.alcoa.com]. Further separation of TAME is accomplished in the distillation column. Methanol from the top product of distillation is recovered by water extraction followed by distillation [Rock, 1992]. Process flow diagram for TAME by conventional process is given in Fig. 16.3. Similar to MTBE, the combine etherification technology for TAME consists of a fixed-bed reactor followed by a catalytic distillation unit which allows the isoamylene conversion in excess of 95%.

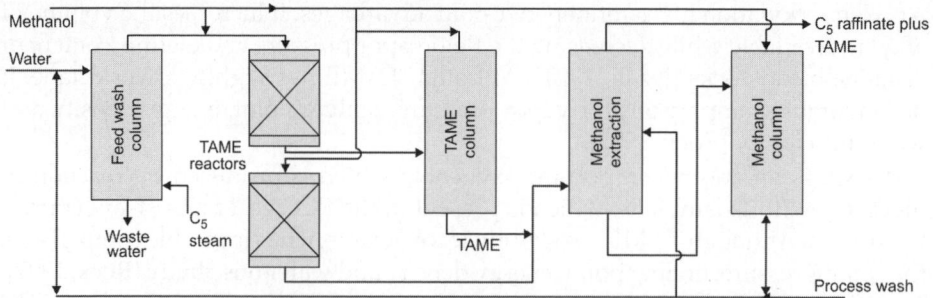

Fig. 16.3 TAME manufacture by conventional fixed-bed technology

16.3 MIXED ETHERS FROM C₅ STREAMS

Processes are available for manufacturing of oxygenates like MTBE, ETBE, TAME and/or higher molecular ether as octane reformulated gasoline components from C_1 to alcohol and reactive hydrocarbons present in C_4–C_8 cuts.

16.4 ETHYL TERTIARY BUTYL ETHER (ETBE)

ETBE has certain advantage over MTBE as $(R + M)/2$ of ETBE is about one number higher than MTBE and lower vapor pressure (blending RVP is 3.5 psi, which allows greater butane absorption) [Sagu et al., 1997].

The manufacture of ETBE is similar to MTBE. ETBE is made by reacting isobutylene with ethanol in presence of catalyst. The process flow diagram is same as MTBE except incase of ETBE, ethanol is used instead of methanol.

$$CH_2 = C(CH_3)_2 + C_2H_5OH \longrightarrow (CH_3)_3–C–O–CH_2CH_3$$

The major difference between MTBE and ETBE process are: (i) Less favorable equilibrium and a lower ethanol/isobutene ratio resulting in lower conversion, (ii) a higher water concentration recycled with ethanol resulting in a higher production of TBA and (iii) a higher level of ethanol impurities resulting in a shorter catalyst life time [Travers, 1998].

16.5 DIISOPROPYL ETHER (DIPE)

DIPE is a high octane low RVP oxygenate. It has high octane, low vapor pressure and excellent gasoline blending properties. DIPE is produced by acid catalyzed hydration of propylene with water. As large amount of propylene is available from refineries, diisopropyl ether and isopropyl ether can significantly increase oxygenated compound production. UOP oxypro process is unique and low cost refinery-based catalytic process for the production of DIPE. The various sections in the process are:

- *Depropanizer:* Separation of propane, thus increasing content of propylene in the feed.
- *Reaction section:* It consists of two reactors: (i) Isopropyl alcohol (IPA) reactor and (ii) DIPE reactor for synthesis of IPA and DIP respectively.
- *Distillation section:* Azeoptropic distillation of DIPE/IPA/water
- *Washing and drying:* Final washing and drying of DIPE product
- *Reactions:*

$$CH_3–CH = CH_2 + H_2O \longrightarrow CH_3–CHOH–CH_3$$
Propylene   Propanol
$$CH_3–CHOH–CH_3 + CH_3–CH = CH_2 \longrightarrow CH_3(CH_3)–O–CH(CH_3)_2$$
Propanol   + propylene   DIPE

The oxyproduct generates high octane number as compared to product from polymerization and alkylation of propylene [Krupa et al., 2004].

16.6 REFORMULATED GASOLINE (RFG)

Reformulated gasoline was coined in June 1989 after its appearance in Clean Air Act and is defined as gasoline whose chemical composition has been changed such that when burnt in vehicle has significant reduction in the air pollution and complies with tighter emission standards and regulations [Birsain et al., 1994]. Reformulated gasoline is now recognized option to reduce hydrocarbon emissions. Reformulated gasoline to reduce emissions is now a recognized option for reducing ozone and other air pollutants [Piel and Thomas, 1990]. Some of the major components of reformulated gasoline are alcohols, methanol, ethanol, tertiary butyl alcohol, ethers (MTBE, TAME, DIPE), alkylate, isomerate, polymers, reformate, FCC gasoline, hydrocracker gasoline, butanes, coker and visbreaker gasoline and pyrolysis gasoline. A typical RFG production scheme in a future refinery is given in Fig. 16.4. Octane range of major refinery gasoline streams is given in Table 16.6.

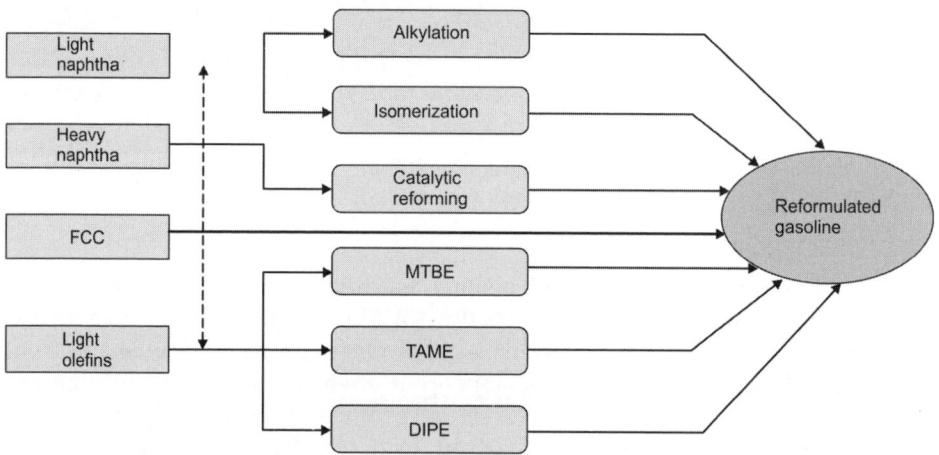

Fig. 16.4 RFG production scheme in a future refinery

Table 16.6   Octane range of major refinery gasoline streams

Stream	Octane range
Butanes	91–93
Straight run gasoline	55–73
Catalytic cracked gasoline	84–88
Hydrocracked gasoline	85–87
Coker gasoline	60–70
Alkylate	90–94
Reformate	81–96
Isomerate	

Source: Bir Sain et al., 1994.

REFERENCES

1. Ancillotti F, Pescarollo E, Szarmari E and Lazer L. MTBE from Butadiene rich C_4's, Product Fuel Blends Rate Solubility Problem. Hydrocarbon Processing, Nov 1987, p50.

2. Brockwell HL, Sarathy PR and Trotta R. Synthesize ethers. Hydrocarbon Processing, Vol 70, No. 7, 1991, p133.

3. Chavel A and Lefebvre G. Treatment of Olefinic C_4 and C_5 Cuts, in Petrochemical Processes Synthesis-Gas derivatives and Major Hydrocarbons. Editions Technip Paris, 1989, p195.

4. Ignatius J, Jarvelin H and Lindqvist P. Use TAME and Heavier Ethers to Improve Gasoline Properties. Hydrocarbon Processing, Feb 1995, p51.

5. Krupa S and Meister J. Huls Ether Process in Handbook of Petroleum Refining Processes, 3rd edition, edited by Meyers RA, McGraw Hills, 2004a, p13.3.

6. Krupa S, Meister J and Luebke C. UOP Ethermax Process for MTBE, ETBE and TAME in Handbook of Petroleum Refining Processes, 3rd edition, edited by Meyers RA, McGraw Hills, 2004b, p13.9.

7. Krupa S, Richardson L and Meister J. OxyProcess in Handbook of Petroleum Refining Processes, 3rd edition, edited by Meyers RA, McGraw Hills, 2004c, p13.19.

8. Jezak A. C_5 Alkylate: A Superior Blending Component. Hydrocarbon Processing, Feb 1994, p47.

9. Rock K. TAME Technology Merits. Hydrocarbon Processing, May 1992, p86.

10. Tsai MJ and Kolodziej Ching D. Consider New Technologies to Replace MTBE. Hydrocarbon Processing, Vol 82, No. 2, 2002, p81.

11. Bader JM and Guesneux S. Increase MTBE Plant Productivity. Hydrocarbon Processing, Oct 2005, p49.

12. Przelj M. Pool Octanes Via Oxygenates. Hydrocarbon Processing, Sep 1987, p68.

13. Refining Processes, 98. Hydrocarbon Processing's Refining Processes, 98. Hydrocarbon Processing, p53.

14. Sagu ML, Gupta AK and Prasada Rao TSR. Role of Oxygenates in Gasoline. Hydrocarbon Technology, Feb 1997, p52.

15. Woo C. Clean Fuel Trends in Asia: Contribution of MTBE Hydrocarbon Asia, July 2007, p86.

16. Travers P. Olefin Etherification in p231, Chapter 8, IFP Petroleum Refining, Part 3. Conversion Processes, edited by Leprince P. Editions Technip, 1998, p291.

17. Piel WJ and Thomas RX. Oxygenates for Reformulated Gasoline. Hydrocarbon Processing, July 1990, p68.

18. Bir Sain, Bhagat SD and Joshi GC. Some Refining Options to Meet Challenges of Reformulated Gasoline. Hydrocarbon Technology, May 15, 1994, p40.

17

Biofuels

Due to increasing cost of crude oil, increasing supply of heavier crude with higher sulfur there have been increasing interest during recent years for the production of biofuel in the form of gashol and biodiesel for improving the quality of environment as well as to provide environmentally friendly fuels. Globally, biofuels developments are primarily driven by few fundamental policy considerations: rural development, energy independence, reduced carbon footprint, to reduce dependency on imported oil and to improve the security to improve supply for transportation fuels [Casone, 2007; Stockle, 2007]. Historically, biofuel was used since the early days of the car industry. Nkolaus August Otto, the German inventor of combustion engine, conceived his invention to run on ethanol while Rudolf Diesel, the German inventor of diesel engine, conceived it to run on peanut oil [http://en.wikipedia.org/wiki/biofuel]. Renewable fuels are expanding worldwide due to increasing petroleum processing, regulations and commitment to reduce green house gases [Holmgren et al., 2007]. First generation fuels refer to biofuels made from sugar, starch, vegetable oil, or animal fats using conventional technology [UN biofuel reports (http://esa.un.org/un-energy/pdf/susdev.Biofuels. FAO.pdf)]. Second generation fuels are made from lignocellulosic biomass using advanced technical process. Fist generation biofuel includes vegetable oil, biodiesel, ethanol, butanol, methanol, alcohols, biogas, and solid biofuels. Second generation biofuels use biomass to liquid technology. The second generation biofuels which are under development are biohydrogen, bio-DME, biomethanol and high temperature upgrading (HTU) diesel, Fischer-Tropsch diesel, mixed alcohols (mixture of mostly ethanol, propanol and butanol with some pentanol, hexanol, heptanol and octanol [tripatlas.com]. Biofuels value chain is a multicomponent industry. Numerous organic feeds can be converted into various fuels and blending streams with established and developing processing technologies. The biofuel value chain is illustrated in Fig. 17.1 from the feedstock through process (biotransformation or thermochemical processing) to the principle biofuel products [Cascone, 2002]. Biomass gasification is getting importance as most potential feedstock. Biomass transformation routes offer numerous processing routes to process various feedstocks into desired transport fuels and blending components. The system has flexibility aided by diversity from varying feeds and processing technologies [Cascone, 2002]. The US and Brazil are two largest biofuel producers in the world.

Renewed interest on use of ethanol gasoline blends and petroleum and biodiesel blend has generated. Because of the availability of alcohol from sugar producing countries, use of ethanol gasoline blend has received momentum during recent years. Possibility of use of ethanol diesel blend has been also explored. New biofuels are emerging in market especially biodiesel which include hydrogenated vegetable oil and biomass-to-liquid diesel produced from cellulosic biomass [DePlan, 2008]. Driving force for biofuels development is rising green house gas emissions from conventional fuels and need for reduction in dependence on imported oil [Stockle, 2007].

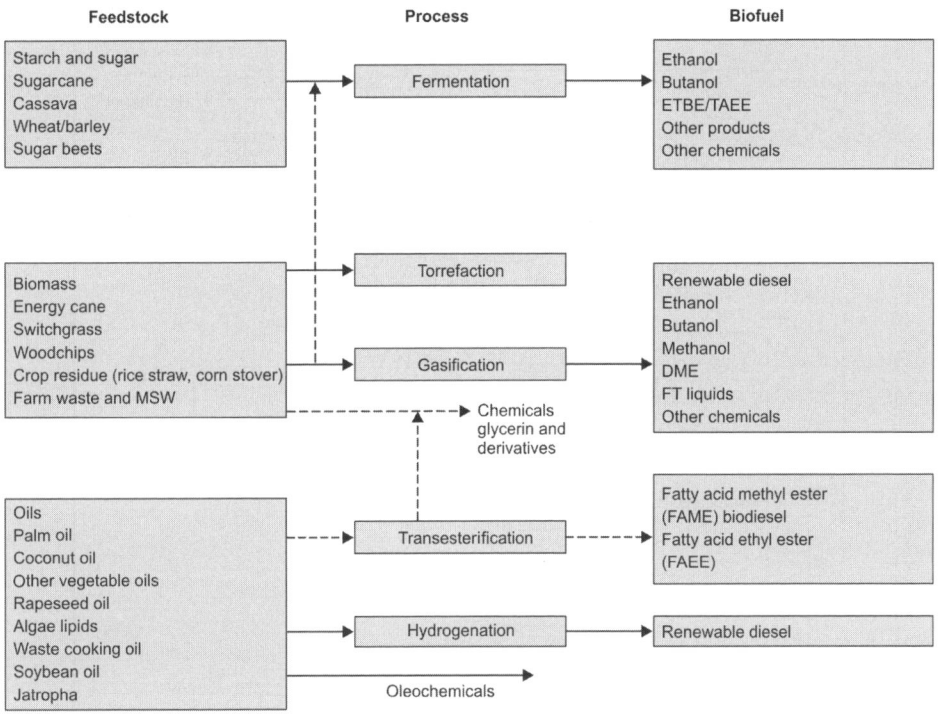

Fig. 17.1 Biofuels value chain

Source: Cascone, 2007.
Courtesy: Hydrocarbon Processing.

17.1 ETHANOL AS AN ALTERNATIVE TO GASOLINE

Alcohol has been globally accepted as alternative to gasoline and alcohol blend gasoline are being used in many countries in varying proportion ranging from 5–20% as ethanol. Even is being sold in Brazil pure ethanol as E-100 fuel. Ethanol accounts for about 40% of the fuel consumed by passenger vehicles [Sehagal, 2006]. Brazil and US together account for about 70% of world's ethanol production. Although ethanol is produced from food based feedstocks like molasses and corn.

However, as demand for ethanol is increasing, continuous efforts are being made all over the world to utilize alternate feedstock like biomass which is sustainable and inexpensive feedstock [Kumar et al., 2010]. Biomass transformation routes to biofuel is shown in Fig. 17.2. Manufacturing of biofuel is shown in Fig. 17.3.

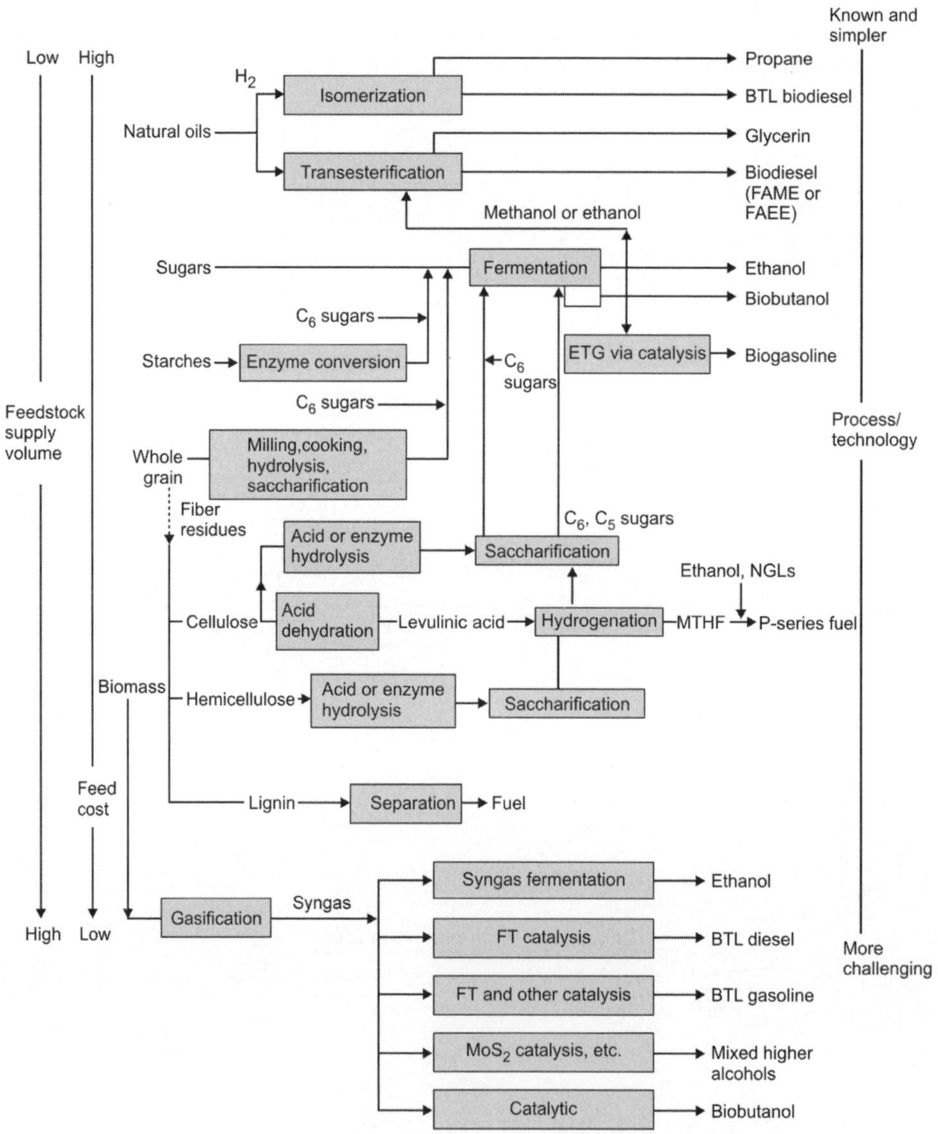

Fig. 17.2 Biomass transformation routes to biofuel

Source: Casone, 2005; Biofuels: What is Beyond Ethanol and Biodiesel, Hydrocarbon Processing, p95–109.
Courtesy: Hydrocarbon Processing.

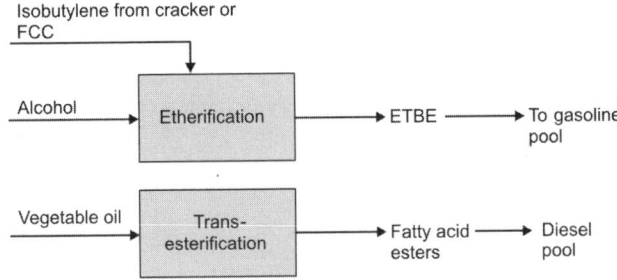

Fig. 17.3 Biofuel manufactures

The estimated total global production of ethanol at present is about 11 billion gallons per annum. IFP estimates that the present production of ethanol can meet just 2.7% of total global fuel demand. Being an oxygen carrier, ethanol helps fuel burn more fully and clearly thereby substantially reducing emissions of CO and CO_2 is resulting in reduction of 78% carbon emission as compared to fossil fuels. Ethanol blended fuels are environmentally friendly and produce 13% less green house gases than fossil fuel [Sehagal, 2006].

Although ethanol is produced from conventional raw materials like sugarcane, sugar beet, corn, wheat, etc., sugarcane being the largest single source for ethanol production. Other route which has been explored is biomass which is available in huge quantity. There are three types of feedstocks for ethanol production [Singh et al., 2008].

Sugars	:	Molasses, sugarcane, beat sweet sorghum and fruits
Starches	:	Corn, wheat, rice, potatoes, cassava, sweet potatoes, etc.
Lignocellulosic	:	Straw, bagasse, other agricultural residues, wood, energy crops
Algae	:	Ethanol production

Molasses is the major route for the manufacture of alcohol. Process flow diagram for the manufacture of alcohol from molasses is shown in Fig. 17.4. The process steps involved are:

Prefermenter	:	Growing of yeast
Fermentation	:	Molasses handling, fermenter feeding and fermentation of molasses
Yeast treatment	:	Treatment of yeast cell generated during fermentation and recycling system
Fermentation	:	Fermentation of molasses after addition of yeast and adjusting pH and alcohol recovery, recovery of alcohol from fermentation section, washing of fermenter alcohol
Distillation	:	Distillation of fermented wash, alcohol separation and separation of spent wash

The current cost of manufacturing ethanol and biodiesel is about ₹ 21/liter. Cost of ethanol production can be reduced by using improved agricultural practices to increase sugarcane production and deploying energy efficient ethanol dehydration methods like pressure swing adsorption and membrane separation [Chauhan, 2012].

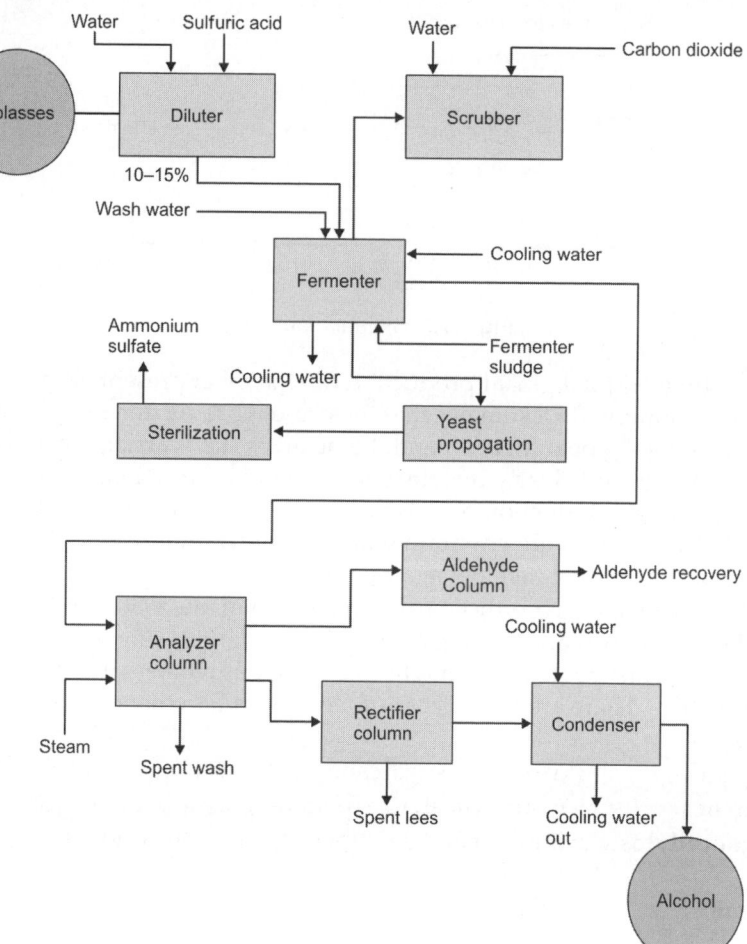

Fig. 17.4 Process of manufacturing of alcohol from molasses

Biofuels industry in India is also poised to make important contribution to meet energy needs by supplying clean environment friendly fuel. Ethanol industry is although mature in India, can benefit from improved agricultural practices, more efficient production processes and use of alternate feedstocks including cellulosic material. On the other hand, biodiesel industry is at incubation stage and large scale jatropha cultivation and the infrastructure for oil seed collection and oil extraction must be established before the industry can be placed on rapid growth track [www.unctad.org] [United National Conference on Trade and Development of the Biofuels Industry in India, April 16, 2011].

Lignocellulosic biomasses are getting importance as bioethanol feedstock. Lignocellulosic biomass refers to organic material such as wood chips, bagasse, straw, corn stalks, grass, etc. Some of the advantages of lignocellulosic material for ethanol are it is renewable, inexpensive, locally and domestically available. Lignocellulosic biomass is composed of cellulose (40–60%), hemicellulose (20–40%) and lignin (10–25%). Potential for ethanol from some cellulosic matters is given in

Table 17.1. The nature and availability of lignocellulosic feedstocks in different parts of the world depend on climate and other environmental factors, agricultural practices and technological developments. Some of the technological barriers which need to be addressed in efficient conversion of biomass to ethanol are:

- Pretreatment
- Sachatification of cellulose and hemicelluloses matrix
- Simultaneous fermentation of hexose and pentose sugars.

The most commonly pretreatment methods used are steam explosion and dilute acid prehydrolysis, which are followed by enzymatic hydrolysis cellulose in the lignocellulosic materials to fermentable reducing sugars, fermentation of sugar into ethanol and downstream processing of ethanol.

Typical flow diagram for manufacture of alcohol from biomass is given in Fig. 17.5.

17.2 BIODIESEL

Biodiesel is renewable, ecofriendly, clean burning fuel. Biofuel can be produced from virgin or used vegetable oils (both edible and non-edible) and animal fats. Many countries all over the world are using biodiesel blended with petroleum diesel in different compositions [www.gisdevelopment.net]. National standard for biodiesel in different countries is given in Table 17.2. Vegetable oil-based diesel can offer better integration within crude oil refineries for fuel blending. Biodiesel consists

Table 17.1   Potential for ethanol from cellulosic matter

Feedstock	Gallons/dry
Bagasee	112
Cornstover	113
Rice straw	110
Forest thinning	82
Hardwood dust	101
Mixed paper	116

Source: Report of The Committee on Development of Biofuel Planning Commission, Govt. of India, 2003.

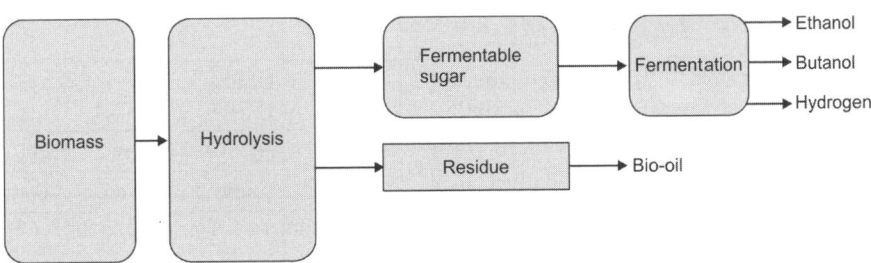

Fig. 17.5 Alcohol from biomass

of the monoalkyl esters formed by a catalyzed reaction of triglycerides in the oil or fat with a simple monohydric alcohol [Gerpen, 2005].

Options currently available for renewable diesels are given below [Ray, 2009]:

Feedstock : Jatropha, rapeseed, soybean, tallow, algal oils, palm oils

Technologies : Biodoesel (FAME), ecofining™ process, stand-alone hydro-processing/isomerization), co-processed hydroprocessing

Some of the major advantages of biodiesel are: Low toxicity, derived from renewable sources, biodegradable, reduced engine due to lubrication, reduced emission level of particulates, carbon oxides, sulfur oxides and under some conditions, nitrogen oxides, reduced petroleum imports [Fukuda et al., 2001; Kaieda et al., 2001]. Compared to petroleum based diesel, the cost of biodiesl is a major barrier to its commecialization as 60–90% is the cost feddstock [Al-Zuhair, 2006] of palm oil is being used for production of biodiesel in Malaysia which is one of the leading nations in palm oil producer and exporter. Al-Zuhair et al., 2005, has reported use of immobilized lipase for the production of biodiesel from waste palm oil as an alternative for reducing cost of virgin palm oil feedstock. Elumbaring-Rayat et al. 2006, has reported the use of the coconut oil fatty acid deodorizer distillate (COFAD) which is about half of the cost of refined oil.

Biodiesel is made through transesterification where generally methyl, ethyl, or higher alcohol esters produced from triglycerides. The reaction of triglycerides with alcohol is catalyzed by either acid or base. Properties of biodiesel and green diesel are given in Table 17.3. Reaction of transesterification process is given below.

CH₂COOR'''			NaOH	CH₂OH		RCOOR'''

$$
\begin{array}{ccccccc}
CH_2COOR''' & & & & CH_2OH & & RCOOR''' \\
| & & & \xrightarrow{NaOH} & | & & \\
CH_2COOR'' & + & 3ROH & \xleftarrow{\hspace{1cm}} & CHOH & + & RCOOR'' \\
| & & & & | & & RCOOR' \\
CH_2COOR' & & & & CH_2OH & &
\end{array}
$$

Vegetable oil (Triglyceride)	**Alcohol**	**Catalyst**	**Glycerol**	**Alkyl ester** (Biodiesel)

Source: Biodiesel: Clean Fuel of the Future; Pramanik T and Tripathi S; Hindustan Petroleum Corp Ltd, New Delhi, India.

Courtesy: Hydrocarbon Processing.

Table 17.2 National standards for biodiesel

	Austria	Czech Republic	France	Germany	Italy	Sweden	US
Standard/ specification	ONC 1191 6507	CSN 65 Official	Journal 51605	DIN V 10635	UNI 155436	SS D6751	ASTM
Date	July 1997	Sep 1998	Sep 1997	Sep 1997	April 1997	Nov 1996	2001
Application	Fame	Rme	Vome	Fame	Vome	Vome	Femae
Density @ 15 (°C, g/cm³)	0.85–0.89	0.87–0.89	0.87–0.90	0.875–0.90	0.86–0.90	0.87–0.90	–

Contd.

Table 17.2 National standards for biodiesel *(Contd.)*

	Austria	Czech Republic	France	Germany	Italy	Sweden	US
Visc @ 40 (°C, mm²/s)	3.6–5.0	3.5–5.0	3.5–5.0	3.5–5.0	3.5–5.0	3.5–5.0	1.9–6.0
Distillation, (95% °C)	–	–	< 360	–	< 360	–	360
Flash point (°C)	> 100	> 110	> 100	> 110	> 100	> 100	> 130
CFPP (Cold filter plug pt) (°C)	0/–15	–5	–	0/–10/ –20	–	–5	–
Pour point (°C)	–	–	< –10	–	< 0/< –15	–	–
Sulfur (% mass)	< 0.02	< 0.02	–	< 0.01	< 0.01	< 0.01	< 0.05
Water (mg/kg)	–	< 500	< 200	< 300	< 700	< 300	500
Cetane No.	> 49	> 48	> 49	> 49	–	> 48	> 47
Neutral No. Mg KOH/g	< 0.8	< 0.5	< 0.5	< 0.5	< 0.5	< 0.6	–
Methanol (% mass)	< 0.2	–	< 0.1	< 0.3	< 0.2	< 0.2	–
Ester content (% mass)	–	–	> 96.5	–	> 98	> 98	–

Courtesy: Hydrocarbon Process [Pramanik and Tripathi, 2005].

Table 17.3 Comparison of biodiesel and green diesel properties

	Mineral ultra low sulfur diesel	Biodiesel (FAME)	Green diesel
%O_2	0	11	0
Density (g/mL)	0.84	0.883	0.78
Sulfur content	< 10	< 10 ppm	< 10 ppm
Heating value (lower) (MJ/kg)	43	38	44
Cloud point (°C)	–5	–5	–5 to –30
Distillation (10–90% pt)	200–350	340–355	265–320
Cetane	40	50	80-90
Stability	Good	Marginal	Good

Source: Ray, 2009.

Comparing between synthetic diesel and FAME manufacturing is given in Table 17.4. Properties for FAME originating from different raw materials are given in Table 17.5. Synthetic diesel product in comparison with other diesel fuels is given in Table 17.6.

Table 17.4   Comparing between synthetic diesel and FAME manufacturing

Feature	BTL-diesel	FAME
Raw material origin	Flexibility to use several vegetable oils and animal fats	Fixed to specific vegetable oil
Additional feeds	Hydrogen	Methanol
Product quality	Excellent blending properties	Limitations with blending, stability issues, cold property issues
By-product	Handled in refinery	Glyserine
NO_x aspects	Reduces NO_x emissions	Increase NO_x emissions
CO_2 balance (kg) CO_2/kg of fuel	0.5–1.5	1.4–2
Logistics	Can use refinery logistics; no restrictions for blending	Requires investment for separate logistics
Investment considerations	Large units utilizing integration with oil refinery	Small independent units

Courtesy: Hydrocarbon Processing, Feb 2006, 61 [Koskinen et al., 2006].

Table 17.5   Properties for FAME originating from different raw materials

Raw material	Melting point (°C)	Cetane no.
Rapeseed oil, soybean oil	−10–0	55–58
Sunflower oil	−12	52
Olive oil	−6	60
Cotton seed oil	−5	55
Corn oil	−10	53
Coconut oil	−9	70
Palm oil	14	65

Courtesy: Hydrocarbon Processing, Feb 2006 [Koskinen et al., 2006].

Table 17.6 Synthetic diesel product in comparison with other diesel fuels

	Synthetic diesel	GTL diesel[1]	FAME	EN590/2005 diesel fuel, summer grade
Density at 15 °C (kg/m^3)	775–785	770–785	~885	~835
Viscosity at 40 °C (mm^2/s)	2.9–3.5	3.2–4.5	~4.5	~3.5
Cetane number	84–99[2]	73–81	~51	~53
10% distillation (°C)	260–270	~260	~340	~200
90% distillation (°C)	295–300	325–330	~355	~350
Cloud point (°C)	–5–30	+5–25	0–5	~ –5
Heating value (lower) (MJ/kg)	~44	~43	~38	~43
Heating value (MJ/L)	~34	~34	~34	~36
Polyaromatics content (%wt)	0	0	0	~4
Oxygen content (%wt)	0	0	~11	0
Sulfur content (mg/kg)	< 10	< 10	< 10	< 10

[1] gas to liquid (GTL).
[2] Blending cetane number.
Source: Pramanik and Tripathi, 2005.
Courtesy: Hydrocarbon Processing, Feb 2006 [Koskinen et al., 2006].

Process

The methanol esterification of vegetable oil can be achieved using a homogeneous catalyst operated in batch or continuous mode. Alkali catalyzed systems use NaOH, KOH or sodium methylate while acid catalyzed system uses HCl, H_2SO_4 or sulfonic acid.

The process steps involved in the manufacture of biodiesel are: Pretreatment of vegetable oil, catalyst synthesis and mixing of vegetable oil with methanol in separate reactors, transesterification, methanol recovery, glycerin separation washing and refining of biodiesel. Block diagram of transesterification process is shown in Fig. 17.6.

Characteristics of transesterified vs raw jatropha oil is given in Table 17.7. Component analysis of jatropha seeds (%wt) is given in Table 17.8.

17.3 BIOFUEL FROM ALGAE

There has been recent interested in algae as one of the promising sources of biofuel for production of alcohol and biodiesel. Compared to the first generation agricultural crops that are grown seasonally, usually from food based sources, algae is a sustainable source of biomass and can be produced and harvested three to four times a day continuously year around. Microalgae grow much faster than land grown plants, often 100 times faster. Microalgae are diverse group of primitive and simply organized plant-like organisms and have been investigated for the production of a number of different biofuels including biodiesel, bio-oil, bio-syngas, methane, ethanol, butanol and biohydrogen [Tuli et al., 2009]. Different technologies

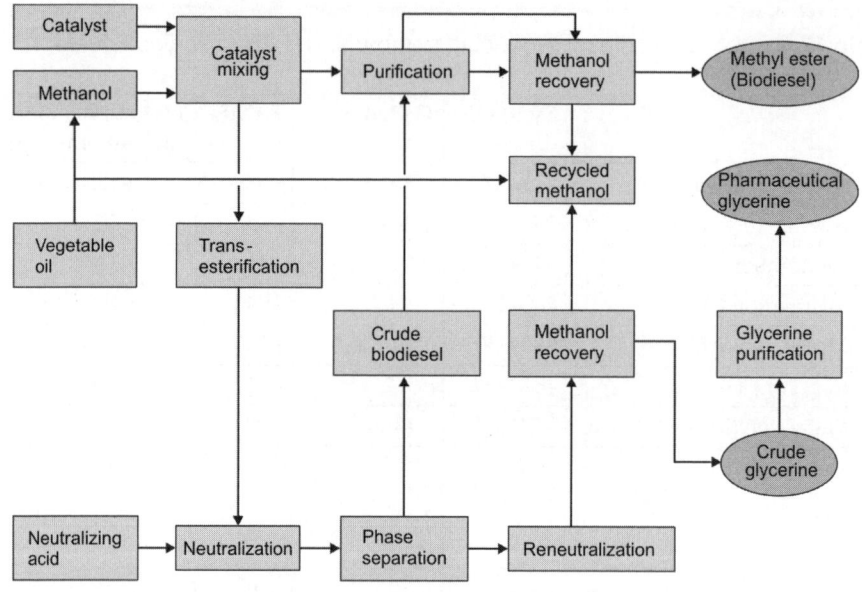

Fig. 17.6 Block diagram of transesterification process

Source: Pramanik and Tripathi, 2005.
Courtesy: Hydrocarbon Processing, Feb 2005.

Table 17.7 Characteristics of transesterified vs raw jatropha oil

Parameter	Jatropha oil, raw	Jatropha oil, transesterified
Density (g cm -3 at 20 °C)	0.920	0.879
Flash point (°C)	236	191
Cetane no. (ISO 5165)	23–41	51–52
Viscosity (mm^2/s at 30 °C)	52	4.84
Neutralization number (mg KOH/g)	0.92	0.24
Total glycerin (%)	–	0.088
Free glycerin (%)	–	0.015
Sulfur content (ppm)		0–13
Sulfated ash (%)	–	0.014
Methanol (%)	–	0.06

Source: Pramanik and Tripathi, 2005.
Courtesy: Hydrocarbon Processing, Feb 2005.

Table 17.8 Component analysis of jatropha seeds (%wt)

Moisture	6.2
Protein	18
Fat	38
Carbohydrate	17
Fiber	15.5
Ash	5.3

Courtesy: Hydrocarbon Processing, Feb 2005 [Pramanik and Tripathi, 2005].

including chemical flocculation, biological flocculation, filtration, centrifugation and ultra-aggregation have been investigated for microbial harvesting [Mittal, 2009]. Microalgae have much faster growth than terrestrial crops. More than 99.5% algae biomass produced worldwide is mainly from seaweeds farmed near seashore. More than one hundred firms across the world are working and harvesting of algae biomass. Biofuel from algae is the third generation biofuel and can be an ideal solution for India's impending fuel crisis [Venkatraman, 2012]. Algae can be used as source for biofuel, bioethanol and biobutanol. Some of the advantages for algae in Indian conditions are country's enormous diversity, vast coastal line, sufficient solar energy [Venkatraman, 2014]. Algae does not compete with food crops and can grow in places away from forests without affecting eco and food chain systems (Fig. 17.7).

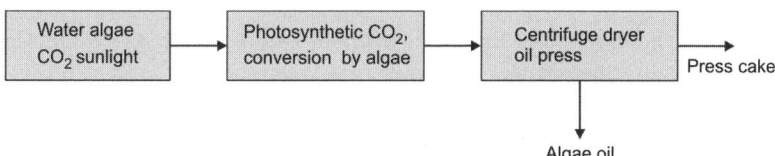

Fig. 17.7 Process flow diagram of biofuel from algae

REFERENCES

1. Al-Zuhair S, Isa D and Nour M. The Use of Immobilized Lipase for the Production of Biodiesel, from Waste Palm Oil: Possibilities and Challenges. Proceedings of The 11th APCChE Congress 2006 27030 August 2006 Kuala Lumpur.

2. Casone R. Biofuels: What is Beyond Ethanol and Biodiesel? Hydrocarbon Processing, Sep 2005, p95.

3. DePlan A. Integrating Biofuels in Refinery Optimization Models. Hydrocarbon Processing, Sep 2008, p337.

4. Elumbaring-Rayat ACM, Fabian B and Agapay RC. Technoeconomic Perspective and Biofuel Prospects of Cocobiodiesel. Proceedings of the 11th APCChE Congress 2006 27030 August 2006 Kuala Lumpur.

5. Fukuda H, Konodo A and Noda H. Fuel Production by Transesterification of Oils. J Biosc Bioengg, Vol 92, No. 5, 2001, p405.

6. Gerpen JV. Biodiesel Processing and Production. Fuel Processing Technology, Vol 86, 2005, p1097.

7. Holmgren J, Gosling C, Marinangeli R, Marker T, Faraci G and Perego C. New Developments in Renewable Fuels Offer More Choices. Hydrocarbon Processing, Sep 2007, p67.

8. http://powerenery.com.

9. http://standeredalcohol.com/biofuel.htm. Hydrocarbon Processing, Feb 2006.

10. Kaieda M, Samukawa T, Kondo A and Fukuda H. Effect of Methanol and Water Contents on Production of Biodiesel Fuel from Plant Oil Catalyzed by Various Lipases in a Solvent Free System. J Biosc Bioengg, Vol 91, No. 1, 2001, p12.

11. Koskinen M, Sourander M and Nurminen M. Apply a Comprehensive Approach to Biofuels. Hydrocarbon Processing, Feb 2006, p81.

12. Kumar R, Puri SK, Tuli DK and Malhotra RK. Exploring Agricultural Biomass for Ethanol Production. Compendium 16th Refinery Technology Meet, Feb 17–19, Kolkata, India, Organized by Center for High Technology and IOCL, Delhi.

13. Mittal A. Algae Farming—Resource Potential and Utilization, Part I. Chemical Industry Digest, Sep 2009, p85.

14. Pelmans L and Papageorgiou A. PREMIA—International AMF Activities: Biofuels in India, Dec 2005.

15. Pramanik T and Tripathi S. Biodiesel: Clean Fuel of the Future. Hydrocarbon Processing, Feb 2005, p50.

16. Ray A. Latest Technology Developments in Biofuels. Lovraj Kumar Memorial Lecture Annual Workshop, 2009. Managing Carbon Footprints in The Process Industry, New Delhi, Nov 26–27, 2009.

17. Report of The Committee on Development of Biofuel Planning Commission, Govt. of India, 2003.

18. Sehagal JM. Emergence of Ethanol as Global Alternative to Gasoline. Chemical, Weekly, Sep 26, 2006, p193.

19. Stockle M. Biofuels are Coming—How will Affect Refiners? Hydrocarbon Processing, Feb 2007, p77.

20. Al-Zuhair Sulaiman, Isa Dino and Nour Mutasim. The use of Immobilized Lipase for the Production of Biodiesel from Waste Palm Oil: Possibilities and Challenges.

21. Singh MP, Tuli DK, Malhotra RK and Kumar A. Ethanol from Lignolcellulosic Biomas: Prospects and Challenges. Journal of the Petrotech Society, June 2008, p39.

22. Tuli DK, Singh MP and Upreti MK. Biofuels from Microalgae. Petrofed, Vol 7, No. 3, 2009, p27.

23. UN biofuels report, (http://esa.un.org/un-energy/pdf/susdev.Biofuels.FAO.pdf).

24. Venkatraman NS. Need to Focus on Renewable Energy. Chemical Industry Digest, Annual, Jan 2014, p71.

18

Hydrogen Production and Management in Petroleum Refinery

Hydrogen is one of the important feedstocks in petroleum refinery. Historically, refinery hydrogen consumption has increased as refiners increase the degree of conversion process heavier and more sour crude. However, there has been tremendous increase in the utilization of hydrogen during recent years due to recent stringent environmental standards for reduction in sulfur, aromatics and improvement in quality of gasoline, diesel and fuel oil. The requirement of cleaner transportation fuels for improving the air environment quality and improved gasoline quality for reducing aromatics and improvement in cetane number of diesel has resulted increased installation of hydroprocessing and hydrocracking processes in modern refinery resulting in increased hydrogen requirement in near future. This will necessitate additional hydrogen supply in refinery. Although refinery gases may be source of hydrogen, however, it will be not sufficient to meet the enhanced requirement of hydrogen due to increased number of hydrocracking and hydroprocessing units.

There has been tremendous increase in hydrogen demand in India also which is largely driven by the ever increasing demand of middle distillate fuels, stringent standard for gasoline and diesel, and legislation necessitating sulfur reduction to meet the demand of ultra low sulfur in gasoline and diesel, cetane increase in diesel, density reduction, etc. This has led to a higher rate of growth of the capacities of hydropressing units in India and requirement of hydrogen.

18.1 HYDROGEN REQUIREMENT IN A REFINERY

Hydrogen requirement refinery has increased during past decade due to the requirement for clean transportation fuels (especially ultralow sulfur fuels), an increase in the application of residue conversion technologies, often accompanied by heavier crude processing, together with a growth in demand for petrochemicals and realization of the potential benefits of refinery/petrochemical integration [www.fosterwheeler.fi, Phillips, 2000]. However, need to abate refinery CO_2 emissions is receiving increasing interest and hydrogen production is likely to come under security as approximately 10 tons of CO_2 is produced per ton of hydrogen produced [Phillips, 2000]. Demand of hydrogen varies from plant to plant depending on crude oil characteristics, processing scheme and products and their characteristics and environmental norms for various products. Balancing hydrogen consumption is a major issue in a modern refinery [Hartmman, 2001]. Pure hydrogen is used in the refinery for the following purposes:

- To increase partial pressure of hydrogen in the systems operating under hydrogen environment and consuming hydrogen such as hydrotreating, hydrocracking, hydrofinishing and reforming.
- To act as reducing agent for catalysts such as steam reforming.
- As high calorific value fuel in a admixture with other hydrocarbons.

18.2 HYDROGEN PRODUCTION TECHNOLOGIES

As hydrogen requirement is increasing, effective management, current hydrogen infrastructure and planning for future requirements require careful selection of best combination of recovery, expansion, efficiency, improvements, purification and new hydrogen supply [www.pewclimate.org] [Davis and Patel, 2004]. Some of the options to meet hydrogen requirement are [Abrardo, 1995]:

- Recover hydrogen contained in various off gas streams that are sent to fuel header
- Expand existing on purpose hydrogen producing facilities (units where hydrogen is the primary product rather than by-product)
- Install new on purpose hydrogen capacity
- Buy pipeline/onsite hydrogen (produced outside the refinery).

There are two major routes to meet the hydrogen requirement of refinery: (1) off gas hydrogen recovery and (2) hydrogen production through various reforming and partial oxidation processes.

18.2.1 Off Gas Hydrogen Recovery

Brief description of these processes is given in Table 18.1. Typical hydrogen content of off gases from a refinery is given in Table 18.2 [Abrardo and Khurana, 1995]. Hydrogen recovery technologies include membranes, adsorption systems and cryogenics. Each technology has its own advantage and disadvantages [Abrardo and Khurana, 1995]. Comparison of hydrogen recovery and purification technology is given in Table 18.3.

Table 18.1   Off gas recovery system

Off gas recovery system	Description of process
Membrane	Typical membrane process consists of: Pretreatment for removal of entrained liquids, preheat feed before gas enters the membrane separators and membrane separator configurations. Typical hydrogen purity is 90–98% and in some cases 99.9%. Membrane separators are compact bundles of hollow fibers contained in a coded pressure vessel. The pressurized feed enters the vessel and flows on the outside of the fibers (shell side). Hydrogen permeates selectively through the membranes to the inside of the hollow fibers (tube side), which is at lower pressure. A common disadvantage of membranes is that

Contd.

Table 18.1   Off gas recovery system *(Contd.)*

Off gas recovery system	Description of process
	hydrogen is produced at pressures much lower than the feed pressure and may require additional recompressing. Off gas membrane recovery system is given in Table 18.4 [Abrardo and Khurana, 1995].
Adsorption	Pressure swing adsorption has been very commonly used in gas separation and produce hydrogen of high purity 99.9%+, however, recoveries are lower than other processes. Feed impurities less of a problem compared to membranes. Heavy hydrocarbons, however, are a problem if they are irreversible. If a hydrogen product with a purity of 99%+ is required, PSA is obvious choice. PSA recovery system is given in Table 18.5.
Cryogenic	Cryogenic technology is the highest cost capital alternative and its application has been very limited. There has been limited to larger capacities when liquids recovery, such as a C_3^+ hydrocarbon cut, is required. Cryogenic unit has the ability to deliver separate hydrocarbon byproduct streams. Cryogenic separation is expensive for small capacities. Recovery of NGL is given in Table 18.6.

Source: Zagoria, 2003; Abrardo and Khurana, 2003; Patel et al., 2005; Gas Processes, 2006.

Table 18.2   Typical hydrogen content of various off gases

Off gas source	Typical hydrogen, concentration, %vol
Naphtha reformer	65–90
Hydroprocessing High pressure Low pressure	75–90 50–75
Fluid catalytic cracking	10–20
Toluene HAD	50–60
Ethylene manufacture	70–90
Methanol manufacture	70–90

Source: Abrardo and Khurana, 1995

Table 18.3   Hydrogen recovery and purification technology characteristics

Characteristics	Membrane	Adsorption	Cryogenics
Hydrogen purity (%)	< 95	99.9+	95–99
Hydrogen recovery (%)	< 90	75–90	90–98
Hydrogen product pressure	<< feed pressure	Feed pressure	Variable
By-products available	No	No	Yes
Feed pressure (psig)	250–1,800	150–800	250–500

Source: Hydrocarbon Processing, (Patel et al., 2005).

Table 18.4   Refinery off gas membrane recovery systems

	Case 1	Case 2
Off gas sources	**Hydrocracker HP purge.**	**Hydrocracker LP purge**
Feed pressure (psig)	1,800	270
Feed flow (MMscfd)	3.2	10.2
Feed composition (mol%)		
H_2	74.6	61.9
C_1		21.1
C_2		10.0
C_3^+	7.0	
H_2 product pressure (psig)		xz17.7
H_2 product flow (MMscfd)	2.3	6.5
H_2 recovery (%)	90.9	1.4
H_2 product purity (mol%)	94.3	450

Source: Abrardo and Khurana, 1995.

Table 18.5   Refinery off gas PSA recovery systems

	Case 1	Case 2
Feed pressure (psig)	395	310
Feed flow (MMscfd)	20	65
Feed composition (mol%)		
H_2	80	66.8
N_2	–	3.2
C_1	15	16.5
C_2	3	7.8
C_3^+	2	5.2
H_2O	–	0.5
H_2 product pressure (psig)	385	300
H_2 product flow (MMscfd)	13.8	37.4
H_2 recovery (%)	86	88
H_2 product purity (mol%)	99.9	99.9

Source: Abrardo and Khurana, 1995.

Table 18.6   Cryogenic hydrogen purification with NGL recovery

Feed pressure (psig)	260
Feed flow (MMscfd)	56.8
Feed composition (mol%)	
H_2	65.3
N_2	2.3
C_1	24.2
C_2	5.4
C_3^+	2.9
H_2 product pressure (psig)	135
H_2 product flow (MMscfd)	35.6
H_2 recovery (%)	92.3
H_2 product purity (mol%)	96.1
C_3^+ product pressure (psig)	30
C_3^+ recovery (%)	78.8

Source: Abrardo and Khurana, 1995.

18.2.2   Hydrogen Production

Steam reforming, autothermal reforming and partial oxidation are three major hydrogen production technologies used in refineries. Some of the other pathways to hydrogen production are gasification of coal or biomass, high temperature nuclear reactor to thermally split water and electrolysis of water with electricity from renewable sources like wind turbines and solar photovoltaic cells. Petrocoke gasification is also getting importance in refineries (Table 18.7).

Table 18.7   Hydrogen production technology

Primary method	Process	Feedstock
Thermal	Steam reforming autothermal reforming	Natural gas/naphtha
	Thermochemical water splitting	Water
	POX/gasification integrated gasification combine cycle	Hydrocarbons/pet residue/coal/biomass
	Pyrolysis	Biomass
	Gasification	Petrocoke, coal, biomass
Electrochemical	Electrolysis	Water
	High temperature electrolysis	Water
	Photoelectrochemical	Water
Biological	Photobiological	Water and algae strains
	Anaerobic digestion	Biomass
	Fermentive microorganisms	Biomass

- *Steam reforming:* Steam reforming of refinery natural gas/refinery off gas, naphtha combined with purification by pressure swing adsorption or membrane process is two major options commonly used in refineries. Various steam reforming technologies are [Patel et al., 2005, Christensen and Pridahi,1994]:
  - o Conventional steam reforming
  - o Partial oxidation (POX)
  - o Catalytic partial oxidation (CPO)
  - o Combined reforming
  - o Combined reforming with prereformer
  - o Gas-heating reforming (GHR)
  - o Autothermal reforming
  - o Combined autothermal reforming (CAR)
  - o Kellogg heat reforming exchanger system (KRES).

Process steps involved in hydrogen production using steam reforming are:
- *Feed purification:* Removal of sulfur compounds by hydrogenation followed by reaction with zinc oxide
- *Steam reforming:* Steam reforming of natural gas/off gases (methane), naphtha
- *CO shift reactor:* Conversion of CO to CO_2
- *Purification:* Separation of hydrogen from CO_2.

Vaporized naphtha or natural gas mixed with recycled hydrogen at about 390 °C is sent to hydrogenation reactor containing Co-Mo catalyst, sulfur compounds thiophenes, and mercaptans are converted to hydrogen sulfide. The hydrogenated feed is sent to series of reactor where hydrogen sulfide reacts with zinc oxide and converted to zinc sulfide. The desulfurized feedstock then goes to super heater to increase the temperature of desulfurized feed to about 500 °C. The preheated feed containing steam then enters the steam reformer containing nickel catalyst. Steam reformer is vertically fired tubular reactor containing nickel as catalyst. Normally reformer operates at exit temperature of about 860°C metallurgy and pressure of 23.5 kg/cm². Reforming reaction is endothermic heterogeneous catalytic reaction. Reformer tubes high alloy following reaction takes place in the reformer. Process flow diagram for production of hydrogen from steam reforming is shown in Fig. 18.1 and typical design condition and composition of naphtha feed to reformer is given in Table 18.8.

- *Steam reforming reaction:*

$$C_nH_m + nH_2O \longrightarrow nCO + (m/2 + n) H_2$$
$$CH_4 + H_2O \longrightarrow CO + 3H_2$$

Reformed gases from the steam reformer after heat recovery goes to high temperature shift converter where CO reacts with steam and is converted to CO_2 in presence of iron catalyst using chrome oxide as stabilizer. The temperature in the high temperature shift reactor is around 315–425 °C. The shift reaction is exothermic and heat of reaction is recovered in series of heat exchanger. Finally the cooled gases are sent to purification section for separation of pure hydrogen.

- *Shift reaction:*

$$CO + H_2O \longrightarrow CO_2 + H_2$$

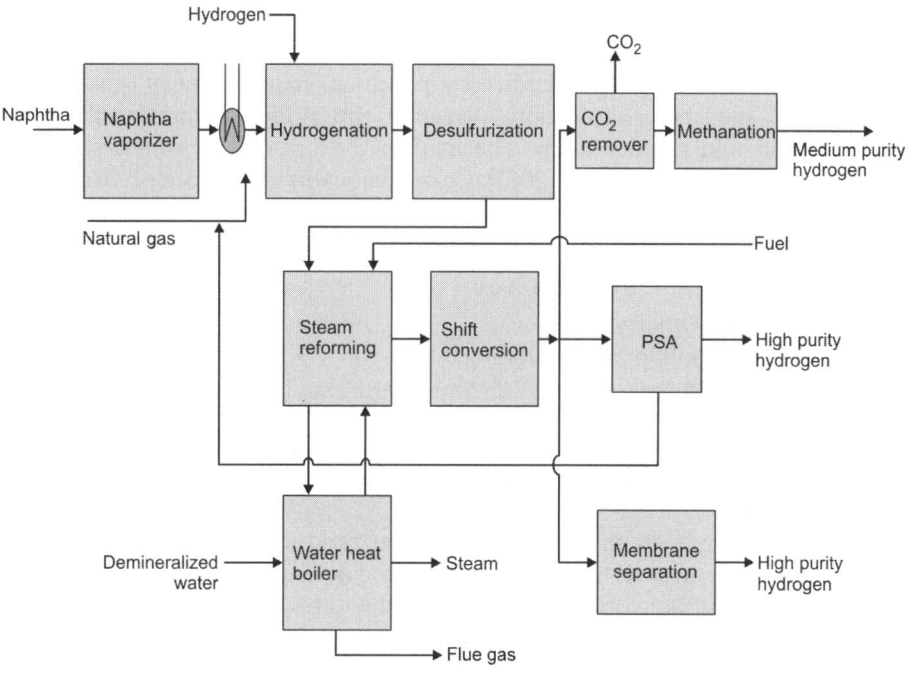

Fig. 18.1 Hydrogen production from steam reforming

Table 18.8   Typical design condition and composition of naphtha feed to reformer

	Design condition for naphtha reformer	Feed to reformer
Specific gravity (15/15)	0.73	0.71
PONA analysis (wt %vol) method		%vol
Total paraffins	54–76	78.4
Naphthenes	19-27	14.6
Aromatics	< 10	7
Sulfur content (wt ppm)	1000	< 0.5
Chlorides (wt ppm)	4	< 0.1
Arsenic (wt ppm)	4	< 0.1
Lead (wt ppb)	100	< 0.1
Nickel (wt ppm)	100	Not indicated
Nitrogen (wt ppm)	4	Not indicated
Vanadium (wt ppm)	100	Not indicated
Copper (wt ppm)	20	Not indicated

Courtesy: CPCL, Chennai.

- *Hydrogen from gasification of heavy liquid residue and petrocoke:* Although gasification for production of hydrogen is more capital intensive, however, as heavy residue and petrocoke are by-products from refinery, gasification technology may be economically attractive option and integrating the liquid residue and coke gasification will be an attractive option for refineries [Sutikno and Turini, 2012; Weiss et al., 2013]. Coke gasification to produce hydrogen for hydroprocessing is more capital intensive than steam methane reforming, but not so vulnerable to natural gas price variations [Sutkno and Turini, 2012].

The process of gasification involves:

o Gasification using oxygen
o Quenching, cooling and scrubbing
o Two-stage shift conversion; high temperature and low temperature
o Acid gas removal unit
o Methanation or nitrogen wash for removal of CO
o Hydrogen purification using pressure swing adsorption or membrane process.

The process needs an air separation unit for producing oxygen. Typical flow diagram for gasification of heavy residue/petrocoke is given in Fig. 18.2. Reliance industries is one of the world's largest refinery will have gasification plant based on Phillip's E-Gas technology which incorporates a gasification system that can be applied with gas and steam turbine combined cycle power generation to produce electric power as well as synthesis gas application for hydrogen production [PTQ 2012] [Petroleum Technology, Quarterly, Q3, 2012, p135].

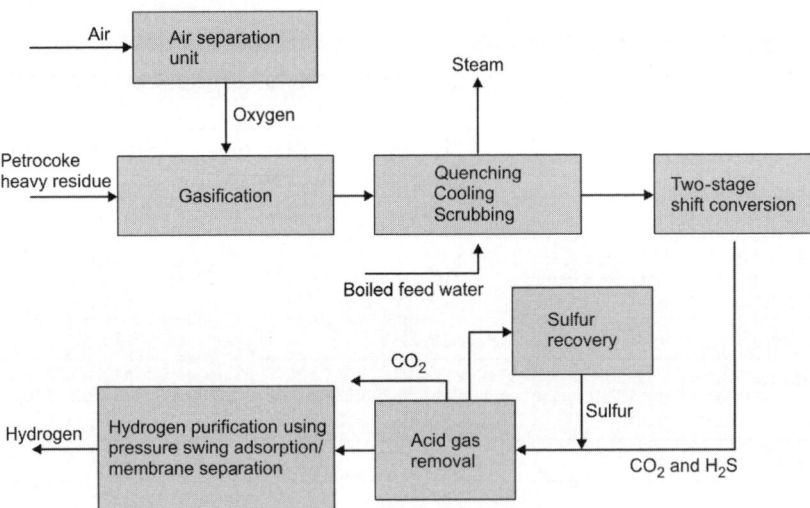

Fig. 18.2 Hydrogen from gasification

18.2.2.1 Purification

Pressure swing adsorption (PSA) or membrane processes are used for purification of hydrogen. Absorption produces low purity hydrogen (95–97%) while PSA unit produces hydrogen of purities above 99.9% [Abrado and Khurana, 1995]. Comparison of hydrogen recovery and purification technologies is given in Table 18.9.

Table 18.9 Comparing hydrogen recovery and purification technologies

Features	Adsorption	Membranes	Cryogenics
H_2 purity	99.9%	90–98%	90–96%
H_2 recovery	75–92%	85–95%	90–98%
Feed pressure	150–600 psig	300–2300 psig	> 75–1100 psig
Feed H_2 content	> 40%	> 25–50%	> 10%
H_2 product pressure	Feed pressure	Feed pressure	Feed/low pressure
H_2 capacity	1–200 MMscfd	1–50+ MMscfd	10–75+ MMscfd
Pretreatment	None	Minimum CO_2	H_2O removal
Requirement			
Multiple products	No	No	Liquid HCs
Capital cost	Medium	Low	Higher
Scale economics	Moderate	Moderate	Good

Source: Sabram et al., 2001.

- *Wet Scrubbing:* In the wet scrubbing plant CO_2 is removed using potassium carbonate or amine. This is followed by methanation where CO and CO_2 are converted to methane.

Pressure Swing Adsorption (PSA) Unit

PSA unit used beds of solid adsorbents in number of absorbers where impure hydrogen stream is purified by separating high purity hydrogen. The impurities are adsorbed and regeneration takes place alternatively by depressurization to low pressure and purging with pure hydrogen. Each absorber operates in following steps: Adsorption, depressurization, purging and pressurization.

In the PSA unit, hydrogen is recovered from the process gas by adsorption of methane, carbon monoxide, carbon dioxide and water vapor on molecular sieves and activated carbon-based adsorbents. Under the high pressure of the feed , the adsorbent attracts the impurities and as a consequence a pure hydrogen product stream exits from the adsorber. Subsequently, the adsorbent is regenerated by bringing the adsorber down to the lowest pressure. The adsorbent releases the impurities which are disposed off as purge gas.

Product Specifications

Gaseous products

Reformer effluent type	Design
Composition	Vol. %
Hydrogen	42.08
Methane	2.54
Carbon monoxide	10.14
Carbon dioxide	7.66
Steam	37.59
Total	100.00

Courtesy: CPCL, Chennai.

Typical Operating Conditions

Flow   : 34,515 kg/hour
Mol. wt   : 14.2
Temperature : up to 860 °C
Pressure   : 24.6 kg/cm^2 g

PSA Outlet

Type	Design
Composition	%vol
Hydrogen	99.9
Methane	Balance
Nitrogen max. ppm V	< 50
Carbon monoxide/carbon dioxide max. ppm V	< 20
Chlorides max. ppm V	Nil
Water max. ppm V	< 50
Total	100.0

Operating Conditions

Pressure   : 22 kg/cm^2 g
Temperature : 40 °C
Flow   : 22,500 Nm^3/hour

Courtesy: CPCL, Chennai.

- *Membrane separation:* This process operates on the principle of selective permeation. Hydrogen and other fast gases which diffuse faster become permeate while the slow gases become the residue stream. Permeate purity and hydrogen recovery are dependent upon the feed gas hydrogen purity.

- *Cryogenic separation:* Cryogenic separation units are operated by cooling of the gas and condensing some of the gas stream. Cryogenic separation technology has highest capital cost [Abrado and Khurana, 1995]
- *Clean hydrogen technology:* This is a new hydrogen production process technology. This hydrogen overage production efficiency (HOPES) process provides emission free 99.9% high purity hydrogen. All CO_2 emissions are removed from the process [www.hydrocarbon processing.com/RS].

18.3 REFINERY HYDROGEN MANAGEMENT

Hydrogen management practices significantly impact on operating costs, refinery margin and CO_2 emission. Therefore, managing hydrogen more effectively has been found to improve refinery profitability by millions of dollars and often enables the refiner to avoid the capital cost of new hydrogen production [Zagoria, 2003]. Many off gas stream from refinery and petrochemical processes contain hydrogen which may vary from 50–90% to as low as 10%. Recovery of hydrogen from off gas is getting importance. Some of the available processes are membrane, adsorption and cryogenics. Comparison of off gas hydrogen recovery and purification technologies is given in Table 18.10 [Patel et al., 2005].

Table 18.10 Comparing off gas hydrogen recovery and purification technologies

Features	Adsorption	Membranes	Cryogenics
H_2 purity (%)	99.9+	< 95	95–99
H_2 recovery (%)	75–90	< 95	90–98
Feed pressure	Feed pressure	<< feed pressure	variable
Feed H_2 content	> 40%	> 25–50%	> 10%
By-product available	No	No	yes
Feed pressure, bar G	10–50	15–125	15–35

Source: Patel et al., 2005.

Use and optimization of high performance catalysts are the important tools to increase hydrogen yield. Steps requiring applications of catalysts in various stages hydrogen production are:
- Hydrodesulfurization using catalyst and adsorbent
- Two-stage steam reforming; pre-reforming and steam reforming stages
- High, medium and low shifts.

Use and optimization of high performance catalysts are the important tools to increase hydrogen supply in a refinery. High activity catalyst is now available to increase the energy efficiency of hydrogen production [Brusnson et al., 2013]. Catalysts like HDMax, ReforMax, Shiftmax and Meth series catalyst developed by Clariant Technology improve the performance in energy efficiency and hydrogen output keeping operating cost down [Brusnson et al., 2013].

A refinery hydrogen management program should be organized to meet the following objectives [Davis and Patel, 2004; Patel et al., 2005]. Hydrogen network in a typical refinery includes following steps:

- Maximize hydrogen utilization through increased recovery
- Decouple catalytic reformer operation with hydrogen production needs
- Take advantage of higher hydrogen purity with specific consumers
- Reduce cost on purpose hydrogen production
- Improve on purpose hydrogen production reliability
- Expand on purpose hydrogen production
- Evaluate refinery hydrogen plant shut down economics
- Integrate with new industrial gas company hydrogen supply [www.pewclimate.org].

Hydrogen pinch techniques have been developed for analysis of hydrogen resources in refinery complexes. This identifies the best opportunities for off gas reuse and purification and helps in quantifying the minimum size of new hydrogen production facilities, if require. Hydrogen pinch technology can be used to define the best hydrogen solution, taking into account the following factors [www.fosterwheeler.fi] [Phillips, 2000]:

- Confirmation of existing network headers
- Compression requirements
- Potential for reuse of off gas streams
- Potential for hydrogen purification.

Hydrogen network operating decisions affect the bottom line every day [zagoria, 2003]. Overview of hydrogen network improvement options is given in Table 18.11.

Table 18.11 Overview of hydrogen network improvement options

	Daily operating decisions	**Engineering improvement**
Hydrogen plant	Steam to carbon ratio Temperature	Catalyst, revamp for different feed debottlenecks
Catalytic reformer	Severity Feed properties	Catalyst change Reduce pressure
Distribution network	Optimize matching of producers and consumers	New compression control system improvement
H_2 purification	Feed rate trade off purity vs recovery vs capacity	Debottleneck revamp for different feed compositions New purifier
Hydroprocessing	Make-up source Purge flow H_2 partial pressure	V compression Add H_2S scrubbing

REFERENCES

1. Abrardo JM and Khurana V. Hydrogen Technologies to Meet Refiner's Future Needs. Hydrocarbon Processing, Feb 1995, p44.

2. Brunson R, Flessner U and Morse P. Catalysts for Hydrogen Management. Catalysis PTQ, 2013, p41.

3. Christensen TS and Primdahi II. Improve Syngas Production Using Autothermal Reforming. Hydrocarbon Processing, March 1994, p39.

4. Davis RA and Patel NM. Refinery Hydrogen Managment. Petroleum Technology, Quarterly, Spring, 2004, p29.

5. Gas Processing, 2006. Hydrogen—PRISM Membrane in Hydrocarbon Processing, Jan 2006, p68.

6. Hartman JCM. Balance Hydrogen Consumption. Hydrocarbon Processing, Dec 2001, p45.

7. Patel N, Ludwig K and Morris P. Insert Flexibility into Your Hydrogen Network, Part 1. Hydrocarbon Processing, Sep 2005, p75.

8. Petroleum Technology Quarterly, Q3, 2012, p135.

9. Phillips G. Hydrogen: Innovative Business Solutions. Petroleum Technology, Quarterly, 2000.

10. Sabram TM, Fairclough DD and Davis RA. Low Cost Hydrogen from Refineries. Nandini Chemical Journal, Dec 2001, p18.

11. Zagoria A. Refinery Hydrogen Management—The Big Picture. Hydrocarbon Processing, Feb 2003, p41.

12. Sutikno T and Turini K. Gasifying Coke to Produce Hydrogen in Refineries. Petroleum Technology, Quarterly, Q3, 2012, p109.

13. Weiss MM, Heurich H, Roma D and Walter S. Petroleum Technology, Quarterly, Q4, 2013, p91.

14. Zagoria A. Refinery Hydrogen Management—The Big Picture. Hydrocarbon Processing, Feb 2003, p41.

19

Petroleum Refinery and Petrochemical Integration

Petrochemical industry has been playing an important role in the social, cultural and economic growth of a nation and providing basic needs of mankind—food, shelter and clothing. Petrochemicals have become an indispensable part of our life and are playing an important role in providing key feedstock for chemical industry for making thousands of different chemicals the general people usually encounter as end or consumer products. Due to increasing cost of crude oil and fuel now trends in the refineries are integration of refineries with petrochemical production to produce value added product, thus improving over all economy of the refiner—petrochemical complex. Figure 19.1 shows integration of refinery with petrochemical complex. Some of the important sources of petrochemicals in a refinery are the fluid catalytic cracking, catalytic reforming and crude distillation unit.

19.1 PETROCHEMICAL COMPLEX

The petrochemical complexes involve one or a combination of the following operations [Mall, 2006]:

- The manufacture of basic raw materials like syngas, methane, ethylene, propylene, acetylene, butadiene, benzene, toluene, xylene, etc. The basic building processes include partial oxidation, steam reforming, catalytic and thermal cracking, alkylation, dealkylation, hydrogenation, disproportionation, isomerization, etc. The commonly used unit operations are distillation, extractive distillation, azeoptropic distillation, crystallization, membrane separation, adsorption, absorption, solvent extraction, etc.

- Manufacture of intermediate chemicals derived from the above basic chemicals by various unit processes like oxidation, hydrogenation, chlorination, nitration, alkylation, dehydrogenation along with various unit operations like distillation, absorption, extraction, adsorption, etc.

Manufacture of target chemicals and polymers that may be used in the manufacture of target products and chemicals to meet the consumer needs. It includes plastics, synthetic fibers, synthetic rubber, detergents, explosives, dyes, intermediates and pesticides. Structure of petrochemical complex is given in Fig. 19.2.

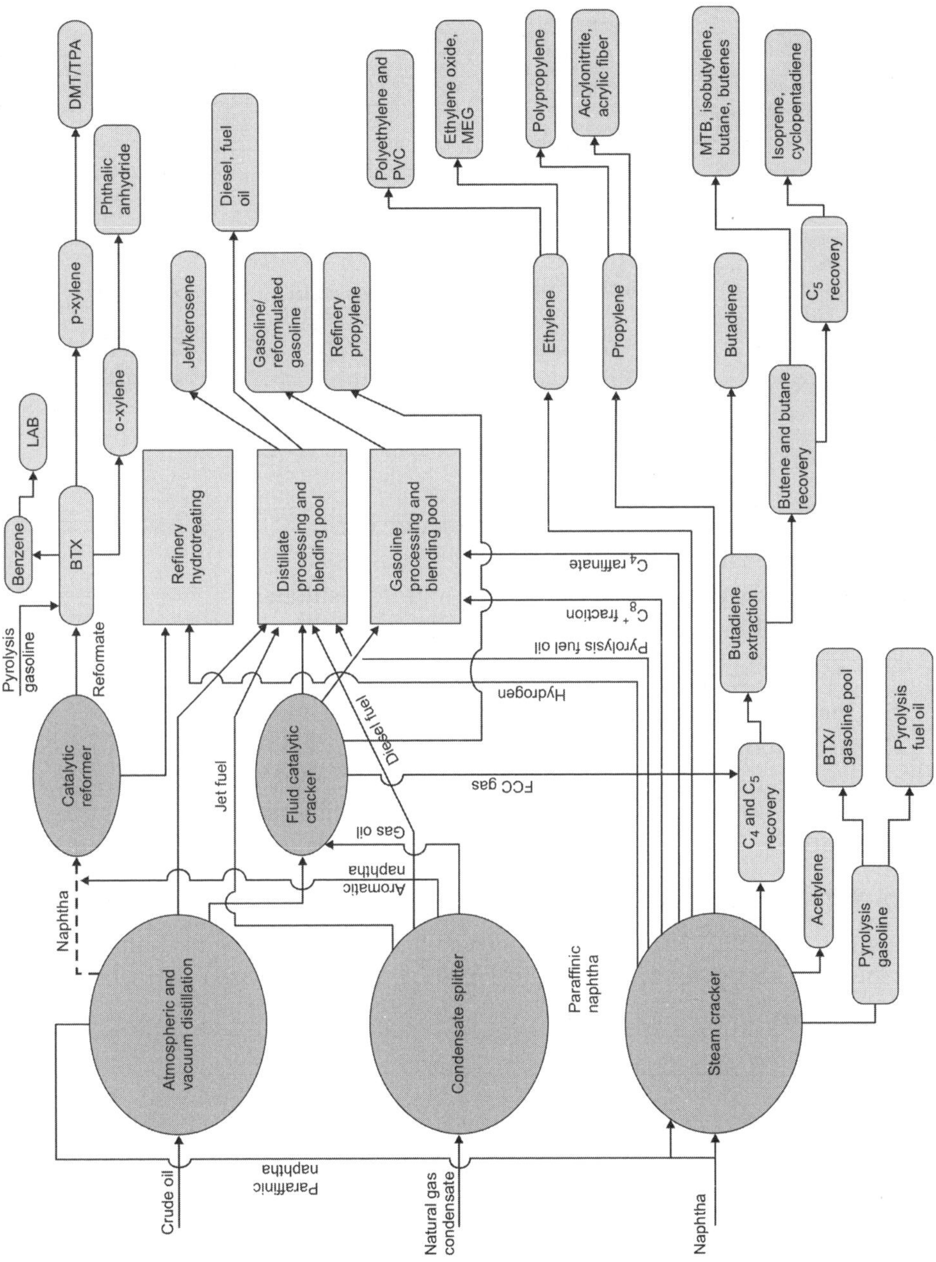

Fig. 19.1 Integration of refinery with petrochemical industry

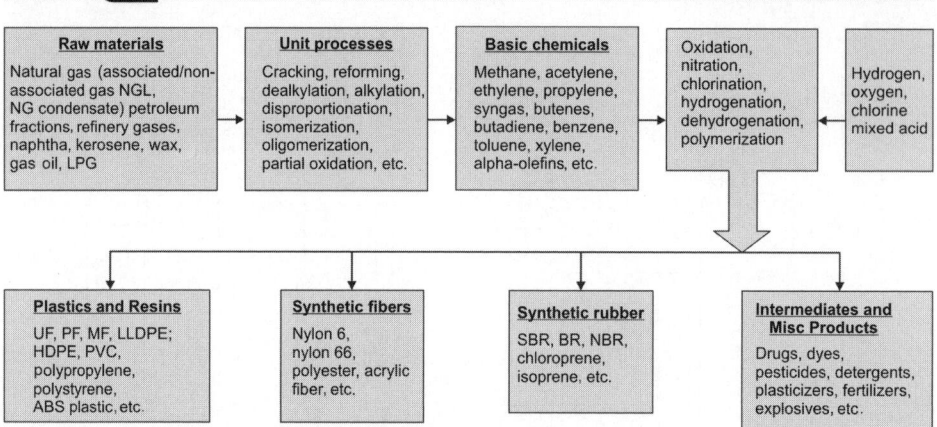

Fig. 19.2 Structure of petrochemical complex

Basic raw materials derived from petroleum are:

Gaseous : Natural gas, condensate, refinery gases

Liquids : Naphtha, solvent extracts, middle distillates

Solids : Coal, coke, wax, residues

The three main routes to petrochemicals are:

Synthetic gas (syngas) : $CO + H_2$

Paraffin olefins : C_2 (ethane, ethylene, acetylene), C_3 (propane, propylene), C_4 and C_5 (butane, butenes, pentane, pentene), butadiene, higher olefins

Aromatics : Ethyl benzene, BTX (benzene, toluene, xylenes)

19.2 PETROCHEMICAL PRODUCT PROFILE

Major petrochemical feedstocks are derived from methane, olefins, butadiene and butenes and aromatics are shown in Fig. 19.3. Aromatic based petrochemical is shown in Fig. 19.4.

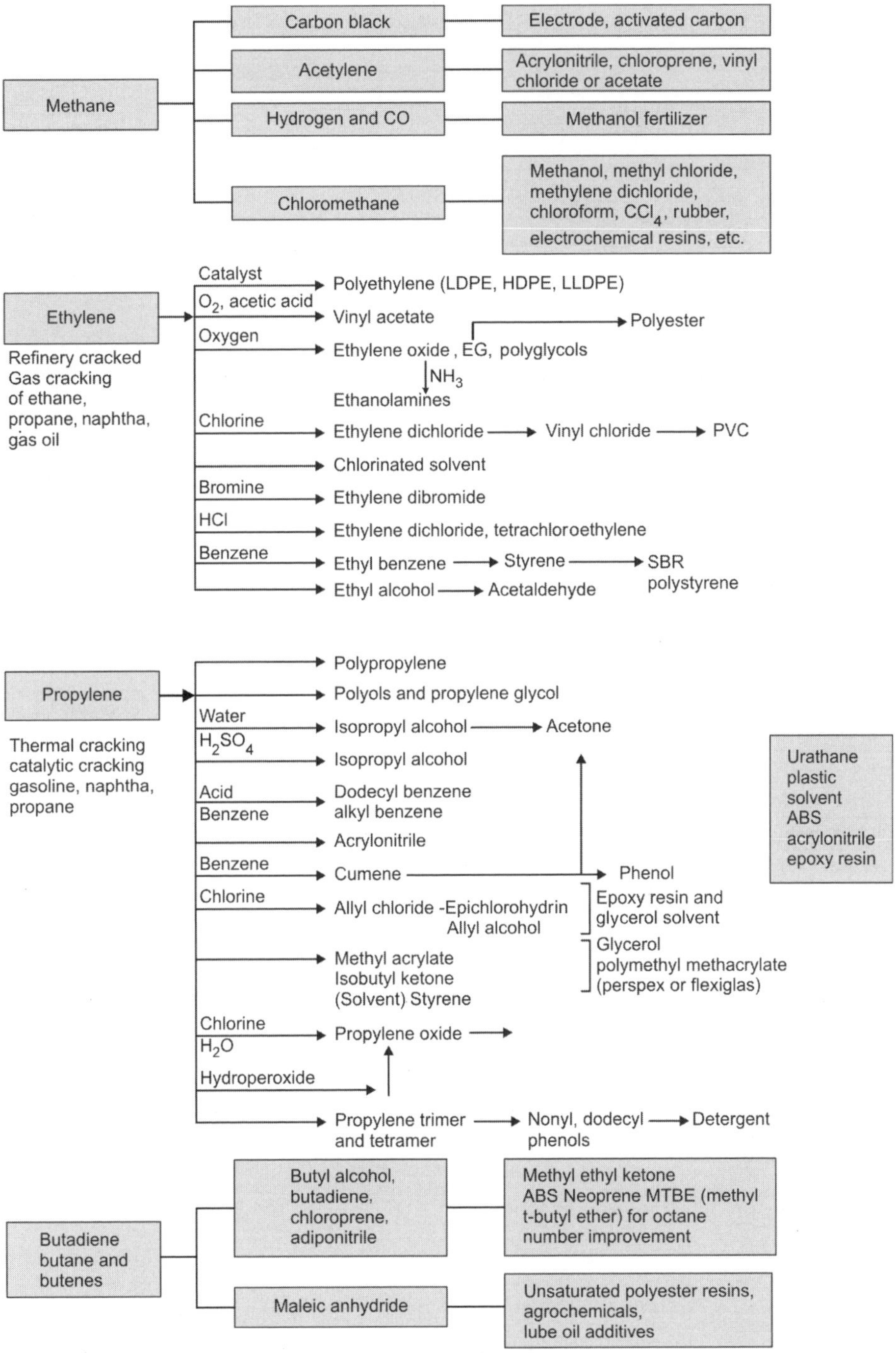

```
Methane ─┬─ Carbon black ──────── Electrode, activated carbon
         ├─ Acetylene ─────────── Acrylonitrile, chloroprene, vinyl
         │                         chloride or acetate
         ├─ Hydrogen and CO ───── Methanol fertilizer
         └─ Chloromethane ─────── Methanol, methyl chloride,
                                   methylene dichloride,
                                   chloroform, CCl₄, rubber,
                                   electrochemical resins, etc.
```

Methane

Carbon black — Electrode, activated carbon

Acetylene — Acrylonitrile, chloroprene, vinyl chloride or acetate

Hydrogen and CO — Methanol fertilizer

Chloromethane — Methanol, methyl chloride, methylene dichloride, chloroform, CCl_4, rubber, electrochemical resins, etc.

Ethylene

Refinery cracked
Gas cracking
of ethane,
propane, naphtha,
gas oil

Catalyst → Polyethylene (LDPE, HDPE, LLDPE)
O_2, acetic acid → Vinyl acetate ────────────→ Polyester
Oxygen → Ethylene oxide , EG, polyglycols
    ↓NH_3
Ethanolamines
Chlorine → Ethylene dichloride ──→ Vinyl chloride ──→ PVC
→ Chlorinated solvent
Bromine → Ethylene dibromide
HCl → Ethylene dichloride, tetrachloroethylene
Benzene → Ethyl benzene ──→ Styrene ──────→ SBR
→ Ethyl alcohol ──→ Acetaldehyde   polystyrene

Propylene

Thermal cracking
catalytic cracking
gasoline, naphtha,
propane

→ Polypropylene
→ Polyols and propylene glycol
Water / H_2SO_4 → Isopropyl alcohol ──→ Acetone
→ Isopropyl alcohol
Acid / Benzene → Dodecyl benzene alkyl benzene
→ Acrylonitrile
Benzene → Cumene ──────────────→ Phenol
Chlorine → Allyl chloride -Epichlorohydrin ─┐ Epoxy resin and
              Allyl alcohol ─────────────────┘ glycerol solvent
→ Methyl acrylate ─┐ Glycerol
  Isobutyl ketone │ polymethyl methacrylate
  (Solvent) Styrene ┘ (perspex or flexiglas)
Chlorine / H_2O → Propylene oxide ──→
Hydroperoxide ─────────────→
→ Propylene trimer ──→ Nonyl, dodecyl ──→ Detergent
  and tetramer   phenols

Urathane
plastic
solvent
ABS
acrylonitrile
epoxy resin

Butadiene
butane and
butenes

Butyl alcohol, butadiene, chloroprene, adiponitrile — Methyl ethyl ketone ABS Neoprene MTBE (methyl t-butyl ether) for octane number improvement

Maleic anhydride — Unsaturated polyester resins, agrochemicals, lube oil additives

Fig. 19.3 Petrochemicals from methane, ethylene, propylene, butadiene and butenes

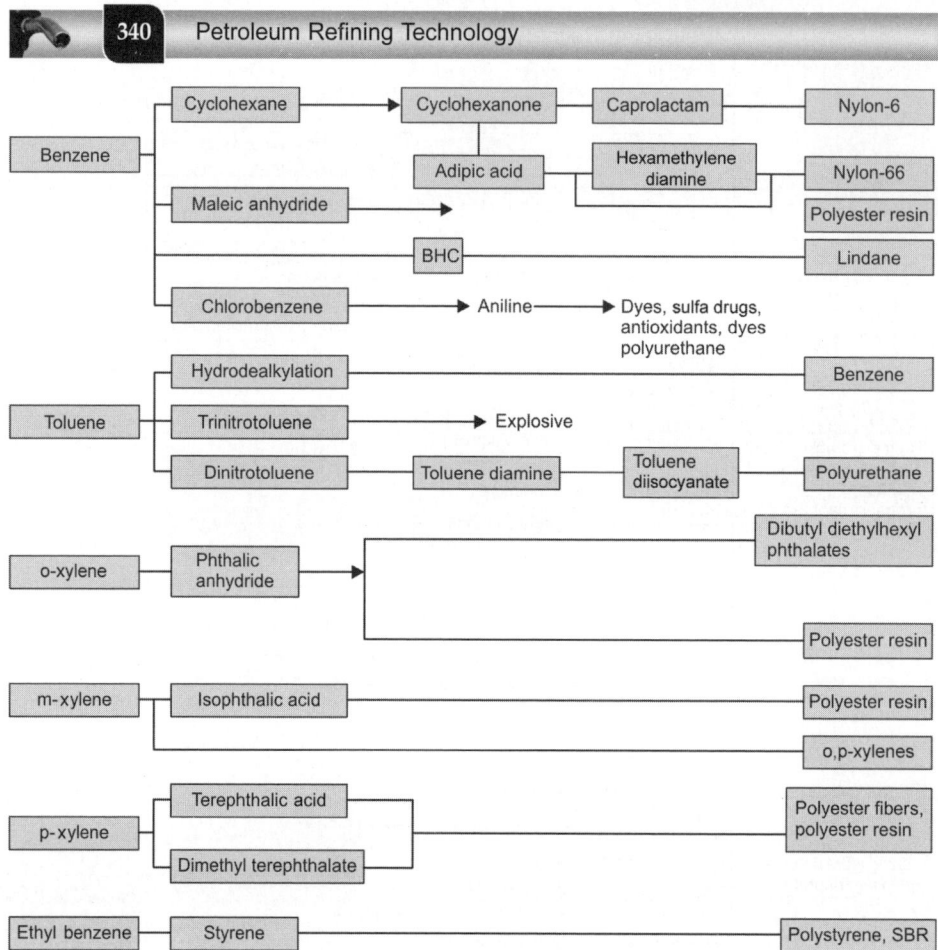

Fig. 19.4 Aromatics based petrochemicals

19.3 PETROCHEMICAL FEEDSTOCKS

The original feedstocks for organic chemical industries are coal, coal tar, fats, oils and molasses. Sources of organic chemicals from various industries are shown in Fig. 19.5.

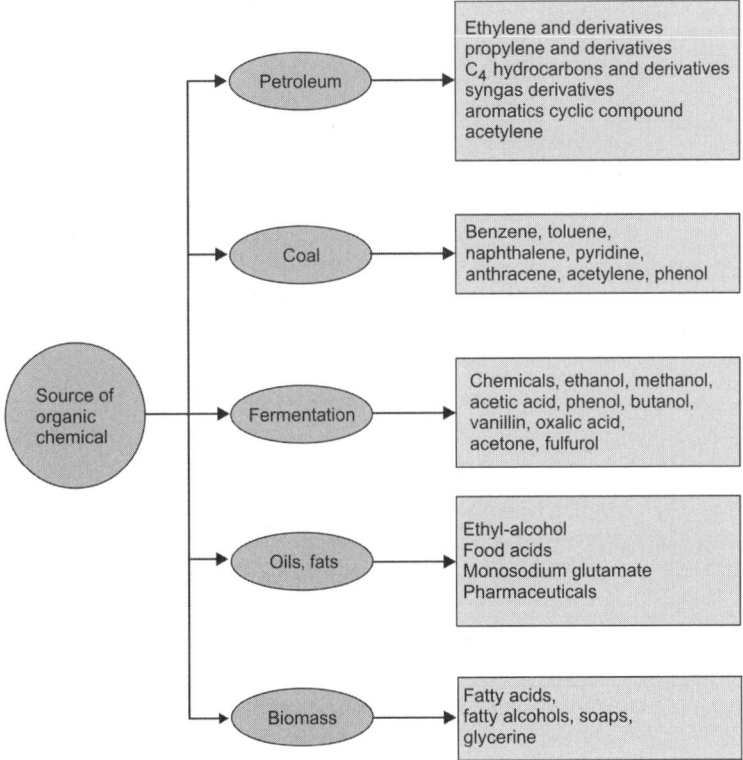

Fig. 19.5 Sources of organic chemicals

World War II gave real impetuous to utilization of petroleum and natural gas petroleum products and quantitatively 95% of the organic chemicals are now based on petrochemicals derived from petroleum and natural gas feedstock. It is estimated that 7–10% of the carbon in hydrocarbon output of the world in terms of gas oil is utilized as feedstock in the petrochemical industry.

A petrochemical feedstock derived from petroleum industry, their sources, composition, intermediate processes and intermediate petrochemical feedstock is given below Table. 19.1. Capacities of Indian petrochemicals are given in Table 19.2.

- **Natural Gas:** Largely methane, ethane (4%)
- **Liquefied petroleum gas:** Propane and butane, extracted from natural gas and produced during various refining processes
- **Refinery off gases:** From catalytic and reforming operations. Paraffinic and olefins used for petrochemicals, H_2 for NH_3

Table 19.1   Petroleum product as petrochemical feedstock

Petroleum fractions and natural gases	Source	Composition	Intermediate processes	Intermediate petrochemical feedstock
Refinery gases	Distillation, catalytic cracking, catalytic reforming	Methane, ethane, propane, butane, BP up to 25 °C	Liquefaction, cracking	LPG, ethylene propylene, butane, butadiene.
Naphtha	Distillation and thermal and catalytic cracking, visbreaking	C_4–C_{12} hydrocarbon, BP 70–200 °C	Cracking, reforming alkylation, disproportionation, isomerization	Ethylene, propylene, butane, butadiene, benzene, toluene
Kerosene	Distillation and secondary conversion processes	C_9–C_{10} hydrocarbon, BP 175–275 °C	Fractionation to obtain C_{10}–C_{14} range hydrocarbon	Ethylene, propylene, butane, butadiene, benzene, toluene
Gas oil	Distillation of crude oil and cracking	C_{10}–C_{25} hydrocarbons BP 200–400 °C	Cracking	Ethylene, propylene, butadiene, butylene
Wax	Dewaxing of lubricating oil	C_8–C_{56} hydrocarbon	C_8–C_{56} hydrocarbon	C_6–C_{20} alkanes
Pyrolysis gasoline	Ethylene cracker	Aromatic, alkenes, dienes, alkanes, cycloalkane	Hydrogenation, distillation, extraction, crystallization, adsorption	Aromatics
Natural gases and natural gas condensate	Gas fields and crude oil stabilization crude oil	Hydrogen, methane, ethane, propane, pentane, aromatics	Cracking, reforming, separation	Ethylene, propylene, LPG aromatics, etc.
Petroleum coke	Visbreaker and delayed coking	Carbon	Residue upgradation processes	Carbon electrode, acetylene, gasification

Source: Mall, 2007.

Table 19.2   Indian petrochemical capacities for building blocks

Product	Installed capacity 000 TPA	Production 000 TPA
Ethylene	2346	1910
Propylene	1522	1350
Butadiene	119	50
Aromatics		
Benzene	669	500
Toluene	260	140
o-xylene	160	110
p-xylene	1680	1312

- **Naphtha:** Olefins, aromatics
- **Fuel oil:** Synthesis gas
- **Kerosene:** For production of n-paraffins used in linear alkyl benzene manufacture
- **Petroleum coke:** Calcium carbide, acetylene, electrodes
- **Heavy residue:** Synthesis gas from gasification.

19.3.1 Feedstock for Olefins

Till few years back, naphtha has been the dominant feedstock for petrochemicals in India, however, recent trends have been more for utilization of natural gas, LPG and pyrolysis gasoline. The major feedstocks for olefins production are gas feedstocks (natural gas, LPG), liquid feedsstocks—light virgin naphtha, full range naphtha, reformer raffinate, atmospheric gas oil, vacuum gas oil and crude oils. Among the various feedstocks, natural gas liquids and naphtha are the most commonly used feedstocks for production of light olefins.

19.3.2 Feedstock for Aromatics

Although coal carbonization was the original source of BTX and still some parts of benzene are manufactured from coke oven plants, however, petroleum fraction are chief source of aromatics. The aromatic yield varies with source of crude oil. Crude oil from Bombay High is rich in aromatics.

19.3.3 Feedstock for Linear Alkyl Benzene (LAB)

The demand for kerosene is likely to increase due to demand of n-paraffin for LAB manufacture which is derived from kerosene.

19.3.4 Feedstock for Higher Olefins

Paraffin wax having C_{25}–C_{30} hydrocarbons is an important source of C_6–C_{20} olefins. The yield of C_6–C_{20} alkene is about 60%. The α-olefins also find wide application in field of plasticizers.

19.4 PETROCHEMICAL BUILDING BLOCKS

Major petrochemical building blocks are: Steam reforming/partial oxidation, production of synthesis gas; steam cracking, production of olefins; catalytic cracking, propylene, C_4 and C_5 hydrocarbons; catalytic reforming, production of aromatics; dehydrogenation of paraffins, oligomerization of light olefins. Detail of petrochemical building block is shown in Fig. 19.6.

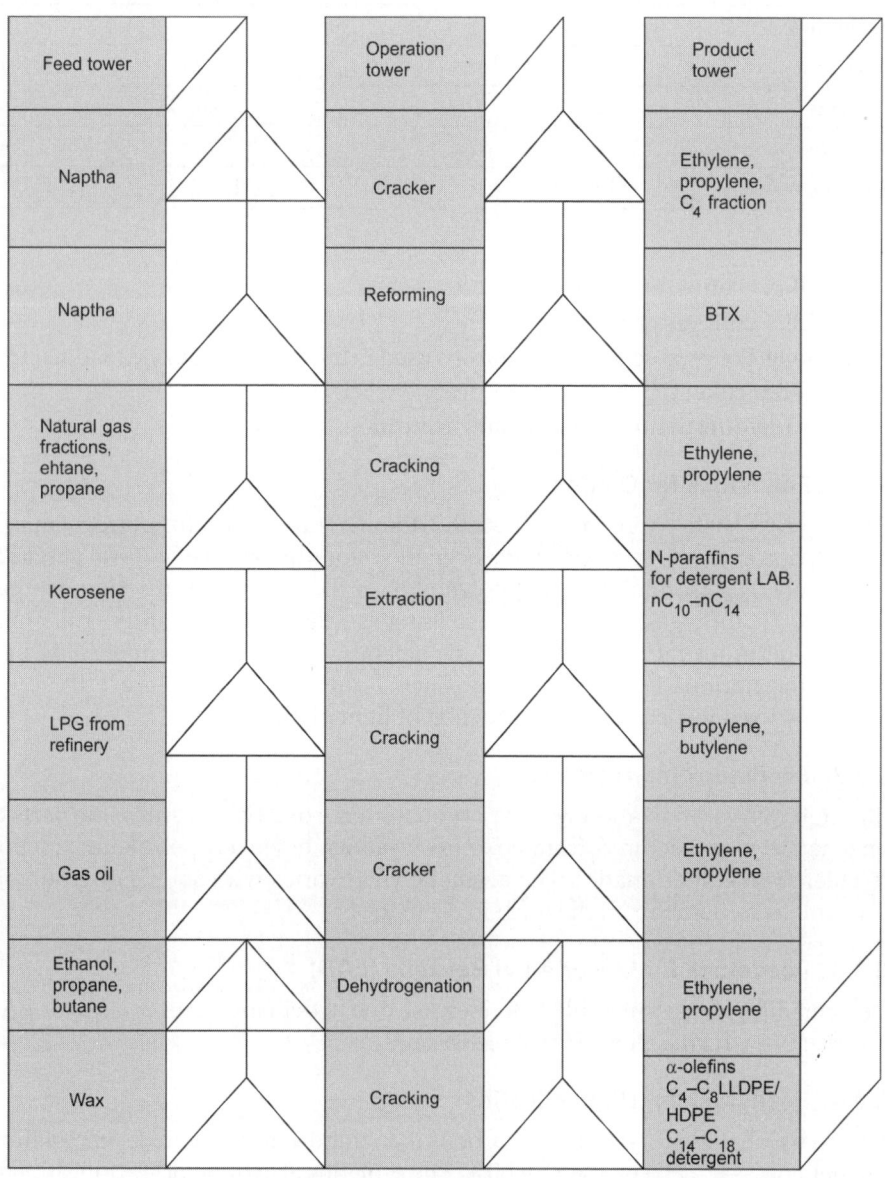

Fig. 19.6 Major petrochemical building blocks

19.4.1 Steam Reforming

Synthesis gas is an important petrochemical feedstock for the production of wide variety of chemicals is given in Table 19.3. Synthesis gas is manufactured by steam reforming or partial oxidation process.

Table 19.3 Synthesis gas requirements for major world scale petrochemicals

Product	Required $H_2:CO$	Typical world-scale capacity, tons per annum.	Syngas required, Nm^3/hour
Methanol	2:1	1,60,000–12,75,000	48,000–1,90,000
Acetic acid	0:1	2,75,000–5,45,000	18,000–36,000
Acetic anhydride	0:1	90,000	3,500
Oxo alcohol	2:1	1,15,000–2,75,000	12,000–25,000
Phosgene	0:1	4,800–1,60,000	3,500–12,000
Formic acid	0:1	45,000	3,500
Methyl formate	0:1	9,000	600
Propionic acid	0:1	45,000–68,000	2,400–3,500
Methyl methacrylate	1:1	45,000	4,700
1, 4-butadiol	2:1	45,000	4,700

Courtesy: Hydrocarbon Processing, April 1999, p87.

Process Technology

Various synthesis gas production technologies are steam methane reforming, naphtha reforming, autothermal reforming, oxygen secondary reforming, and partial oxidation of heavy hydrocarbons, petroleum coke and coal. Various technologies for production of synthesis gas is shown in Fig. 19.7.

Various steps involved in synthesis gas production through steam reforming are:

- Desulfurization of gas/vaporized naphtha which involves hydrogenation of sulfur and chloride to H_2S and HCl in first reactor followed by absorption in second reactor and sorption of H_2S.
- Steam reforming which involves steam reforming of natural gas
- Separation of CO_2

Various synthesis gas generation schemes available are:
- Conventional steam reforming
- Partial oxidation
- Combined reforming
- Parallel reforming
- Gas heated reforming.

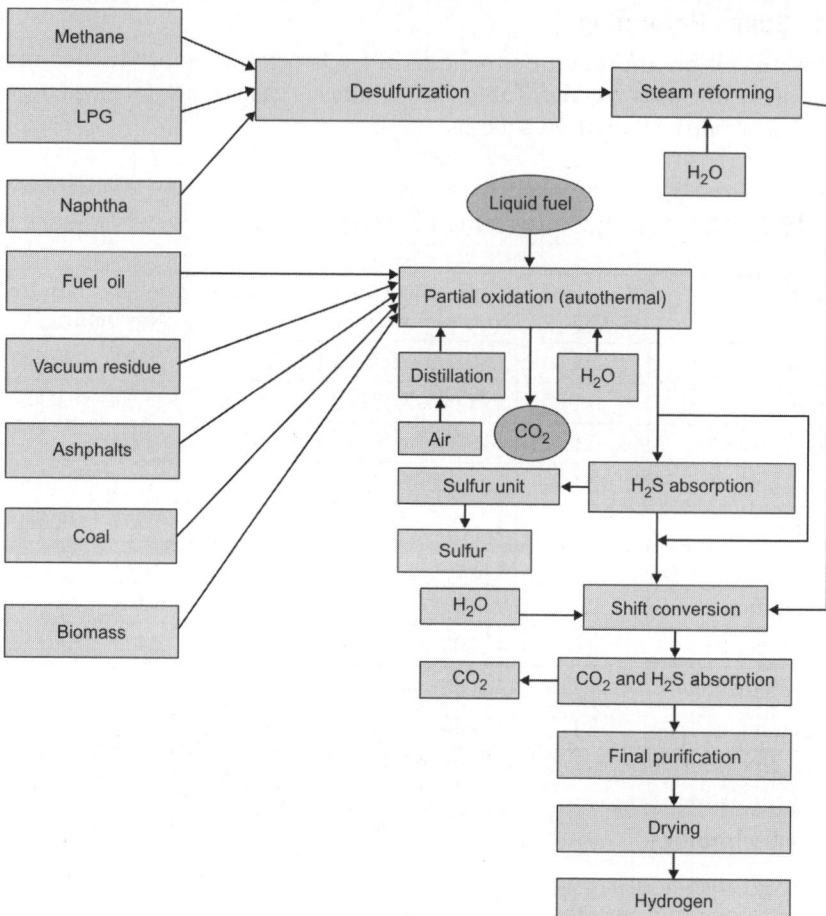

Fig. 19.7 Processing of carbonaceous material for synthesis gas production

19.4.2 Steam Cracking

Steam olefins such as ethylene, propylene and butylenes are one of the building blocks for petrochemicals. They are the major intermediates of petrochemical activities derived from the cracking of hydrocarbon products from oil or natural gas. Their reactivity and versatilities have led to the development of many petrochemical products such as chemical intermediate, solvents thermoplastics, elastormers and detergent. Ethylene is the kind of various petrochemicals and enters into the production of wide variety of petrochemicals which find applications in the production of plastics, antifreeze, apparel, solvents and many other uses. Tremendous growth in ethylene consumption as a chemical intermediate is one of the real success stories of the petrochemical industry and ethylene is the world's most important building block.

Modern ethylene plants incorporate following major process steps: Steam cracking compression and separation of the cracked gas by low temperature fractionation. The nature of the feedstock and the level of pyrolysis severity largely determine the operating conditions in the cracking and quenching sections. Flow diagram for steam cracking of naphtha is given in Fig. 19.8. The steam consumption in cracker plant varies with types of the feedstock being processed and the cost of the cracking plant also varies accordingly. A typical steam ratio in cracking is given in Table 19.4. The relative cost of the cracker plant is given in Table 19.5. Various steps involved in naphtha gas/gas cracking for olefin production is mentioned in Table 19.6.

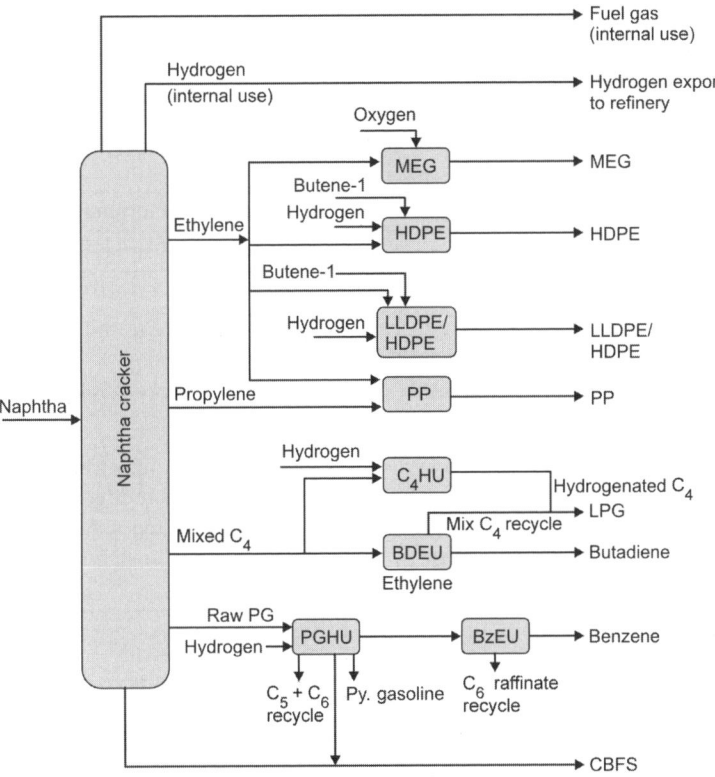

Fig. 19.8 Naphtha cracker unit

Courtesy: IOC, Panipat Refinery.

Table 19.4 Typical steam ratios in cracking

Feed	kg steam/kg hydrocarbon
Ethane	0.2–0.4
Propane	0.3–0.5
Naphtha	0.4–0.8
Gas oil	0.8–1.0

Table 19.5   Relative capital costs of crackers

Feedstock	Relative capital cost
Ethane	1.0
Propane	1.2
Naphtha	1.4
Gas oil	1.5

Table 19.6   Various steps involved in naphtha gas/gas cracking for olefin production

Hot section	Feedstock is pyrolized and the effluent conditioned
Cold section	Product formed are separated and purified
Hot section	
Convection zone	Hydrocarbon feedstock is preheated and mixed with steam that is also preheated
Radiation zone	Rapid rise in temperature. Pyrolysis reaction takes place.
Quench	
I	To avoid subsequent reaction the effluents are fixed in their kinetics development by sudden quench.
II	Indirect quench by water to 400–500 °C generation of high pressure steam. Direct quench by heavy residue by-product of pyrolysis.
Primary fractionation column	Separation of light products of pyrolysis as top and bottom as pyrolysis product
Compression	Compression of light products
Caustic scrubbing and drying	Scrubbing with caustic followed by molecular sieve adsorption to remove sulfur compounds, mercaptan, etc.
Cold section	• Hydrogen separation • Ethylene separation 99.9% • Propylene separation • A C_4 cut containing 25–50% butadiene • Complementary fraction of pyrolysis gasoline rich in aromatic hydrocarbons
Demethanizer	Methane condensed at top around –100 °C; pressure, 32 Pa
Deethanizer separation of ethylene	Separation of C_2 cut (ethane and ethylene). Acetylene eliminated by selective hydrogenation catalyst: Palladium or nickel, temperature 40–80 °C; pressure, 3 kPa

Contd.

Table 19.6 Various steps involved in naphtha gas/gas cracking for olefin production *(Contd.)*

Depropanizer removal of propane from propylene	Column with 110–120 trays, 1.9 kPa, –35 °C C_3^+ cut from bottom of deethanizer is fractionated. C_3 cut from top of depropanizer is selectively hydrogenated to remove methyl acetylene and propadiene. Propylene content 95%.
Debutanizer	Heavier hydrocarbons from bottom of the depropanizer treated in debutanizer to produce 2, 3-butadiene rich C_4 cut at the top.

Typical reactions involved in naphtha cracking are given in Table 19.7.

Table 19.7 Basic reactions involved in the cracking involve free radical reactions

Primary cracking (I)	Cracking of saturated aliphatic hydrocarbon in paraffin and olefin
Secondary cracking (II & III)	Lighter product rich in olefins produced composition and yield depend on operating conditions

19.4.2.1 Operating Severity and Selectivity

- Severity describes depth of cracking or extent of conversion.
- Maximum severity represents acceptable optimum yield of ethylene.
- At low and medium severity: Primary cracking and dehydrogenation reaction predominate. They cause sharp increase in yield of methane, ethylene, propylene and C_4 hydrocarbons. Reduction in C_5 cut.
- At very high severity: Methane and ethylene yield levels off while those of propylene and C_4 cut reach a peak and then decline.
- Ratio of ethylene and propylene yield increases with the severity, which hence favors the formation of ethylene.

Operating variables: Major operating variables in naphtha cracking are:
- Reaction temperature
- Residence time
- Hydrocarbon partial pressure and steam
- Feedstock.
  Operating variables in refinery are given in Table 19.8.

Table 19.8   Operating variables

Reaction temperature	700–900 °C depending upon feedstock ethane 850 °C
Residence time	Longer residence for heavier feedstock than light feedstocks. 0.2 second (lower limit) shorter residence time improved yield
Hydrocarbon partial pressure and role of steam	Pyrolysis reaction more advanced at lower pressure. Ethylene yield decreases as the partial pressure of hydrocarbon increases. i. Reduces overall reaction rate ii. Enhances the selectivity of pyrolysis in favor of light olefins desired. Increases temperature of feedstock.
Steam	0.5–0.64 ton of steam per ton of naphtha
Feedstock	Relative production of ethylene decreases as the feedstock becomes heavier ratio of propylene yield (C_2/C_3 ratio) decreases steadily from ethane to gas oils while pyrolysis gasoline (C_5 – 200 °C) increases. Butadiene yield varies slightly.

19.4.2.2 Trend in Technological Developments of Steam Crackers for Production of Ethylene

- Lower energy consumption per ton of the ethylene. The present specific energy consumption figures have reduced by more than 50% as compared to the mid 70 figure. Major developments have been in fire heaters, compressors, heat exchangers and fractionating column.
- Improved overall yields of ethylene by short residence time, higher severity.

19.4.2.3 Upgrading Performance of Furnace

There has been continuous upgradation in the cracker furnaces which includes:
- Increasing furnace capacity
- Increasing cracking severity
- Improving ethylene selectivity
- Improving thermal efficiency
- Reducing downtime for decoking
- Reducing maintenance.

Increase in thermal efficiency can be achieved by:
- New conversion heat transfer surface
- A combustion air preheater
- Cogeneration
- Combustion air preheating, short residence time, small diameter coils, increase dilution steam use of booster compressor to reduce furnace outlet pressure can increase efficiency and ethylene selectivity.

- Increased feedstock flexibility, i.e. possibility of cracking more than one type of feedstock in the same furnace. The advanced cracking reactor process offers total liquid feedstock flexibility from light naphtha through vacuum gas oils in the same production unit.
- Development in the decoking processes to have higher run length with use of additive.
- Development in the pyrolysis furnace alloy material to provide efficient and reliable performance with minimum maintenance.
- Catalytic pyrolysis of naphtha for increasing plant flexibility as well as reducing the required temperature and coke deposition for a given conversion.
- Improved flue gas heat recovery.

19.4.2.4 Coke Formation

Coke formation is due to:
- Because of sequence of catalytic and non-catalytic reactions
- Gas phase and on the solid surface of the reactor
- Nickel and iron catalyzes coke formation
- Tar droplets are precursors of part of non-catalytic coke.

Rate of coke formation is highest at the start of pyrolysis run when SS tube surfaces are clean. Coke formed during steam cracking results in following undesirable effects.
- Coke reduces the cross-section tube necessitates higher pressure.
- Reduces the heat transfer across the wall necessitates higher metal wall temperatures to main cracking temperature.
- Higher temperature applications create problems with creep, accelerated by carburization.
- Low cycle fatigue.

19.4.2.5 Decoking

Decoking is the process of removal of the coke from furnace tube. The time between two decoking operations is called run length. Decoking is done by different operating methods which have been mentioned in Table 19.9. The C_4 stream composition is given in Table 19.10.

Table 19.9 Decoking operation

Mechanical decoking	Use of high pressure decoking water jets 30–70 kPa longer time
Steam-air decoking	600–800 °C cooler steam cools spalling and thermally shops the coke and breaks away from tube surface and packed up by steam.
Burning	Higher time 30 hours.
Decoking without air	Higher temperature 950 °C

Table 19.10 Typical steam cracker mixed C$_4$ stream composition

Feedstock	Naphtha			Gas oil	Propane	Ethane
Severity	ISV	SSV	HSV	HSV	HSV	HSV
Propylene/ethylene ratio	0.54	0.52	0.48	0.60	0.35	0.02
Mixed C$_4$ yield (% on feed)	10.8	10.2	9.2	9.0	4.5	2.5
Mixed C$_4$ composition (%)						
C$_3$	0.5	0.5	0.5	0.5	0.5	0.5
n-butane	4	3	2	2	3	3
Isobutane	2	1	1	1	2	2
Isobutene	23	23	22	23	4	3
1-butene	14	14	14	17	12	9
2-butene	10	11	11	13	9	7
1, 3-butadiene	46	47	49	43	69	75
Heavies	0.5	0.5	0.5	0.5	0.5	0.5
Total	100	100	100	100	100	100

Courtesy: Chemistry and Industry, 2nd Feb, 1998, p90.

19.4.3 Fluid Catalytic Cracking as Source of Petrochemical

FCC off gases are of considerable importance as feedstock for petrochemicals. Some of the important building blocks which can be derived from FCC are: Dilute ethylene stream, LPG, propylene, C$_4$ streams containing butanes/butylenes.

19.4.3.1 Propylene Recovery from Fluid Catalytic Cracking (FCC)

The FCC process is widely used in the petroleum refining industry to convert vacuum and heavy atmospheric gas oil into lower boiling distillate, gasoline and LPG products. The ethylene, propylene and butenes are available in the fuel gas and LPG products from the FCC units which are being recovered and are providing excellent feedstock for large number of chemicals. Typical composition of FCC gas stream is given in Table 19.11.

Table 19.11 Typical composition of FCC gas stream

Products	Yield weight (%)
Dry gas (including ethylene)	12.7
Propane	6.5
Propylene	21.0
Butene	35.8

Improved FCC technologies can effectively maximize propylene yield from traditional FCC feedstocks and selected naphtha [Niccum et al., 2001]. While FCC operates typically produce less than 6 %wt propylene, specialized fluidized catalytic processes can further raise propylene yields as high as 20% or more from FCC feedstocks [Niccum et al., 2001].

Deep catalytic cracking technology is commercially proved fluid catalytic process to selectively crack a wide variety of hydrocarbon feedstocks into light olefins— particularly propylene and isobutylene [Dharia et al., 2004; Zaiting et al., 2000; Peiling, 1997].

Various options for a refiner to produce and recover light olefins from a catalytic cracking unit are [Dharia et al., 2004]:

- High severity FCC with additives like ZSM-5
- Deep catalytic cracking with high severity type I
- Deep catalytic cracking with high lower severity operation type II
- Naphtha recycle.

It is estimated that 1 million tons capacity in conventional FCC unit of Indian refineries would produce 25–31 tons per annum of propylene depending on the catalyst and process conditions. Some of the emerging technologies for increasing the yield of olefins in the FCC gases are: Deep catalytic cracking (DCC), non-regenerative catalytic cracking (NRCC), and use of ZSM-5 additive in FCC catalyst. DCC yield from Thai petrochemical industry is given in Table 19.12. Operating conditions of DCC types I and II are given in Table 19.13. DCC is the promising technology for producing C_2–C_4 olefins from the heavy feedstock and total yield of gaseous olefins can be as high as 40% by weight of the feedstock as against 10–15% weight obtained in the conventional FCC [Badoni, 1996]. Comparison of propylene yield from various CC options are given in Table 19.14.

Table 19.12 Thai petrochemical industries DCC yield

	1998	1997
Feedstock	HT Arabian light VGO plus 6.5 DAO and 11% wax	HT Arabian light VGO plus wax
Catalyst	CRP-1	20% CRP-1 and 80% CRP-S
Reactor temperature		
Yields %wt		
C_2 minus	11.6	10.3
$C_3 + C_4$	41.5	40.0
C_5 + Naphtha	35.7	31.9
RON	98.5	99.3
MON	85.3	85.3
Olefin yields %wt		
Ethylene	5.3	5.1
Propylene	18.5	17.4
Butylenes in which	13.3	11.0
Isobutylene	5.9	4.8

Table 19.13 DCC types I and II operating conditions

	Type I	Type II
ROT (°C)	530–575	505–555
Cat/Oil (w/w)	8–15	7–12
Steam dilution rate %wt feed	20–30	10–15
Type of cracking	Riser bed	Riser

Table 19.14 Comparison of propylene yield from various catalytic cracking options

Option	Propylene yield (%wt)
Typical FCC	2–4
High severity FCC	3–5
High severity FCC plus ZSM-5	5–10
DCC II	10–18
DCC I	15–25
Naphtha recycle	+2–3

19.5 C_4 AND C_5 HYDROCARBONS

With increasing demand of C_5 hydrocarbons and oxygenates, upgrading of C_4 and C_5 streams from steam crackers and FCC is important to the economic performance of the above processes. It also provides a rich resource of reactive molecules which form the backbone of the synthetic rubber industry [Morgan, 2010]:

- *C_4 hydrocarbons:* Butenes, butadiene, butane, pentane, methyl butenes, solvents, chemical intermediate, MTBE and plasticizers.
- *C_5 hydrocarbons:* Isoprene (important petrochemical feedstock for synthetic rubber), TAME, rubber chemicals, herbicides, lube oil additives and pharmaceuticals.

19.5.1 Sources of C_4 And C_5 Hydrocarbons

C_4 and C_5 hydrocarbons are important petrochemical feedstocks and finding applications in the manufacture of large number of chemical intermediates, MTBE, TAME and feedstocks for synthetic rubber and agrochemicals.

- Steam cracking
- Catalytic cracking
- Product distribution from FCC depends on:
  - Reactor temperature
  - Feed preheat temperature
  - Catalyst activity
  - Catalyst circulation rate
  - Catalyst activity
  - Recycle rate

- Naphtha feed gives higher yield of C_4 (8–10%) than ethane feed (2–3%).
- Upgrading of C_4 olefins:
  - The production of chemical intermediates
  - Butene-1, isobutylene, mixed n-butene
  - Production of motor fuel component (alkylate, dimerate and MTBE).

19.5.2 Processing of C_4 stream of FCC

Process flow diagram for treatment of C_4 cut from steam cracker and FCC is shown in Fig. 19.9 [Briggs et al., 1987; Convers, 1987; Vermilion and Niclaes, 1977; Chavel and Lefebvre, 1989]. Isobutene recovery includes either hydration of the C_4 stream and subsequent decomposition or etherification with methanol to yield MTBE which is cracked to give isobutene. Separation of 1-butene is done by selective hydrogenation followed by adsorption for separation of 1-butene and further processing for separation of isobutene and 2-butene by distillation. Separation of 2-butene involves hydroisomerization and subsequent distillation for separation of isobutene and 2-butene.

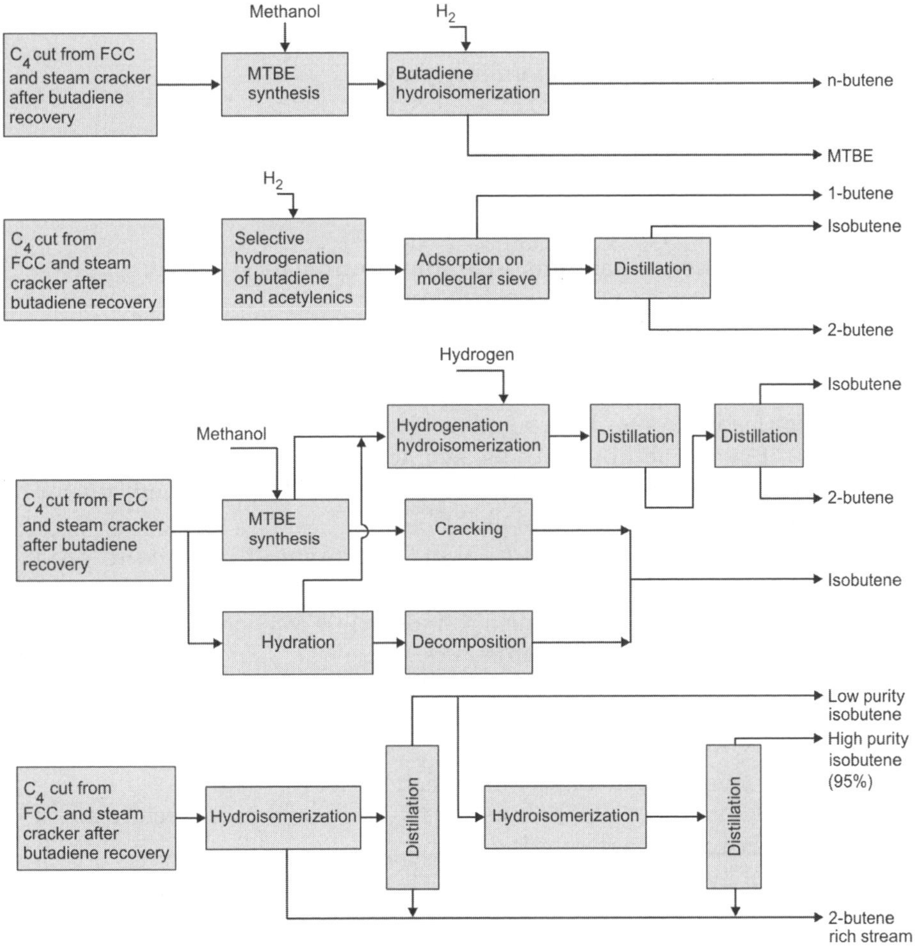

Fig. 19.9 Processing of C_4 stream of FCC

19.6   DEHYDROGENATION OF PARAFFIN (OLIGOMERIZATION OF LIGHT OLEFINS)

Dehydrogenation of paraffins has received considerable interest during recent years due to availability of C_3, C_4 and C_5 hydrocarbons from steam cracking and FCC.

- Dehydrogenation of propane to propylene
- Dehydrogenation of butenes/butanes for butadiene
- Dehydrogenation of isobutane for butadiene
- Dehydrogenation of isoamylenes (2-methyl 1- butene and 2-methyl 2-butene)
- Dehydrogenation of paraffins to olefins (LAB) manufacture

19.7   THERMAL CRACKING OF WAXES

Thermal cracking of waxes is some of the other processes used for production of valuable petrochemicals like α-olefin.

19.8   CATALYTIC REFORMING

Catalytic reforming of naphtha consists of pretreatment and naphtha reforming in reforming details of which has been discussed in Chapter 9.

19.8.1   Reactions in Catalytic Reforming

Reactions involved in catalytic reforming are:

Dehydrogenation:
Methyl cyclohexane $\rightarrow$ toluene + H_2
Isomerization:
n-hexane $\rightarrow$ neohexane
MCP $\rightarrow$ benzene + H_2
Dehydrocyclization of paraffins, i-paraffins to aromatics:
n-heptane $\rightarrow$ toluene + H_2

19.8.2   Aromatic Production

Aromatic hydrocarbons especially benzene, toluene, xylene (BTX) and ethyl benzene are the major feedstocks for large number of intermediates which are used in the production of synthetic fibers, resins, synthetic rubber, explosives, pesticides, detergent, dyes, intermediates, etc., styrene, linear alkyl benzene and cumene are the major consumers of benzene [Mall, 2007].

Aromatic production involves catalytic reforming of naphtha to produce reformate containing BTX.

- First step in making BTX is to distil off a suitable fraction rich in naphthenes which served as precursors for aromatics.
- Catalytic reforming or steam cracking to produce an aromatic gasoline
- Preliminary treatment of this cut: Fractionation and/or selective hydrogenations (essentially pyrolysis gasoline)
- Solvent extraction to eliminate non-aromatics
- Distillation to produce pure benzene and toluene, and in case of reformates used alone or blended with a pyrolysis gasoline
- Distillation aromatic C_8 to yield by super fractionation of ethyl benzene and o-xylene, after passage through a separation column in a light cut and a heavy cut (splitter)
- Production of p-xylene at low temperature with mother liquor by-product rich in m-xylene.

19.8.3 p-xylene Production

p-xylene is one of the important feedstocks for manufacture of terephthalic acid/ DMT which is used for the manufacture of polyester. Manufacturing of p-xylene is shown in Fig. 19.10.

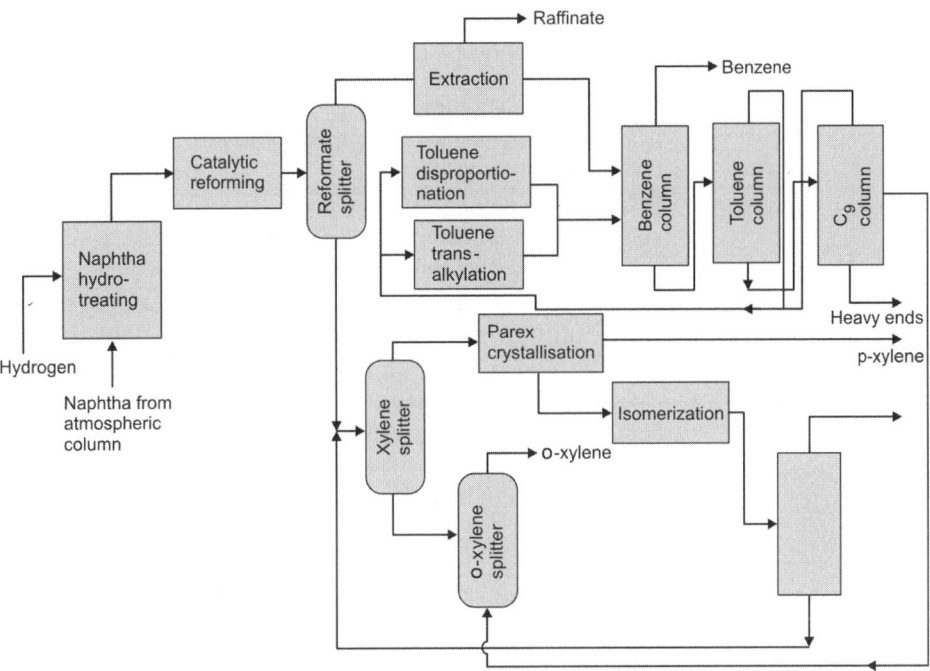

Fig. 19.10 Process flow diagram of p-xylene production

The p-xylene plant consists of following five units.

Pretreatment unit: This unit is used for reducing sulfur content to 5 ppm (max.) by dehydrodesulfurization which takes place at 330–370 °C and 24 kg/cm^2 pressure in presence of cobalt molybdenum catalyst.

Reformer unit: To get maximum amount of C$_8$ aromatics by reforming process (process similar to described earlier).

Fractionation units: For separation of o-, m- and p-xylenes from combined C$_8$ reformate and isomerisate from isomerization unit (after clay treatment).

Parex unit: This unit is for the separation of p-xylene by selective adsorption using molecular sieve followed by desorption. Other method for separation of p-xylene is by crystallization process.

Isomerization: Isomerization of C$_8$ stream from parex unit rich in m- and o-xylenes and ethyl benzene to p-xylene which is sent to fractionation unit for separation of high component. The bottom of the column is recycled for further recovery of xylenes.

Aromatic conversion processes: Various aromatic conversion processes are given in Table 19.15.

Table 19.15 Various aromatic conversion processes

Hydrodealkylation	Hydrodealkylation is the process of producing benzene from toluene by catalytic hydro using chromium oxides based catalyst delkylation.
Isomerisation	The process involves isomerization of ethylene benzene to xylenes or m-xylene to p-xylene using both liquid phase and vapor phase . Various types of catalysts used are Friedel craft type of catalyst, ZSM 5 catalyst, platinum deposited on silica alumina. m-xylene isomerization is also carried out to convert m-xylene into p-xylene.
Toluene dismutation	The process involves toluene disproportionation using vapor or liquid phase using solid catalyst, Friedel craft type, silica aluminas and dual functional or zeolites.
Disproportionation	The process selectively converts toluene to mixed xylenes and high purity benzene. The reaction occurs in vapor phase.
Cyclar process	The process is used for the production of petrochemical grades benzene, toluene, xylenes via aromatization of propane and butane from natural gas. The process consists of reactor section consisting of stacked radial flow reactor, catalyst regeneration and product recovery section.

19.9 LINEAR ALKYL BENZENE (LAB)

Linear alkyl benzene (LAB) is a basic raw material for production of most widely used detergent. Linear alkyl benzene was introduced as a substitute for non-biodegradable branched alkyl benzene. Globally, LAB accounts for one-third of the active ingredients used in detergents. Some of the LAB manufacturers in India are Reliance Industries, Nirma, Chennai Petroleum and Petrochemical Limited. Various steps of manufacturing LAB are given in Table 19.16 and Fig. 19.11 illustrates the process of manufacturing of LAB by HF alkylation.

Table 19.16 Various steps in the LAB manufacture

Prefractionation	To obtain C_{10}–C_{14} range hydrocarbons from kerosene. (n-C_{10} to n-C_{13} for light LAB and n-C_{11} to n-C_{14} for heavy LAB).
Hydrotreater	To remove sulfur compounds
Paraffin separation	To remove n-paraffins from kerosene by selective adsorption using molecular sieve
Dehydrogenation	Dehydrogenation of n-paraffins to olefins
Alkylation	Alkylation of benzene with olefins to obtain LAB. Earlier HF was used as catalyst, however, HF has been replaced with solid acid catalyst in the newer process.

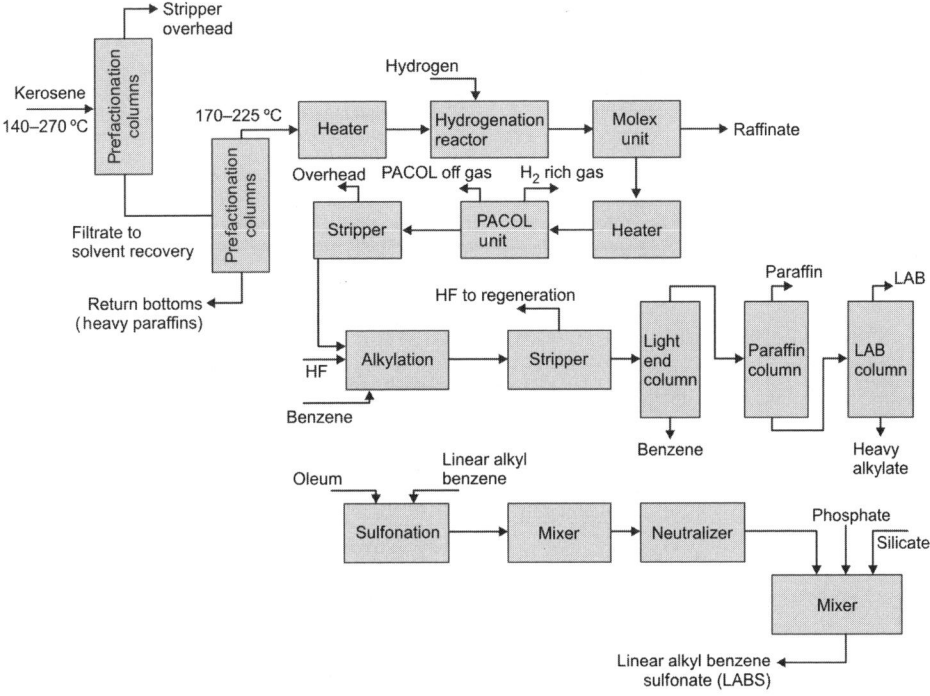

Fig. 19.11 Linear alkyl benzene by HF alkylation

19.10 POLYMERS

The polymer industry plays an important role in the economic development, and this industry is one of the fastest growing sectors all over the world because of the availability of ethylene, propylene from cracker plants. A wide variety of polymers in various forms are being manufactured. During recent years consumption of polymers has been increasing and it has substituted the conventional material in many applications. Plastics have penetrated all sectors of activity and have become essential in daily needs. Various types of polymers and their applications are given in Fig. 19.12.

Polyolefins are high polymers produced by polymerization of olefins. Polyethylene (ldpe, hdpe, lldpe) and polypropylene (pp) are main polyolefins produced and used extensively in areas covering: Automobiles, communication, electric machinery, food packaging, drink containers, civil engineering, building materials, medical machines and apparatus and agricultural materials. Polyethylene (LDPE, LLDPE, HDPE) and polypropylene are two major polyolefins finding large scale application in various sectors. Demand for polypropylene, LDPE, LLDPE, HDPE in India is 1.9 MMTPA, 0.2 MMTPA, 0.8 MMTPA, 1 MMTPA respectively [Bansal, 2009]. Capacities for polymers and elastomers in India are mentioned in Table 19.17.

Polyvinyl chloride is the second largest (after polyethylene) and most versatile in all thermoplastics. Global production and consumption of polyvinyl chloride (pvc) in 2007 was approx. 34 million metric tons. Global capacity utilization was 81% in 2007. It is one of the most widely used thermoplastic polymers with a total

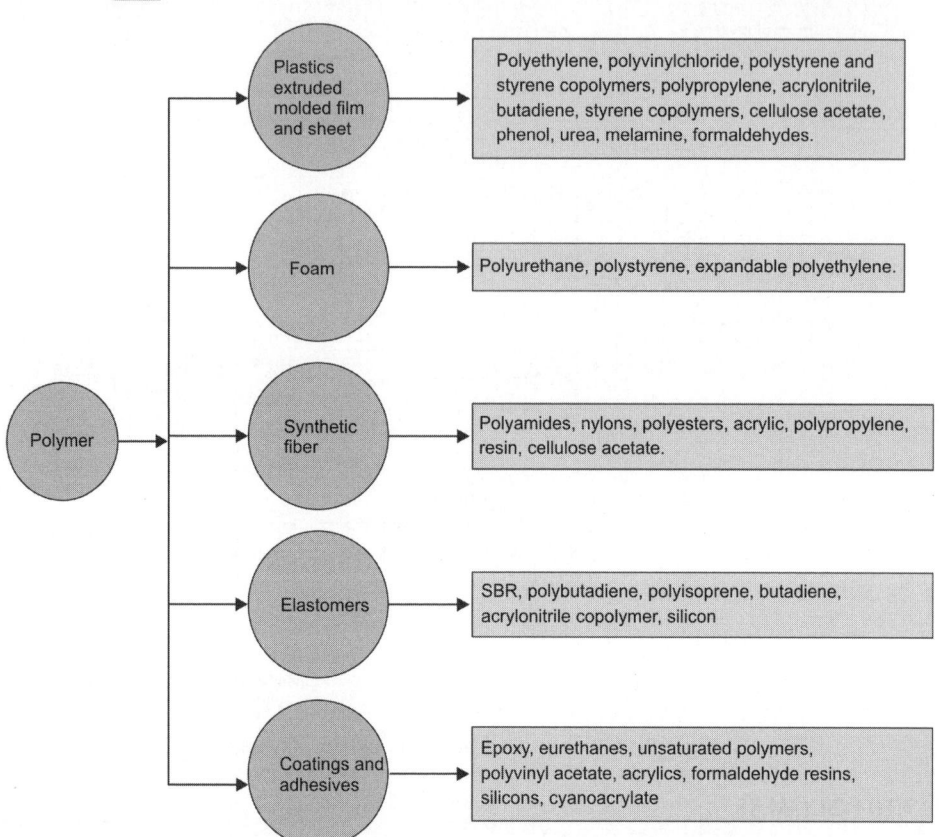

Fig. 19.12 Various types of polymers and their applications

Source: Mall, 2007.

Table 19.17 Indian petrochemical capacities for polymers and elastomers

Product	Installed capacity '000' TPA	Production '000' TPA
Polyethylene	1,625	1,317
Polypropylene	1,391	1,170
PVC	790	760
SBR	21	14
PBR	50	41

world consumption of about 30 million tons. It contains about 56.8% chlorine and balancing hydrocarbon.

Application of various types of polymers is shown in Fig. 19.12. Most of the petrochemical complexes have polyethylene, polypropylene, PVC plants with availability of ethylene, propylene as feedstocks from cracker plant.

19.11 SYNTHETIC FIBER INDUSTRY

There has been phenomenal growth to the synthetic fiber industry during recent years due to growth of petrochemical industry. The synthetic fiber industry based on petrochemicals consist of polyamides, polyester, acrylics and polypropylene fibers and enjoy major share in international business and marketing. Viscous rayon based on natural fibers is also one of the important synthetic fibers. Some of the important applications are staple fibers, textile grade filament yarns, and industrial and high tenacity yarns. Acrylic, polyamide, polyester and polypropylene fibers account for 98% of the world synthetic fiber market.

With increasing demand of polyester, polyester plant has been integrated with petroleum refinery due to availability of naphtha for catalytic reforming to produce aromatics. With availability plant of propylene from cracker plant, there has been continuous increase in the production of acrylic fiber. Detail of fiber is shown in Fig. 19.13 and detail of feedstock and capacity of fiber are given in Tables 19.18 and 19.19.

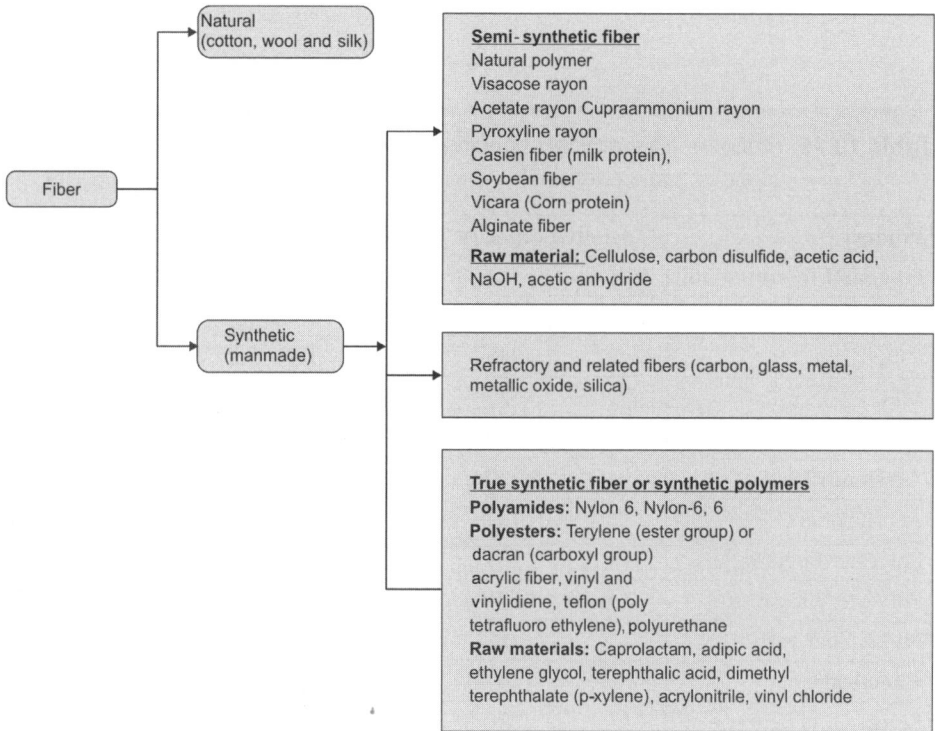

Fig. 19.13 Various types of fibers

Table 19.18   Various synthetic fiber and their monomers and intermediate feedstocks

Name of the synthetic fiber	Monomer	Intermediate feedstocks
Nylon 6	Caprolactam	Benzene, toluene, ammonia
Nylon-6, 6	Adipic acid HOOC–$(CH_2)_4$–COOH, hexamethylene diamine H_2N–$(CH_2)_6$–NH_2	Benzene, butadiene, propylene, phenol, adiponitrile
Polyester	Dimethyl terephthalate and purified terephthalic acid (PTA)	p-xylene, monoethylene glycol
Acrylic fiber	Acrylonitrile	Propylene
Modified acrylics	Acrylonitrile, vinyl chloride, vinylidene chloride	Propylene, ethylene dichloride from ethylene
Polypropylene	Propylene	Propylene

Table 19.19   Indian petrochemical capacities for fibers, intermediates and important basic organic chemicals

Product	Installed capacity '000' TPA	Production '000' TPA
Fiber and intermediates		
DMT	270	190
PTA	1445	556
MEG	520	494
Caprolactam	150	149
Acrylonitrile	30	32
Acrylic fibers	114	99
Polyester fibers	564	563
Polyester fill yarns	1,091	818
Nylon fiber yarn	27	26
Chemicals		
Linear alkyl benzene	370	359
Ethylene oxide	87	66
Phenol	35	80
Phthalic anhydride	227	180

REFERENCES

1. Briggs BA, Simpson SO, Lemerk CA and Ward DJ. Fluid Catalytic Cracker as a Source of Petrochemical Olefins. Chemical Age of India, Vol 38, No. 1, 1987, p21.

2. Badoni RP, Kumar Y, Shanker U and Prasada Rao TSR. Emerging Technologies for Light Olefins Production. Chemical Engineering World, Vol 31, Dec 1996, p105.

3. Chavel A and Lefebvre G. Treatment of Olefinic C_4 and C_5 Cuts, in Petrochemical Processes. Synthesis, Gas Derivatives and Major Hydrocarbons, Editions Technip, Paris, 1989, p195.

4. Convers A. Make Chemicals from C_4 Olefinic Fractions. Chemical Agre of India, Vol 38, No. 1, 1987, p31.

5. Dharia D, Letzsch W, Kim H, McCue and Chapin L. Increase Light Olefins Production. Hydrocarbon Processing, April 2004, p61.

6. Kothary NC, Mulchandani HK, Jain SK, Gomkale and Khilnani S. Technoeconomics of Utilization of Refinery C_4 Streams. Chemical Engineering World, Vol 13, No. 12, 1988, p31.

7. Mall ID. Petrochemical Process Technology, Macmillan India, 2007.

8. Morgan M. The C_4 Industry Beyond 2000. Chemistry and Industry, Feb 1998, p90.

9. Niccum PK, Gilbert MF and Tallman MJ. Consider Improving Refining and Petrochemical Integration as Revenue—Generation Option. Hydrocarbon Processing, Nov 2001, p47.

10. Peilling Z. Integration of a DCC Unit into a Refinery. Hydrocarbon Engineering, Oct 1997.

11. Vermilion WL and Niclaes HJ. Petrochemicals from the FCC Unit. Hydrocarbon Processing, Sep 1977, p193.

12. Zaiting LJ, Fukang X, Cahogang X and Youhao X. DCC Technology and Its Commercial Experience. China Petroleum Processing and Petrochemical Technology, No. 4, Dec 2000.

20

Product Quality Analysis, Specification and Standards

Better refinery stream distillation and contaminant data ultimately improves the accuracy of various refinery decision making tools [Golden et al., 1995]. The assessment of the quality of product is of equal importance. Some of the important parameters are specific gravity, smoke point, pour point, viscosity, freezing point, stability test, corrosion tests, flammability tests, acidity and alkalinity, ash, sediment, wax, sulfur, inorganic constituents and contaminants, organic constituents, RON and MON tests, etc.

20.1 PETROLEUM PRODUCT QUALITY ANALYSIS

Petroleum product analysis is important for producing fuel of desired specifications. Some of the major analytical instruments used are gas chromatograph, HPLC, surface area and pore volume analyzer, TBP distillation unit (ASTMD 2892), high temperature simulated distillation, FIA apparatus, atomic absorption spectrophotometer, UV/visible spectrophotometer, inductively coupled plasma atomic emission spectroscopy (ICP-AES), MSCP, wiped film evaporators potentiograph, Wickbold sulfur apparatus, automatic distillation apparatus, vacuum distillation apparatus, etc. Some of the engine/ATF special tests include residues—ASTM D 4530, asphaltene IP-143, ASTM D 3279 and ASTM D 4124. Table 20.1 shows requirement of various parameters for evaluation of petroleum product quality [Mall, 2007].

Table 20.1 Requirement of various parameters for evaluation of crude oil, petrochemical feedstocks and product quality

Product	Information required
LPG	Copper strip test, total volatile sulfur, vapor pressure at 65 °C, volatility, dryness, odor, composition, free water content, gross calorific value, composition of C_2–C_5 hydrocarbon, unsaturated hydrocarbon
Naphtha	Specific gravity, paraffins, naphthenes and aromatic content, RON and MON, sulfur, nitrogen, chloride, metals, nitrogen, RVP, aromatics/olefins, gross calorific value

Contd.

Table 20.1 Requirement of various parameters for evaluation of crude oil, petrochemical feed stocks and product quality *(Contd.)*

Product	Information required
Kerosene	Acidity, burnability quality, color, copper strip corrosion test, distillation range flash point, smoke point, total sulfur, density
Vacuum gas oils	Specific gravity, viscosity, cetane number/diesel index, nitrogen, carbon residue, oxidation stability, flash point, aromatics, polycyclic aromatics, acidity, sediment, RVP, moisture, RON/MON
Reformulated gasoline	RVP, sulfur, aromatics, benzene olefins, oxygenates, T90 °C, T50 °C, sp. gravity, RON/MON, oxidation stability, gums, antiknock index, lead content, vapor lock index, copper corrosion test, color
Catalytic reforming feed and reformate	Specific gravity, RVP, pour point, flash point, aniline point, total sulfur, CCR% specific gravity, distillation (IBP °C/FBP °C), benzene, toluene, xylene, total sulfur
Residues	Specific gravity, sulfur, nitrogen, metals, asphaltenes, viscosity, aniline point
Lube cuts	Specific gravity, viscosity, nitrogen, metals, asphaltenes, viscosity, mid distillate vol percentage curves, viscosity yield curves, CCR, cloud point, solvent treating susceptibility
Greases	Penetration, softening point
Detergent alkylate	Specific gravity, bromine index, saybolt color, doctor test, water, sulfonation, biodegradability, paraffins, n-alkyl benzene, average molecular weight
Fuel oil	Acidity, ash, carbon residue, copper strip corrosion test, sediment, flash point, smoke point, sulfur
Bitumen	Specific gravity, penetration, softening point, ductility, flash point, matter soluble in carbon disulfide

20.1.1 Density and API Gravity, Viscosity, Water, Sediments and Salts (See Chapter 4)

20.1.2 Foaming Characteristics of Lubricating Oils

Foaming of lubricating oils in applications involving turbulence, high speed gearing or high volume pumping can cause inadequate lubrication, cavitations, overflow and premature oxidation [koehlerinstrument.com].

20.1.3 Water Separability of Petroleum Oils and Synthetic Fluids

The ability of a lubricating oil to separate from water and resist emulsification is an important performance characteristic for applications involving water contamination and turbulence [koehlerinstrument.com].

20.1.4 Refractive Index

Refractive index is a fundamental physical property that is used in conjunction with other properties to characterize pure hydrocarbons and their mixtures. It is useful property for concentration measurements, purity determinations and chemical identification [www.clarksonlab.com].

20.1.5 Reid Vapor Pressure and Light End Analysis

Reid vapor pressure is an important parameter and is a measure of the vapor pressure of petroleum and petroleum products and indicates the relative percentage of gaseous and lighter hydrocarbons present in crude oil and the fuel. Vapor pressure is a critical factor in the handling and performance of liquids petroleum and liquefied gas (LPG) products. The vapor pressure of automotive gasoline is subject to environmental regulations for pollution control [koehlerinstruments.com].

20.1.6 Pour Point

Pour point is an indicator of the lowest temperature of utility for petroleum products. Pour point is an important parameter which gives an idea of the lowest temperature at which a fuel can be used and is approximated indication of its pumpability. For estimating the relative amount of wax present in the crude oil, pour point is the lowest temperature at which movement of the oil is observed. Oils used for lubricants must have a low pour point. Pour points of crude oils generally vary in the range –60 to +30 °C.

20.1.7 Cloud Point

The cloud point is the highest temperature at which first trace of wax starts to separate out, causing it to become turbid or cloudy and haze is observed. Cloud point, usually range may be –10 °C to 0 °C.

20.1.8 Aniline Point

Aniline point indicates the lowest temperature at which the oil is completely mixed with an equal volume of aniline. High aniline point indicates that the fuel is paraffinic, and hence has a high diesel index and very good ignition quality. Aromatic hydrocarbons exhibit the lowest aniline points. Cycloparaffins and olefins exhibit values that lie between those paraffins and aromatics.

20.1.9 Wax Content

Wax present in the crude oil affects the flow behavior of the crude and affects the product quality of gas oil, vacuum oil, asphalt and lubricating oils. Wax content is determined by Engler Holde Method.

20.1.10 Penetration

Penetration test is important for measuring consistency to describe the hardness or softness of grease. Penetration tests are performed to determine consistency and shear stability (lubricating greases) for design, quality control and identification purposes. In penetration test, a standard cone or needle is released from a penetrometer and allowed to drop freely into the sample for 5 seconds or at different (specified interval) or constant temperature. The depth of penetration of the cone or needle into sample is measured in tenths of a millimeter by the penetrometer [www.clarksonlab.com]. Microprocess based digital penetrometer is available. Different types of instruments are available.

20.1.11 Asphaltene, Carbon Residue and Ash

Carbon residue and asphaltenes indicate the presence of heavier hydrocarbons in the crude. Asphaltenes are pole nuclear condensed aromatic hydrocarbons having high molecular weight. Ash in the crude oil is due to presence of metallic constituents. Metal present in the crude oil has detrimental effect on the catalyst activity and life.

20.1.12 Carbon Residue

It is the measure of thermal coke forming property and is the tendency of a fuel to form carbon residue under high temperature conditions in an inert atmosphere. It is determined by conradson carbon residue and Ram's bottom carbon residue method. Carbon residue value gives an indication of the combustibility and deposit forming tendencies of the fuels.

20.1.12.1 Conradson Carbon

Conradson carbon provides an indication of relative coke forming properties of petroleum oils. The sample after weighing is placed in a crucible and subjected to destructive distillation where the residue undergoes cracking. After the end of specified heating period, the residue is cooled and weighed. The residue remaining is reported as a percentage of the original sample.

20.1.12.2 Ramsbottom Carbon Residue

It determines the carbon residue left after evaporation and pyrolysis of a sample oil in a special glass bulb having a capillary opening placed in Ramsbottom furnace at 550 °C for 20 ± 1 minutes after which it is cooled and weighed. The residue remaining is reported as percentage of original sample as carbon residue. It provides an indication of the deposit forming tendencies of fuels and guidelines for the processing of refinery products.

20.1.13 ASTM D 1289 Conradson Carbon Residue

ASTM D 4530 microcarbon residue: This procedure determines the carbon residue left after evaporation and pyrolysis of an oil sample under prescribed conditions and is rough indicator of oil's relative coke forming tendency or contamination of a lighter distillate fraction with a heavier distillate fraction or residue [home.earthlink.net] [Golden et. al., 1995].

Carbon residue and atomic H-to-C ratio can be correlated by

H/C = 171 − 0.015CR (conradson)

20.1.14 Flash Point

Flash point is the lowest temperature at which application of test flame causes the vapor and air mixture above the sample to ignite [www.sersc.org]. It is a guide to the fire hazard associated with fuel. During its storage, transportation, and use of liquid petroleum products in either closed or open vessels.

Some of the instruments used are Pensky-Martens closed cup flash points test, automatic tag closed cup flash point tester and automatic Cleveland open cup flash point tester.

Flash point of the diesel can be represented by following empirical relationship

FP (flash point, °C) = IP (distillation initial pointing, °C) – 100

20.1.14.1 Pensky-Martens Closed Cup Flash Tester

Pensky-Martens closed cup flash tester determines flash point of a wide range of products by a closed cup method with two option speed stirring of the sample.

20.1.15 Fire Point

Fire point is the lowest temperature at which the oil ignites and continues to burn. This is used as an index of fire hazards.

20.1.16 Water and Sediments

These cause irregular behavior in the distillation and cause blocking and fouling of heat exchanger and result in corrosion. Centrifuging provides a convenient means of determining water content in crude oil, fuel oils and middle distillate fuels. Also used in determining the precipitation number, demulsibility characteristics, trace sediments, and insoluble in used lubricating oils [labequip.com].

Water content is determined by Dean and Starck. Sediment and water is determined by centrifuging a mixture of crude oil and toluene. Salt content is determined by titrating the water extract with $KCNS/AgNO_3$.

20.1.17 Smoke Point

It is an indication of smoking tendency of fuel. It is used for evaluating the ability of kerosene to burn without producing smoke. It is the maximum flame height in mm at which the fuel will burn without smoking. Presence of higher aromatics gives higher smoke.

Smoke volatility index (SKI) = Smoke point + 0.42 × Recovery at 204 °C.

20.1.18 Acidity

Acidity is a measure of corrosive properties of products which include organic and inorganic acidity. Acids in the sample are extracted in neutral alcohol and titrated against standard alcoholic potassium hydroxide.

20.1.19 Copper Corrosion Test

This test serves as a measure of possible difficulties with copper, brass, bronze part of the fuel system. In copper corrosion test a polished copper strip is immersed in the sample for three hours at 50 °C and then removed and washed. The condition of the copper surface is qualitatively rated by comparing it to standards [www.ctnor.com].

20.1.20 Autoignition Temperature

Autoignition temperature determines the lowest temperature at which the vapors of a liquid or solid chemical sample will self-ignite under prescribed laboratory conditions [www.alpe.net.au].

20.1.21 Freezing Point

Freezing point is the temperature at which a hydrocarbon passes from a liquid to a solid state, ASTM 2386, IP16 [Speight, 2001].

20.1.22 Melting Point and Congealing Point

Melting point of wax is the temperature at which the wax becomes sufficiently fluid to drop into liquid phase. Congealing point is the temperature at which melted petroleum ceases to flow when allowed to cool under prescribed condition.

20.1.23 Reid Vapor Pressure and Light End Analysis

Reid vapor pressure is the measure of the vapor pressure of petroleum and petroleum products and indicates the relative percentage of gaseous and lighter hydrocarbons present in crude oil and the fuel. Vapor pressure is a critical factor in the handling and performance of liquids petroleum and liquefied gas (LPG) products. The vapor pressure of automotive gasoline is subject to environmental regulations. Reid vapor pressure is the vapor pressure exerted by a liquid under the specific conditions of test, temperature, vapor:liquid ratio and air and water saturation.

20.1.24 Gum

It is an indication of gum at the time of test and amount of deposition during service time and presence of gum olefins which have very poor storage stability.

Extinct gum is the amount of non-volatile heptane insoluble residue left when the sample is evaporated under specified condition.

Potential gum is the amount of gum formed after sample is aged in an oxidation stability bath and evaporated under specified condition.

20.1.25 Color

It is an indication of the thoroughness of the refining process. This is determined by saybolt chromometer or by Lovibond Tintometer. Saybolt color of petroleum products test is used for quality control and product identification purposes on refined products having an ASTM color of 0.5 or less.

ASTM color of petroleum products applies to products having ASTM color of 0.5 or darker, including lubricating oils, heating oils and diesel fuel oils [www.clarksonlab.com].

Pale = 4.5 ASTM color or lighter

Red = Darker than 4.5 ASTM

Dark = Darker than 8.0 ASTM

20.1.26 Antiknock Quality (Octane Number)

Knocking is a characteristics property of motor fuels that governs engine performance and is expressed in terms of octane number. It depends on the properties of hydrocarbon type and nature. Octane number is the percentage of iso-octane in the reference fuel which matches the knocking tendency of the fuel under test. Research octane number (RON) and motor octane number (MON) are two methods used and are measured with a standardized single cylinder, variable compression ratio engine. For both octane numbers, same engine is used, but operated at different conditions. The distinction between two octane number (RON and MON) measurement procedures are engine speed, temperature of admission and spark advance [Chatila, 1995]. The motor method captures the gasolines at high engine speeds and loads, and the research octane method at low speed depending on the fuel characteristic, the MON is normally 8–10 point lower than the RON. A high tendency to auto-ignite, or low octane rating, is undesirable in a gasoline engine, but desirable in a diesel engine [www.dmperf.com].

Anti-knock index (AKI) = (RON + MON)/2

For paraffins, the higher the branching of hydrocarbon compound, the higher is its octane value. Olefins also have high octane number, while naphthenic hydrocarbons have mid range octane numbers. N-paraffins have lowest octane numbers, while aromatic compounds have highest octane numbers. Isoparaffins have octane number at varying levels [Rao, 2007]. The octane numbers of hydrocarbon mixtures are not additives based on individual hydrocarbon octane number [Rao, 2007]. Three new terms octane gravity, type factor, boiling point gravity product have been given by Rao, 1997 for estimating octane numbers of hydrocarbon mixtures.

20.1.27 Cetane Number and Calculated Cetane Index

Cetane number is used to measure the ignition qualities of diesel fuels. It is the percentage of cetane which must be mixed with heptamethyl nonane to give the same ignition performance as the fuel in question. The higher the cetane's number the better the quality and will facilitate easy starting of engine and lessen engine roughness. The cetane number is determined in a single cylinder cooperative fuel research (CFR) engine by comparing its ignition quality with that of reference blends of known cetane number [www.sersc.org].

Cetane number = 0.72 diesel index × 10

Caculated cetane index (CCI) can be calculated by four variables:

$$
\begin{aligned}
CCI = 45.2 \; &+ (0.0892) \, (T_{10} \, N) \\
&+ [0.131 + 0.901 \, (B)] \, [T_{50} \, N] \\
&+ [0.0523 - (0.420)B)] \, [T_{90} \, N] \\
&+ [0.00049][\, (T_{10} \, N)^2 - (T_{90} \, N)^2 \\
&+ 107 \, B \; + 60 \, B^2
\end{aligned}
$$

where T_{10} = 10% distillation temperature °C
T_{50} = 50% distillation temperature °C
T_{90} = 90% distillation temperature °C
B = $e^{-3.5DN} - 1$
D = Density @15 °C
DN = $D - 0.85$

20.1.28 Diesel Index (DI)

It is an indication of ignition quality of a diesel. Higher diesel is an indication of better ignition quality.

Diesel index = (Aniline point in °F × API)/100

Diesel index = (Cetane number – 10)/0.72 or Cetane index = 0.72 DI + 10

Cetane index = AP – 15.5

20.1.29 Thermal Stability of Jet Fuels

Jet thermal oxidation test is the most common method for estimating the thermal stability of jet fuel.

20.1.30 Oxidation Stability

It is used for the evaluation of storage stability and resistance to oxidation as most of the oils, when exposed to air over time, react with oxygen and degraded. Oil with poor oxidation stability, form corrosive acids at high temperature condition in the engine.

20.1.31 Carbon-Hydrogen Ratio

C-H ratio = 74 + 15d/26 – 15d, where d is sp. gr. at 15 °C/15 °C

20.1.32 Weathering Test for LPG

This test shows the volatility of the LPG.

20.1.33 Corrosion Inhibition Properties of Greases

It measures the ability of grease to protect a bearing against corrosion in the presence of water.

20.1.34 Lubricating Ability of Greases

It measures the ability of grease to lubricate under various speeds and at various temperatures, by recording the number of running hours before the grease ceases to lubricate and causes the bearing to fell [www.koehlerinstrument.com].

20.1.35 Evaporation Loss of Lubricating Greases and Oils

It evaluates the potential for evaporation loss of lubricant components in high temperature service.

20.1.36 Dropping Point of Lubricating Greases

Dropping points are used for identification and quality control purposes, and can be an indication of the highest temperature of utility for some applications. This is the temperature at which grease passes from a semisolid to a liquid state under prescribed conditions [Speight, 2001].

20.1.37 Apparent Viscosity of Lubricating Greases

It is used to evaluate pumpability and handling characteristics of greases and is also suitable for analysis of adhesives, sealants and other semisolid products.

Cone penetration of lubricating grease: Penetration of lubricating grease is the depth in tenths of millimeter that standard cone penetrates the sample under prescribed conditions of weight, time temperature ["Standards", Journal of Synthetic Lubrication, 10/1984].

20.1.38 Grease Mobility Test

It determines the resistance of lubricating greases to flow under prescribed condition.

20.1.39 Oil Separation from Lubricating Greases

It determines the tendency of oil and lubricating grease to separate at elevated temperature.

20.1.40 Frass Breaking Point

This determines the breaking point of solid and semisolid bitumens. This is the temperature below which the bitumen tends to break rather than flow.

20.1.41 Bromine Number and Bromine Index

Bromine number is an indication of amount of saturation in the sample. Bromine number is defined as the number of grams of bromine consumed by 100 g of sample when reacted under specified condition. Bromine index is defined as the number of milligrams of bromine that will react with 100 g of sample [Novak et al., 1989].

20.1.42 Ductility of Bituminous Materials

This measures the distance of elongation of a bitumen sample when a briquette specimen is pulled apart at a specified speed and temperature.

20.1.43 Odor Point of Petroleum Wax

Odor of wax is the numerical rating corresponding to the odor scale description that fits the sample being tested. This property is important where wax is used for packaging purpose.

20.1.44 Silver Corrosion Test

This test is used to measure the corrosiveness of aviation turbine fuels towards silver. A polish silver strip test is immersed in a fuel sample at elevated temperature and specified time. After a specified test period, the strip is removed from the sample, washed and evaluated for corrosion [koehlerinstruments.com].

20.1.45 Antirust Properties of Petroleum Products Pipeline Cargos

This test is used to control corrosion in product pipeline caused by moisture condensed from gasoline and distillate fuels. Antirust properties are determined by immersing a polished steel test specimen in a stirred mixture of the sample and distilled water held at constant temperature [koehlerinstruments.com].

20.1.46 Oxidation Stability

Gasoline contains cracked components having tendency to form gum materials during storage and handling which affect performance. Oxidation stability provides an indication of the tendency of gasoline and aviation fuels to form gum in storage. In this test the sample is oxidized inside a stainless steel pressure vessel initially charged with oxygen at 689 kPa and heated in a boiling water bath. The amount of time required for a specified drop in pressure (gasoline) or the amount of gum and precipitate formed after specific aging period (aviation fuel) is determined [koehlerinstruments.com].

20.1.47 Sulfur

Sulfur is always present in the crude in the form of organic sulfur compounds, dissolved hydrogen sulfide and sometimes even in suspended form with increasing use of heavier crude sulfur content is increasing in the fuel. The sulfur content may be in the range from as low as 0.05 %wt to as high as 5 %wt. The heavier crude oils contain higher sulfur. Presence of sulfur in the crude and the products undesirable not only from environment point of view but also from its impact on process in the form of corrosion, poisoning of catalyst and degradation of the product. This is determined by lamp method or wickbold.

20.1.48 Nitrogen

Crude oil contains nitrogen compounds in the form of quinoline, isoquinoline, and pyridine and causes result in poisoning of reforming catalyst.

20.1.49 Loss on Heating in Bitumen

As the bitumen is heated on application or mixing, this test is devised to prevent the inclusion of exccessive amounts of volatile matter.

20.1.50 Identification and Structural Group Analysis [Speight, 1999]

The crude oil is a complex mixture of saturated hydrocarbons, saturated hetero-compounds, and aromatic hydrocarbons, olefinic hydrocarbons and aromatic heterocompounds. With the advancement of the instrumental analysis techniques like chromatography and spectroscopic methods, now it has been possible to go indepth study of the identification and structural group analysis. Some of the major analytical instruments used are gas chromatography, ion exchange chromatography, simulated distillation by gas chromatography, absorption chromatography, gel permeation chromatography, high-performance liquid chromatography and supercritical fluid chromatography. The application of spectroscopic methods includes infrared spectroscopy, nuclear magnetic resonance spectroscopy, mass spectroscopy, electron spin resonance, X-ray diffraction, inductively coupled plasma emission spectroscopy, X-ray absorption spectroscopy and atomic absorption spectrophotometer.

20.1.51 Specification of Diesel and Gasoline

There has been continuous upgradation of fuel specification to meet the various problems arising from changing crude oil quality from lighter to heavier crudes and sweet to sour quality of crude. Specification of the diesel and gasoline as per the Bharat-II, Euro-III and Euro-IV norms is mentioned in Tables 20.2 and 20.3. IS specification of various petroleum products is given in Tables 20.4 to 20.16.

Table 20.2   Roadmap for vehicular emission norms for new vehicles (except 2 and 3 wheelers)

Entire country	
Bharat stage-II	1st April 2005
Euro-III equivalent	1st April 2010
Metros and major cities (Bengaluru, Hyderabad, Ahmedabad, Pune, Surat, Kanpur and Agra)	
Bharat stage-II	Already introduced in metro cities and other major cities as on 1st April 2003
Euro-III equivalent	1st April 2005
Euro-IV equivalent	1st April 2010
Bharat stage-II	1st April 2010
Euro-III equivalent	Preferably from 1st April 2008, but not later than 1st April 2010

Source: Mehroton RK, Petrotech Society, New Delhi, July 3, 2006.

Table 20.3   Key specifications of gasoline (MS)

Specification		BIS 2000	BS-II	Euro-III eqv.		Euro-IV eqv.	
				Regular	Premium	Regular	Premium
Sulfur, [ppmw (max.)]	↓	1000	500	150	150	50	50
RON (min.)	↑	88	88	91	95	91	95
MON (min.)		No spec	No spec	81	85	81	85
AKI (min.)	↓	84		No spec	No spec	No spec	No spec
Benzene [%vol (max.)]	↓	5		1	1	1	1
Aromatics [%vol (max.)]	↓	No spec	No spec	42	42	35	35
Olefins	↓	No spec	No spec	21	18	21	18

Source: Gopal S Ray. Petroleum Refining Petrochemicals, June 6–10, 2011.

Table 20.4 Key specifications of HSD

Specification		BIS 2000	BS-II	Euro-III eqv.	Euro-IV eqv.
Density @ 15 °C kg/m³	↓	820–860	820–860	820–845	820–845
Sulfur content [ppmw (max.)]	↓	2500	500	350	50
Cetane no. (min.)	↑	48	48	51	51
Distillation 95% vol (°C, max.)	↓	370	370	360	360
Polycyclic aromatics hydrocarbons (PAH) (% mass, max.)	↓	No spec	No spec	11	11

Source: Gopal S Ray. 6th Summer School as Petroleum Refining Petrochemicals, June 6–10, 2011.

Table 20.5 Requirements of motor gasoline [IS: 2796 (2008)]

Sl. No.	Characteristics	Test requirements				GR typical value
		Unleaded regular	Unleaded premium	Regular	Method	
1.	Color, visual	Orange	Red	Orange	–	Orange
2.	Copper strip corrosion, for 3 hours, at 50 °C	Not more than no. 1	Not more than no. 1	Not more than no. 1	P:15	1
3.	Density at 15 °C, (kg/m³)	720–775	720–775	720–775	P:16	782.4
4.	Distillation:					
	a. Initial boiling point (°C)			–	P:18	40
	b. Recovery up to 70 °C (%v, min.)	10–45	10–45	10–45	–	28
	c. Recovery up to 100 °C (%v, min.)	40–70	40–70	40–70	–	58
	d. Recovery up to 150 °C (%v, min.)	75	75	75	–	–
	e. Final boiling point, (°C, max.)	210	210	210	–	160
	f. Residue (%v, max.)	2	2	2	–	1
5.	Octane requirements: a. Anti-knock index (min.)	84	88	84	P:26/ P:27	84
6.	Potential gum [g/m³, max. (4 hours test)]	50	50	50	P:147	30

Contd.

Table 20.5   Requirements of motor gasoline [IS: 2796 (2008)] *(Contd.)*

Sl. No.	Characteristics	Test requirements				GR typical
		Unleaded regular	Unleaded premium	Regular	Method	value
7.	Existent gum (g/m³, max.)	40	40	40	P:29	12
8.	Sulfur (total, %wt max.)	0.10	0.10	0.20	P:34	0.065
9.	Lead content (as pb) (g/L, max.)	0.005	0.005	0.15	P:38	0.003
10.	Reid vapor pressure at 38 °C (kPa, max.)	60	60	60	P:39	55
11.	Benzene content (%vol)	1	1	1	P:104	0.90
12.	Vapor lock index (max.) a. Summer b. Other months	750 950	750 950	750 950	– –	746 –
13.	Water tolerance of gasoline alcohol blend, temperature for phase separation a. Summer (°C, max.) b. Winter (°C, max.)	10 00	10 00	10 00	– –	– –

Table 20.6   Requirements for mineral turpentine oil (MTO) (IS: 1945–1978)

Sl. No.	Characteristics	Requirement for method of		GR test	Typical value gr. I
1.	Density at 15 °C (g/mL)	Not limited, but to be reported	Not limited, but to be reported	P:16	0.7875
2.	Color (saybolt) (min.)	+21	+20	P:14	+30
3.	Flash point (abel) °C (min.)	30	35	P:20	–
4.	Distillation: a. IBP °C (min.) b. 50% vol Rec @ (°C) c. 95% vol Rec. @ (°C) d. Final boiling point (°C, max.)		240	205	–228
5.	Aromatic content (%vol, max.)	40	40	P:23/ P:48	20.6

Contd.

Table 20.6   Requirements for mineral turpentine oil (MTO) (IS: 1945–1978) *(Contd.)*

Sl. No.	Characteristics	Requirement for method of		GR test	Typical value gr. I
6.	Copper strip corrosion @ 50 °C for 3 hours	Not worse than no. 1	Not worse than no. 1	P:15	Passes
7.	Residue on evaporation (air jet) mg/100 mL (max.)	5.0	5.0	P:29	1.0

Table 20.7   Requirements for aviation turbine fuels, kerosene type (JET A-1) IS: 1571 (2008)

Sl. No.	Characteristics	Requirement	Method test value	GR typical
1.	Appearance	Clear bright free from solid matter and undissolved water at normal ambient temperature	–	Passes
2.	Composition:			
	a. Acidity total (max., mg, KOH/g)	0.015	P:113	0.008
	b. Aromatics [% v, max. 22 (defence)]	25 (civil),	P:23	13.9
	c. Olefins contents (% v, max.)	5	P:23 and 4	1.0
	d. Sulfur total (% mass, max.)	0.30	P:34 (method B)	0.03
	e. Sulfur, mercaptan (% mass, Max.)	0.003	P:109	–
	or doctor text	Negative	P:19	Negative
3.	Volatility:			
	a. Distillation:			
	1. Initial boiling point (°C)	Report	–	150
	2. Fuel recovered:			
	10% by vol (°C, max.)	205	–	163
	20% by vol (°C)	Report	–	168
	50% by vol (°C)	Report	–	180
	90% by vol (°C)	Report	–	208

Contd.

Table 20.7   Requirements for aviation turbine fuels, kerosene type (JET A-1) IS: 1571 (2008) *(Contd.)*

Sl. No.	Characteristics	Requirement	Method test value	GR typical
	3. Final boiling point (°C, max.)	300	–	230
	4. Residue (%vol, max.)	1.5	–	1.0
	5. Loss (%vol, max.)	1.5	–	1.0
	b. Flash point (abel) (°C, min.)	38	P:20 (Method B)	40
	c. Density at 15 °C (kg/m^3, range)	775–840	P:16	0.7866
4.	Fluidity:			
	a. Freezing point (°C, max.)	–47	P:11	–49
	b. Kinematic viscosity at 20 °C (mm^2/s, max.)	8	P:25	3.0
5.	Combustion:			
	a. Specific energy MJ/kg (min.)	42.8	ASTM 4529	43.2
	b. Calorific value (net) (cal/g, min.) or	10225	P:6	–
	Product of API gravity and aniline point (min.)	4800	P:3	7055
	c. Smoke point (mm, min.) or	25	P:31	26
	Smoke point (mm, min.) and	19	P:31	–
	naphthalene (%vol, max.)	3.0	P:118	–
	or luminometer no. (%vol, min.)	45.0	ASTM 1740	–
	d. Hydrogen content (% mass)	13.8	IP-338	Not mandatory
6.	Corrosion:			
	a. Copper strip corrosion for 2 hours at 100 °C	Not worse than 1	P:15	Passes
	b. Silver strip corrosion classification (4 hours °C)		IP:227/82	Passes
7.	Stability:			
	Thermal stability (JFTOT)			
	a. Filter pressure differential (mm Hg, max.)	25	P:97	Nil

Contd.

Table 20.7   Requirements for aviation turbine fuels, kerosene type (JET A-1) IS: 1571 (2008) *(Contd.)*

Sl. No.	Characteristics	Requirement	Method test value	GR typical
	b. Tube rating (visual) or	Less than 3, no peacock or abnormal color deposits	–	1
	TDR spun (max.)	15, no peacock or abnormal color deposits	–	Zero/NPC
8.	Contaminants:			
	a. Existent gum (mg/ 100 ml, max.)	7	P:29	1.0
	b. Water reaction			
	1. Interface rating (max.)	1 b	P:42	Passes
	2. Separation rating (max.)	Sharp separation, no emulsion or precipitate, within or upon either layer 85	P137 and P:142	80
	c. Water separometer index (WSIM) (min.)	85	P:137 and P:142	80
9.	Conductivity: Electrical conductivity (at the point, time and temperature of delivery to the purchaser (ps/m, range)	50–450 (250–350)	IP:274/82	250

Table 20.8   Requirements for kerosene (IS: 1459–1974) reaffirmed in 2001

Sl. No.	Characteristics	Requirement	Method of test ref. to (P:) method of IS: 1448	GR typical value
1.	Acidity, inorganic	Nil	P:2	Nil
2.	Burning quality:			
	a. Char value of oil consumed (mg/kg, max.)	20	P:5	12
	b. Bloom on glass chimney	Not darker than grey	–	Passes
3.	Color (saybolt) (min.)	10	P:14	+16
4.	Copper strip corrosion for 3 hours at 50 °C	Not worse than no. 1	P:15	Passes

Contd.

Table 20.8   Requirements for kerosene (IS: 1459–1974) reaffirmed in 2001 *(Contd.)*

Sl. No.	Characteristics	Requirement	Method of test ref. to (P:) method of IS: 1448	GR typical value
5.	Distillation:			
	a. Percent recovered below 200 °C (max.)	20	P:18	24
	b. Final boiling point (°C, max.)	(295) 300	–	275
6.	Flash point (abel) (°C, min.)	(38) 35	P:20	37
7.	Smoke point (mm, min.)	20	P:31	20
8.	Total sulfur (% mass, max.)	0.25	P:34	0.18
9.	Density (g/mL)	–	–	0.8152

Table 20.9   Requirements for high speed diesel oil (HSD) (IS: 1460–2005)

Sl. No.	Characteristics	Requirement	Method of test	GR typical value
1.	Acidity (total, mg KOH/g, max.)	0.20	P:2	0.2
2.	Ash % by mass (max.)	0.01	P:4	0.002
3.	Carbon residue, (Ramsbottom)	0.30	P:8	0.02
4.	Cetane number (min.) or cetane index (min.)	5146	P:9 ASTM D4731	–54
5.	Pour point (°C, max.) (see remarks 1)	3 °C for winter 15 °C for summer	P:10	0–18
6.	Copper strip corrosion for 3 hours at 100 °C	Not worse than no. 1	P:15	1
7.	Distillation:			
	a. 85% rec. @ temperature (°C, max.)	350	–	345
	b. 95% rec. @ temperature (°C, max.)	370	–	365
8.	Flash point (°C, min.)	35.0	P:20	34
9.	Kinematic viscosity at 40 °C (cSt, range)	2.0–5.0	P:25	3.5

Contd.

Table 20.9   Requirements for high speed diesel oil (HSD) (IS: 1460–2005) *(Contd.)*

Sl. No.	Characteristics	Requirement	Method of test	GR typical value
10.	Sediment (% by mass, max.)	0.05	P:30	0.04
11.	Total sulfur (% by mass, max.)	0.25	P:33	0.4
12.	Water content (%vol, max.)	0.05	P:40	0.01
13.	Total sediments (mg/ 100 mL, max.)	1.5	Appx. 'A'	1.0
14.	Cold fitter plugging point (CFPP) (°C, max.)	6 for winter and 18 for summer	P:110	4–18
15.	Density at 15 °C (kg/m^3)	820–870	P:16	834

Table 20.10 Requirements for light diesel oil (LDO) (IS: 1460–2005)

Sl. No.	Characteristics	Requirement	Method test	GR typical value
1.	Acidity, inorganic	Nil	P:2	Nil
2.	Ash %mass (max.)	0.02	P:4	0.01
3.	Carbon residues (Ramsbottom) (%mass, max.)	1.5	P:8	1.3
4.	Pour point (°C, max.)	12 for winter 21 for summer	P:10	6
5.	Copper strip corrosion 3 hours @ 100 °C	Not worse than no. 2	P:15	Passes
6.	Flash point, Pensky-Martens (closed) (°C, min.)	66	P:21	69
7.	Kinematic viscosity @ 40 °C (cSt)	2.5 to 15.7	P:25	10.1
8.	Sediment (%mass, max.)	0.10	P:30	0.04
9.	Total sulfur (%mass, max.)	1.8	P:33/IP:336	0.77
10.	Water content (%vol, max.)	0.25	P:40	0.08

Table 20.11 Requirements for army diesel (DHPP-A)

Sl. No.	Characteristics	Requirement for DHPP-A	Method of test	DHPP-AGR typical value
1.	Color (max.)	3.5	D-1500	–
2.	Flash point (Pensky-Martens closed) (°C, min.)	45	P:21	52
3.	Kinematic viscosity at 37.8 °C (cSt, range)	2–7.5	P:25	3.1
4.	Carbon residue (Ramsbottom) on 10% residue of distillation (%mass, max.)	0.2	P:8	0.08
5.	Cetane number (min.)	45	P:9	60
6.	Distillation:			
	a. Percent of recovery up to 350 °C (%vol, min.)	90	–	95
	b. Final boiling point (°C, max.)	385	–	349
	c. Residue (%vol, max.)	2	–	1
7.	Copper strip corrosion for 3 hours at 100 °C	Not worse than No. 1	P:15	Passes
8.	Water content (%vol, max.)	0.05	P:40	Traas
9.	Sediment (%wt, max.)	0.05	P:30	0.01
10.	Ash (%wt, max.)	0.01	P:4	< 0.01
11.	Acidity, inorganic (mg KOH/g)	Nil	P:2	Nil
12.	Acidity, total mg KOH/g (max.)	0.50	P:2	0.25
13.	Sulfur, total (%mass, max.)	0.50	P:33	0.17
14.	Pour point (°C, max.)	6	P:10	0.17
15.	Specific gravity at 60 °F/ 60 °F range	0.800 to 0.860	P:16	0.8264
16.	Olefins (%vol, max.)	5	P:23	–
17.	Aromatics (%vol, max.)	20	P:23	–

Table 20.12 Requirements for light aluminums rolling oil (LARO) (revised)

Sl. No.	Characteristic	Requirement	Test method	GR typical value
1.	Acidity, inorganic (mg/KOH/g)	Nil	IP:1	Nil
2.	Acidity, total (mg KOH/g, max.)	0.01	IP:1	0.01
3.	Aromatic hydrocarbons (%vol, max.)	25	D-1019/1319	17.0
4.	Ash content (%wt)	Not more than traces	D-482	–
5.	Bromine number (max.)	1.0	D-1159	0.70
6.	ASTM color (max.)	1	D-1500	0.5
7.	Distillation (°C, range) 5% recovery @ temperature (°C) 95% recovery @ temperature (°C)	240 to 280 245 275	D-98	240–280
8.	Recovery (%vol, min.)	98	D-86	99
9.	Density 15 °C (g/mL)	To be reported	P:16	0.831
10.	Flash point (°C, min.)	115	D-92	120
11.	Viscosity @ 40 °C (cSt, range)	2.3–2.5	D-445	34
12.	Sulfur (%wt, max.)	0.06	D-1266	0.03
13.	Water crackle test	Negative	–	–

Table 20.13 Requirements for fuel oils (IS: 1593–1982)

Sl. No.	Characteristic	Requirement for gr. MV1 and MV2	Method of test (ref. to P:) of IS: 1448	GR typical value
1.	Acidity inorganic	Nil	P:2	Nil
2.	Ash (%wt, max.)	0.1	P:4	0.06
3.	Density @ 15 °C (g/mL)	To be reported	P:16	0.9438
4.	Gross calorific value (cal/g)	Not limited, but to be reported	P:6	10520
5.	Flash point Penskey-Martens (closed) (°C, min.)	66 °C	P:21	70
6.	Kinematic viscosity at 50 °C (cSt, max.)	125	P:25	124 winter 170 summer
7.	Sediment (%wt, max.)	0.25	P:30	0.17

Contd.

Table 20.13 Requirements for fuel oils (IS: 1593–1982) *(Contd.)*

Sl. No.	Characteristic	Requirement for gr. MV1 and MV2	Method of test (ref. to P:) of IS: 1448	GR typical value
8.	Sulfur (total %wt, max.)	3.5	P:33 for reference and P:35 for routine	2.1
9.	Water content (%vol, max.)	1	P:40	Nil
10.	Pour point (°C, max.)	+15 for winter +21 for summer	P:10	6 - 21

Table 20.14 Recommended chemical composition of feed water and boiler water for high pressure boilers (IS: 4343–1967) requirements for light aluminums rolling oil (LARO)

Sl. No.	Characteristic	Limits recommended for boiler pressure		Test method IS: 3025–1964
		Up to 70 kg/cm²	Above 70 kg/cm²	
1.	Feed water:			
	a. Hardness (as $CaCO_3$) (mg/L)	Less than 5	Nil	16.1
	b. pH	8.5–9.5	8.5–9.5	8
	c. Dissolved oxygen (max., mg/L)	0.005	0.002	A-2, App. 'A'
2.	Boiler water:			
	a. Hardness (as $CaCO_3$) (mg/L)	Nil	Nil	16.1
	b. Caustic alkalinity (as $CaCO_3$) (mg/L)	10% of dissolved solids with min. of 50	10% of dissolved solids with a min. of 20	–
	c. Silica (as SiO_2) (max., mg/L)	50	5	30.1 and 30.2
	d. Phosphates (as PO_4) (mg/L)	40–80	20–40	22
	e. Hydrazine (mg/L)	0.05–0.1	0.02–0.05	–
	f. Sodium sulfite (if used) (mg/L)	20–30	21	–
	g. Total dissolved solids (max., mg/L)	500	500 (up to 100 kg)	A-4, App. 'A'
	h. Chlorides (as NaCl) (max., mg/L)	10% of TDS	10% of TDS	24

Contd.

Table 20.14 Recommended chemical composition of feed water and boiler water for high pressure boilers (IS: 4343–1967) requirements for light aluminums rolling oil (LARO) *(Contd.)*

Sl. No.	Characteristic	Limits recommended for boiler pressure		Test method IS: 3025–1964
		Up to 70 kg/cm^2	Above 70 kg/cm^2	
3.	Condensate:			
	a. pH (min.)	8.5	8.5	8
	b. Hardness (as CaCO$_3$) (mg/L)	Nil	Nil	16.1

Table 20.15 Proposed BIS specification for food grade hexane first revision of IS: 3470

Sl. No.	Characteristics	Requires	Method of test ref. to	
			Annex	(P:) of IS: 1448
1.	Color (saybolt) (min.)	+30	–	P:14
2.	Density at 25 °C	0.687	–	P:32
3.	Refractive index (n = 20)	1.381–1.384	*	–
4.	Distillation:			
	a. IBP (°C, min.)	63	–	–
	b. Dry point (°C, max.)	70	–	–
	c. Temperature range of final 10% (OC, max.)	2	–	–
5.	Non-volatile residue (g/100 mL, max.)	0.0005	A	–
6.	Reaction of non-volatiles residue	Neutral to methyl orange indicator	B	–
7.	Sulfur content (ppm, max.)	5	C	–
8.	Lead as Pb (mg/kg, max.)	0.2	**	–
9.	Aromatic hydrocarbons percent (vol/vol, max.)	–	E	–

Contd.

Table 20.15 Proposed BIS specification for food grade hexane first revision of IS: 3470 *(Contd.)*

Sl. No.	Characteristics	Requires	Method of test ref. to	
			Annex	(P:) of IS: 1448
10.	Polycyclic aromatics hydrocarbon, UV absorbance per optical path length, max. at wavelength (nm)			
	280–289	0.15	–	–
	290–299	0.13	–	–
	300–359	0.08	–	–
	360–400	0.02	–	–

Additional test requirements can be met according to customer's specifications.

* (Expressed as percentage of item 5).

** Carbon tetrachloride or trichloroethylene may also be used instead of carbon disulfide, the method of test being the same. This is as per Amendment No. 2, March 1983.

Table 20.16 Various grades of paving bitumen conforming to IS: 73–2006

Sl. No.	Characteristics	Requirement for grades					Method GR typical value		
		S.35 30/40	S.45 40/50	S.65 60/70	S.90 80/100	S.200 180/200	Of test	S.65 60/70	S.90 80/100
1.	Specific gravity at 27 °C g/mL (min.)	0.99	0.99	0.99	0.98	0.97	IS: 1202–1958	1.1014	1.0147
2.	Water (%wt, max.)	0.2	0.2	0.20	0.2	0.2	IS: 1211–1958	0.1	0.1
3.	Flash point (PMCC) (°C, min.)	220	220	220	220	175	IS: 1209–1958 (Method A)	205	200
4.	Softening point (R & B) (°C)	40	45	47	50	30 to 45	IS: 1205–1958	45	42
5.	Penetration at 25 °C, 100 g	80–100	60–80	50–70	40–60	175–225	IS: 1203–1958	65	90
6.	Ductility at 25 °C (cm, min.)	75	50	40	25	–	IS: 1208–1958	88	80
7.	a. Loss on heating, (%wt, max.)	1	1	1	1	2	IS: 1212–1958	0.5	0.5
	b. Penetration of residue (%wt min.)	60	60	60	60	60	IS: 1203–1968	62	60
8.	Matter soluble in carbon disulfide (%wt, min.)	99	99	99	99	99	IS: 1216–1958	99.6	99.2

REFERENCES

1. Chatila SG. Evaluation of Crude Oils in IFP Petroleum Refining Part 1: Crude oil Petroleum Products. Process Flow Sheets; Edition Technip, 1995, p315.

2. Golden SW, Craft S and Villaalati DC. Refinery Analytical Techniques Optimize Unit Preparations Performance. Hydrocarbon Processing, Nov 1995, p1.

3. http://www.petrotechnical.com/pti/crudechar.html.

4. Mall ID. Petrochemical Process Technology, MacMillan India. 2007.

5. OSHA Technical Manual, Section IV, Chapter 2, http://www.osha.gov/dts/osta/ otm/otm iv/otm_iv 2.html.

6. Rao PK. Relation Between Knock and Physical Properties Explored. Hydrocarbon Processing, March 2007, p89.

7. Speight JG. Handbook of Petroleum Analysis. Wiley Interscience, 2001.

8. Speight JG. The Chemistry and Technology of Petroleum, 3rd edition, Marcel Dekker Inc, New York, 1999, 300.

21

Separation Processes in Petroleum Refinery

Petroleum refining process is a complex process involving large number of separation processes and conversion processes. There has been continuous development in the separation process in order to increase cost effectiveness, boost energy efficiency, increase productivity, improve the quality of product and prevent pollution demand. Technological and used maturities of separation process is shown in Fig. 21.1 [Banik, 2006]. Various separation and conversion processes involved are given in Table 21.1. Separation equipment in most of the chemical and petrochemical manufacturing plants account for the major part of the investment in the plant.

21.1 DISTILLATION

Distillation has been the king of all the separation processes and most widely used separation technology and will continue as an important process for the foreseeable

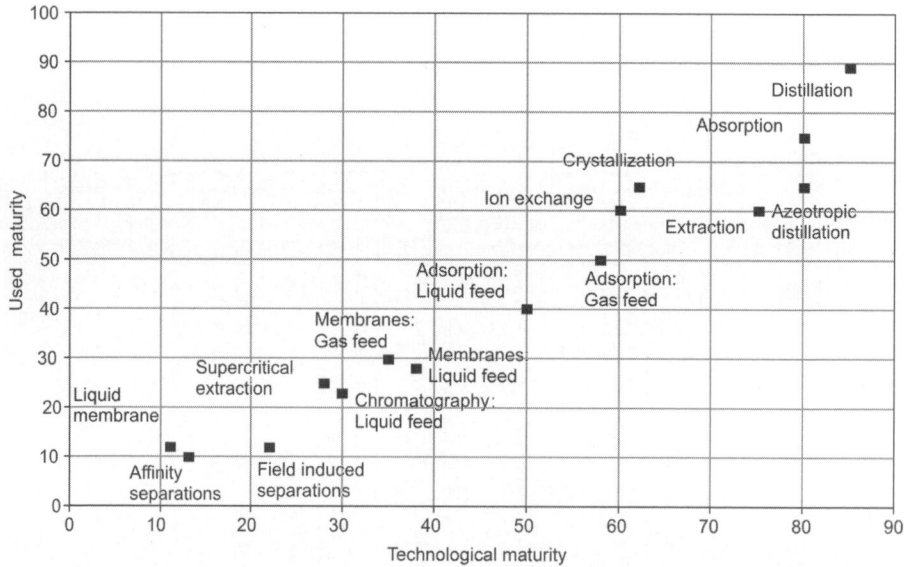

Fig. 21.1 Technological and used maturities of separation processes

Table 21.1 Separation processes in petroleum refining

Sl. No.	Petroleum process	Unit operation	Technology	Purpose
1.	Crude desalting	Membrane separation		To remove oil from waste water obtained after crude washing
2.	Primary separation processes	Distillation		Various fractions are obtained from desalted crude oil like naphtha, kerosene, diesel, etc.
		Liquid extraction	(1) Rose process (2) UOP SDA	To remove asphalt from distillation residue
			Furfural/NMP Solvent extraction	Remove aromatics and impurities in streams from vacuum distillation
3.	Alkylation and polymeriza-tion	Distillation	UOP alkylene	To separate light ends, propane, butane and alkylate
			Stratco effluent refrigerated H_2SO_4 alkylation process	To separate butane, isobutane and alkylate
			UOP HF alkylation	(1) As iso-stripper to recover HF (2) Depropanizer
			Linear alkylbenzene (lab.) manufacturer	To separate benzene and recover HF
			Q-Max process for cumene production	Separate benzene and cumene from product streams
			ConocoPhillips reduced volatility alkylation process (revap.)	Recover HF and propane
		Absorption	ConocoPhillips HF alkylation process	Waste treatment of vent gases
4.	Catalytic reforming and aromatics production	Distillation	Platforming	Stabilize reformate by separating light gases
				(1) Separate benzene, toluene and heavy aromatics, (2) separate xylenes (i.e. para, meta and ortho), (3) separate olefins
			Sulfolane unit	Separate sulfolane from stream after extraction of aromatics

Contd.

Table 21.1   Separation processes in petroleum refining *(Contd.)*

Sl. No.	Petroleum process	Unit operation	Technology	Purpose
		Adsorption	Parex	Separates p-xylene from its isomers
		Liquid-liquid extraction	Sulfolane unit	Recovers aromatics from feed
		Absorption	Sulfolane unit	Separates light hydrocarbons from sulfolane
5.	**Catalytic cracking**	Distillation	UOP FCC	(1) To produce separate streams of naphtha, gases, cycle oil, etc. (2) In GCU it is used to separate olefins from gasoline.
			Hydrocracking	Separates cracked product into various useful products
		Absorption	UOP FCC	(1) Separates C_2 and C_3 fractions (2) Separates H_2S from product stream
6.	**Desulfurization**	Distillation		Amine regeneration
			UOP sulfur-X	Solvent recovery with the help of steam
		Absorption		Absorption of H_2S by amine
			Belco Edv wet scrubbing system	Absorption of SO_2 by caustic solutions
		Filtration	Belco Edv wet scrubbing system	Separation of catalyst fines and sodium sulfite, sodium sulfate
		Liquid-liquid extraction	UOP merox	Mercaptans are dissolved in caustic solutions and hence removed
			UOP sulfur-X	Thiophenes and mercaptans are extracted from feed stream
7.	**Isomerization**	Distillation	Bensat	Separates light ends from saturated benzene
			Butamer	Separates light ends from isomerized n-butane
			Penex	Separates light ends from isomerate
		Adsorption	UOP tip and O-T zeolitic isomerization processes	Separation of n-paraffins from its branched isomers

Contd.

Table 21.1   Separation processes in petroleum refining *(Contd.)*

Sl. No.	Petroleum process	Unit operation	Technology	Purpose
		Molecular sieve	UOP butamer	Dry water from feed so that it does not poison the catalyst
			UOP penex	Dry water from feed
			UOP penex	separate the stabilized penex product into a high-octane isoparaffin stream and a low-octane normal paraffin stream.
8.	**Hydrotreating**	Liquid-liquid extraction	KLP process	Separates butadiene from product stream
		Distillation	In many technologies	Fractionates product into various streams
		Membrane separation	RCD unionfining	Removes light gases like methane formed in reactor
			Chevron LC-fining	Produces pure H_2 to recycle
		Absorption	UOP unisar	Removes H_2S from stream
			Chevron LC-fining	Removes H_2S from stream
			Chevron Lummus Global Rds/Vrds hydrotreating	Removes H_2S and NH_3
		Filtration	Chevron LC-fining	Separates coke precursors to prevent coking
			Chevron Lummus Global Rds/Vrds hydrotreating	Removes reservoir mud and iron sulfide scale
			RCD unionfining	removing scale particulates

future [Olujie et al., 2003]. Distillation is used in petroleum refining and petrochemical manufacture. Distillation is the heart of petroleum refining and all processes require distillation at various stages of operations. Crude oil distillation is the first step in any refinery in order to separate the crude oil into value added product. Crude oil is separated in various fractions by distillation in atmospheric and vacuum distillation columns.

Distillation columns are among the biggest energy consumers in the chemical industries, particularly in oil refineries. Retrofit project in refineries mostly aim at reducing energy consumption and increasing throughput to increase profit and meet market demands.

Major variables in a continuous distillation column are: Pressure, distillate flow rate, reflux flow rate, residue flow rate, reboiler flow rate, level of top and bottom drums [Gourila, 1998]. Distillation column design criteria is given in Fig. 21.2 [Majumder 2000].

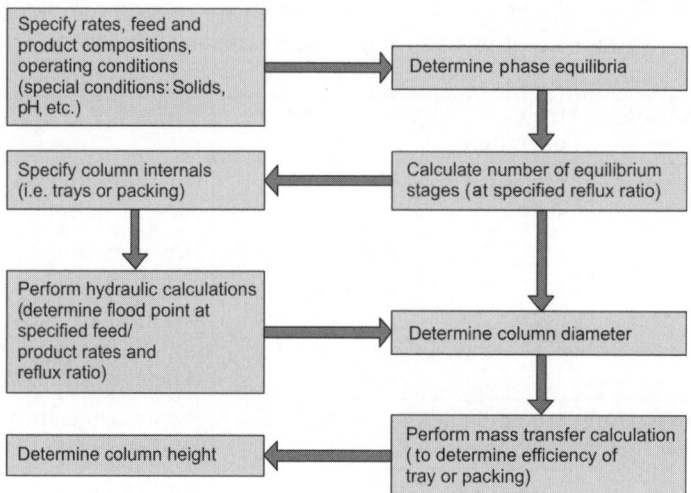

Fig. 21.2 Distillation column design

Source: Majumder, 2000.

21.1.1 Distributed Distillation

Distributed distillation has been practiced, to a limited extent, in natural gas processing, olefin production and refinery units, for every 30 years. Its utilization can allow a 10–30% reduction in net separation energy usage at the cost of adding more equipment [kolmetz.com].

(Ref; IOCL, Distributed distillation with heat integration; compendium on energy management and conservation in hydrocarbon industry, November, 2003.)

Distributed distillation system can be viewed as a separation sequence that provides some or more parallel flow paths for the production of a boiling fraction or component. The basic concept in distributed distillation is to minimize the number of sharp separations used to fractionate a multicomponent mixture. Distributed distillation system frequently increases opportunities for heat integration by shifting heat duty temperature levels [Reid, 2000] of different purity products.

21.1.2 Extractive Distillation

Extractive distillation combines the effect of a phase change by energy input with the effect of solvent addition to separate a component on the basis of volatility and its chemical nature [Rojey, 1998]. In extractive distillation first a close boiling distillate cut is produced; this is then mixed with a solvent to increase the boiling differences between components, thus increasing the relative volatility. Extractive distillation is preferred over extraction when prefractionation heart cut is necessary.

21.1.3 Azeotropic Distillation

Azeotropic distillation involves addition of a third component to a binary system to facilitate the separation by distillation of the added component modifies VLE in a favorable direction. The third component and the energy input to the reboiler are two different separating agents.

21.1.4 Reactive Distillation

Reactive distillation is the hybrid operation which applies simultaneous implementations of reaction and distillation without countercurrent column. During the last decade, there has been a rapid upturn in interest in reactive distillation. The chemical industry recognizes the favorable economics of carrying out simultaneously with distillation for certain classes of reacting system and many new processes have been built based on this technology [Bokade, 1998]. The main advantages of reactive distillation are use of the exothermal heat of reaction for the distillation process, achievement of higher yields for reversible reactions, simplification of reactor effluent separations, ability to overcome distillation boundaries, e.g. azeotropes and lower capital cost [Stichlmair and Frey, 1999]. The conversion pass is high, reduces the operating cost. Some of the benefits of reactive distillation are process simplification, improved conversion, enhanced selectivity, longer catalyst life, reduced catalyst requirement, better handling of difficult separation, control of highly toxic reaction and heat integration increased reliability [Amte and Rao, 2009].

Area of Application of Reactive Distillation

Etherification
- Methyl tert-butyl ether (MTBE)
- Ethyl tert-butyl ether (ETBE)
- Tert-amyl methyl ether (TAME).

Selective hydrogenation
- C_4 butadiene selective hydrogenation
- C_5 pentadiene selective hydrogenation
- C_6 hexadiene selective hydrogenation
- Benzene saturation.

Advantage
- Reduction in capital investment
- Higher reaction rate
- Better utilization of heat of reaction.

21.1.5 Short Path Distillation

Short path distillation is a continuous fractionating process without steam injection, using wiped film evaporator with an internal condenser. A special feature of the short path still is the use of very high vacuum at least 10–3 mm Hg. Large scale short path distillation plant facilities for processing vacuum residue at a feed rate of 2500 kg/hour (20000) and generating heavy distillate to the extent of 60% of feed have been created at Digboi Refinery [Banik, 2003]. In short path distillation, the feed is spread into film on the evaporating surface for quick residence time.

21.1.6 Divided Wall Columns

Divided wall column is the another development in the distillation where a single column is used to separate hydrocarbon mixture into three high purity fractions and there is wide field of application especially in petroleum refinery and petrochemical industry. Divided column found first application in 1985 [Asprion and Kaibel, 2010]. Dividing wall column is the most daring variation of the so called Petlyuk column. It allows substantial energy saving, while separating in a single body of a three-component mixture into pure products [Oluje et al., 2003]. Compared with conventional fractionation using two columns, the divided wall column entails lower investment and operating cost as it requires one overhead condensing system and one reboiling system [Ennenbach and Kolbe, 2000]. Typical application of divided wall column may be in the separation of low benzene gasoline fraction and benzene containing C_6 fraction, which is suitable for recovery of high purity benzene [Ennenbach and Kolbe, 2000]. Dividing wall columns offer significant advantages with respect to investment costs and energy consumption [Asprion and Kaibel, 2010]. For batch distillation, Schmidt et al., 2008, have reported lower batch time, i.e. higher capacity, lower specific energy demand, high product quality and yield, low thermal stress in case of divided wall column compared to conventional batch distillation.

Divided wall column (DWC) system consists of reflux splitter, which separates the liquid products from the column overhead into two liquid steams: One to the prefractionation section and the other to the main column section of the DWC [Shin et al., 2013]. A new configuration with introduction of divided wall distillation with reactive distillation exhibits high degree of integration. Typical divided wall column is shown in Fig. 21.3.

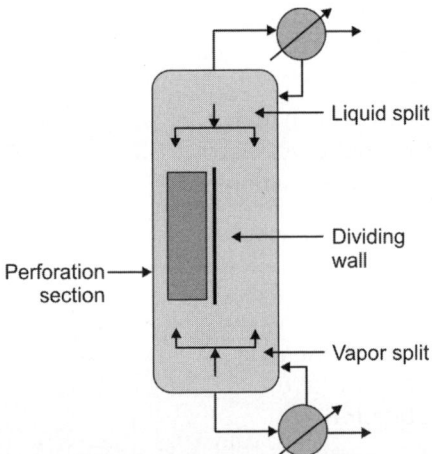

Fig. 21.3 Divided wall column

Courtesy: Hydrocarbon Processing [Shin et al., 2013].

21.1.7 Distributed Distillation

Distributed distillation is one of the several basic tools in the development of modification of a distillation sequence [Reid, 2000]. Distributed distillation frequently increases opportunities for heat integration. Some of the areas where distributed distillation can be used are crude oil distillation, hydrotreating, alkylation, isomerization and reformer units, LPG and NGL processing, olefin production. Figure 21.4 shows a typical application of distributed distillation system using conventional two towers and a three-tower distributed distillation system for separation of a three-component mixture [Reid, 2000].

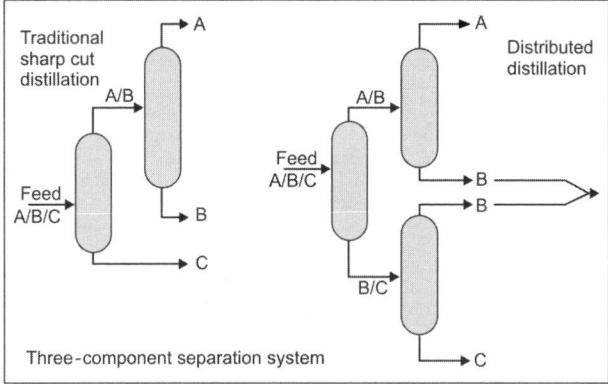

Fig. 21.4 Distributed distillation system

Source: Petroleum Technology, Quarterly, Autumn, 2000.

21.1.8 Design of Distillation Column

Distillation column design criteria is shown in Fig. 21.2.

21.1.9 Column Internals

Tray and packed columns are commonly used for distillation because they guaranteed excellent countercurrent flow and permited a large overall height [Kashani et al., 2005; Waals, 1995]. Various types of column internals are tray, packing and distributors. Distributors: Good distributors are critical to the performances of packings. Vapor maldistribution depends on velocity head of vapor entering column and pressure drop across packed bed. Performance of column internal is mentioned in Table 21.3.

21.1.9.1 Trays and Packing

Historically, trays have been dominant tower internals because of their reliability and column internal is shown in Fig. 21.5, ease of installation and low cost relative to other devices [Kunesh et al., 1995]. There are wide varieties of trays in use and some of them are shown in Figs. 21.5 to 21.7 [Majumder, 2000].

The overall objective is to maximize vapor/liquid contact and throughput with minimum pressure drop, weeping and entrainment [Kunesh et al., 1995]. Bubble cap tray was most commonly used trays during 1940s and early 1950s. However, valve trays and sieve trays have dominated for the last 40 years. Sieve trays are constructed with large number of apertures. Valve trays are constructed with valve covers mounted above the apertures. Valve trays covers may be fixed or floating. Fixed valves are integrally attached to the tray deck [Fan and Rubin, 2000]. Valve trays always utilize a smaller number of larger apertures than sieve trays because of the relatively high cost of valve construction, especially floating valves [Fan and Rubin, 2000]. Valve trays offer greater operating range because of the movable valve units, minimize weeping at low vapors rate and serve to deflect entrainment at high vapor rates [Kunesh et al., 1995].

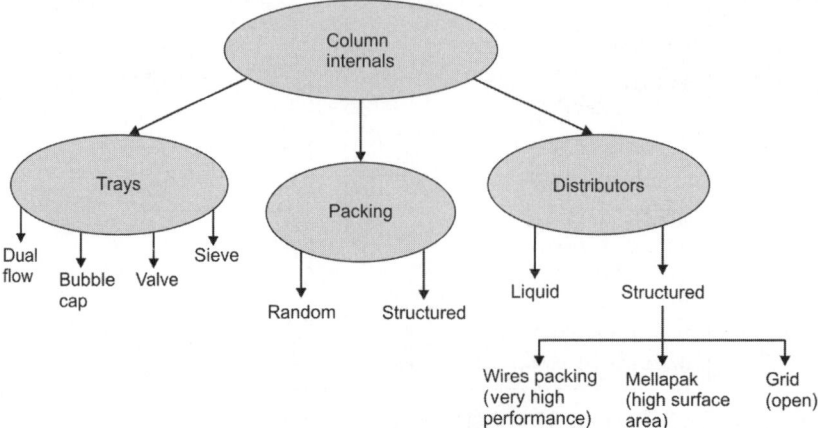

Fig. 21.5 Column internals

Source: Majumder, 2000.

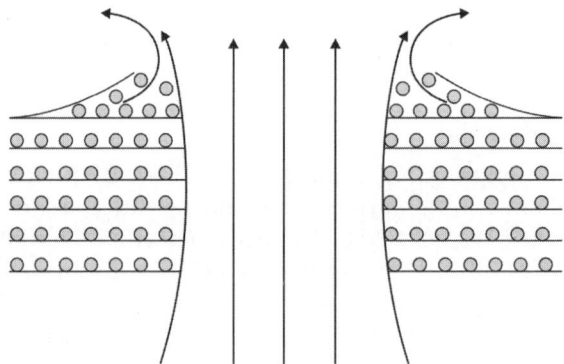

Fig. 21.6 Sieve tray

Source: Majumder, 2000.

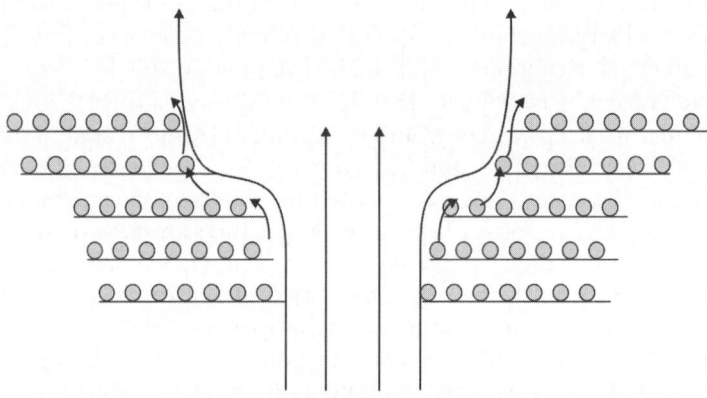

Fig. 21.7 Valve tray

Source: Majumder, 2000.

The basic objective of a packing is to maximize specific surface area, spread surface area uniformly, maximize void space, minimize friction and minimize cost. Some of the commonly used packings are random packing, structured packings, grids and sandwich packing. In sandwich packing, partitions of the packing operate selectively at the flooding point [Kashani et al., 2005]. Earlier, random packings were crushed rock and jack chain followed by pall packings, however, the third generation has high open area with complex surfaces in an attempt to maximize vapor/liquid contact and minimize pressure drop. Random packings are made up of metals, plastics and ceramics. Various types of random and structured packings are given in Figs. 21.8 and 21.9. Trays and packing character are mentioned in Table 21.2.

Because of the low resistance to vapor flow, structured packing offers considerably more capacity than random packing having same HETP. Compared to trays, structured packing provides a pressure drop/theoretical stage that is an order of lower magnitude [Kashani et al., 2005] and is given in Tables 21.3 and 21.4. Capacity and separation gains due to the lower pressure drop of packing. Random packings pressure drop is typically one-third to one-fifth that of trays and structured packings is even lower [Kashani et al., 2005]. In most cases lower pressure drop raises relative volatility, and therefore, improves separations [Kashani et al., 2005].

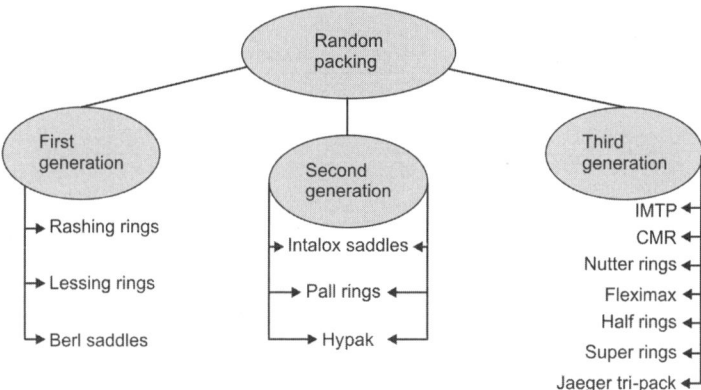

Fig. 21.8 Random packing

Source: Majumder, 2000.

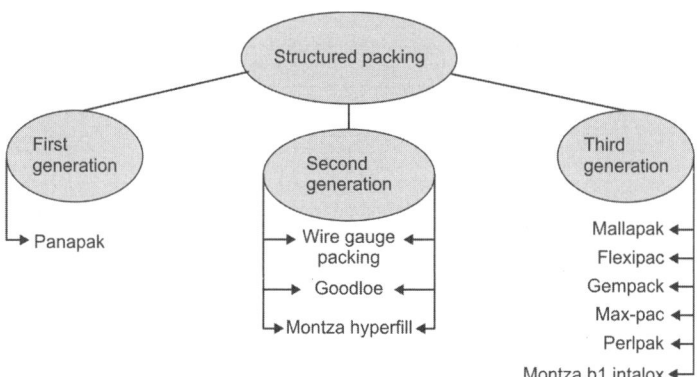

Fig. 21.9 Structured packing

Source: Majumder, 2000.

Table 21.2   Trays and Packing

	Trays	Random packing	Structured packing
Effect of scale on HETP	Predictable	Difficult to predict	Predictable
Pressure drop	High	Low	Low
Established design technique	Yes	*	
Cost	Low	Low–medium	High
Suitability for fouling service	Yes	No	No
Feed point flexibility	Easy	Difficult	Difficult

* Only for Capacity, Not for HETP.
Source: Majumder, 2000.

Table 21.3   Performance of column internals

Characteristics	Trays	Random packing	Structured packing
F-factor	0.25–2.0	0.25–2.4	0.1–3.6
C-factor	0.03–0.25	0.03–0.3	0.01–0.45
Pressure drop, mm Hg/theoretical stage	3–8	0.9–1.8	0.01–0.8
Mass transfer efficiency (HETP), inches	24–48	18–60	4–30

Source: Majumder, 2000.

Table 21.4   Comparison of common tray types

	Sieve tray	Valve tray	Bubble cap tray
Capacity	High	Very high	Mod. high
Efficiency	High	High	Mod. high
Turn down	2:1	5:1	Excellent
Entrainment	Moderate	Moderate	High
Pressure drop	Moderate	Moderate	High
Cost	Low	Moderate	Very high
Maintenance	Low	Moderate	High
Fouling tendency	Low	Moderate	High
Effect of corrosion	Low	Moderate	High

Source: Majumder, 2000.

21.1.10 Deep Cut Technology

Objective of deep cut technology is to:

- Recover as much HVGO from vacuum residue as possible
- Increase TBP cut point of vacuum gas oil from 530 °C to as high as 625 °C
- Maintain vacuum gas oil quality within permissible limits of FCC/HCU
- Expected additional yield of vacuum gas oil by 5–9% on crude.

Features

- Feed characterization
- Process design and unit modeling
- Analysis of crude preheat train
- Equipment design: Furnace, ejector system, transfer line, column; flash zone, wash zone, stripping section.

21.2 ADSORPTION

Adsorption is a proven and reliable pollution control technology that has the added benefit of recovering valuable materials for reuse [Ruhl, 1993]. Principle of adsorption process is shown in Fig. 21.10.

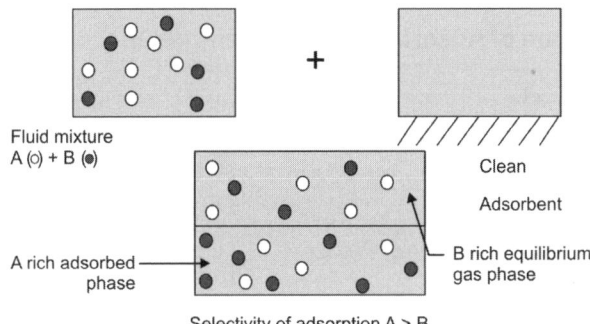

Fig. 21.10 Selectivity of adsorption

Although adsorption was conventionally used for the removal of contaminants from waste water stream, however, adsorption has received much attention in petroleum and petrochemical industry for important separation process. Adsorption technology is now used very effectively in the separation and purification of many gas and liquid mixtures in chemical, petrochemical, biochemical and environmental industries and is often a much cheaper and easier option than distillation, adsorption or extraction.

Adsorption process is widely used in process because of its versatile applications in process industries such as separation of CO_2–CH_4 mixtures, in H_2 purification, in production of H_2 from steam reformer off gas and from coke oven gas, paraffin separation from kerosene (molex process), in p-xylene separation (parex) process, in removal of VOC, AOX and other refractory organics, in recovery of phenols and chlorinated phenols, in recovery of 1, 1-dichloro-1-fluoroethane. It has also

wide application in gas masks and desiccators for drying. Adsorption is one of the most effective methods for controlling volatile organic compounds [Ruhl, 1993].

The important attributes sought in an adsorbent are capacity, selectivity, regenerability, kinetics, compatibility and cost. Mechanism of selectivity adsorption is shown in Fig. 21.8. Apart from physicochemical properties of adsorbent, some of the important characteristics commonly measured in case of adsorbents are particle size, surface area, pore size, pore volume, pore volume distribution, density, adsorption capacity—iodine number, molasses number, abrasion number. Various factors affecting adsorption are nature of adsorbents and adsorbates, pH of the solution, contact time, initial concentration, temperature, and degree of agitation, molecular size and molecular configuration of the adsorbate.

- Adsorption increases as particle size decreases
- Larger surface area, greater adsorption capacity
- Effective adsorbents, large volume of very small pores
- Pore volume is related to surface area distribution.
- Abrasion number represents ability of carbon to withstand handling.
- Bulk density for determining volume occupied by a given weight of carbon.
- Adsorption occurs on active site of solid surface.
- Active sites are homogeneous when they all contain same energy potential.
- With different energy potentials the sites are said to be heterogeneous.

21.2.1 Regeneration of Adsorbents [Crittendent,1988]

Thermal Swing Cycle

Most commonly used cycle. Any cycle which employs different temperature levels for adsorption and desorption. It is good for a strongly adsorbed species.

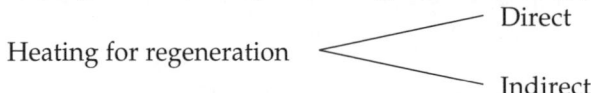

Heating for regeneration $\left\{\begin{array}{l}\text{Direct}\\\text{Indirect}\end{array}\right.$

Pressure Swing Cycle

Adsorption processes operated at a constant temperature by utilizing power pressure for regeneration than for adsorption. This is good for weekly adsorbed species required in high purity.

Purge Gas Stripping Cycle

Purging an adsorbent with an inert gas reduces partial pressure of the adsorbate in vapor phase and this causes desorption.

Displacement Cycle

Displacement cycle in which more strongly adsorbed adsorbate is used to displace the original adsorbate, and used when original adsorbate is easily decomposed. It is good for strongly held species.

Combination Cycles

Steam for regeneration is a most common way since steam acts to create a thermal swing, purge gas and displacement cycle.

21.2.2 Transfer Mechanism

Transfer mechanism involved in adsorption process are:

- Mass transfer from bulk of gas to particle surface
- Diffusion through the passage within the particles
- Through macropores
- Through the micropores
- Adsorb at an appropriate site, i.e. adsorption on an internal particle surface.
- Time for adsorption to take place is seen at the times for each step.

Each of the process depends on the system operating conditions, physical and chemical characteristics of the gas stream and solid adsorbent [Cheremisinoff, 2002].

The slowest of these steps will determine the rate for the whole process and is called rate determining step.

Adsorbent capacity is the most important property, if adsorbent as higher the adsorption capacity, lower is the requirement of adsorbent, lower the volume of adsorber. The adsorption capacity depends mainly on fluid phase concentration, temperature, and initial condition of adsorbent [Knabel, 1995]. Selectivity is the ratio of the capacity of an adsorbent for one component to its capacity for another at a given fluid concentration. Smaller the value of selectivity, the larger is the required equipment.

Some of the commonly used adsorbents are activated carbon, carbon molecular sieves, silica gel, activated alumina, zeolite molecular sieves, silicalite, polymer adsorbents, irreversible adsorbents and biosorbents [Keller II,1995]. Adsorbents are available in different shapes—powdered, granular and cylindrical.

Activated carbon: Activated carbon is one of important adsorbents and is available in granular and powdered forms. Activated carbons are used to adsorb impurities in gaseous and liquid phases.

Molecular sieves: Molecular sieves are crystalline alumino silicates. It can be molecular sieves which can be microporous, mesoporous or macroporous material. The porous crystalline structure and large surface area of molecular sieves exhibit a high electronic activity which gives molecular sieves outstanding adsorption properties [Terrgoel, 2013]. In molecular sieve principle of difference in molecular size of components is used for separation. The larger molecules are trapped in specially formulated fixed-beds or molecular sieves. The smaller molecules cannot be trapped in the adsorbing-beds (molecular sieves) and are separated from the larger molecules at stipulated pressures. These trapped molecules are then released when pressure is reduced. Molecular sieves are being used as adorbent in petroleum and petrochemical industries for removal of moisture, acid gases, sulfur and mercury, separation of olefins, paraffins, xylene, etc.

Type of selectivity involved is equilibrium, steric or kinetic selectivity. The process may be regenerative or non-regenerative. Various regeneration processes are: Temperature increase (TSA), pressure decrease (PSA), by stripping, by displacement with an adsorable compound or by combination of several techniques. The adsorber may be batch or continuous. Commonly used continuous systems are moving bed, simulated moving bed. Both PSA and TSA systems employ two identical, parallel adsorption beds. One of them adsorbs, while other is regenerated by heating with a regenerative gas or by pressure lowering the pressure. Separation of gas or liquid components by adsorption is widely used in chemical, petroleum and petrochemical industries. Some of the applications of adsorption process are given in Table 21.5.

Table 21.5　Application of Adsorption process

In gas bulk separation in process industries	Removal of organics from vent streams; removal of SO_2 from vent streams; removal of water vapor from air and other gas streams. Removal of CO_2 from natural gas; separation of normal paraffins and isoparaffins, aromatics; separation of nitrogen from natural gas. In gas bulk separations like N_2 and O_2; H_2O and ethanol, acetone and vent streams, C_2H_4 and vent streams; CH_4, CO_2, N_2, A, NH_3 and H_2. Removal of H_2O from organic solutions; removal of organics from H_2O; removal of sulfur compounds from organic solutions; decolorization of solutions.
Chemical process industries	Removal of traces of chlorine from brine in chlor-caustic industries. For reduction of contaminants in pulp prior to ozone bleaching.
In petroleum and petrochemical industries	CO_2 separation from natural gas; normal paraffins/iso-paraffins, normal paraffins/olefins; xylene and other C_8 aromatics; p- or m-cymene/other cymene isomers; p- or m-cresol/other cresol isomers; for adsorbing arsenic compounds from hydrocarbon; treatment of lubricating oils; adsorptive storage of natural gas.
In air pollution control	Removal of sulfur from gas streams; removal of solvents and odors from air; removal of NO_x from N_2; removal of CO_2 from power plant flue gas; removal of mercury vapor; automotive canister, sorption and catalytic destruction of chlorinated VOCs.

21.2.3 Pressure Swing Adsorption (PSA)

PSA is one of the important gas separation processes. In PSA, the adsorbent is regenerated by reducing the partial pressure of the adsorbed component. By lowering the total pressure or using a purge gas reduction in partial pressure can be accomplished. PSA is useful for bulk separation, applied to a concentrated feed stream, as well as for purifications. Principle of PSA is given in Fig. 21.11.

Some of the applications of PSA processes in the separation of gas mixture are as follows [Keller II, 1995; Kiran and Chakraborty, 2002]:

  a. Air drying

  b. Hydrogen purification

  c. Production of H_2, CO, CO_2 and syngas from steam reformer off-gas

  d. Air fractionation to produce O_2- and N_2-rich streams

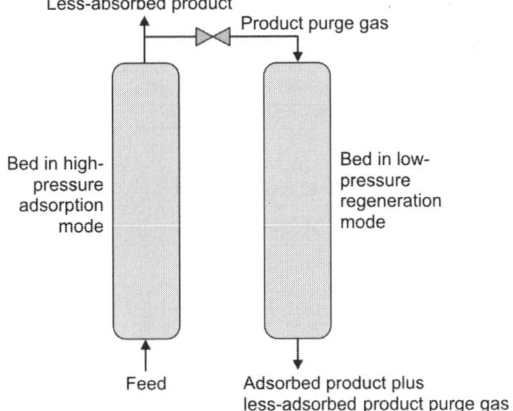

Fig. 21.11 Principle of pressure swing adsorption

e. Acid gas removal: Separation of CO_2 from CH_4
f. Flue gas desulfurization—removal of SO_2 and removal of CO_2 from flue gases
g. Bulk separation of normal paraffins
h. Hydrogen recovery from coke oven gas
i. Solvent vapor recovery by PSA.

Various important characteristics of the adsorbents are pore volume, pore distribution, surface area. Other characteristics include molasses number, iodine number, bulk density, compressive strength and abrasion strength.

UOP's sorbex simulated moving bed (SMB) (Fig. 21.12) has achieved large commercial success. Some of the commercially available commercial UOP sorbex processes based on adsorption using molecular sieves are given in Table 21.6. In SMB by changing the positions of feed and product removal points as function of time simulate of a moving adsorbent bed by using a large rotary valve [Keller II, 1995]. Adsorber design consideration criteria is given in Table 21.7 [Knaebel, 1999].

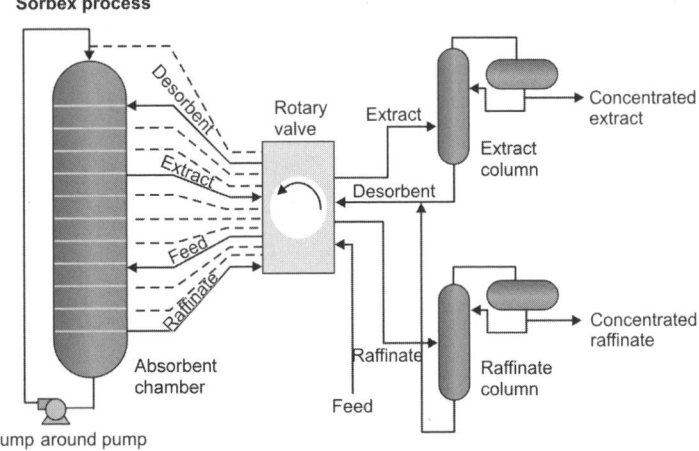

Fig. 21.12 UOP sorbex simulated moving bed process

Table 21.6 Commercial adsorption processes

Sorbex process	Application
Parex	Separation of para-xylene from mixed C_8 aromatics isomers
MX sorbex	Meta-xylene from mixed C_8 aromatics isomers
Molex	Linear paraffins from branched and cyclic hydrocarbons
Olex	Olefins from paraffins
Cresex	Para-cresol or meta-cresol isomers
Cymex	Para-cymene or meta-cymene from cymene isomers
Sarex	Fructose from mixed sugar
UOP ISOSIV processor	Separation of normal paraffins from hydrocarbon mixture
Kerosene Isoiv process	For separation of straight chain normal paraffins from the kerosene range (C_{10}–C_{18}) used for detergent industry

Table 21.7 Adsorber design considerations

Parameters	
Basic adsorbent properties	
Isotherm data	Uptake, release measurements; hysteresis observed, pre-treatment conditions, aging upon multiple cycles, multi-component effects
Mass transfer behavior	Interface character; intraparticle diffusion, film diffusion, dispersion
Particle characteristics	Porosity, pore size distribution, specific surface area, density, particle size distribution, particle shape, abrasion resistance, crush, strength, composition, stability, hydrophobicity
Application considerations	
Operating conditions	Flow rate, feed and product concentrations, pressure, temperature, density recovery, cycle time, contaminants
Regeneration	Thermal: Steam, hot fluid, kiln chemical: Acid, base, solvent pressure shift regenerant, adsorbate recovery or disposal
Energy requirement	
Adsorbent life	Attrition, swelling; aging, fouling
Equipment, flow sheet	
Contactor type	Fixed, axial, radial flow pulsed, fluidized bed
Geometry	Number of beds, bed dimensions, flow distribution, dead volumes
Column internals	Bed support, ballast, flow distribution, insulation
Miscellaneous	Instrumentation, materials of construction, safety, maintenance, operation, start-up, shut down

21.3 LIQUID-LIQUID EXTRACTION

Liquid extraction is the second most widely used unit operation in petroleum and petrochemical industries. The first commercial liquid-liquid extraction process in the petroleum refinery was due to pioneering work of Edeleanu in 1907. In Edeleanu process, liquid sulfur dioxide is the solvent used and the feed may be any one of the distillates from heavy naphtha to diesel [Ghosh, 1993].

Extraction application in petroleum refinery: The main applications of extraction in petroleum refinery [Fig. 21.13] are:

- Extraction of aromatic compounds from lube oil stocks to produce lubricants to improve viscosity index
- SO_2 extraction from kerosene fractions to improve smoke point using:

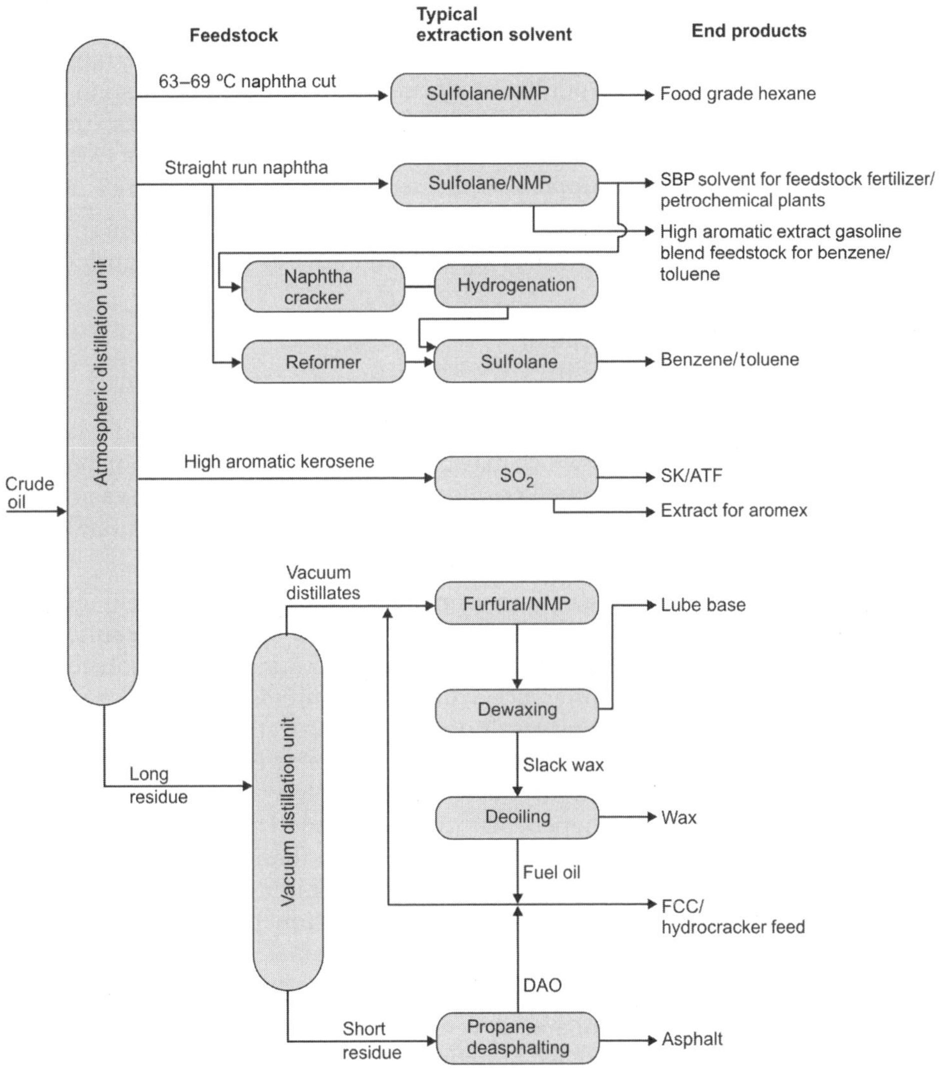

Fig. 21.13 Solvent extraction processes in refinery

o Production of aromatics from reformate for separating aromatic compounds from lighter non-aromatics

o Solvent deasphalting using propane (propane deasphalting) and deasphalting using supercritical solvent techniques (UOP/FW USA) solvent deasphalting for producing deasphalted oil and asphalts

o Extraction of food grade hexane based on sulfolane and NMP

o Extraction of aromex from medium kerosene stream using sulfolane

o Centrifugal extractor (differential).

Liquid-liquid extraction is an important separation process for separation of closed boiling mixture. Separation is based on the different distribution of the components to be separated between two liquid phases and depends on the mass transfer of the component to be extracted from a first liquid phase to the second one (solvent phase) [Freund, 2011]. Equipment for extraction must be therefore, capable of providing intimate contact between the two phases so as to transfer of solute between these and also of ultimately effecting a complete separation of the phases. The solvent extraction requires two main steps: The solvent extraction and solvent regeneration. The extraction process essentially consists of extractor, extractive stripper, solvent stripper and raffinate washer.

The solvent extraction process can be carried out as a batch or continuous process industrially important extractors may be divided into four major categories:

• Mixer-settlers (stagewise)
• Unagitated columns (differential)
• Agitated columns (differential).

Many types of extractors/contactors are available for achieving mass transfer and there is no single contactor which may be best for all solvent extraction process, either technically or economically. The selectivity, the more effectively/easily, will be the separation and consequently requirement of simpler equipment. Sieve tray extraction column is given in Fig. 21.14.

Ease of regeneration is another important characteristic as the solvent must be regenerated by separation from the extract and the raffinate streams. Low miscibility of feed and solvent is required in order to minimize the solvent that leaves with raffinate. Larger difference in density requirement is required to have large driving force for the settling resulting in rapid separation resulting in higher capacities. As high viscosity of either phase, reduces mass transfer. Low viscosity solvent will minimize resistance to mass transfer. Low viscosities are preferred for rapid settling of phases. Higher stability is required to have to avoid thermal, chemical degradation. Melting point should be lower than ambient temperature for ease of handling. High interfacial tension allows rapid settling due to easier coalescence. Low interfacial tension facilitates dispersion, but too low interfacial tension leads to emulsification and coalescence problems.

In the extraction process important variables are: Temperature, temperature gradient, extract recycle, use of anti-solvent, use of secondary solvent, use of co-solvent, extraction and extractive distillation combination. Extractive distillation and extraction processes for aromatic separation is given in Table 21.8.

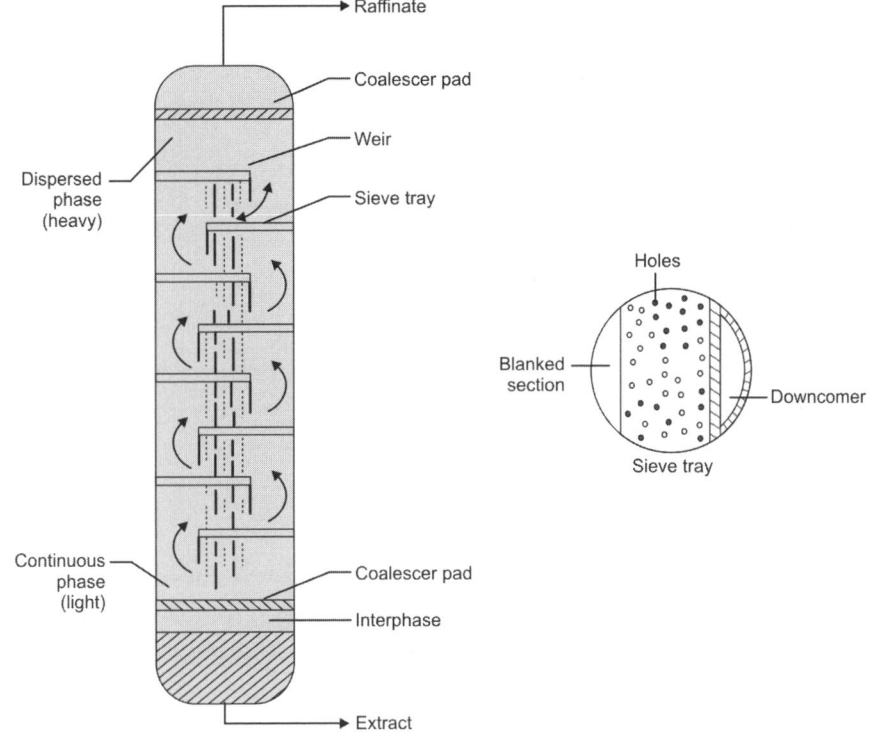

Fig. 21.14 Sieve tray extraction column

Source: Nanoti, 2005.

In the lube extraction, furfural and NMP are being used successfully given in Table 21.9. NMP has received considerable interest because of lower solvent circulation because of higher solvent power, lower utility requirement, lower solvent losses, reduced equipment fouling and lower solvent toxicity.

Commercial application of Liquid-Liquid extraction

Liquid-liquid extraction has been commonly used in petroleum and petrochemical industries for separation of close boiling hydrocarbons. Extraction unit in India is given in Table 21.10. Some of the major applications are:

- Removal of sulfur compound from liquid hydrocarbons
- Recovery of aromatics from liquid hydrocarbon
- Separation of butadiene from C_4 hydrocarbons
- Extraction of caprolactam
- Separation of homogeneous aqueous azeotropes
- Extraction of acetic acid
- Removal of phenolic compounds from waste water
- Manufacture of rare earths
- Separation of asphaltic compounds from oil
- Recovery of copper from leach liquor
- Extraction of glycerides from vegetable oil

Table 21.8   Extractive distillation and extraction processes for aromatic separation

Separation process	Solvent
Extractive distillation	
IFP	Dimethyl formamide
Krupp-koppers (octenal)	N-formylmorpholine
Lurgi (distapex)	N-methylpyrrolidine
SNAM progetti (formex)	N-formylmorpholine
UOP	Sulfolane
Union carbide (tetra)	Tetraethylene glycol
Toray (stex)	Dimethyl acetamide
Extraction process	
Shell UOP	Diethylene-triethylene glycol
UOP Udex	Tetra-ethylene glycol
Union carbide (tetra)	Tetra-ethylene glycol
IFP	Dimethyl sulfoxide
Snamprogetti (formex)	N-formylmorpholine
Lurgi process (arosolvan)	N-methylpyrrolidine + ethylene glycol or water
Howe-baker (aromax)	Diglycol amine
ARCO	Sulfolane
Edeleann process	Liquefied sulfur dioxide
Leunawerke (molex)	Methyl formamide
Koppers (aromex)	N-formylmorpholine

Table 21.9   Comparison of furfural, MP and phenol as extraction processes

Characteristics	Furfural	MP	Phenol
Adaptability	Excellent	Very good	Good
Emulsibility	Low	Moderate	High
Solvent-to-oil ratio	Moderate	Very low	Low
Extraction temperature	Moderate	Low	Intermediate
Refined oil yield	Excellent	Very good	Good
Product color	Very good	Excellent	Good
Corrosiveness	Intermediate	Low	Moderate
Energy cost	Moderate	Low	Intermediate
Investment cost	Intermediate	Low	Moderate
Maintenance cost	Low	Low	Moderate
Operating cost	Intermediate	Low	Moderate

Source: Hydrocarbon Technology; 15 Aug 1993; Issue No. 26.

Table 21.10 Liquid-liquid extraction units in Indian petroleum industry

	Service	Extractor
BPCL, Bombay	Sulfolane Food grade hexane LPG amine treating	Sieve tray – Do –
CRL, Cochin	Sulfolane – BTX LPG – Amine	Sieve tray
HPCL, Bombay	Phenol extraction (LUBE)	EXXON tray
HPCL, Vizag	LPG – Amine	Sieve tray
IPCL, Baroda	Sulfolane – BTX Butadiene extraction	Sieve tray
MRL, Madras	Furfural extraction (LUBE) NMP extraction (LUBE) Food grade hexane LPG – Amine	RDC Packed column Sieve tray
IOCL, Haldia	Furfural extraction PDA extraction	RDC Baffle tray
Digboi	Kerosene treating	Sieve tray
Barauni	Kerosene treating (SO$_2$) Phenol extraction (CBFS)	Packed column Packed column
Gujarat	Tetra process LPG – Amine	Sieve tray
Mathura	LPG – Amine	Packed column
BRPL, Bongaigaon	Kerosene treating	Packed column
NOCIL, Bombay	BTX extraction butadiene extraction	RDC
GSFC, Baroda	Caprolactam extraction	
FACT, Cochin	Caprolactam extraction	

Source: Nanoti, 2005, Nanoti, 2000, Rawat, 1995.

21.4 MEMBRANE PROCESSES

Membrane processes has emerged one of the major separation processes during the recent years and finding increasing application in desalination, waste water treatment and gas separation and product purification. Different types of membrane processes are available depending on the driving force. Various types of membrane processes are mentioned in Table 21.11. Membrane technology is vital to the process intensification strategy as membrane processes are more energy intensive [Sridhar, 2009]. Principle of membrane separation is given in Fig. 21.15 [Joshi et al., 2010].

Membrane systems have become a tried and accepted natural gas treading with distinct technology with distinct advantage in a variety of processing applications.

Table 21.11 Membrane processes

Membrane process	Driving force
Reverse osmosis	Pressure difference
Ultrafiltration	Pressure difference
Microfiltration	Pressure difference
Nanofiltration	Pressure difference
Dialysis	Concentration difference
Pervaporation	Concentration difference
Liquid membrane	Concentration difference
Electrodialysis	Electrical potential
Gas permeation	Concentration difference
Thermo-osmosis	Temperature difference

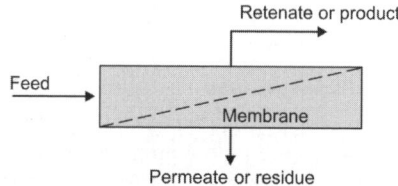

Fig. 21.15 Principle of membrane separation

Based on lower operating costs, comparable capital cost and only slightly product loss (including fuel), membranes have demonstrated as a flexible, cost effective alternative to amine treating for some natural gas processing applications [Cook and Losin, 1945]. Material used for membrane manufacturing is given in Table 21.12.

Membrane technology has become an established technology for CO_2 removal since their first use in application in 1981. Membranes are being widely used for two major applications in CO_2 removal application: Natural gas sweetening, CO_2 removal from NGL and enhanced oil well recovery using CO_2 [Gall and Sanders, 2002; Dortmundt and Doshi, 2003]. Many comprehensive onshore and offshore natural gas pretreatment facilities exist for removal of CO_2. These systems exploit the reliability and minimum manpower requirements of membranes which has led to the development of a robust and comprehensive pretreatment scheme to ensure extended membrane life to improve overall economy of the process. This pretreatment and continuing development of advanced membranes, have enhanced the reliability and performance of membrane technology and resulted CO_2 removal technology of choice in a variety of processing conditions using membrane separation. Gas separation membranes are manufactured in one of two forms: Flat sheet and hollow fiber [Dortmundt and Doshi, 2003] (Table 21.13).

Although PSA is being used commonly for separation of hydrogen, membrane process is also gaining importance in hydrogen production from synthesis gas.

Detail of application of membrane process in natural gas processing and hydrogen production is given in Chapters 15 and 18. Commercially viable membranes for CO_2 removal are polymer based cellulose acetate, polyamides, polysulfone, polycarbonate and polyetherimide (Table 21.14).

Table 21.12 Membrane material

Modified natural product	Cellulose acetate, cellulose autobutyrate, cellulose nitrate
Synthetic products	Polyamide, polybenzimidazole, polysulfone, vinyl polymers, polyfuran, polycarbonate, polyethylene, polypropylene, polyvinyl acetate, polyacronitrile, polytetrafluoroethylene
Miscellaneous	Polyelectrolyte complex, porous glass, graphite oxide, ZrO_2-polyacrylic acid, ZrO_2-carbon, etc.

Table 21.13 Gas membrane application area

Common gas separation	Application
O_2/N_2	Oxygen enrichment, inert gas generation
H_2/hydrocarbons	Refinery hydrogen recovery
H_2/CO	Syngas ratio adjustment
H_2/N_2	Ammonia purge gas
CO_2/hydrocarbons	Acid gas treatment, landfill gas upgrading
H_2O/hydrocarbons	Natural gas dehydration
H_2S/hydrocarbons	Sour gas treating
He/N_2	Helium recovery
Hydrocarbon/air	Hydrocarbon recovery, pollution control
H_2O/Air	Air dehumidification

Source: Spillman, 1989.

Table 21.14 Types and characteristics of various membrane processes

Separation process	Separation mechanism	Feed stream
Microfiltration	Sieving	Liquid or gas
Ultrafiltration	Sieving	Liquid
Dialysis	Sieving and sorption-diffusion	Liquid
Reverse osmosis	Sorption-diffusion	Liquid
Pervaporation	Sorption-diffusion	Liquid
Gas and vapor permeation	Sorption-diffusion	Gas or vapor

Source: Koros, 1995.

21.4.1 Membrane Contactors

Among large variety of membrane operations, membrane contractors (MC) represent relatively new membrane based devices that, because of their potential advantages are gaining consideration both in industrial and scientific fields. Membrane contactors are gaining importance in chemical, petroleum and petrochemical, pharmaceutical and galvanic industries. In a membrane contactor, the membrane separation is completed and integrated with a extraction or absorption [Klaassen et al., 2005]. In membrane contactor conventional phase contacting operation like all traditional stripping, scrubbing, absorption and liquid-liquid extraction operations, as well as emulsification, crystallization and phase transfer catalysts can be carried by membrane contactors [Drioli et al., 2005]. Both gas liquid contact as well as liquid-liquid contact can be performed in membrane contactor. important features of membrane contactors are flexibility and compactness, reduced size, absence of foaming, loading or flooding, besides no requirement of separation of the two independent phases after operation [Sridhar, 2009].

21.5 ABSORPTION

Absorption is one of the most commonly used separation techniques for gas cleaning purpose for removal of various gases like H_2S, CO_2, SO_2 and ammonia. Cleaning of solute gases is achieved by transfer into liquid solvent by contacting the gas stream with liquids that offers specificity or selectivity for the gases to be recovered. Some of the commonly used solvents are:

12.5.1 Chemical Absorption

Amine processes: Monoethanol amine (MEA), diethanol amine (DEA), triethanol amine (TEA), diglycol amine (DGA), methyldiethanol amine (MDEA)

Carbonate process: K_2CO_3, K_2CO_3 + MEA, K_2CO_3 + DEA, K_2CO_3 + arsenic trioxide

12.5.2 Physical Absorption

Polyethylene glycol dimethyl ether (selexol), N-methylpyrrolidone (purisol), methanol (rectisol), sulfonane mixed with an alkanolamine and water (sulfinol). Factors affecting gas absorption are mentioned in Table 21.15 [Speight, 1993].

Table 21.15 Various factors affecting gas absorption

Zone	Factor
Gas	Gas velocity, molecular weight, and sizes of solvent and solute gases, temperature, concentration gradient
Solvation liquid	Solubility, molecular weight, and sizes of solvent and solute gases, temperature, viscosity of liquid, liquid surface velocity, surface renewal rate, concentration gradient of solute gas, concentration gradient of neutralizing reagent

Absorption processes are used in petroleum and petrochemical industries for sweetening of natural gas, CO_2 removal, SO_x removal and NO_x removal. Various commonly used absorber equipments are packed towers, spray towers, tray towers and ventury scrubbers.

Some of the commonly useed chemical solvents for acid gas removal are ethanol amines and potassium carbonates. In chemical solvents, regeneration is achieved by application of heat. New brid of physical solvents used for acid gas removal are dimethyl ether of polyethylene glycol (DEPG), methanol, N-methyl-2-pyrrolidone (NMP) and propylene carbonate (PC). In physical solvents regeneration can be achieved by sampling reducing pressure [BR and E PTQ1, 2013]. Gas solubilities of these solvents for hydrocarbon and acid gases are given in Table 21.16.

Although membrane processing is commonly compared with conventional amine processing, membrane/amine hybrid system in a combination of membrane with amine absorption has been found economical for CO_2 removal from natural gas at lower concentrations [Mckee et al.,1991].

21.6 CRYSTALLIZATION PROCESS

Crystallization processes are used in petroleum industry for separation of wax. The process involves nucleation, growth, and agglomeration and gelling. Some of the applications of crystallization are in the separation of wax, separation of p-xylene from xylenes stream. Typical process of separation of p-xylene involves cooling of the mixed xylene feedstock to a slightly higher than that of eutectic followed by separation of crystal by centrifugation or filtration. Some of the commonly used crystatllization techniques for separation of p-xylene are Phillips consisting of vertically pulsed crystallizer/purifier and ARCO one-stage and two-stage crystallization processes [Chauvel and Lefebvre, 1989]. In solvent dewaxing, the parrafinic wax is separated by crystallization as paraffinic materials present in lubricating oil separate out lowering temperature.

Table 21.16 Physical solvents for acid gas removal

Gas component	DEPG at 25 °C	PC at 25 °C	NMP at 25 °C	Methanol at 25 °C
H_2	0.013	0.0078	0.0064	0.0054
Methane	0.066	0.038	0.072	0.051
Ethane	0.42	0.17	0.38	0.42
CO_2	1.0	1.0	1.0	1.0
Propane	1.01	0.51	1.07	2.35
n-butane	2.37	1.75	3.48	–
COS	2.30	1.88	2.72	3.92
H_2S	8.82	3.29	10.2	7.06
n-hexane	11.0	13.5	42.7	–
Methyl mercaptan	22.4	27.2	34.0	–

Source: BR and E. PTQ1, 2013.

REFERENCES

1. Amete V and Rao E. Recent Developments in Reactive Distillation. Chemical Industry Digest, Dec 2009.

2. Asprion N and Kaibel G. Dividing Wall Columns: Fundamentals and Recent Advances. Chemical Engineering and Processing: Process Intensification. Chemical and Processing, Vol 49, 2010, p139.

3. Banik S. Short Path Distillation at AOD, Digboi. National Workshop on Energy Management and Conservation in Hydrocarbon Industry, Nov 11–12, 2003, organized by Lovraj Kumar Memorial Trust and Indian Institute of Chemical Engineers, NRC, New Delhi.

4. Banik S. Distillation Options in Refinery. Petrotech Society's Summer School Program on Advances in Petroleum Refining Industry, July 3–8, 2006.

5. Bokade VV. Catalytic Distillation. Chemical Engineering World, Vol 33, No. 6, June 98.

6. Chauvel A and Lefebvre G. Petrochemical Processes, Part 1. Editions Technip, 27, Paris, 1989.

7. Cook PJ and Losin MS. Membranes Provide Cost Effective Natural Gas Processing. Hydrocarbon Processing, April 1995, p80.

8. Crottenden B. Selective Adsorption. Chemical Engineer, Sep 1998, p21.

9. Dortmundt D and Doshi K. CO_2 Removal Membrane Technology—Recent Developments. Chemical Engineering World, Sep 2003, p55.

10. Drioli E, Curcio E and Di Profio G. State-of-the-art and Recent Progress in Membrane Contactors. Chemical Engineering Research and Design, Vol 3, No. A3, March 2005, p233.

11. Ennenbach F and Kolbe B. Divided Wall Columns—A Novel Distillation Concept. PTQ, Autumn, 2000, p97.

12. Fan L and Rubin D. MaxFlow^{TM} High Performance Valve Tray. Hydrocarbon Engineering, March 2000, p55.

13. Gourila JP. Distillation, Absorption and Stripping. in 2000 in IFP Petroleum Refining, Vol 2, Separation Processes edited by Wauquier JP. Editions Technip, 2000.

14. Handbook of Solvent Extraction, edited by Baird MHI and CX.

15. Joshi MK, Goyal GD, Deshpande A and Sarkar DK. Membrane Process for Natural Gas Sweetening. Petrofed Journal, January–March, 2010, p20.

16. Kasani N, Siegert M and Sirch T. A New Kind of Column Packing for Conventional and Reactive Distillation—The Sandwitch Packing. Chem. Eng. Technol, Vol 28, 2005, No. 5, p549.

17. Keller II GE. Adsorption: Building upon a Solid Foundation. Chemical Engineering Progress, Oct 1995, p56.

18. Klaassen R, Feron PHM and Jansen AE. Membrane Contactors in Industrial Application. Chemical Engineering Research and design, Vol 83 (A3), p234.

19. Knaebel KS. For Your Next Separation—Consider Adsorption. Chemical Engineering, Nov 1995, p92.

20. Knaebel KS. The Basics of Adsorber Design. Chemical Engineering, April 1999, p92.

21. Koros WJ. Membranes: Learning from Nature. Chemical Engineering Progress, Oct 1995, p69.

22. Kunesh JG, Kister HZ, Lockett MJ and Fair JR. Distillation Still Towering Over Other Options. Chemical Engineering Progress, Oct 1995, p43.

23. Mazumder K. Advances in Distillation. AICTE, Staff Development Program on Hydrocarbon Engineering, Jan 3–22, 2000, IIT Roorkee.

24. McKee RL, Changela MK and Reading GJ. CO_2 Removal: Membrane Plus Amine. Hydrocarbon Processing, April 1991, p64.

25. Nanoti SM. Industrial Extraction, Equipment Selection and Process Design. PhD Thesis, IIT Roorkee, 2005.

26. Nanoti, 2000, AICTE. Staff Development Program on Hydrocarbon Engineering. Department of Chemical Engineering, IIT Roorkee, Jan 3–22, 2000.

27. Olujie Z, Kaibel B, Jansen H, Rietfort T, Zich E and Frey G. Distillation Column Internals/Configurations for Process Intensification. Chem. Biochem. Eng, Vol 7, No. 4, 301–309 (20030).

28. Rawat BS. Advances in the Production of Petrochemicals. Department of Chemical Engineering, IIT Roorkee, June 21–July 5, 1995.

29. Reid JA. Distributed Distillation with Heat Integration. PTQ, Autumn, 2000, p85.

30. Rojey A. Distillation, Absorption and Stripping in IFP Petroleum Refining, Vol 2, Separation Processes, edited by Wauquier JP. Editions Technip, 2000.

31. Ruhul MJ. Recover VOCs Via Adsorption on Activated Carbon. Chemical Engineering Progress, July 1993, p37.

32. Schmidt W, Hiltmann T and Herbrecht D. Batch Distillation in Trendwandkolonnen, Jahrestreffen Derprocess Net-Fachausschusse Adsorption und Fluidverhrestechnik, Bingen, Deutschiand, 13–14.03.2008.

33. Shin J, Lee J, Lee S, Lee B and Lee M. Enhance Operation and Reliability of Dividing Wall Columns. Hydrocarbon Processing, May 2013, p85.

34. Spillman RW. Economics of Gas Separation Membranes. Chemical Engineering Progress, Jan 1989, p41.

35. Sridhar S. Membrane Technology in Process Intensification. Chemical Industry Digest, Feb 2009, p62.

36. Stichlmair J and Frey T. Reactive Distillation Processes. Chem. Eng. Technol, 22(1999)2, p95.

37. Terrigoel A. Molecular Sieves in Gas Processing: Effects and Consequences by Contaminants. Hydrocarbon Processing, Jan 2013, p51.

22

Energy Management in Petroleum Refinery

Economic growth of a nation is dependent upon its energy resources and energy consumption is one of the yard sticks of industrialization and urbanization of a nation. Rapid pace of economic growth in developing and developed countries has led to an ascending demand of energy and energy resources are becoming more and more scare and costlier. The energy which mankind is using today took million of years to reach the earth; however, its utilization efficiency has been at very low profile at the earlier stage of industrialization. Energy is the most essential input for sustaining and for speedy economic development of a country [Ghosh and Narayana, 2010].

The global energy crisis during 70s has brought increasing energy conservation awareness through use of benchmarking tools, periodic energy audit, promoting renewable energy sources and general energy conservation measures through good house keeping and installation of energy efficient devices and processes. Energy audit has been made mandatory by most of the developing and developed countries. Industrial and transportation sectors are two major energy consumer sectors. Some of the keys for shaping global energy supply and their use in the future are population growth, economic and social development, financial institutional conditions, efficiency of energy supply and use, technological innovation and development, access to sufficient modern energy resources in the developing world [www.ciigbc.org] [World Energy Council, 2000].

Petroleum industries are energy intensive and the energy consumption is affected by refinery configuration, type of feedstock processed, severity of operation, vacuum system employed, steam and power balance, process technology, yield pattern, mandate in product specifications, environmental regulations, flexibility in operation [Dey et al., 2003]. Energy in a typical refinery is 60% of variable costs. A 1% improvement in energy equals to roughly $6,50,000 year saving for a 1,00,000 bpsd refinery [Krishnan, 2003]. Some of the major approaches for energy conservation are energy efficient design, improving energy efficiency, energy efficient operation, benchmark energy performance and continuous energy improvement.

22.1 WORLD ENERGY SCENARIO

Energy demand is growing and with increase in population which is projected to increased by 1.4 billion over the next 20 years. World primary energy consumption is likely to increased by approximately 40% over the next 20 years. Oil and gas are still expected to play a significant role [BP Annual Report, 2011]. Global energy demand will be about 30% higher in 2040 compared to 2010. As economic output more than doubles and prosperity expands across a world whose population will grow to nearly 9 billion people [Vasudeva, 2012]. The global energy is expected to reach 17,100 million tons oil equivalent (MTOE) by 2030 as compared to 11,468 MTOE at present [Bhatia, 2008].

22.2 INDIAN ENERGY SCENARIO

India is the fourth largest consumer of energy in the world and with current growth, India will become the third largest consumer of energy by 2030. Currently, Indian energy is dominated by coal 53%, followed by oil 30%, natural gas 10.5%, hydropower 5% , nuclear and renewable 1% each. To maintain 8% growth rate from 2012 to 2032, primary energy consumption requires to be scaled up from 537 million tons oil equivalent in 2011–12 to 1856 million tons oil equivalent (MTOE) in 2031–32. This oil consumption would rise from 1,666 million tons per annum and gas would go up from 44 MTOE to 197 MTOE [Vasudeva, 2012]. Total energy requirement in India is given in Table 21.1 [2011 India Energy Handbook August, 2010]. Per capita consumption energy has increased from 1,204 kWh in 1970–71 to 4,861 kWh in 2010–11. Total carbon dioxide emissions have been gradually increasing over 50% of the total CO_2 emissions from the power sector. Coal is expected to continue to meet India's energy needs in significant power generation and other industrial processes. As the industry is growing, the consumption of energy is also increasing which has direct impact on the cost of fuel. Rising cost scenario of coal and oil is shown in Fig. 22.1 and rising cost scenario of electric power is shown in Fig. 22.2.

Table 21.1   Total energy requirements (MTOE)

Year	Hydro	Nuclear	Coal	Oil	Gas	Toal
2016–17	18	31	375	241	64	729
2021–22	23	45	521	311	97	997
2026–27	29	71	706	410	135	1,351
2031–32	35	98	937	548	197	1,815

Source: IEPN Report, p28, Table 2.10.

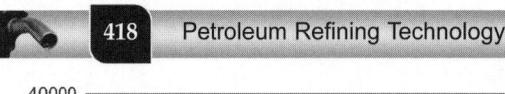

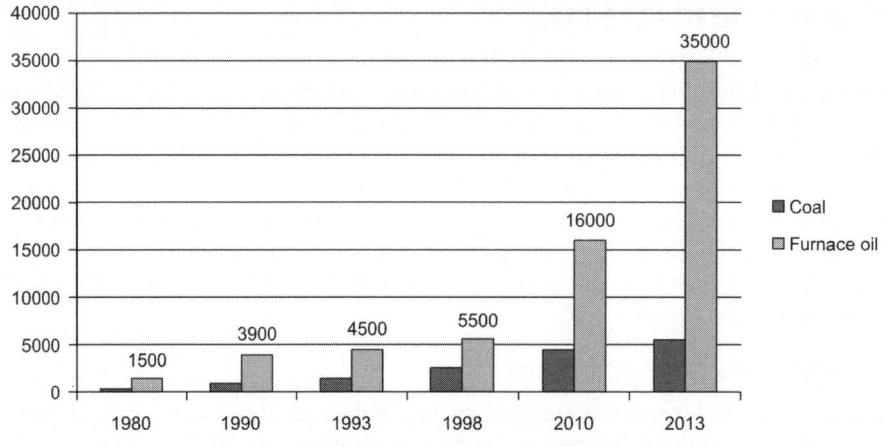

Fig. 22.1 Rising cost of coal and furnace oil

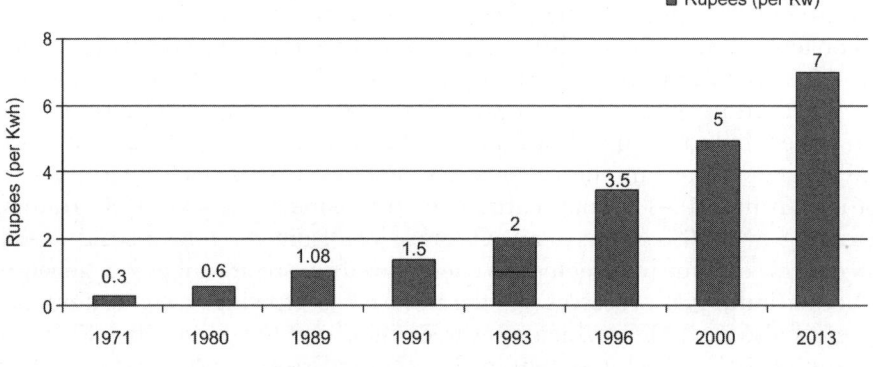

Fig. 22.2 Rising cost of electricity

22.3 ENERGY MANAGEMENT IN PETROLEUM REFINERY

Petroleum refineries are one of the major energy consuming industries and contribute major portion of the cost of production. The energy consumption in refineries varies widely from refinery to refinery depending on design and age of the plant, types of crude processes, energy management policy, and product profile fuel and petrochemical. Typical energy consumption in refinery is shown in Fig. 22.3 [Chopra, 2003]. Energy is by far the largest operating expenses for petroleum refineries and it accounts for over 40% of refinery operating expenses and over 60% of olefin plant operating expenses [Birchfield, 2002].

Energy is consumed in the petroleum refinery in the following manner:

1. As direct fuel in process heaters/boilers/GTGs
2. As indirect fuel for rising steam or generating power
3. Hydrocarbon losses during handling/storage of crude oil/products, processing, handling during dispatch and loading and unloading of products; flare losses and leakages, etc.
4. Steam and power for drives
5. Circulating cooling water.

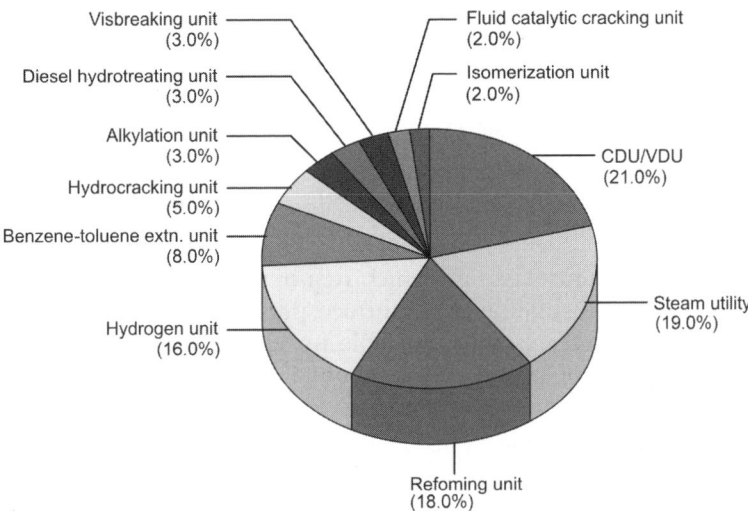

Fig. 22.3 Breakup of energy consumption in different sections of refinery

Source: Chopra, 2003.

Total energy consumed in the refinery depends upon the types of crude processed; the processing schemes; plant capacity; equipment and systems; the degree of heat in the system; mode of product dispatches; the general housekeeping and maintenance practices; the complexity of operation and the general concern for the energy efficient operation.

Energy is required to operate refining processes and to produce products to required specifications (Table 22.2). Thus optimizing refinery energy systems require an integrated approach as some energy expenditures are independent of process operations and on the other hand, there are numerous process/energy interactions where yields and energy should be considered simultaneously. Thus optimizing refinery energy systems requires:

Table 22.2   Energy system that must be operated and maintained

To use energy	Furnaces, boilers, reboilers, motors, steam and gas turbines
To recover energy accounting systems	Exchangers, waste heat boilers, economizers, expanders, insulation
To distribute energy	Natural gas, fuel gas, fuel oil, steam, electric power
To remove energy	Cooling towers, recirculating water, once-through water system, air fan/fin coolers, condensers (to water)

Source: Birchfield GS, Aspen Tech, 2002.

- Energy balancing
- Rigorous energy economics
- Process analysis
- Steam power system analysis
- Equipment level efficiency assessment
- Analysis of process/energy interactions
- Use of advance methods/optimization tools
- Efficiency monitoring.

Apart from good housekeeping which requires little investment in repairing leaking joints, proper insulation of hot surface, periodical maintenance of bearing, pump seals, etc; the major point available for energy conservation is through modifications/retrofit on process equipment namely:

- Fired heaters
- Steam/power generation systems
- Heat exchangers
- Distillation/absorption/extraction/stripping/evaporation system
- Power/steam drives

Total cost of energy: This includes purchased fuels, steam and power, plant produced, energy consumed, distributing energy in plant, operating energy equipment, boilers, furnaces, exchangers, drivers, removing waste energy by air or water cooling [www.batchplus.com].

Energy policy and energy information system: An effective energy policy and energy information system and cost accounting system are vital for a sound energy management.

22.3.1 Energy Policy

Energy policy is one of the major steps in energy conservation program which includes:

- Adopting energy efficient and environment friendly technologies
- Replacement of energy inefficient equipment with energy efficient equipment
- Maximizing the recovery of waste energy
- Benchmarking performance with the best in the world and endeavoring to be ahead
- Promoting use of renewable sources of energy
- Creating awareness amongst the employees
- Maximizing the recovery of waste energy
- Fostering a culture of participation and innovation, improvement in energy conservation
- Propagating the message of avoiding wastage of energy [www.ciigbc.org].

22.3.2 Energy Information Systems

A good energy information system and cost accounting system is vital for a sound energy management. Energy consumption in Indian refineries is given in Table 22.3. The objectives are:

Table 22.3 Typical specific power and thermal energy consumptions in Indian refineries

Year	Sp. power consumption per ton kWh/Mt of intake			Thermal consumption Mkcal/Mt of intake		
	Reliance	Haldia	Mathura	Reliance	Haldia	Mathura
2002–03	73.8	57.13		0.668	0.88	
2003–04	73.49	59.75	36.4	0.661	0.82	0.5
2004–05	73.1	57.73		0.662	0.74	

Standard refinery fuel (SRF) is defined as the fuel with net value of 10,000 Kcal/kg
1 SRFT = 13T steam + 36000 kWh = 20,000 m^3 cooling water.

- To motivate operating personals and others to minimize fuel and electricity cost
- To provide information essential to the allocation of costs to specific processes and products
- To facilitate identification of fuel and electricity cost reduction investment opportunities
- To provide information which may be required for statutory requirement.

22.3.3 Distillation Unit

Crude oil distillation unit in a refinery is biggest energy consumer in a refinery and energy intensive distillation systems are fundamental in a refinery for improving energy efficiency. The performance of distillation column is very closely linked with the refinery processes. The efficient operation of column can result in better yield, improved product quality and overall reduction in utilities and energy consumption. Many energy saving measures adopted in refineries are [Krisnan and Verma, 1993; Verma, 1991; EIL 1992]:

- Optimizing reflux ratio to have use of more trays and lower reflux ratio and low column pressure which results in fewer trays and/or lower reflux ratio for a given separation
- Use of high efficiency/low pressure drop packing for better separation and lower energy consumption
- Use of split tower arrangement
- Use of divided wall column technology which offers better separation and energy saving
- Multiple effect heat cascading for distillation
- Increased heat recovery from side draw off and pump around
- Use minimum stripping steam/dry vacuum towers
- Better control system so as to reduce reboilers and overhead condenser loads
- Possible reduction in reflux ratio
- Minimize overflash
- Installation of absorption heat pump and mechanical vapor compressor using screw compressor

- Use of structured packing which results in improved fractionation efficiency, higher capacity and reduced energy consumption
- Better instrumentation and control of distillation column
- Optimization of insulation thickness.

Distributed distillation allows 10–30% reduction in separation energy [Reid, 2000]. A distributed distillation system frequently increases opportunities for heat integration by shifting heat duty temperature levels, producing different purity products [kolmetz.com] [Reid, 2000; Gadalla, 2003] has suggested an optimization framework for the existing distillation system and its heat exchanger network simultaneously, lowering energy consumption and freeing up capacity at a minimum capital investment.

22.3.4 Optimization of Refinery Steam/Power System

The efficient usage of utilities, optimization of steam power balance and optimization of heat exchanger train have been the prime concern in energy conservation areas [bee-indiqa.nic.in]. The major areas where energy efficiency can be improved are:
- Heat integration
- Fired heaters efficiency
- Optimizing steam and power system.

Reducing fouling/surface cleaning/surface coating in heat exchanger/furnace: Some of the measures taken for reducing fouling/surface cleaning/surface coating in heat exchanger/furnace are:
- Preventing solids from forming
- Preventing solids from adhering to themselves and to the heat transfer surfaces
- Removing solids from surface
- Efficient desalting.

22.3.5 Pumping System

Some of the measures which should be taken during design stage of a project [Harindranath and Kamath, 2007] are:
- Ensure that the actual system head requirement of the installed pump and rated heads are as close as possible to avoid mismatch in heads leading to inefficient operation
- Proper sizing of pumps with respect to flow head as per process demand
- Incorporation of speed control mechanisms like variable speed drives for variable flow requirements instead of throttling of the delivery valves of the pump
- Incorporation of high quality impellers for corrosive liquids
- Incorporation of high capacity pumps instead of parallel operation of pumps wherever possible
- Use of low friction pipes whenever required
- Minimizing the number of bends and using the shortest routes, when installing pumps.

22.3.6 Electric Drives

Electric drives account for major portion of auxiliary electricity consumption. Be it pumps, mixers or fans, a variable speed or frequency drive serves significantly by allowing soft starts and matching motor torque and speed to the load. This saves energy and accounts for longer life [Hydrocarbon Asia, 2000]. The following points should be considered while selecting drive [Harindranath and Kamath, 2007].

- Adequate sizing of the motor
- Use of high efficiency motor over standard conventional motors
- Inventory
- Incorporation of dual speed motors, wherever applicable
- Incorporation of highway from dusty environment
- Providing proper protection against rain/wind efficiency belt and transmission belts
- Installation of motors

22.3.7 Furnace System

In the refinery large number of furnaces are used for heating the feed. Forced draft burners are increasingly used as they offer important advantages like significantly higher combustion efficiency, reduced particle emission, lower consumption of atomizing steam, better control of flame shape and ability to create more compact furnace [Verma, 1991].

For enhancing the efficiency following measures can be taken.

- Monitoring of CO and excess air to reduce rejected energy and improve efficiency, installation of oxygen analyzer
- Installation of improved burner, economizer and air preheaters
- Preheating boiler feed water with available low temperature process streams
- Maximizing use of heat transfer surface by optimizing soot blowing frequency and decoking of tubes
- Flash blow down to produce low pressure steam [Woods, 2005]
- Reduce radiation loss
- Reduce flue gas temperature
- Choose right auxiliaries
- Increasing convection section duty/area
- Generating steam heating boiler feed water
- Providing additional cleaning facilities

Possible improvement in furnaces is shown in Table 22.4.

Table 22.4   Possible improvements in furnaces

Equipment system	Purpose	Improvement possible
Burners	To burn oil	To keep excess air as low as possible. Simultaneously, promoting flame conditions that result in complete combustion of fuel at lower level of excess air with minimal NO_x and CO emissions.
Air preheater	Recovering energy from stack gases by heating the combustion air	2.5% increase in energy improvement for each 55 °C drop in stack gases temperature.

Contd.

Table 22.4   Possible improvements in furnaces *(Contd.)*

Equipment system	Purpose	Improvement possible
Economizer and waste heat boiler	Transfer energy from stack gases to raise steam or heat water	1% increase in boiler efficiency for each 55 °C increase in feed water temperature.
Combustion control systems	Regulate flow of fuel and air for desired fuel-air ratio	0.25% efficiency improvement for each 1% decrease in excess air
Soot blowers	Remove deposits that cover heat transfer surface	To keep heat transfer surface clean and available for effective heat transfer both in the convection bank as well as in APH system

Source: EIL, 1982.

22.3.8 Heat Exchanger

Factors that are relevant to effective utilization of heat exchangers are:
- Optimum heat exchanger train configurations to increase crude preheat temperature
- Closer approach temperature between hot and cold streams in heat recovery
- Use of welded plate exchanger in feed-effluent service
- Optimum recovery of waste heat from hot product draws off streams
- Integration of heat exchangers between various processes and utility units to economize on heat inputs, and optimal heat recovery
- Lower pressure drop in heat exchangers and simplified piping arrangements
- Use of antifoulants for control of scaling in heat exchangers
- Use of high flux tubes for low T boiling
- Effective utilization of low level heat
- Online cleaning system in certain heat exchangers.

22.3.9 Compressed Air

Compressed air is an important utility in refinery operation and has a tremendous impact on production processes and overall cost of production. Many commercial and industrial compressed air users are improving air system energy efficiency, reducing maintenance costs and lowering noise levels with rotary screw compressors and incorporating variable speed (VSDs) [newequipment.com]. Variable speed drivers lowered compressed air costs. Using advanced VSD technology to provide improved efficiency, flexibility and noise control, state-of-art rotary screw compressors can save users 20–35% on electricity in situations where they have variable loads [Perry, 2002].

22.3.10 Flare System

Garo liquid ring compression systems have proved to be cost effective and profitable solution for the recovery of flare gas and other refinery off gases since they act as a sort of scrubber, feeding the process with clean gas. Further, these liquid ring compressors allow plant operators to optimize gas treating capacity and increase profits without new capital equipment [Ines Milanesi, Garo Hydrocarbon Engineering, June 1999, p79].

22.3.11 Steam Traps

An effective, but modest maintenance and replacement program can reduce number of broken traps to 5–10%, saving about 10% of the energy consumed in a steam distribution system. Despite improvements in steam traps are not invincible. Most common cause of failure is dirt which blocks the flow of condensate through the orifice in the steam trap and causes it to fail, close. Accumulated dirt on the seat of a steam strap prevents the valve from closing the orifice resulting in failure and leakage of steam. Dirt also responsible for corrosion, erosion and pitting. New steam traps are taking these problems into consideration and steam straps with built-in dirt handling capabilities are available [Hairston, 2003].

22.4 BENCHMARKING AND TARGETING

Benchmarking and targeting is the diagnostic tool for energy optimization in a refinery operation. Benchmarking is a tool for ongoing search for best practices that produce superior performance when adapted and implemented. Targeting is a set point for energy consumption which can be achieved in a refinery taking into consideration the modifications/corrective engineering to be carried out under the constraints of existing configurations. Benchmarking encourages savings for excellence, opens up thinking and promotes innovation, provides external perspective, identification and rapid implementation of best practices [Bokare, 2003].

- *Specific energy consumption (SEC):* It is expressed in terms of Mbtu/bbl/NRGF
- *NRGF (energy factor):* Each process unit is assigned an energy factor based on the ratio of its energy consumption to that of CD.
- *Energy intensity index (EII)*

  EII = (actual energy consumed in study year × 100)/[Sum (U offsites unit actual capacity × unit standard energy) + (standard sensible heat + standard offsite energy)]
- *Shell corrected energy and loss index (CEL)*

  CEL index = actual energy consumed in study year (SRFT) × 100/theoretical allowance of units.

22.4.1 Benchmarking in Refinery [Sil et al., 2005]

Benchmarking has become one of the important practices that produce superior performance when adapted and implemented. Benchmarking is recognized as an effective approach towards improving efficiency, productivity, quality, profitability and other such dimensions of performance that determines competitiveness. Benchmarking may or may not be achievable because a large number of parameters affect the energy consumption depending on age and capacity of the plant, type of crude oil processed. Target energy consumption may be more realistic figure. Targeting means setting an intermediate attainable energy consumption level for process units/steam power plant/offsite keeping in view the present configuration. Feedstock available, product pattern and local constrains of each refinery and identifying potential energy saving areas. Table 22.5 gives the comparison of actual

Table 22.5   Comparison of actual energy consumption for major process units with benchmark consumption

S. No.	Unit	Actual energy consumption		Benchmark energy consumption	
		(BTU/BBL)	Kcal/M³	(BTU/BBL)	Kcal/M³
1.	CDU (stand alone)	74,640–1,23,900	1,18,305–1,96,383	73,600–78,650	1,16,657–1,24,661
2.	VDU (stand alone)	86,200–1,98,400	1,36,628–3,14,466	65,330	1,03,549
3.	C & VDU (combined)	1,04,900–1,55,700	1,66,267–2,46,786	88,000–1,09,000	1,39,481–1,72,766
4.	Naphtha splitter	1,02,660–2,36,740	1,62,717–3,75,235	1,02,150	1,61,909
5.	FCCU (with coke)	2,56,675–5,05,000	4,06,832–8,00,430	2,50,400	3,96,886
6.	Delayed coker (LR)	3,70,100–4,21,140	5,86,612–6,67,511	3,16,710	5,01,988
7.	Aromatics recovery	6,54,175	10,36,873	5,05,840	8,01,761
8.	Hydrocracker unit	4,33,300	6,86,784	2,62,320	4,15,780
9.	Hydrogen unit	87,387–1,10,850	1,38,509–1,75,698	66,930	1,06,085
10.	Propane deasphalting	4,54,380–5,73,255	7,20,196–9,08,614	2,61,640	4,14,702

Source: Sil et al., 2003.

energy consumption for major process units with benchmark consumption and energy saving potential in the refineries [Sil et al., 2005].

22.5 ENERGY SAVING POTENTIAL

Energy consumption in Indian refineries varies widely from refinery to refinery. Although after the implementation of Energy Act 2001, Indian process industries and refineries have taken number of initiatives to reduce their energy bills, however, there is a lot of scopes for energy saving. Energy saving potential in Indian refineries is given in Table 22.6.

Thus we can see that there is lot of opportunity for reducing the energy consumption through reduction in fuel and hydrocarbon losses. Reduction in fuel and hydrocarbon losses by 1% would mean a staggering annual saving of around ₹ 700 crore which besides, the increased profits to the refineries.

Table 22.6   Energy saving potential in the refineries

Refinery units	Energy saving potential between actual and target	Energy saving potential between actual and benchmark
Process units	8–10%	20%
Steam and power plants	8–10%	15–43%

Source: Sil et al., 2003.

22.6 ENERGY CONSERVATION MEASURES IN INDIAN REFINERIES

Some of the major energy conservation measures taken in Indian refineries are:
- Provision of high efficiency burner
- Heat integration through pinch technology in crude distillation unit
- Use of pressure gas
- Installation of soaker technology
- High emissivity refractory coating in crude distillation furnace
- Residuum oil supercritical extraction process which consumes less energy than conventional method
- Replacement of existing refractory with ceramic fiber in crude distillation furnace
- Replacement of metallic fans with FRP blades
- Effect on condensate recovery
- Installation of more efficient steam traps
- Replacement of old low efficiency reciprocating compressor with high efficiency centrifugal compressors
- Installation of blow down recovery system
- Use of oxygen analyzer for regular checking of excess air and efficiency in furnace
- Installation of lakos filter
- Hydrocarbon loss reduction by the way of strict control and monitoring of flare (by the use of physical acoustic leak detector for detection of leakage of fuel gas from safety valves, control valves and bypass valves of flare system) handling losses, tank form monitoring
- Installation of shell and tube type condenser, primary condenser using refinery cooling water as cooling media and two-stage ejector system in vacuum column
- Installation of high efficiency boilers
- Installation of hollow FRP blades in cooling towers in place of GRP/aluminum blades
- Preheat improvement in visbreaker unit by exchanger cleaning sequence to avoid fast fouling of preheaters which use very heavy residue as feedstock containing high wax, asphaltene and metal contents
- Use of steam turbine for boiler feed pump
- Chemical treatment of cooling water, for improving exchanger, fouling factor
- Steam trap auditing
- Better insulation management
- Increasing of LP burner trip

- Improvement in burner performance by nitrogen injection in sulfur recovery unit
- Installation of autocombustion control loop
- Installation of fugitive emission monitor
- Monitoring hydrocarbon losses
- Use of energy efficient mechanical seals
- Use of slop oil for quenching
- Installation of autosampler in crude oil pipeline
- Fuel loss measurements
- Routing vacuum distillation hot-well off gases to furnaces
- Optimizing hydrogen management.

22.7 ENERGY AND CO_2 EMISSIONS IN REFINERY

Energy costs typically represent 50–60% of non-feedstock refinery operating costs. Energy consumptions in a refinery can be classified into: (i) thermal energy consumption [fired heaters, incinerators, boilers and coke burn in fluid catalytic cracking (FCC)] (ii) electrically energy consumptions: Motors, lighting and power supply, etc., sound energy management have direct impact on CO_2 emissions from refinery. Various sources for CO_2 emissions in a typical refinery are distillation columns, hydrodesulfurization processes, catalytic reforming, hydrogen production, FCC, delayed coker, alkylation sulfur block, captive power plants, flares. About 65–70% of the total CO_2 emission from a refinery is contributed by captive power plant and hydrogen production units alone. About 10 tons of CO_2 is produced per tons of hydrogen. Crude distillation and FCC are the other prominent CO_2 generators among the process units. Three major areas where CO_2 reduction opportunity are: Efficiency improvements of various process fired heaters, steam rising and power generation; hydrogen production and coke burn-off from FCC. Switching over from refinery fuel oil to natural gas can significantly reduce CO_2 emissions. About 15% reduction in total CO_2 emissions from a refinery can be achieved by switching the fuel from regular refinery fuel oil to natural gas. Various hydrogen treatment processes increase the CO_2 emission from refinery. Better hydrogen management through reoptimizing of the refinery hydrogen balance and strategies to minimize hydrogen losses, use of more selective catalysts in hydrotreaters, fuel substitution with fuel having higher hydrogen content and lower carbon with improved thermal efficiency, maximizing the H_2 production in catalytic reforming efficiency [Vardhrajan et al., 2009].

22.8 ENERGY CONSERVATION ACT 2001

Considering the economic growth rate needed for the country to emerge as a leader in the Asian region, the present and future energy scenario and its impact on the overall economy, Government of India has enacted the Energy Conservation Act 2001. The Act provides the legal framework and institutional arrangement for embarking on an energy efficiency drive. A brief description of the Energy Conservation Act 2001, role of the Bureau of Energy Efficiency [energy.esansar.com] and its plan of actions are given in the following pages.

22.8.1 Establishment of Bureau of Energy Efficiency (Sections-3 to 12)

a. Establishment of Bureau of Energy Efficiency (BEE) directed by a Governing Council with minimum 20 and maximum 26 members appointed by the Government.

b. Director General appointed by the Government shall be the Chief Executive Authority of the Bureau.

c. The Bureau shall, within 6 months from the date of commencement of this Act, constitute Advisory Committees for the efficient discharge of its functions.

d. The Bureau may constitute Technical Committees of experts for the purpose of formulation of energy consumption standards or norms in respect of equipment and processes.

e. By this Act, the assets, liabilities and employees of the Energy Management Center is transferred to BEE.

22.8.2 Powers and Functions of the Bureau (Section-13)

- Recommend to the Govt. norms for the processes and energy consumption standards
- Recommend particulars required to be displayed on label on equipment or on appliances and manner of their display
- Recommend to the Govt. for notifying any user or class of users of energy as designated consumer
- Take suitable steps to prescribe guidelines for energy conservation building codes
- Take measures to create awareness and disseminate information for efficient use of energy and its conservation
- Arrange and organize training of personnel and specialists in the techniques of efficient use of energy and its conservation
- Strengthen consultancy services
- Promote R & D
- Develop testing and certification procedures and promote testing facilities for certification and testing for energy consumption of equipment and appliances
- Formulate and facilitate implementation of pilot and demonstration projects
- Promote use of energy efficient processes, equipment, devices and systems
- Promote innovative financing of energy efficient projects
- Give financial assistance to institutions for promoting energy conservation
- Levy fee for services provided for promoting efficient use of energy
- Maintain a list of accredited energy auditors
- Specify qualification for energy auditors
- Specify periodicity and the manner in which energy audit to be conducted
- Specify certification procedures for energy managers
- Prepare educational curriculum on efficient use of energy for educational institutions and coordinate with them for inclusion in the syllabus
- Implement international cooperation programs relating to efficient use of energy as directed by Central Govt.
- Perform such other functions as may be prescribed [ceikerela.gov.in].

22.8.3 Powers of the Central Government (Section-14)

The Central Govt. may, by notification, in consultation with the Bureau:

- Specify norms for processes and energy consumption standards for any equipment which generates/transmits/supplies energy.
- Specify equipment/appliances or class of equipment/appliances [ceikerela.gov.in].
- Prohibit manufacture/sale/purchase of import of specified equipment/appliances not conforming to standards. Provided that no notification prohibiting manufacture/sale or purchase/import of equipment/appliances shall be issued within 2 years from the notification of norms for processes and standards for equipment/appliances [energymanagertraining.com].
- Direct display of particulars on label on equipment/appliances
- Specify designated consumers
- Alter the list of energy intensive industries in the schedule
- Establish and prescribe norms/standards for designated consumers
- Direct the energy intensive industries in the schedule to get energy audit conducted by accredited auditors in a manner and interval specified by regulations
- Direct the designated consumer to get energy audit conducted by accredited auditors
- Direct any designated consumer to furnish information with regard to the energy consumed and action taken on the recommendations of the accredited energy auditor
- Direct any designated consumer to appoint energy manager and submit an annual report on the status of energy consumption in the prescribed form
- Prescribe minimum qualification for the energy manager
- Direct every designated consumers to comply with the energy consumption norms and standards
- Direct any designated consumer, who does not fulfill the norms and standards, to implement schemes for energy conservation keeping in view its economic viability
- Formulate codes for energy conservation building/building complex
- Amend the energy conservation building codes to suit the regional and local climatic conditions
- Direct owner or occupier of the building/building complex, being a designated consumer to comply with the provisions of energy conservation building codes
- Direct the designated consumer referred to in (r) to get energy audit conducted by accredited auditors
- Take all measures to create awareness and disseminate information for efficient use of energy [ceikerela.gov.in].
- Arrange and organize training of personnel in efficient use of energy
- Take steps to encourage preferential treatment for use of energy efficient equipment or appliances.

22.8.4 Powers of the State Government (Section-15)

The State Govt. may, by notification, in consultation with the Bureau:

- Amend the energy conservation building codes to suit the regional and local climatic conditions and may specify and notify the codes [ceikerela.gov.in].
- Direct owner or occupier of the building/building complex, being a designated consumer to comply with the provisions of energy conservation building codes
- Direct the designated consumer referred to in (b) to get energy audit conducted by accredited auditors
- Designate any agency as designated agency to coordinate, regulate and enforce the provisions of this Act.
- Take all measures to create awareness and disseminate information for efficient use of energy
- Arrange and organize training of personnel in efficient use of energy
- Take steps to encourage preferential treatment for use of energy efficient equipment or appliances
- Direct any designated consumer to furnish to the designated agency information on energy consumption
- Specify matters to be included for the purposes of inspection.

22.8.5 State Energy Conservation Fund (Section-16)

The State Government will constitute a fund for promotion of energy conservation. Grants/loans made by the State/Central Govt., any other organization, individual will be credited to this fund.

22.8.6 Power to Inspect (Section-17)

The designated agency may appoint inspecting officers, after 5 years from the date of commencement of the Act, to inspect consumers for checking compliance with the norms.

An inspecting officer shall have the power to:

a. Any operation carried on or in connection with the equipment/appliance specified under clause (b) of Section 14 or in respect of energy standards under clause (a) of Section 14.

b. Enter any place of designated consumer to inspect equipment/processes

An inspecting officer shall not move or cause to move any equipment/appliance, book of account and other documents [ceikerela.gov.in].

22.8.7 Power of Central/State Government to Issue Directions (Section-18)

The Central and the State Governments may issue directions to any person, officer, authority, any designated consumer. This includes the power to direct regulation of norms for processes in any industry, building and building complex.

22.8.8 Penalty (Section-26)

Penalty provision has also been provided in the Act. If any person fails to comply with the provisions of clause c, d, h, i, k, I, n, r, s of Section 14, or clause b, c and h of Section 15, he shall be liable to a penalty which shall not exceed ₹ 10,000 for each such failure and in case of continuing failure, with an additional penalty which may extend to ₹ 1,000 for every day during which such failure continues [nlsenlaw.org].

22.8.9 Power to Adjudicate (Sections-27 to 29)

State Electricity Regulatory Commission, after giving a reasonable opportunity of being heard, may appoint one of its members to be an adjudicating officer. If he is satisfied that the person has failed to comply with the provisions given in Section 8, he may impose suitable penalty considering the amount of disproportionate gain or unfair advantage made out of default.

22.8.10 Responsibilities and Duties of Energy Manager

Responsibilities

- Prepare an annual activity plan and present to management concerning financially attractive investments to reduce energy costs
- Establish an energy conservation cell within the firm with management's consent about the mandate and task of the cell
- Initiate activities to improve monitoring and process control to reduce energy costs
- Analyze equipment performance with respect to energy efficiency
- Ensure proper functioning and calibration of instrumentation required to assess level of energy consumption directly or indirectly
- Prepare information material and conduct internal workshops about the topic for other staff
- Improve disaggregating of energy consumption data down to shop level or profit center of a firm
- Establish a methodology how to accurately calculate the specific energy consumption of various products/services or activity of the firm
- Develop and manage training programme for energy efficiency at operating levels
- Coordinate nomination of management personnel to external programs
- Create knowledge bank on sectoral, national and international developments on energy efficiency technology and management system and information denomination
- Develop integrated system of energy efficiency and environmental upgradation
- Wide internal and external networkings
- Coordinate implementation of energy audit/efficiency improvement projects through external agencies
- Establish and/or participate in information exchange with other energy managers of the same sector through association [bee-india.com].

Duties

- Report to BEE and State level Designated Agency once a year. The information with regard to the energy consumed and action taken in the recommendation of the accredited energy auditor, as per BEE format.
- Establish an improved data recording, collection and analysis system to keep track of energy consumption.

- Provide support to Accredited Energy Audit Firm retained by the company for the conduct of energy audit.
- Provide information to BEE as demanded in the Act, and with respect to the tasks given by a mandate, and the job description.
- Prepare a scheme for efficient use of energy and its conservation and implement such scheme keeping in view of the economic stability of the investment in such firm and manner as may be provided in the regulations of the Energy Conservation Act [irieen.com].

22.8.11 Responsibilities and Duties of Energy Auditor

- Conduct internal audit of individual equipment/system once a year
- Submit copy of reports to energy manager with recommendation on action
- Keep record of calibration status of all energy measurement instruments/devices
- Maintain portable tools/instruments required for audit
  - ○ Keep abreast of all codes of practices for energy efficiency testing
  - ○ Training of measurement staff on use of instruments and codes
  - ○ Be a team member of the external audit team
- For ESCO performance contract projects be verified for M & V system and baseline and savings.

REFERENCES

1. Birchfield GS. Energy Management—Now is the Time. Aspen Tech, 2002.
2. Birchfield GS. Aspen Tech, 2002. Workshop on Energy Management and Conservation, Lovraj Kumar Memorial Trust, New Delhi, November 11–12, 2003.
3. Bhatia S. Carbon Capture and Storage: Solution or a Challenge. Hydrocarbon Processing, Nov 2008, p99.
4. Bokare UM. Best Practices on Energy Conservation. Presentation of Petrochemical and Refinery Sectors Task Force Program on Energy Conservation at Reliance Industries Ltd, Navi Mumbai, Aug 29, 2003.
5. Chopra SJ. Refinery for Future. QIP Short-term Course on Advances in Hydrocarbon Engineering, June 23–July 4, 2003, IIT Roorkee, India.
6. EIL, 1982. Energy Conservation in Process Industry, 1982.
7. Energy Conservation in Process Industry. Engineers India Limited, 1992.
8. Energy Information Administration/International Energy Outlook, 2003.
9. Gadalla M, Jobson M and Smith R. Chemical Engineering Progress, April 2003, p44.
10. Ghosh BD and Narayana CSS. Energy and Environment Performance—Two Sides of Same Coin: Outlook and Options for Optimization. Compendium 16th Refinery Technology Meet, organised by Center for High Technology and Indian Oil Corporation, India, held at Kolkata, Feb 17–19, 2011.
11. Hairston D. Trapping Steam. Chemical Engineering, Jan 2003, p23.
12. Harindranath N and Kamath M. Integration of Energy Efficiency Concepts at the Design Stage in Process Industries. Chemical Industry Digest, July 2007, p76.
13. Krishnan V and Verma RP. Energy Conservation in Refineries. Hydrocarbon Technology, Aug 15, 1993, p47.

14. Perry W. Variable Speed Drives Lower Compressed Air Costs. Hydrocarbon Processing, July 2002, p52.

15. Reid JA. Distributed Distillation with Heat Integration. Petroleum Technology, Quarterly, Autumn, 2000, p85.

16. Sil K, Dey GK and Naryana CSS. Benchmarking in Refinery Area. Workshop on Energy Management and Conservation, Lovraj Kumar Memorial Trust, New Delhi, Nov 11–12, 2003.

17. Vardarajan K, Handa SK and Goyal GD. Lovraj Kumar Memorial Annual Workshop, 2009. Managing Carbon Footprints in The Process Industry, organized by EIL India, Petrotech Society, Petrofed, IIT Delhi, IICHE and PTQ international, New Delhi, Nov 26–27, 2009.

18. Vasudeva S. Dr HL Memprial Lecture 2012, Chemcon 2012, organized by NIT Jalandhar, Dec 27, 2012.

19. Verma RP. Energy Conservation in Refineries. Hydrocarbon Technology, Special Issue, Jan 1991, p45.

20. World Energy Council, 2000.

23

Corrosion and Material of Construction in Petroleum Industries

Corrosion is a global phenomenon. It is recognized as one of the most serious problems damaging the economic growth of both developed as well as developing countries. Resulting losses each year are in hundreds of billion dollars. The global corrosion costs is estimated to be roughly USD 2.2 trillion annually which is over 3% of the world GDP [Ramkrishanan, 2013]. India also losses a staggering figure of over USD 45 billion per year due to corrosion [Ramkrishanan, 2013]. The loss in the US is an estimated $300 billion a year. India is losing 80,000 crores per annum on account of corrosion in various sectors. US is loosing more than $276 billion on account of corrosion (Economics Times, Sep 30, 2007). As per Times of India, in India it is ₹ 1.25 lakh crores [Times of India, Oct 2, 2007]. The figure could be 10 times more, because many things go unaccounted. Corrosion is like a cancer cell that often slowly develops for many years and remains undetected and when observed it is too late to mend the damage caused by it. Every year the chemical industries suffer huge loss which is estimated to the tune of 4000 crores.

Corrosion problems have always presented a severe challenge to oil and gas producing operations as it encompasses a broad range of conditions and chemicals. Crude oil contains many corrosive materials like chlorides, organic liquids, water, sulfur compounds and corrosive environment like high temperature. Petroleum industry is incurring huge losses in terms of unscheduled downtime resulting in production losses, reduced equipment life, high maintenance and corrosion control costs, replacement costs, huge loss of properties and sometimes even injury or death. Cost of design, manufacturing and construction and cost of management are two major corrosion costs. The cost of designing and manufacturing includes material selection, such as stainless steel to replace carbon steel, additional material, such as increased wall thickness for corrosion allowance, material used to mitigate or prevent corrosion such as coatings, sealants, corrosion inhibitors, and cathodic protection, and cost of labor and equipment. Corrosion related inspection, corrosion related maintenance, repairs due to corrosion, replacement of corroded parts, inventory components, rehabilitation, and loss of productive time are the cost of management [Koch G, http://www.corrosioncost.com/news/2002/corrosion costs.htm]. Some of the major categories of investment on corrosion are: Coatings, corrosion inhibitors, cathodic protection, non-metallic, corrosion resistant alloys and testing and analyses.

Major areas of concerns in petroleum sector are: Oil and gas exploration and production, petroleum refining, gas and liquid transmission pipelines, gas distribution, hazardous material transport. Environmental concern due to release of pollutants to air, soil or water caused by corrosion leaks are also becoming high consequence events [corrosinoncost.com]. Some of the major corrosives in petroleum industries are naphthenic acids, sulfur compounds, ammonia, carbon monoxide, carbon dioxide, oxygen, chlorides, cyanides, hydrogen, hydrochloric acid, sulfuric acid, nitric acid, phenols, and various organic chemicals, etc. are given in Table 23.1.

Corrosion is a very complex phenomenon. There is no single solution to minimize corrosion. Knowledge of the mechanism of corrosion, factors causing corrosion, variables affecting corrosion as well as compatibility of the conditions of the environment by proper choice of material of construction and good design for greater corrosion protection and corrosion monitoring are of great importance for corrosion control. Preventive measures are important steps in corrosion control.

Table 23.1   Major corrosives in petroleum industry

Plant	Corrosives pollutants
Corrosion in drilling process	Oxygen, moisture, CO_2, H_2S, chloride, salts, organic acid
Natural gas processing	Moisture, CO_2, H_2S, chloride, amine degradation products, impurities in amine system.
Crude oil processing and refining	Naphthenic acid, sulfur compounds, high temperature, chloride, sulfuric acid, HF, caustic soda, MEA, K_2CO_3
Residue upgradation process (Visbreaking, delayed coking, FCC, hydrocracker, thermal cracking processes)	High temperature, naphthenic acid, cyanide, chloride, sulfide, hydrogen sulfide.
Olefin plant (naphtha/gas cracker units)	Inorganic sulfides, mercaptans, CO_2, soluble hydrocarbons, polymerized product, phenolic compounds, cyanide, coke, spent caustic, SO_x, NO_x, hydrocarbons, high temperature and chloride.
Aromatic production unit catalytic reforming	Dissolved organics, chlorides, sulfides in naphtha, high temperature, benzene and its homologous, presence of acids, bases and/or salts in the aromatic hydrocarbons.

23.1 FUNDAMENTALS OF CORROSION

Corrosion is the destruction or deterioration of material (metal or non-metal) because of its interaction with its environment and is derived from the Latin word corrodere which means "to gnaw to pieces". Deterioration of the material is caused by chemical or electrochemical reaction with the environment whereas destruction is mainly because of mechanical wear or abrasion. Corrosion is also sometimes called extractive metallurgy in reverse.

23.1.1 Theory of Corrosion

Mechanism of corrosion has been explained on the basis of free energy, thermodynamics, reaction kinetics and extractive metallurgy [Fontana, 1987]. Corrosion occurs in metals, because most of the metals are not in their natural state until they return to the ore form in which they are found. Corrosion in the plastic occurs because different environments can affect the bonds between the organic molecules making up plastic. The free energy changes determine whether a corrosion reaction will start spontaneously.

23.1.2 Classification of Corrosion

Corrosion of the equipment takes place because of environment condition and fluid used in the equipment. Classification of the corrosion is given in Table 23.2.

23.1.3 Forms of Corrosion

Corrosion may be uniform are localized which may macroscopic or microscopic. Various forms of corrosion are uniform corrosion, pitting corrosion, crevice corrosion, galvanic corrosion, stress corrosion cracking, intergranular corrosion, dealloying or selective corrosion, caustic embrittlement corrosion, fatigue, under deposit attack, filiform or underfilm corrosion, high temperature corrosion, microbiological corrosion, stray current corrosion, erosion corrosion, and fretting corrosion. Brief descriptions of various forms of corrosion are given in Table 23.3.

Table 23.2 Classification of corrosion

Based on mechanism of corrosion process	Chemical corrosion: Uniform or non-uniform Electrochemical corrosion: Galvanic corrosion or bimetallic corrosion, galvanic microcell within a metal, differential aeration cells, uniform corrosion, crevice corrosion, deposit corrosion, water line corrosion.
Based on nature of corrosion	Corrosion can be wet or dry. A liquid moisture is necessary for the wet corrosion while dry corrosion usually involves reaction with high temperature gases.
Based on nature of corrosion	Atmospheric, marine, underground corrosion, biological corrosion, high temperature corrosion, hydrogen cracking, hot sulfide corrosion, metal salt corrosion, carburizing.
Based on corrosion deterioration	General corrosion, localized corrosion, intergranular corrosion.
Based on mechanical factors	Stress corrosion cracking, fatigue, cavitations.

Table 23.3   Types of corrosion

Type of corrosion	Description
Uniform corrosion	This is a surface phenomenon and occurs through uniform attack of all surfaces of the metal exposed to acidic, alkaline, humid or moisture laden environment and is normally characterized by a chemical or electrochemical reaction. Uniform corrosion can be prevented or reduced by proper material of construction, use of inhibitors, surface coating and cathodic protection.
Galvanic corrosion	This type of corrosion occurs when two dissimilar metals are brought into contact in a corrosive or conductive solution resulting corrosion of the metal which is less noble. This type of corrosion can be prevented by selection of combination of metals as close as possible, by insulating two dissimilar metals, applying coating, addition of inhibitors and cathodic protection, etc.
Pitting corrosion	Pitting is a localized form of corrosion and is characterized by surface cavities which can have different shapes. The process of pitting is slow and results in homogeneity on the metal surface, local loss of passivity, mechanical/chemical rupture of protective oxide film. Pitting is accelerated by more acidic or higher temperature conditions. Prevention can be done by proper selection, material of construction (using metals showing fewer tendencies to pitting) and addition of inhibitors.
Crevice corrosion	Intensive localized attack in crevices which exist at lap joints, bolts, rivets and gaskets. Prevention can be done by proper design and operating procedures.
Stress corrosion cracking (SCC)	SCC is caused by the simultaneous presence of tensile stress, a specific corrosive environment. Chloride SCC and sulfide SCC are two forms of SCC. In SCC stressed regions undergo localized attack resulting in hairline cracks. Suitable environment, tensile strength, a sensitive metal, appropriate temperature and pH are important conditions necessary for stress corrosion cracking. Stress corrosion cracking can be prevented or reduced by lowering stress, eliminating the critical environmental species, changing the alloy, applying cathodic protection, adding inhibitors, coating and shot penning.
Caustic embrittlement	This is a form of corrosion occurring in metals in contact with caustic under certain condition by an alkaline environment. The cracks result from the combined action of tensile stress and corrosion.
Hydrogen damage	Hydrogen blistering, hydrogen embrittlement, decarburization, and hydrogen attack are various forms of hydrogen damage. Hydrogen blistering is caused by the diffusion of atomic or nascent hydrogen in the crystal lattice and collection in fissures or cavity. Hydrogen embrittlement is caused by penetration of atomic or nascent hydrogen through the metal structure, resulting in loss of ductility. Carburization and hydrogen attack occur at high

Contd.

Table 23.3   Types of corrosion *(Contd.)*

Type of corrosion	Description
	temperature. Decarburization is the removal of carbon from steel by moist hydrogen and high temperature.
Corrosion fatigue	This is due to tendency of metals and alloys to fracture under repeated cyclic stressing. Fatigue life which is number of cycles needed for failure, is dependent on the stress level. Corrosion fatigue is the cracking of metals resulting from combined action of a corrosive environment and repeated or alternate stress. Corrosion fatigue can be prevented or reduced by eliminating or reducing the stress by use of inhibitors and by use of coating.
Exfoliation	Exfoliation is a severe form of intergranular corrosion. It occurs mostly in heavily rolled or extruded products where the grains are flattened and elongated in the direction of hot working.
Intergranular corrosion	This is a localized form of attack occurring at grain or adjacent to grain boundaries with little or no attack on grain boundaries themselves resulting in loss of strength and ductility.
High temperature corrosion	Selective oxidation of chromium when exposed to low oxygen atmosphere at high temperature. Some of the other form of high temperature corrosion may be oxidation-reduction, sulphidation, carburization and nitriding.
De-alloying or selective on corrosion	Selective removal of one constituent of a metal from the alloy, e.g. dezincification, graphitization.
Stray current corrosion	A form of attack caused by electrical currents.
Filiform of corrosion or Underfilm corrosion	This is a form of corrosion initiated electrolytically due to presence of moisture, oxygen and corrosive ions and results in form of fine trenches under paint, enameled or lacquered surfaces.
Microbiological corrosion	Deterioration of a metal by caused directly or indirectly as a result of the activity of living organism. Living organisms as a result of their influence on anodic and cathodic reactions develops a corrosive condition. This type of corrosion is commonly used in water storage tanks pipe lines and in integral fuel tanks. This can be prevented or substantially reduced by use of biocides.
Erosion corrosion	Deterioration or attack on metal by contact with high velocity liquids resulting in pitting type of corrosion and is characterized in appearance by grooves, gullies, waves, rounded holes and valleys. Resistance to erosion and corrosion both play a part in providing erosion corrosion resistance. Hardness is considered to be a measure of material's erosion resistance Prevention can be accomplished by reducing velocity, using material with better resistance to erosion, proper design and coatings. The second generation duplex alloys, with their relatively highly abrasive solid resistances, provide very good erosion resistance protection.

Contd.

Table 23.3   Types of corrosion *(Contd.)*

Type of corrosion	Description
Cavitation	Cavitation is the formation of growth and collapse of bubbles or cavities in a liquid. This is corrosion of material removed by the formation and collapse of vapor bubbles in a liquid near a metal surface. Various forms of cavitation are traveling cavitation, fixed cavitation, vortex cavitation, vibratory cavitation.
Fretting corrosion	It is a form of wear or damage of material occurring at contact between materials under load subjected to vibration and slip. Fretting destroys the dimensional accuracy of closely fitted parts and increases the susceptibility to fatigue failure.

Source: Fontana, 1987; Schweitzer, 1998; Shreir, 1976; Uhlig, 1963; Chawla and Gupta, 1996; Mall, 1988; Mall, 2007; Das, 1998; Raja Gopalan, 1998; Govinda Ram, 1998; Rayner, 1994; Charkabarty, 2013.

23.1.4 Factors Affecting Corrosion

Various factors affecting corrosion are mentioned below Table 23.4.

Table 23.4   Factors affecting corrosion

Factors pertaining to metals	Place of metals in the e.m.f series, purity of metals, physical state of metals, inclusions, oxide surface behavior of metals.
Factors pertaining to oxidizing medium	
Atmospheric corrosion	Primarily due to moisture, but is accelerated by contaminants such as sulfur compounds, chlorine, sodium chloride, carbon dioxide, oxides of nitrogen, temperature, pH, velocity, concentration, impurities.
Corrosion of metal in immersed liquids	Oxidizing and reducing conditions.
External factors	Differential aeration, effect of light, presence of colloids, presence of bacteria, presence of cathodic metals, existence of stray current, existence of stresses.

Source: Fontana, 1987.

23.1.5 Corrosion Cells

Principal kinds of corrosion cells are classified according to the sources of the driving voltage. These are [Fontana 1987, most.gov.mn]: bimetallic cells, concentration cell, oxygen concentration (differential aeration cells), stress cells and temperature cells.

- *Bimetallic cells:* In bimetallic cell two different metals are coupled; sometime one electrode may be nonmetal such as mill scale, graphite or an oxide film.
- *Concentration cell:* Concentration cell exist when two electrodes of the same metal lie in electrolytes of different compositions or concentrations.
- *Oxygen concentration (differential aeration cells):* In diiferential aeration cell there is difference in the amount of oxygen reaching the surfaces of the two electrodes.

- *Stress cells:* In stress cells, metal under stress is coupled to unstressed metal in a uniform electrolyte.
- *Temperature cells:* Temperature cell is developed when one electrode is at a higher temperature than the other.

Basic methods of combating corrosion for external pipe corrosion are as follows:
- Substitution of another material for the metal attacked
- Modification of the environment
- Separation of the metal from the environment: Protective coating cathode protection
- Cathodic protection: Combining coating with cathodic protection makes the best method for protecting a pipeline.

23.2 CORROSION MONITORING

Corrosion monitoring which involves the direct application of corrosion measurement techniques to industrial plants and structures for the purpose of diagnosis or corrosion control are very important aspects of corrosion prevention and control. The most important corrosion monitoring methods are: Coupon method, electric resistance method, potential measurements, polarization resistance method, impedance method and hydrogen permeation method. On destruction techniques such as ultrasonic, eddy current and radiography have also been used for monitoring corrosion. In coupon method, weighed specimens are exposed to the environment for a specified period and the loss of metal is measured. Corrosion rate is commonly expressed in terms of mm per year.

23.3 VARIOUS FORMS OF CORROSION IN PETROLEUM INDUSTRIES

Corrosion in oil and gas processing is one of the serious problems in the oil and gas sector. The complexity of the problem is increasing due to use of more and heavier feedstocks containing higher sulfur and impurities. Although use of heavier crude and high acid crudes improve profitability, however, presence of sulfur and naphthenic acid cause corrosion problem at various stages of refinery operation and refineries are investing huge amount in advanced in upgrading materials of construction. Corrosion in pipelines, storage vessels, pumps, valves, heat exchangers, condensers and structures, etc., are the causes of major concern in petroleum industries due to the presence of chloride, organic and inorganic sulfur compounds, naphthenic acids and moisture (which are always present in crude oil and natural gas), elevated temperature, high pressures, temperature cycling during normal operation, startup and shut down [Mall, 2007]. Although various contaminants are treated for removal in various stages, however, traces of these remain and they split, combine or convert into numerous corrosive combinations. Further the corrosion problems are compounded because of high temperature and pressure at various stages of operation. In refining and petrochemical industries, it is necessary to guard against carburizing, nitriding, oxidation, sulphidation, hydrogen attack, intergranular corrosion and stress corrosion cracking of stainless steels and other high alloys [Chawla and Gupta, 1996]. Stress corrosion is a common phenomenon in petroleum and petrochemical complexes. Stress corrosion occurs when certain metals are exposed under a tensile stress to specific environments and failure can occur rapidly without warning [Setterlund, 1991].

23.3.1 Sulfide Corrosion

Organic sulfur compounds such as mercaptans, polysulfides, thiophenes as well as elemental sulfur are present in all crudes in various concentrations and can cause heavy corrosion even at lower temperature, but are especially aggressive in the refining and petrochemical operations above 260–280 °C. Hot sulfide corrosion and wet sulfide corrosion which are also common in petroleum due to presence of hydrogen sulfide and it continues to be a major concern that can impede efficient refinery operation. It is formed either by thermal decomposition of organic sulfur compounds or is present originally in the crude and natural gases. The sulfur compounds can cause corrosion at lower temperature, but are especially aggressive in refining operation above 260 °C. At temperature above 500–550 °C, H_2S becomes corrosive in vapor phase and reacts directly with iron. During wet sulfide corrosion, H_2S is absorbed in the water where it dissociates to react with iron. Typically sulfur contents in crude oils are in the range of 0.5–2.5 %wt, but it can be as high as 4% [Batra et al., 1993]. Mercaptans and organic sulfides comprise the majority of sulfur species. H_2S is most active corrosive sulfur compound. H_2S can corrode steel below the dew point of water. H_2S is absorbed into water where it dissociates and reacts with iron according to following mechanism [Batra et al., 1993].

$$H_2S \Rightarrow H^+ + HS^-$$
$$Fe^{++} + HS^- \Rightarrow (FeSH)^+$$
$$(FeSH)^+ + HS^- \Rightarrow HSFeSH$$

Dissolved iron is produced through acidic corrosion in water. The polymeric iron sulfide formed is not soluble in hydrocarbons and will lie down on the metal surface to provide protection film. At high temperature, especially in furnaces and transfer lines, the presence of NA increases severity of sulfide corrosion.

23.3.2 Naphthenic Acid Corrosion

Naphthenic acid corrosion (NAC) is an another problem in petroleum. The corrosive action of organic acid is complicated because as per rule these acids are not handled in isolation, but rather as process mixtures with inorganic acids, organic solvent and salts as well in combination with other organic acid [Schillmoller, 1997]. Naphthenic acid (NA) are most active at their boiling points. ASTM total acid number (TAN) and naphthenic acid number (NAT) are measurements of NA. There are more than 1,500 types of naphthenic acids in atypical crude oil with average molecular weight ranging from about 200 to approximately 400. Most are believed to have the chemical formula R $(CH_2)nCOOH$, where R is cyclopentane ring and n is typically greater than 12 [Nagi-Hanspal et al., 2013]. Crude oils with TAN higher than 0.5 and cuts with a TAN higher than 1 potentially corrosive between the temperatures 450 °F and 750 °F. Velocity and more importantly wall shear stress is main parameter affecting NAC [http://setlaboratories.com/nac.htm]. The corrosion of the naphthenic acids begins at 225–250 °C and increases with higher temperatures until about 320 °C. Corrosion is reduced at higher temperatures [Danllov, 1981].

Organic acids may be present in crude oils, naphthenic or saturated ring acids. Although naphthenic acid corrosion is a major source of corrosion in atmospheric and vacuum distillation units, other units in naphthenic acid corrosion occurs are:

- Furnace coils usually at exit
- Transfer pipe
- Vacuum columns
- Site stream cooler
- Pumps.

At higher temperature naphthenic acid may increase the severity of the sulfide corrosion. Presumably these organic acids disrupt sulfide film thereby promoting sulfide corrosion on alloys that would normally be resistant to this attack [Kane and Cayard, 1995]. Velocity is a key parameter in naphthenic acid corrosion and in some cases very high corrosion rate even at very low levels of naphthenic acid content is possible when combined with high temperature and velocity. Naphthenic acid and sulfur content are two most important variables which affect the naphthenic acid corrosion. Austenite stainless steels containing molybdenum have provided good resistance to naphthenic acid corrosion. Velocity in the furnace tubes should be kept to a minimum to reduce the risk of naphthenic acid corrosion. Low chromium alloys (12–17% Cr SS) and type 304 SS alloys are generally not reliable for this purpose. Naphthenic acid corrosion and chloride corrosion are affected by the presence of hydrogen sulfide. Naphthenic acid corrosion is worse for low H_2S concentrations. There are four options for managing corrosion: Upgrading material of construction, blending high TAN crudes with low TAN crudes, process control and use of inhibitors [Nagi-Hanspal et al., 2013].

23.3.3 CO_2 Corrosion

Although CO_2 in natural gas stream is not corrosive, if the gas is moisture free, however, in the presence of water CO_2 reacts to form carbonic acid. CO_2 corrosion is directly related to the partial pressure of CO_2. General rules of thumb relating to the CO_2 partial pressure to corrosion are: (i) Low corrosion rates with CO_2 partial pressure below 7 psia; (ii) possible high corrosion rates with a CO_2 partial pressure between 7–15 psia; (iii) high corrosion rates of CO_2 partial pressure above 15 psia; (iv) CO_2 corrosion is directly related to gas temperature [Kresse, 1987].

Dissolved CO_2 in water or aqueous solution causes severe corrosion of pipeline steel and process equipments used in the extraction, production and transportation of oil and gas in the petroleum industry [Das and Khanna, 2004].

CO_2 is not corrosive, if gas is clean and dry; however, in the presence of water CO_2 reacts to form carbonic acid:

$$CO_2 + H_2O \longrightarrow H_2CO_3$$

The corrosion rate of iron increases rapidly in an acidic atmosphere. Iron reacts with the carbonic acid to form iron carbonate.

$$2Fe^{++} + H_2CO_3 \longrightarrow Fe_2CO_3 + 2H^+$$

CO_2 corrosion is directly related to the partial pressure of CO_2 and increases with increase in partial pressure of CO_2. CO_2 corrosion is also directly related to the gas temperature. Higher the gas temperature greater is the CO_2 corrosion rate. Internal corrosion can be controlled by removing one of the active ingredients, water and air or by adding an interior material which will make the steel inactive [Lyons, 2004]. Dehydration is an effective system for controlling internal pipe

corrosion. Another method commonly used for control of internal corrosion of pipelines is the use of inhibitory materials.

23.3.4 Microbial Corrosion

Microbiologically influenced corrosion occurs in storage and operating systems for metal exposed to various waters, process chemicals, crude and distillate fuels [Little et al., 1998]. Various parameters that influence microbiologically influenced corrosion within an operating system include environmental temperature, nutrients, chloride concentration, operating procedure particularly those related to flow, construction materials and system design. Some of the important organisms responsible for microbiologically influenced bacteria are sulfate reducing (*Desulfovirodesulfurisericans*), sulfur oxidizing (*thio bnacillus thio parns*) and iron bacteria (*Crenothriux and Leptothrix*) [Little et al., 1998].

23.3.5 Stress Corrosion Cracking

Some of the causes of stress corrosion cracking are environmental factors: Aqueous medium, chloride, carbonate, bicarbonate, ammonia, NaOH, acids, H_2S, sea water, nitrate solution; metallurgical factors: Average chemical composition, percentage orientation of grains, composition and distribution of precipitates, dislocation, interactions, progress of phase transformation.

23.3.6 Corrosion Due to Moisture/Water

In the crude oil, handling the presence of water is the major cause of corrosion. A film of liquid water adheres to the pipeline surface and oxygen is available from dissolved air in the product. The solubility of air in products varies, but some oxygen is always present [most.gov.mn]. There are wide varieties of experiences with corrosion in pipelines carrying crude oil. Most of crude oils contain oil-well brine which is an excellent electrolyte to promote corrosion. Some oils are paraffinic in nature and deposit a protective layer of paraffin on the pipe wall.

The amount of corrosion is often related to the amount and composition of sediment. High velocity of flow tends to sweep sediment out of pipeline while low velocity allows it to settle on the bottom of the pipe. When the sediment settles it shields the pipe and pitting tends to take place beneath the sediment. When the oil is sour, the sediment contains iron sulphate. This aggravates the corrosion because in addition to the shielding effect the iron surface is strongly cathodic to steel. Thus the iron surface and steel produce a battery type action which destroys the steel at a rapid rate.

Amount of sediment and water should be controlled. Most cross country pipelines have sufficient flow velocity to keep a moderate amount of sediment under suspension. Low velocity gathering lines may give a great deal of trouble especially when they are collecting sour crudes. Chloride, sulfur compounds, naphthenic acids, O_2, CO_2 and H_2 are the main corrosive components of crude oil and natural gas.

23.3.7 Chloride Corrosion and Chlorine Corrosion

Chloride corrosion is an another problem in petroleum industry. Chloride corrosion is caused by hydrogen chloride which is formed from hydrolysis of chloride salts present in crude oil. The majority of chloride salts present are magnesium, calcium and sodium salts. Sodium chloride is thermally and hydrolytically stable to about 428 °C and it does not contribute significantly to the hydrogen chloride. However, hydrolysis of calcium chloride and magnesium chloride is significant and it begins at temperatures above 121°C and accelerates at temperatures above 177–215 °C [Athar, 2002].

Halogen especially dissociated chlorine (chloride ions) frequently contaminate refinery and petrochemical streams and can lead to serious corrosion, e.g. chloride stress corrosion cracking of austenite steel in aqueous solution containing quite low concentration of chloride [Tillack and Guthrie, 1999]. Gaseous chlorine at low temperature and in the absence of moisture is not severely corrosive. Dry HCl behaves similar as chlorine. Hydrochloric acid is typically reducing acid through its entire concentration range. Its strong acidic character is harmful to steel. During the manufacture of vinyl chloride; monomer, chlorine, hydrogen chloride and hydrochloric acid are present at various stages of operation and result in severe corrosion, if proper material of construction is not there.

23.3.8 High Temperature Corrosion

High temperature corrosion is another type of corrosion which is common in various stages of operation especially in steam cracking, reformer, reactors, distillation column, fluid catalytic cracking units, furnaces, heat exchangers and condensers operating at high temperature. Some of the constituents like carbon, hydrogen, nitrogen, halogens, sulfur, ash and molten ash can work separately or synergistically to increase the severity of corrosion by a number of degradation mechanisms like carburization, hydrogen attack, nitriding, hot ash and salt corrosion. Carburization can occur when metals are exposed to carbon monoxide, methane, ethane or other hydrocarbons at elevated temperature [Tillack and Guthrie, 1999; Kane and Cayard, 1995]. A major problem with carburization is that it is non-uniform and unpredictable. Carburization has been the major cause of ethylene furnace tube failure in the thermal cracker units because of high temperature and high carbon potential associated with ethane, propane, naphtha and other hydrocarbons [Tillack and Guthrie, 1999]. The rate of carburization is a process-driven and is a function of tube metallurgy and roughly doubles for every 37.8 °C increase in tube metal temperature. Carbon penetration during cracking and decoking reduces durability and making the tube much more susceptible to stress damage from either thermal cycling or bending moments. During carburization, carbon picks up increases the volume of metal and the coefficient of expansion resulting in strong internal stresses that cause premature failures of the tubing. Carburization also results in a change of mechanical properties mainly a reduction of creep strength and ductility and aging effect [Schillimoller, 1984].

23.4 CORROSION IN OIL AND GAS DRILLING, PRODUCTION AND PETROLEUM REFINERIES

23.4.1 Corrosion in Oil and Gas Drilling and Production

Corrosion in oil and gas production is one of the serious problems and resulting in huge loss in form of corrosion. Major corrosion costs come in the form of either

premature deterioration or failure resulting in maintenance, repair replacement of damaged equipment and pipelines, cost of corrosion inhibitors, protective coatings of offshore structure, vessels, and cathodic protection of pipelines. Major corrosives are CO_2, H_2S, moisture, chloride, sediments, organic acids, salts and microorganisms. Corrosion in various costs is the major cause of drill pipe failures which significantly add to drilling cost. The recent trends toward drilling of deeper wells, use of higher strength steel, presence of higher stresses and use of lower pH drilling fluids contribute to increased susceptibility of metal failures [Bertness and Chilingarian, 1981]. Corrosion in oil and gas production units may be in three categories: (i) Internal corrosion: Caused by the produced fluids and gases, (ii) external corrosion: Caused by exposure to ground water or sea water and (iii) atmospheric corrosion: Caused by salt spray and weathering offshore.

Downhole tubing, surface pipelines, pressure vessels, and storage tanks in oil and gas production are subject to internal corrosion by water, which is enhanced by the presence of CO_2 and H_2S. Internal corrosion is a major cost factor [waterjetting.com]. Hydrogen embrillment (sulfide cracking) and corrosion fatigue are two types of cracking which are associated with drilling and producing environment [Bertness and Chilingarian, 1981].

23.4.2 Gas and Oil Pipeline Corrosion

Corrosion in pipeline carrying crude oil and natural gas has been also the cause of major concern in petroleum and petrochemical industries resulting in loss of million dollars all over the world. Large amount of natural gas and crude oil and petroleum products are being transported through pipelines. Corrosion in pipeline can be either internal or external. Based on the laboratory analysis of different pipelines coatings presently available, it is found that the pipeline coatings offer the optimum protection in minimizing the risk of external pipeline corrosion. While hydrocarbons alone are not corrosive impurities such as sulfur compounds, CO_2, bicarbonates, chlorides and chemicals added during processing can cause corrosion [Setterlund, 1991]. Corrosion in pipeline may be either internal or external. Carbonate-bicarbonate corrosion has been identified as the environmental species, responsible for stress corrosion cracking in pipeline carrying natural gas [Christman, 1987]. External pipeline corrosion takes place by either direct soil corrosion or by stray current attack. Direct soil corrosion involves the existence of corrosion cells on the surface of the pipe. Principal kinds of corrosion cells are bimetallic cells, concentration cells, oxygen concentration or differential cells and temperature cells. In stray current corrosion externally driven electric currents are responsible for corrosion. Corrosion of sour crude increases with temperature and increased sulfur content.

Large amount of natural gas and crude oil and petroleum products are being transported through pipelines. Corrosion in pipeline carrying crude oil and gases has been a cause of major concern due to loss in millions of dollar each year. Corrosion in pipeline can be either internal or external. Based on the laboratory analysis of different pipeline coatings presently available, it is found that the pipeline coatings offer the optimum protection in minimizing the risk of external pipeline corrosion.

23.4.3 Crude Oil Desalting

Crude oil containing bottom sediments, salts, water which is treated for removal of salts. The effluent water high amount of dissolved salts which is at about 150 °C and there is always risk of stress corrosion cracking and pitting. Carbon steel is not suitable for such corrosive environment.

23.4.4 Crude Oil Distillation Column

Corrosion in crude oil distillation units represents significant portion of refining costs as a result of lost production, inefficient operation and high maintenance and corrosion control chemicals costs [Batra, 1993]. Corrosion in the crude unit overhead is primarily due to acid attack at the initial water condensation point [Payne, 2012]. Chlorides, sulfur compounds, naphthenic acid are the major corrosives. Magnesium chloride present in crude oil easily hydrolyzes at temperature above 120 °C to form HCl. Electrochemical corrosion occurs due to presence of H_2S-HCl. The presence of naphthenic in crude oil further complicates the corrosion problem in distillation. H_2S-HCl present at the overhead vapors system of fractionation towers, affecting the condensers, coolers, reflux tanks, vapor lines and gasoline reflux pipes and pumps [Danllov, 1981]. The naphthenic acid corrosion is present in the tubes and return bends of the furnace. The attack continues into the flash section of the main fractionation tower. And on trays shell corresponding to the kerosene—gas oil section of the fractionation tower [Danllov, 1981].

Some of the majors for corrosion prevention and control in crude oil desalting, caustic injection, neutralization of crude and vacuum column overhead vapors and condensed water, water washing in crude column overheads and use of corrosion inhibitors [Batra, 1993]. Controlling corrosion in the overhead condensing of crude atmospheric distillation is a big challenge for refiners—a typical treatment strategy involves a delicate balancing act of neutralizing acids in condensed waters with ammonia and amines while avoiding the formation of corrosive salts via a vapor phase reaction with hydrogen chloride [Lack and Harrell, 2013].

23.4.5 FCC Corrosion

Efficiency of fluid catalytic cracking vapor recovery units can be significantly reduced by corrosion and fouling. Steel and copper alloys corrosion, hydrogen blistering of steel in a FCC fractionator and vapor recovery section are traceable mainly to contaminants generated in the reactor from the nitrogen and sulfur bearing compounds in the feedstock [Walker, 1984]. Corrosion in fluid catalytic cracking unit is mainly because of the presence of nitrogen and sulfur bearing compounds which are converted to ammonia and cyanide in case of nitrogen compounds and to hydrogen sulfide in case of sulfur compounds [Walker, 1984]. Apart from ammonia, cyanide, hydrogen sulfide, some CO_2, chloride and organic acids may be also present. The cracked nitrogen compounds are converted to ammonia and cyanides and the cracked sulfur compounds form hydrogen sulfide. The severity of corrosion and whether it can progress to the point of hydrogen blistering depends very much on the temperature and the activity of the catalyst within the reactor and the feedstock contaminants levels [Walker, 1984]. Some of the corrosion reactions in FCC are [Walker, 1984]:

Steel corrosion reactions:

$$Fe + 2HS^- \longrightarrow FeS + S^{-2} + 2H^0$$
$$Fe + 6CN^- \longrightarrow Fe(CN)_6^{-4} + S^{-2}$$

Copper corrosion reaction:

$$Cu^{+2} + 4NH_4^+ \longrightarrow Cu(NH_3)_4^{+2} + 4H^+$$

Effective water wash is a key to any corrosion control program. Ideally, an effective water wash will at least move vapor phase corrosive into liquid phase, where they are more easily treated [Strong and Wihelm, 1991]. Use of polysulfide and corrosion inhibitors can reduce FCC corrosion [Strong and Wihelm, 1991; Walker, 1984]. Either sodium or ammonium polysulfide can be used. However, ammonium polysulfide is more common. Polysulfide converts cyanides to thiocyanates by the reaction [Walker, 1984].

$$S_x^{-2} \text{ (polysulfide)} + CN^- \longrightarrow S_{x-1}^{-2} + SCN^-$$

Most FCC inhibitors are filming which provide a thin barrier of organic material on the inside surface of equipment at risk [McNab and Treseder, 1971].

23.4.6 Amine Absorber Corrosion

Corrosion and cracking in gas processing systems using amines is very common phenomenon and corrosive conditions in gas treating system can result from poor design and operating practices. Corrosivity of steel in amine absorber is related to the specific amine chemicals used along with specific operating conditions, i.e. temperature, velocity, amine and gas loading [Kane and Cayard, 1995]. Monoethanol amine has been used since long for CO_2 and H_2S removal. Corrosion in alkanolamine gas treatment plants results in unscheduled down time, production losses, reduced equipment life and even injury [DuPart, 1993]. Both acid gas types, ratio of H_2S to CO_2 have impact on corrosion rate [McNab and Treseder, 1971]. Corrosion in mono-ethanol amine is a serious problem in amine absorbing columns. However, corrosion and cracking in gas processing monoethanol amine for CO_2 and H_2S removal has been a cause of major concern because of severe corrosivity to carbon alloy steels commonly used material of construction. Corrosive condition in gas treating system can result from poor design, operating practices (too high flow rate or changes in flow direction) and from contamination of amine solution with sulfide and CO_2 [Kane and Cayard, 1995]. Three major parameters which affect the carbon steel corrosion in amine system are pH, temperature and velocity. Corrosivity increases with decrease in pH over the pH range of 9–12. With decrease in pH localized corrosion increases. With increase in temperature and velocity, corrosion rate is increased [Richert, 1988]. Corrosion is most severe in rich amine systems in the area of high velocity and/or turbulence where there is rich amine solution [Kane and Cayard, 1995]. Corrosivity of steel in amine absorber units is related to the specific amine chemicals used along with specific operating conditions (temperature, velocity, amine and gas loadings) [Kane and Cayard, 1995].

Hydrocarbon oxidation plants are especially prone to acid formation since many organic species are readily oxidized to organic acids [Bennion, 1977]. In aqueous solution of organic acids the presence helps to maintain passivity and keep corrosion rules low [Bennion, 1977]. The presence of chloride increases the liabilities to pitting, stress corrosion cracking which are generally favored by acidic conditions.

23.4.7 Steam and Water Line Corrosion

Corrosion of steam and condensate line is also a serious problem in petrochemical industries. The two major causes of corrosion in steam and condensate line—oxygen which results in pitting and a low pH which gives rise to generalized thinning of piping. Formation of low pH condensate is due to carbonic acid which is formed by reaction of CO_2 with condensed steam. CO_2 formation takes place due to break down of bicarbonate and carbonates. Deposit and corrosion occur throughout the entire steam generating system. Boiler tubes with the highest heat transfer are the common location for deposit of contaminants from feed water and returning condensate [Huchler, 1998].

Cooling water corrosion is also common in petroleum and petrochemical complex as huge amount of cooling water is used at various stages of operation in chemical and petrochemical industries. Cooling water both once through and recirculation is responsible for both corrosion and scaling of piping and exchangers. Crude refining operation in the coastal area forces additional problem of seawater corrosion, pitting, uniform corrosion, stress corrosion cracking, and tuberculation. Important factors which influence corrosion in cooling water system are dissolved oxygen, temperature, velocity, pH and dissolved solids.

23.4.8 Corrosion in Storage Tanks

Corrosion of storage tank components—the roof, shell, bottom plates and foundation may occur due to atmospheric corrosion, underside corrosion due to electrochemical process. Unlike atmospheric corrosion of visible equipment parts and accessories, undesirable corrosion often remains hidden and unnoticed until leaks developed [Habiby et al., 2003]. Electrochemical corrosion occurs at the bottom of tanks due to anodic and cathodic reactions.

23.4.9 Corrosion Under Insulation

Corrosion under insulation is one of the costliest avoidable problems facing the hydrocarbon processing industry and has been also a cause of major concern [Hanratty, 2013]. Corrosion under insulation afflicts refineries specifically the steel piping, storage tanks, container vessels and other process equipment [Hanratty, 2013].

Corrosion takes place under insulation can be caused either by the insulation itself or by improper insulation. Various types of corrosion under insulation can be—acidic corrosion, chloride corrosion and galvanic corrosion.

23.4.10 Pipeline Corrosion

23.4.10.1 Internal Pipeline Corrosion

Although dry crude oil and gases with normal additives are non-corrosive, however, due to presence of water and air they become corrosive. Carbonate-bicarbonate has been identified as the environmental species responsible for the stress corrosion cracking in the pipeline [Christman and Beavers, 1987].

23.4.10.2 External Pipeline Corrosion

An operating pipeline has a variety of metal surfaces exposed to corrosion. The attack in metal surface in contact with the soil takes place. This may be due to either direct soil corrosion resulting from the existence of corrosion cells on the surface of pipe or a stray current attack.

23.5 CORROSION PREVENTION AND CONTROL

Although corrosion is unavoidable, however, its scope and severity can be minimized. Some of the basic approaches for corrosion control are:

- Selecting corrosion resistant material of construction
- Use of corrosion inhibitor
- Isolating the material from the corrosive atmosphere
- Coating and linings
- Altering the environment through process changes
- Employing electrochemical control—cathodic and anodic protection
- Designing and fabrication to minimize localized corrosion
- Corrosion monitoring.

In almost situations, corrosion can be managed, slowed or even stopped with the use of corrosion prevention measures [Ramkrishanan, 2013]. Corrosion control can be achieved by effective corrosion management. This can be achieved by using preventive strategies which include: Increase in awareness of large corrosion costs, change in misconception that nothing can be done about corrosion, improving education and training of staff regarding risk and hazards of corrosion and importance of corrosion control, applying advance design practices for better corrosion management, better material of construction without compromising with cost, advance life prediction, performance assessment methods and implementation of advance corrosion technology through research, development and implementation [waterjetting.com].

23.5.1 Selecting Corrosion Resistant Material of Construction

Metallurgy choices have expanded significantly for process equipment and pumps used for handling difficult corrosive fluids [Rayner, 1994]. Selection of material of construction is one of the very important aspects of corrosion mitigation and should be done based on sustained maximum normal operating conditions (stream composition, temperature, pressure and velocity) with due considerations for short-term transient conditions or alternate operations (e.g. catalyst regeneration, no flow, power outage, steam out, chemical cleaning, etc.) such as start up, shut down, upset and emergency conditions [Narain, 2000]. Material selection and corrosion control requirements should be based on specified design life. Material selection should be based on anticipated corrosion or material degradation rate, material availability, fabricability, maintainability, maintenance cost and availability of spare parts [Narain, 2000]. Material selection is often guided by standards which promote safety, reliability, productivity and efficiency throughout industry. ASME codes and standards provide technical definitions, guidelines and instructions for designers, manufacturers and users [Picciotti and Picciotti, 2006].

There is an increasing number of materials available to meet today's corrosion needs, the selected material may be metal, alloy, plastic, elastomeric, ceramic or combination of two or more metals. It may be noted that corrosion resistance control may not involve the use of corrosion resistance alloy materials. Often adequate life can be obtained in corrosion services with carbon steel piping in conjunction with control of process and operating variables [Setterlund, 1991]. The principal selection criteria for material of construction in petroleum industries are mechanical properties, corrosion resistance, stability of properties, sensitization, fabricability, availability, cost, effective service life [Brown, 1997].

Some of the major corrosion resistance alloys available are—Fe based 18–8 austenitic SS alloys, high performance austenitic alloys, Ni based general purpose alloys, 6% Mo super austenitic SS alloys, alloy 31, Ni based special alloys, nickel based high performance alloys and chromium based high performance wrought super austenitic alloys [Agarwal, 1999]. Super austenitic stainless steels with 6% molybdenum offer enhanced resistance to chlorides and non-oxidizing acids such as sulfuric acid and phosphoric acid. These alloys also exhibit enhanced resistant to sulfide stress cracking (hydrogen embrittlement) and chloride SSC in severe sore brine service sometimes to cracking. Costs are well-below those of competing high nickel alloys [Hibner and Fende, 1999]. These alloys exhibit enhance resistance to sulfide stress cracking and chloride stress corrosion cracking. Some of the alloys in this group are 25–6 Mo, Al-6XN (NO8307), 254 SMO (S31254) and 1925 Nn Mo. Some of the major alloying element in these alloys are Ni (17–26%), Cr (19–22%), Mo (6–7%), Cu (0.5–1.55%), N (0.15–0.25%).

Ni-Cr-Fe alloys are extensively used in petroleum refining and petrochemicals for both liquid and gaseous low temperature corrosion resistance. The austenite stainless steel has good resistance to oxidation in high temperature services and may be used in conditions where steel would rapidly be converted to oxide scales and the structure would loose its integrity [Bennion, 1977]. Alloys for oil field environment must resist sulfur stress corrosion cracking in presence of hydrogen sulfides and water in combination with a tensile strength [Brown, 1997]. Some of the established heat resistant material alloys are 304H, 316H, 321H, 347H, 309H, 310H, 85H, 253MA, 330, 800H, 600, 601, 617, 625, in which major alloying elements are chromium, nickel with small amount of some of the alloying elements like molybdenum, titanium, cesium, nitrogen and niobium. Newer heat resistance wrought alloys for refining and petrochemical applications are 803, HK4M, HPM, 55M, 602CA with chromium and nickel as major alloying elements [Tillack and Guthrie, 1999]. Table 23.5 shows the established heat resistant alloys for refining/petrochemical applications [Tillack and Guthrie, 1999]. Table 23.6 shows the newer heat resistant wrought alloys for refining/petrochemical applications [Tillack and Guthrie, 1999]. These alloys also contain some of the alloying elements like cobalt, zirconium, tungsten, niobium, nitrogen, cerium and boron. Chromium and molybdenum improve localized corrosion whereas along with nitrogen enhances the chlorides stress corrosion cracking. Presence of nitrogen improves localized corrosion resistance. Increased chromium and molybdenum improve the resistance of stainless steels to pitting and crevice attack. Stainless steel with increased molybdenum has provided improved resistance to chloride corrosion cracking in environment such as aqueous sodium chloride [Brown, 1997]. Sulfide stress corrosion is a major cause in petroleum and petrochemical industries due to presence of hydrogen sulfide. High strength, highly alloyed materials such as nickel base alloys are susceptible to sulfide stress corrosion cracking when coupled to iron, indicating that failures are the result of hydrogen charging of metal during the corrosion process [Brown, 1997]. Alloy 600, a high nickel alloy offers excellent resistant to caustic and chlorides in presence of sulfur compounds, even at high temperature and pressures [Picciotti and Picciotti, 2006].

Table 23.5   Established heat resistant alloys for refining/petrochemical applications

Alloy	Cr	Ni	Fe	Co	C	Si	Mn	Al	Other
304H	19	9	Balance	–	0.07	0.75	–	–	Mo 2.5
316H	17	12	Balance	–	0.07	0.75	2	–	Ti × C + N
321H	18	10.5	Balance	–	0.07	0.75	2	–	Nb8 × C
347H	18	10.5	Balance	–	0.07	0.75	2	–	–
309H	23	13.5	Balance	–	0.07	0.75	2	–	–
310H	25	20.5	Balance	–	0.07	0.75	2	–	–
85H	19	15	Balance	–	0.20	3.5	0.8	1	–
253MA	21	11	Balance	–	0.08	1.7	0.6	–	N O.17, Ce 0.04
330	19	35	Balance	–	0.05	1.2	1.5	–	–
800H	20	31	48	–	0.08	0.3	0.8	0.3	Ti 0.3
600	16	76	7	–	0.04	0.2	0.2	–	–
601	23	61	14	–	0.04	0.2	0.2	–	–
617	22	52	2	12	0.06	0.5	0.5	1.2	Mo 9, Ti 0.5
625	22	61	2	–	0.05	–	0.2	–	Mo 9, Nb 3.6

Courtesy: Chemical Engg. Progress (CEP, Feb 1999, p59 and Nickel Development Institute, Toronto, Canada).

Table 23.6   Newer heat resistant wrought alloys for refining/petrochemical application

Alloy	Cr	Ni	Fe	Co	C	Si	Ti	Al	Other
803	27	34	Balance	–	0.08	0.3	0.4	0.4	–
HK4M	25	25	Balance	0.25	0.75	0.4	0.4	–	B 0.004
HPM	25	38	Balance	0.15	1.7	0.4	–	–	Mo 2, Zr 0.05, B 0.01
HR120	25	37	Balance	1	0.05	0.6	0.1	0.1	W 2, Mo 2, Nb 7, B 0.004, N 0.2
HR160	28	37	2	29	0.05	2.7	0.45	–	–
AC66	27	32	41	–	0.05	–	–	–	Nb 0.8, Ce 0.06
617LCF	22	52	1.5	12.5	0.08	0.2	0.3	1.2	Mo 9
45TM	27	47	23	-	0.08	2.7	-	-	N 0.08, Re 0.10
602CA	25	63	9.5	-	0.18	-	0.15	2	Y 0.8, Zr 0.07

Courtesy: Chemical Engg. Progress (CEP, Feb 1999, p59 and Nickel Development Institute, Toronto, Canada).

Some of the measures for control of chloride corrosion in crude oil processing units are proper selection of material of construction, crude desalting with water, neutralization of crude and vacuum column overhead, water washing in crude column overhead system and use of corrosion inhibitor [Baton et al., 1993]. Filming amine corrosion inhibitors has been recommended for chloride corrosion control. Alloy steels containing chromium have good corrosion resistance to high temperature sulfur and hydrogen attack.

Sulfuric acid is widely used in petroleum and petrochemical industries in various processes. Resistance of alloys to concentrated sulfuric acid increases with increasing chromium, molybdenum, copper and silicon contents. Corrosiveness of sulfuric acid is highly dependent upon concentration, temperature and acid velocity. Alloy 20 and similar alloys can handle sulfuric acid at ambient and slightly elevated temperatures throughout the concentration range including oleum. High nickel-chromium-molybdenum alloys such as C-276, C-22 and alloy 59 can handle sulfuric acid at all concentration. Tetrafluoroethylene (TFE), fluorinated ethylene propylene (FEP) and perfluoro alkoxy polymer materials such as teflon [Muller, 3006] has resistance to all acid concentration. Nickel, chromium, molybdenum, copper and silicon are the most important of the elements that enhance the corrosion resistance of alloys in sulfuric acid service [Scillmoller, 1997].

Resistance of alloys to concentrated sulfuric acid increases with increase in chromium, molybdenum, copper is highly dependent upon concentration, temperature, acid velocity and acid manufacturing. Alloy 20 and similar alloys can handle sulfuric acid at ambient and slightly elevated temperatures throughout the concentration range including oleum. High nickel-chromium-molybdenum alloy such as C-276, C-22 and alloy-59 can handle sulfuric acid at all concentration. Alloy 20Cb-3 an enriched grade of stainless steel can be used with dilute sulfuric acid (up to 30%) at elevated temperature [Picciotti and Picciotti, 2006].

Resistance to high temperature halide increases with increasing level of both nickel and chromium. Minimizing Mo and W helps the resistance of alloy to oxidizing halogen corrosion [Tillack and Guthrie, 1999]. Zirconium is used in many process applications due to its resistance to corrosion, including pitting, crevice corrosion and stress corrosion cracking. Zirconium can withstand in corrosive environment due to organic acid [Yau and Bird, 1992]. Titanium has good resistance in SO_2, H_2S and nitrogen and typical services in refinery include regenerators, reboilers and overhead condensers. In commercial high temperature operation up to 750 °C the most commonly used material of construction are carbon steel (1% Cr) and Cr-Mo steels (1–12% Cr and 0–1% Mo) [Kane and Cayard, 1995]. For high temperature application up to 1,500 °C, it is common to use Cr-Ni-Mo steel which contains 12–25% Ni. These materials offer better high temperature strength by virtue of their austenitic structure and high nickel contents. Increase in chromium and molybdenum improves the resistance of stainless steels to pitting and crevice attack. Resistance to chloride stress corrosion cracking is strongly affected by nickel content and is increased from a residual value to about 85%, and then increases with further nickel additions to about 45%. Alloys with more than 32% nickel can be considered for more severe applications [Brown, 1997]. Stainless steel with increased molybdenum results in improved resistance to chloride stress corrosion cracking [Brown, 1977].

Carburization has been a cause of major concern in high temperature application in petroleum and petrochemical industries. As per guideline the relative carburization resistance of a group of Ni-Cr alloys can be sum of the percent content plus nine times the percent of silicon content. Silicon is probably the most potential alloying element in reducing the depth of carburization [Schillomer and Vanden Bruck, 1984]. Some of the measures for corrosion control in FCC are water wash, injection of polysulfide and use of organic filming inhibitors [Strong et al., 1991; Walker, 1984]. Using incorrect welding materials can also cause serious problem. This can be overcome by quality control procedures during construction/ fabrication followed by positive material identification of alloy steel welds to ensure that the correct filler metal is used [Setterlund, 1991].

Alloy 400 which contains two-thirds nickel and one-third copper, finds application for caustic, chloride salts because of resistant to stress corrosion cracking. It is an excellent material for fluorine, hydrogen fluoride and hydrofluoric acid [Picciotti and Picciotti, 2006].

Iron, nickel and cobalt based high performance super alloys (containing specific concentration of niobium, tungsten, tantalum, titanium, chromium and molybdenum) show good strength at high temperature (up to 2500 °F), excellent corrosion and oxidation resistance and unusual degree of creep resistance [Picciotti and Picciotti, 2006].

The commonly used materials for gaskets are [Picciotti and Picciotti, 2006]:
- Polytetrafluoroethylene (teflon): Temperature up to 350 °C,
- Self-sealing O-rings made of fluoroelastomer (FFKM): Good resistance up to 225 °C] or perfluoroelastomer (FFKM): Extremely broad chemical resistance up to 300 °C
- Flexible graphite gaskets (grafoll): Operating temperature above 350 °C. Graphite gaskets are flexible, but their lives are shorter than that of PTFE gaskets.
- Metal gaskets (above 350 °C).

23.5.2 Corrosion Inhibitors

Corrosion of the metallic surfaces can be reduced or controlled by use of corrosion inhibitors which form protective film on the surface of the metal. Various types of corrosion inhibitors are passivating inhibitors, organic inhibitors and precipitation inhibitors. Various types of corrosion inhibitors used in petroleum and petrochemical industries are given in Table 23.7.

Table 23.7   Use of inhibitors in petroleum and petrochemical industries

System	Medium	Corrosion inhibitor	Remarks
Oil well operation	Acidic condition, chloride.	Mercaptans and glycol—xanthates, oleic and naphthenic acid, derivatives of amine, diamines, zinc metaphosphate, biocides: Biomin–I & II, Scale inhibitors: Scalemin-I & II.	

Contd.

Table 23.7   Use of inhibitors in petroleum and petrochemical industries *(Contd.)*

System	Medium	Corrosion inhibitor	Remarks
		Stable free flowing oil line corrosion inhibitor, corrosion inhibitor: corromin -OLC, corromin-H_2S, corromin-A3	
Condensate well	CO_2, organic acid	Cronox film -Plus	
Crude oil pipe-line	Crude oil, water, air, salts, sulfur compounds, naphthenic acid.	Water-soluble and oil-soluble inhibitors. Amines and nitrites.	Corrosion occurs due to presence of water and air in the crude oil. CO_2 and H_2S present are highly corrosive.
Natural gas pipeline	Natural gas and condensate, CO_2, H_2S.	Filming inhibitor.	Presence of CO_2, H_2S and water makes the corrosive environment. Dehydration is effective means of reducing corrosion.
Oil storage	H_2S, chlorides.	Phosphates, nitrites, imidazolines, oleic acid, salts of amines, Filming amine.	
Atmospheric and vacuum distillation columns	Crude oil, chlorides, H_2S, sulfur compounds, nitrogen compound, naphthenic acid.	Aminoalkyl aryl phosphate.	Crude desalting, caustic injection, neutralization of vacuum column overhead vapors and condensed water, water washing in crude column can reduce corrosion.
Fluid catalytic	Naphthenic acid, H_2S, chlorides, water/steam, CO_2, oxygen.	Diamide, imidazoline, quaternary amine, polysulfide, thiocyanate.	Water wash followed by use of inhibitor. Water wash dilutes and scrubs the corrosive matter, H_2S, ammonia, chloride and cyanide. Most FCC corrosion inhibitors or filming. Some of the FCC inhibitors are oil-soluble amide, quaternary amine, oil-soluble imidazoline inhibitor.
Steam and condensate line	Water, CO_2	Oxygen scavenger: inorganic sulfite, hydrazine,	Addition of catalyst to the hydrazine mixture ensures completion of the oxygen

Contd.

Table 23.7   Use of inhibitors in petroleum and petrochemical industries *(Contd.)*

System	Medium	Corrosion inhibitor	Remarks
		carbohydrazine, hydroquinone, and ascorbic acid. Neutralizing type amines: Morpholine, diethylaminoethanol, cyclohexyl amine, ammonia.	scavenging and metal passivating reactions. Filming amine reacts with carbonic acid in condensate to form neutral amine salts, thus raising the pH.
Cooling water	Dissolved oxygen, hardness, chloride, sulfate, etc.	Filming type amine: Ethyloxylated soya amine, diadodecylamine, and tridodecylamine. Addition of sodium silicate, chromates, polysulfide, sodium molybdate. Use of biocides (oxidizing and non-oxidizing) oxidizing: Chlorine, ClO_2 (calcium hypochlorite). Non-oxidizing: Isothiozolines, dimethyl bisthiocyanate.	Octadecylamine, biocides are used for removal of undesirable formation of biological film.

Source: Fontana, 1987; Schweitzer, 1998; Shreir, 1976; Uhlig, 1963; Chawla and Gupta, 1996; Muralidharan et al., 1997; Mall, 2007; Thermex.
(http://www.thermaxindia.com/v2 ProductPage.asp?levelno=2&divid=3&pageno=1&objecti).

23.5.3 Design Consideration

Many corrosion failures are due to poor design of equipment/instruments, poor fabrication of joints/piping/equipment and due to lack of precautions to be employed to reduce environmental damage. Corrosion can be controlled or minimized to a greater extent by good design which can be obtained by considering surface conditions, smoothness of surface, cleanliness of surface, dissimilar material control, elimination of crevices, proper design of supports, by providing adequate drainage, avoiding electrical contact between dissimilar metals, avoiding sharp bends in piping system, providing thicker structures to take care of impingement effects, by properly considering the relevant codes and standards, by properly designing against excessive vibration, selecting plant site upwind from other polluting industries, by avoiding vapor spaces, uneven and stress distributions from control of metal surface temperature, environment, stress, by welding procedure, eliminating or minimizing the presence of impurities, reducing velocity of fluid, by providing easy access to the structure for periodic inspection, maintenance and replacement of damage parts, eliminating shielding area, etc. Selection of welding process is also very important. Material of construction and corrosion allowances should be selected based on anticipated corrosion or material degradation rates under most severe combination of process parameters and also long-term material performance experiences of similar types of services and facilities.

23.5.4 Coating and Lining

Apart from selection of proper material of construction, sometimes surface treatments like coating, cladding, heat treatment, diffusion treatment and surface finish are also helpful in reducing corrosion by isolating the material from the corrosive environment. Corrosion prevetion by coating can be achieved by barrier protection, inhibition protection, sacrificial protection [Ramkrishanan, 2013]. Various types of coatings may be either metallic or inorganic coatings like ceramic and glass. Commonly used organic coatings are oil base, alkyd, chlorinated rubber, coal tar epoxy, catalyzed epoxy, silicon aluminum, vinyl, urethane, etc. Although exterior coating is very common, however, internal coating of pipelines, storage tanks and other equipments are also very common. For petrochemical/chemical plants and refineries, protecting heated surfaces found on stacks, breaching, boilers, heat exchangers, reactors, stills, crackers, furnaces and engine exhaust manifolds have traditionally required the upper limits of coating [Mogul, 1999]. Once installed these coatings are subjected to corrosive fumes, cyclic heating operations, thermal expansion and atmospheric weathering [Mogul, 1999]. While going for selection of paint or coating in chemical process industries, one must consider a myriad of requirements which include adhesion to the substrate, resistance to anticipated corrosives and chemicals, flexibility, hardness and stability during temperature cycling [Finzel, 1991]. Some of the commonly used coatings are oil base, alkyd short oil, alkyd long oil, chlorinated rubber, epoxy, coal tar, amine, polyamide, ketamine, urethane aliphatic, vinyl, etc. Some of the requirements for effective pipeline corrosion coating are—ease of application, good adhesion to pipe, good resistance to impact, flexibility, resistance to soil stress, resistance to flow, water resistance, electrical resistance, chemical and physical stabilities, resistance to soil bacteria, marine organism and cathodic disbondment [McConkey, 1982]. Alkyd resins provide hard, durable films with very good weathering resistance along with good glass and color retention. Medium oil type alkyd resins show excellent important abrasion and more resistance. Long oil type offers increased flexibility and exterior durability.

Coating

Coating has been commonly used in the process industries for protection of structures, pipeline, equipments, etc. Coatings are the first line of defence against exterior pipeline corrosion supplemented by cathodic protection [Punj, 1993]. There has been continuous development in the area of coating material and method of application. Coating may be metallic and inorganic or organic. Various methods of metal coating are electrodeposition, cladding, hot dipping and vapor deposition and chemical reduction of metal-salt solution. Various types of metallic coatings are—nickel coating, zinc coating, cadmium coating, tin coating, aluminum coating and vitreous enamel. Various methods for inorganic coating are spraying, diffusion or chemical conversion. Various types of commonly used organic coatings are epoxy amine, epoxy amide, epoxy ester, epoxy coal tar, furan, phenolic unmodified, phenolic oil, polyester, polyurethane, silicon, acrylic, asphaltic and coal tar, polyamide, polyethylene fluorocarbon, vinyl, vinylidiene chloride, chlorosulfonated polyethylene, neoprene, etc. Coating must be resistant to the corrodant and free of pine holes through which the corrosive could penetrate and reach the surface of material being coated.

The essential properties in a pipe coating system are ease and fast application, good adhesion to pipe surface, chemical stability, electrical stability, flexibility, high stress and impact resistance, resistance due to cathodic displacement, stability at elevated temperature, high soil stressing resistance, water absorption resistance and resistance to soil bacteria, marine organisms, low hydrocarbon solubility [Punj, 1993]. Internal coating of pipeline results in reduced pipeline flow friction and intercorrosion, resulting in lower operating cost and installation cost, higher product purity and increased throughput [Singh and Samdal, 1988].

Coating stability

Coating stability means least practical change with time in those practical properties which affect cathodic protection system and its ability to maintain adequate pipeline corrosion control. Such pipeline coating should be selected which has high effective coating resistance. Matching the right coating to the right condition can become a feat in itself. It means the resistance per average square feet between a pipeline surface and earth when the pipeline is in place under operating conditions. This effective resistance is dependent on factors like:

- Basic resistivity of the coating materials
- Coating thickness
- Deleterious effect of environment on coating
- Resistivity of the conducting environment in which the pipeline is buried
- Bond between pipe surface and coatings.

Important criteria which must be met during coating are proper cleaning of pipe, correct priming and application at correct temperature, even distribution, correct flood coating and positioning of wraps. Various types of coatings commonly used are: Coal tar enamel, coal tar polyester, extruded polyethylene, fusion bond epoxy, polyethylene tape (PS), PVC tape, polyurethane, alkyd short oil, chlorinated rubber, polyamide, zinc-rich primers (on top coated) inorganic, ketimine. Some of the factors which affect the performance of the coatings are: Type of coating, service requirement and environmental conditions, surface preparation and profile, solids content of coatings and number of coats in application method.

Pipeline coating failure

One of the major problems facing pipeline operations everywhere has been pipeline coating failures. A coating may perform superbly under environmental and operating conditions at one location, yet it may fail miserably because of different parameters at another location. External heating required pumping, normally heavy crudes (82.2 °C) combined with soil stress, pipe movement, gravel moisture, cathodic protection currents, etc. results in conditions that many present coating systems cannot withstand. High temperature invariably causes problems in adherence of coating to steel pipe. Disbondment results in partial or complete failure of the coating.

23.5.5 Cathodic Protection

Cathodic protection has been very common in petroleum and petrochemical industries, especially in case of pipelines, storage tanks and steel structure exposed to soil. Cathodic protection is done either by an external power supply or by appropriate galvanic coupling. Anodic protection is based on the formation of a protective film on metals by externally applied current.

23.6 CORROSION MONITORING

Corrosion monitoring is one of the important aspects of overall corrosion prevention and control as it helps in diagnosis and corrosion control. Corrosion monitoring has been found to improve output, increase of life of plant, improve product quality and reduce capital and operating cost. It helps in monitoring effectiveness of solutions to the problem in providing information for solution of specific corrosion problem and helps in assessing the behavior of new materials and helps in reducing downtime. It helps in the safe operation of any equipment. Significant benefits of a corrosion monitoring program are: Improved safety, reduced down time, early warning before costly serious damages sets in, reduced maintenance cost, reduced pollution and contamination risks, longer intervals between scheduled maintenance, reduced operating costs and life extension [Khanna, 2006]. Various methods used for corrosion monitoring have been already discussed.

REFERENCES

1. Agarwal DC. Defy Corrosion With Recent Nickel Alloys. Chemical Engineering Progress, Jan 1999, p62.

2. Batra B, Borchert CA, Lewis KR and Smith AR. Design Process Equipment for Corrosion Control. Chemical Engineering Progress, May 1993, 93.

3. Bennion D. Hazards Associated with the Use of Austenitic Stainless Steel in Process Plant. Advances in Petrochemical Technology. The Institutional Chemical Engineers Symposium, Series No. 50, 1977.

4. Bertness TA and Chilingarian GV. Corrosion in Drilling Operations, in Drilling and Drilling Fluids by Chilingarian GV and Vorabutr P, Elsevier Scientific Publishing Company, 1981, p559.

5. Brown RS. Select Alloys for Severely Corrosive Environment. Chemical Engineering Progress, March 1997, p74.

6. Charkrabarty C. Corrosion of Metals. Chemical News, Feb 2013, p31.

7. Chawla SL and Gupta RK. Material Selection for Corrosion Control. ASTM International, Materials Park, OH 44073, 1996.

8. Christman TK and Beavers JA. Cause of Stress Corrosion Cracking in Pipe. Oil and Gas Journal, Jan 5, 1987, p40.

9. Danllov B. Examples of Corrosion Control. Hydrocarbon Processing, Feb 1981, p95.

10. Das GS and Khanna AS. Corrosion Behavior of Pipeline Steel in CO_2 Environment. Trans India Inst Met 37, June 2004, p277.

11. DuPart MS, Bacon TR and Edwards DJ. Understanding Corrosion in Alkanolamine Gas Treating Plants. Hydrocarbon Processing, April 1993, p75.

12. Finzel WA. Use of Low VOC Coatings. Chemical Engineering Progress, Nov 1991, p50.

13. Fontana MG. Corrosion Engineering, 3rd edition, 1987, McGraw Hill Book Company.

14. Habiby F, Imtiaz RR and Mutairi AH. Reduce Underside Corrosion in Aboveground Storage Tanks. Hydrocarbon Processing, Jan 2003, p59.

15. Hibner EL and Fende DS. Conquer Chloride and Alloy Cost. Chemical Engineering Progress, April 1999, p63.

16. Hanratty T. Corrosion Under Insulation is a Hidden Problem. Hydrocarbon Processing, March 2013, p51.

17. Huchler LA. Select the Best Boiler Water Chemical Treatment Program. Chemical Engineering Progress, Aug 1998, p45.

18. Kane RD and Cayard MS. Improve Corrosion Control in Refining Process. Hydrocarbon Processing, Nov 1995, p129.

19. Kane RD and Cayard MS. Select Materials for High Temperature. Chemical Engineering Progress, March 1995, p83.

20. Khanna AS. Corrosion Monitoring and Control Methods for Chemical Process Industry. Chemical Industry Digest, Oct 2006, p49.

21. Kresse TJ. CO_2 Removal Reduces Pipeline Corrosion at Two Storage Tank. Oil and Gas Journal, Oct 12, 1987, p77.

22. Lack J and Harrell B. Reduce Salt Corrosion with Stronger Base Amines. Hydrocarbon Processing, Sep 2013, p67.

23. Little BJ, Ray RI and Wanye PA. Tame Microbiologically Influenced Corrosion. Chemical Engineering Progress, Sep 1998, p51.

24. McNab AJ and Treseder RS. Materials Requirement for Gas Treating Process. Material Performance, Vol 10, No. 1, 1971.

25. McConkey SE. Fusion Bonded Epoxy Pipe Coatings are Economical, Practical. Oil and Gas Journal, July 19, 1982, p148.

26. Mogul MG. Reduce Corrosion in Amine Gas Absorption Columns. Hydrocarbon Processing, Oct 1999, p47.

27. Muralidharan S, Venkatachari G and Rengaswamy NS. Inhibitors and Chemical Treatments for Cooling Water Systems. Chemical Industry Digest, SPl Issue, Maintenance, Coating and Corrosion, May, p87.

28. Nagi-Hanspal, Subramaniyam M and Shah P. Corrosion Control with High Acid Crudes. Petroeum Technology, Q3, 2013, p115.

29. Narain S. Material Science and Corrosion Control. Hydrocarbon Processing, Oct 2000, p76.

30. Payne B. Minimize Corrosion While Maximizing Distillate. Petroleum Technology, Quarterly, Q3, 2012, P75.

31. Picciotti M and Picciotti F. Selecting Corrosion Resistant Materials. Chemical Engineering Progress, 2006, p45.

32. Punj A. Pipe Coating Selection. Chemical Engineering World, Vol 28, No. 10, Oct 1993.

33. Ramakrisnan B. Preventing Corrosion Through Coatings. Chemical Engineering World, Dec 2013, p52.

34. Rayner RE. Better Metallurgy or Process Equipment. Hydrocarbon Processing, 1994, p53.

35. Richert JP, Bbadasarian AJ and Shagay CA. Stress Corrosion Cracking of Carbon Steel in Amine System. NACE Material Performance, Jan 1988, p9.

36. Schillmoller CM. Control Organic Acid Corrosion with These Metals and Alloys. Chemical Engineering Progress, Feb 1997, p66.

37. Schllimoller CM and Vanden Bruck. Furnace Alloys Update. Hydrocarbon Processing, Dec 1984, p55.

38. Schweitzer PA. Encyclopedia of Corrosion Technology, Marcel Dekker, 1998.

39. Setterlund RB. Selecting Process Piping Materials. Hydrocarbon Processing, Aug 1991, p93.

40. Setterlund RB. Preheat Piping Failures. Hydrocarbon Processing, Oct 1997, p47.

41. Singh G and Samdal OR. Internal Coating Justified by Operating Costs. Oil and Gas Journal, April 4, 1988, p50.

42. Shreir LL. Corrosion, Vol 1–2, Newnes-Butterworths, 1976.

43. Strong RC, Majeshi VK and Whilelm SM. Basic Steps Lead to Successful FCC Corrosion Control. Oil and Gas Journal, Sep 30, 1991, p81.

44. Tillack DJ and Guthrie JE. Select the Right Alloys for Refiners and Petrochemical Plants. Chemical Engineering Progress, Feb 1999, p59.

45. Uhlig HU. The Corrosion Handbook, John Wiley & Sons, 1963.

46. Walker HB. Reduce FCC Corrosion. Hydrocarbon Processing, Jan 1984, p81.

47. An T and Bird KW. Know Which Reactive and Refractory Metals Work for You. Chemical Engineering Progress, Feb 1992, p65.

24

Environmental Management in Petroleum Industry

Petroleum industry is a backbone of industrial development and survival without petroleum product is itself a big question mark. In spite of its significance contribution, petroleum industry is amongst the top highly polluting industries. All steps starting from oil exploration to refining, and application of petroleum product from automobile to petrochemical production, hydrocarbon sector produces large variety of pollutants in form of liquid air, solid waste, noise and thermal land pollution. To defend and improve the environment for present and future generations has become an imperative goal of mankind (Stockholm Conference Proclamation, 1972). Sound environmental management to mitigate the environmental pollution has become one of the major issues. Use of heavy crude oil and changing stringent environmental standards for fuel have further complicated the environmental control strategies requiring huge investment.

24.1 ENVIRONMENTAL POLLUTION CONTROL ACTS, REGULATIONS AND STANDARDS

For the prevention and control of environmental pollution and protection of environment for conservation of natural resources several Acts, rules and standards have been instituted in India and other parts of the world. Various environmental protection Acts, rules and various amendments are given in Table 24.1. Every industry has to take environmental clearance for their project; list of projects has been mentioned in Table 24.2. National ambient air quality standards, ambient air noise standards, minas for petroleum and petrochemical complexes and general standards for discharge of effluent are given in Tables 24.3 to 24.5. General standards for discharge of effluents as per the Environmental Protection Rules, 1986 [Scheme VI] are given in Table 24.6.

Table 24.1   Environmental protection Acts, rules and various amendments

- Indian Boiler Act, 1923
- Mines and Mineral Act (Regulation and Development Act), 1947.
- Wildlife Protection Act, 1972.

Contd.

Table 24.1   Environmental protection Acts, rules and various amendments *(Contd.)*

- Forest (Conservation) Act, 1980
- The Water (Prevention and Control of Pollution) Act, 1974, as amended to date
- The Water (Prevention and Control of Pollution) (Procedures for Transaction of Business) Rules, 1975.
- The Water (Prevention and Control of Pollution) Cess Act, 1977, as amended to date
- The Water (Prevention and Control of Pollution) Cess Rules, 1978, as amended to date
- The Air (Prevention and Control of Pollution) Act, 1981, as amended to date
- The Air (Prevention and Control of Pollution) Rules, 1982, as amended to date
- The Air (Prevention and Control of Pollution) (Union Territories) Rules, 1983
- The Environmental (Protection) Act, 1986
- The Environmental (Protection) Rules, 1986

Table 24.2   List of projects requiring environment clearance in India

- Nuclear power and related projects such as heavy water plants, nuclear fuel complex and rare earths.
- River valley projects including hydropower, major irrigation and their combination including flood control.
- Ports, harbours, airports (except minor ports and harbours).
- Petroleum refineries including crude and product pipelines.
- Chemical fertilizers (nitrogenous and phosphatic other than single superphosphate)
- Pesticides (technical)
- Petrochemical complexes (both olefinic and aromatic) and petrochemical intermediates such as DMT, caprolactam, LAB, etc. and production of basic plastics such as LDPE, HDPE, PP and PVC.
- Bulk drugs and pharmaceuticals.
- Exploration for oil and gas and their production, transportation and storage
- Synthetic rubber
- Asbestos and asbestos products.
- Hydrocyanic acid and its derivatives.
- (a) Primary metallurgical industries (such as production of iron and steel, aluminum, copper, zinc, lead and ferro alloys).
- Electric arc furnaces (mini steel plants)
- Chlor-alkali industry
- Integrated paint complex including manufacture of resins and basic raw materials required in the manufacture of paints.
- Viscose staple fiber and filament yarn.
- Storage batteries integrated with manufacture of oxides of lead and lead antimony alloy [www.envfor.nic.in].

Contd.

Table 24.2 List of projects requiring environment clearance in India *(Contd.)*

- All tourism projects between 200–500 m of high tide line or at locations with an elevation of more than 1000 m with investments of more than ₹ 5 crore.
- Thermal power plants
- Mining projects (major minerals) with leases more than 5 hectares
- Highway projects (except projects relating to improvement work including widening and strengthening of roads with marginal land acquisition along the existing alignments provided it does not pass through ecologically sensitive area such as national parks, sanctuaries, tiger reserves, reserve forests.)
- Tarred roads in Himalyas and/or forest areas
- Distilleries
- Raw skins and hides
- Pulp, paper and newsprint
- Dyes
- Cement
- Foundries (individual)
- Electroplating [www.elaw.org]

Table 24.3   National ambient air quality standards (NAQS), 1994

Sl. No.	Pollutant	Time weighted average	Concentration in ambient air		
			Industrial, residential, rural and other area	Ecological sensitive area (notified by central government)	Method of measurement
1	2	3	4	5	6
1	Sulfur dioxide (SO_2) $(\mu g/m^3)$	Annual*	50	20	• Improved West and Gaeke
		24 hours**	80	80	• Ultraviolet fluorescence
2	Nitrogen dioxide (NO_2)	Annual*	40	30	• Modified Jacob and Hochheiser (Na-Arsenite)
		24 hours**	80	80	• Chemiluminescence
3	Particulate matter (size less than 2.5 ppm) or PM_{10} $(\mu g/m^3)$	Annual*	60	60	• Gravimetric
					• TOEM
		24 hours**	100	100	• Beta attenuation
4	Particulate matter (size	Annual*	40	40	• Gravimetric

Contd.

Table 24.3   National ambient air quality standards (NAQS), 1994 *(Contd.)*

Sl. No.	Pollutant	Time weighted average	Concentration in ambient air		
			Industrial, residential, rural and other area	Ecological sensitive area (notified by central government)	Method of measurement
1	2	3	4	5	6
	less than 10 µm) or $PM_{2.5}$ (µg/m³)	24 hours**	60	60	• TOEM • Beta attenuation
5	Ozone (O_3) (µg/m³)	8 hours** 1 hour**	100 180	100 180	• UV photometric • Chemiluminescence • Chemical method
6	Lead (Pb) (µg/m³)	Annual* 24 hours**	0.5 1	0.5 1	• AAS/ICP method after sampling on EPM 2000 or equivalent filter paper • ED-XRF using teflon filter
7	Carbon monoxide (CO) (mg/m³)	8 hours** 1 hour**	2 4	2 4	• Non-dispersive infrared (NDIR) spectroscopy
8	Ammonia (NH_3) (µg/m³)	Annual* 24 hours**	100 400	100 400	• Chemiluminescence • Indophenol blue method
9	Benzene (C_6H_6) (µg/m³)	Annual*	5	5	• Gas chromatography based continuous analyzer • Adsorption and/ desorption followed by GC analysis
10	Benzo (o) Pyrene (BaP) particulate phase only (ng/m³)	Annual*	1	1	• Solvent extraction followed by HPLC/ GC analysis
11	Arsenic (As), (ng/m³)	Annual*	6	6	AAS/ICP method after sampling on EPM 2000 or equivalent filter paper
12	Nickel (Ni)	Annual*	20	20	AAS/ICP method after sampling on EPM 2000 or equivalent filter paper

Source: MoEF Notification, 2009.

Table 24.4   Ambient air noise standards

Area code	Category of area	Limits in dB (A) Leg	
		Day time	Night time
A	Industrial area	75	70
B	Commercial area	65	55
C	Residential area	55	45
D	Silence zone	50	40

Exposure time (hours/day)	Limit in dB (Leg)
8	90
4	93
2	96
1	99
1/2	102
1/4	105
1/8	108
1/16	111
1/32 (2 minutes or less)	114

Table 24.5   Minas (minimal national standards) for oil refineries, petrochemical, organic chemical industries

Oil refineries	Maximum permissible limit mg/L	Maximum permissible quantity kg/100 kg of crude oil
Oil and grease	10	7
Phenol	1	0.7
Sulfide	0.5	0.35
BOD (3 days at 27 °C)	15	10.5
Suspended solid	20	14
pH	6–8.5	

Table 24.6   General standards for discharge of effluents

Sl. No.	Parameter	Standards			
		Inland surface water	Public sewers	Land for irrigation	Marine coastal areas
1	2	3			
		(a)	(b)	(c)	(d)
1.	Color and odor	See Note 1	–	See Note 1	See Note 1
2.	Suspended solids (mg/L, max.)	100	600	200	(a) For process waste water—100 (b) For cooling water effluent—10 percent above total suspended matter of influent cooling water
3.	Particle size of suspended solids	Shall pass 850 micron IS sieve			(a) Floatable solids max 3 mm (b) Settleable solids max 850 microns
4.	Dissolved solids (inorganic) (mg/L, max.)	2,100	2,100	2,100	–
5.	pH value	5.5–9.0	5.5–9.0	5.5–9.0	5.5–9.0
6.	Temperature (°C, max.)	Shall not exceed 40 in any section of the stream within 15 m downstream from the effluent outlet	45 at the point of discharge	–	45 at the point of discharge
7.	Oil and grease, (mg/L, max.)	10	20	10	20
8.	Total residual chlorine (mg/L, max.)	1.0	–	–	1.0
9.	Ammonical nitrogen (as N), (mg/L, max.)	50	50	–	50
10.	Total Kjeldahl nitrogen (as N), (mg/L, max.)	100	–	–	100

Contd.

Table 24.6   General standards for discharge of effluents *(Contd.)*

Sl. No.	Parameter	Standards			
		Inland surface water	Public sewers	Land for irrigation	Marine coastal areas
1	2	3			
		(a)	(b)	(c)	(d)
11.	Free ammonia (as NH$_3$) (mg/L, max.)	5.0	–	–	5.0
12.	Biochemical oxygen demand (5 days at 20 °C) (max.)	30	350	100	100
13.	Chemical oxygen demand (mg/L, max.)	250	–	–	250
14.	Arsenic (as As) (mg/L, max.)	0.2	0.2	0.2	0.2
15.	Mercury (as Hg) (mg/L, max.)	0.01	0.01	–	–
16.	Lead (as Pb) (mg/L, max.)	0.1	1.0	–	1.0
17.	Cadmium (as Cd) (mg/L, max.)	2.0	1.0	–	2.0
18.	Hexavalent chromium (as Cr^{+6}) (mg/L, max.)	0.1	2.0	–	1.0
19.	Total chromium (as Cr) (mg/L, max.)	2.0	2.0	–	2.0
20.	Copper (as Cu) (mg/L, max.)	3.0	3.0	–	3.0
21.	Zinc (as Zn) (mg/L, max.)	5.0	15	–	15
22.	Selenium (as Se) (mg/L, max.)	0.05	0.05	–	0.05
23.	Nickel (as Ni) (mg/L, max.)	3.0	3.0	–	5.0
24.	Boron (as B), (mg/L, max.)	2.0	2.0	2.0	–
25.	Percent sodium (max.)	–	60	60	–

Contd.

Table 24.6 General standards for discharge of effluents *(Contd.)*

Sl. No.	Parameter	Standards			
		Inland surface water	Public sewers	Land for irrigation	Marine coastal areas
1	2	3			
		(a)	(b)	(c)	(d)
26.	Residual sodium carbonate (mg/L, max.)	–	–	5.0	–
27.	Cyanide (as CN) (mg/L, max.)	0.2	2.0	0.2	0.2
28.	Chloride (as Cl) (mg/L, max.)	1000	1000	600	–
29.	Fluoride (as F) (mg/L, max.)	2.0	15	–	15
30.	Dissolved phosphates (as P) (mg/L, max.)	5.0	–	–	–
31.	Sulfate (as SO_4) (mg/L, max.)	1000	1000	1000	–
32.	Sulfate (as S), (mg/L, max.)	2.0	–	–	5.0
33.	Pesticides	Absent	Absent	Absent	Absent
34.	Phenolic compounds (as C_6H_5OH) (mg/L, max.)	1.0	5.0	–	5.0
35.	Radioactive materials (a) Alpha emitters (μ/mL, max.)	10^{-7}	10^{-7}	10^{-8}	10^{-7}
	(b) Beta emitters (Ci/mL, max.)	10^{-6}	10^{-6}	10^{-7}	10^{-6}

Notifications [Under the Environmental (Protection) Rules, 1986]

- Authorized officers/agencies to enter the premises for inspection
- Officers/agencies authorized to take samples
- Delegation of powers to the state governments under Section 5 of the The Environmental (Protection) Act, 1986.
- General standards for effluents, noise standards, emission standards for vehicles, etc. (Schedule-IV) [cpcbenvis.nic.in]
- Emission standards for pollutants from various industries
- Various forms
- General standards (Schedule-VI)
- National Ambient Air Quality Standards
- Notification on Doon Valley
- Prohibition on the Handling of Azodyes
- Notification regarding Antop Hill, Bombay, Notification on Dahanu Talukam Maharashtra
- Environmental Impact Assessment Notification
- Environmental Impact Assessment Notification regarding thermal power plants [www.cleantechindia.com]
- The Manufacture, Use, Import, Export and Storage of Hazardous Micro-organisms, Genetically Engineered Organisms or Cell Rules, 1989.
- The Hazardous Wastes (Management and Handling) Rules, 1989, as amended to date
- The Manufacture Storage and Import of Hazardous Chemical Rules, 1989, as amended to date [www.cpcb.nic.in]
- The Manufacture, Use, Import, Export and Storage of Hazardous Chemical Rules
- Scheme of Labeling of Environmental Friendly Products (Eco-Marks)
- The Bureau of Indian Standards (Certification) (Amendment) Regulations, 1997
- Restricting certain activities in special Specified area of Aravalli range
- The Chemical Accidents (Emergency Planning, Preparedness and response) Rules, 1996
- The Biomedical Waste (Management and Handling) Rules, 1998, as amended to date [www.lexisnexis.in]
- The Recycled Plastics Manufacture and Usage rules, 1999
- The Municipal Solid Wastes (Management and Handling) Rules, 2000
- The Noise Pollution (Regulation and Control) Rules, 2000
- The Ozone Depleting Substances (Regulation) Rules, 2000
- The Batteries (Management and Handling) Rules, 2001
- Notification (Miscellaneous)
- The Public Liability Insurance Act, 1991, as amended to date
- Public Liability Insurance Rules, 1991, as amended to date [www.ecacwb.org]
- The Motor Vehicle Act 1988 and amended in 1988
- The National Environmental Tribunal Act, 1995
- The National Environmental Appellate Authority Act, 1997
- The National Environmental Appellate Authority (Appeal) Act, 1997

24.2 ENVIRONMENTAL POLLUTION IN PETROLEUM INDUSTRY

The petroleum industry is a complex and an integrated industry that includes a large variety of processes and products starting from oil exploration, crude oil and natural gas processing, because of a large number of processes, use of wide variety of crude oil starting from low to heavy crudes and low sulfur to high sulfur crude oils, catalysts, additives, chemicals and presence of explosives and hazardous materials, the environmental pollution problem from petroleum industries is quite complex. The large scale oil and gas exploration, drilling and production of oil and gas, integration of petroleum refinery with petrochemical has further complicated the issues. Wide variety of pollutants are discharged into water stream and emitted in the environment. General standard for effluent discharge is mentioned in Table 24.6. Environmental pollution from production of crude oil and gas, processing of crude oil includes following major areas:

- Oil and gas exploration, drilling, production, pipeline and marine transportation
- Storage and handling emissions
- Process emissions
- Fugitive emissions
- Secondary emission.

Exploration and production and crude oil processing units before refinery operation are also major sources of pollution. Typical sources of emissions in exploration and production are given in Table 24.7. Offshore activities and transportation of crude oil are also the major sources of marine pollution and there are several statutory regulations regarding prevention and pollution of the sea by oil and protection of the marine environment which must be followed as the coastal zones in India are generally very high population density areas and the well-being of the coastal population linked with ecological balance of the land sea system. At the planning and design stage of each project, therefore, specific provision must be made which clearly indicates control measures to combat the adverse impact on the marine environment. Oil discharges from offshore drilling and production can be controlled by installing equipment and treatment facilities and observance of prescribed procedures. Safety and pollution control on platforms are inter-independent.

Storage tanks, truck, railway and marine terminals are the sources of VOC emission and oil spillage problems. Storage and handling emissions are functions of construction and size of storage tank, the vapor pressure of the stored organic liquid and the ambient quality at the tank location. Process emission occurs from reactors, distillation columns, purification equipments, fire heaters, condensers, reformers, crackers, filters, sulfur recovery units, recovery and control equipments, stacks, vent, etc [Mall, 2007]. There are two types of fugitive emissions: Low level leaks from process equipment, episodic fugitive emissions where an even such as equipment failure results in sudden large release [Shen, 2004]. Fugitive emission occurs from pumps, valves, flange, mechanical seal, relief valves, tanks, instrument connections, sample connections and open-ended lines, etc [Allen and Rosselot, 1997]. Secondary emission results from the waste water treatment unit, cooling tower, boilers, process sewers, etc. In almost cases fugitive emission for equipment leaks is the largest source typically accounting for 40% to 60% of total VOC emission [Siegell, 1997].

Table 24.7   Typical sources of emissions in exploration and production and control strategies

Source	Contaminant	Control system
Well-drilling (test, completion and work over operations)	Hydrocarbons, H_2S, CO_2, SO_2, NO_x, particulate matter, fumes	Incineration combustion improvement with special burners
Well-exploration Pumping triphasic separation	Light hydrocarbons Light hydrocarbons	Good maintenance (joints) Emissions discharge in gas circuit
Oily water treatment Storage tank	Hydrocarbons Hydrocarbons, H_2S	Covering and blanketing Tank pressure control (floating roof)
Oil and fuel leaks Leading for exploration	Hydrocarbons Hydrocarbons, H_2S	Good house keeping Vapour recovery, discharge to storage tank, incineration
Well-draining Open drain system Cold vent Flares	Hydrocarbons, H_2S Hydrocarbons Hydrocarbons, H_2S Hydrocarbons, H_2S, CO, CO_2, NO_x, unburnt hydrocarbons	Closed circuit, Hydraulic seals, siphons, Good dispersion Flares tip design (improvement of flare nozzles)
Gas turbine exhaust, other combustion units	CO, CO_2, NO_x, SO_2, CO, CO_2, NO_x, SO_2, particulate matter	High stack adjustment, high stack burner adjustment, selection of combustibles, dust removal (system depending on requirements)

Source: Perspective Plan on Environmental Management, 1985–86 to 1989–90. Safety and Environnent Management. Oil and Natural Gas Commission, India.

Some of the major sources of pollutants in petroleum refinery are crude oil and natural gas processing, crude oil storage, crude processing—desalting and distillation, secondary processes—thermal, catalytic cracking, hydrocraking units, reforming, catalyst regeneration units, hydroprocessing units, lube refining and lube treatment processes, boiler blow downs, power plants, effluent treatment plant, etc. Major pollutants are free and emulsified oils, phenols, cyanides, inorganic salts, naphthenic acids, heavy metals, sulfide, spent catalyst; tars; H_2S, NH_3, NO_x, SO_x, CO, CO_2 hydrocarbons, volatile organic chemicals, particulate matter, fine catalyst dusts, solvents like phenol, NMP, sulfonate, etc. Typical sources of environmental pollution are given in Table 24.8 and Fig. 24.1.

Middle East crude which is high in sulfur content is being processed by many refineries in India which is a major source of H_2S and SO_x emission. Fuel oil manufacture from these refineries when used in captive fuel emits SO_x. In refineries and petrochemical complexes SO_x is emitted from a number of sources which include process heaters and boilers, FCCs, sulfur recovery units, blow down system, vpour recovery and flares, process gas flares, fluid coking, etc. Flue gases leaving

Table 24.8 Waste water generation in an integrated refinery complex

Unit	Type of waste	Waste water generation (m³/hour)
Crude oil distillation I	Oily/chemical water	8
Crude oil distillation II	Oily/chemical water	8
Kerosene treating unit	Oily/chemical water	15
Delayed coker unit I	Oily/chemical water	12
Delayed coker unit II	Oily/chemical water	12
Pump house	Oily water	2
Tankages	Oily water	4
Laboratory	Oily water	2
Sanitary	Sanitary water	10
Floor cleaning/washing, etc.	Oily water	50
	Total	123
Sanitary sewer generation from township sewer system		

the FCC regenerator is a major source of SO_x emission. Hydrotreating units are the largest source of SO_x emission. CO emission takes place from regenerator where the coke from the catalyst is burnt off. In hydrocarbon industry process heaters are used at number of places in the process, which are major source of NO_x emission in hydrocarbon industries. NO_x is formed during combustion through thermal fixation of atmospheric nitrogen introduced with the fuel or through the oxidation of nitrogen introduced through the fuel.

Potential sources of liquid pollution may be reactors, equipment water overflows, sample blow downs, distillation column, absorber, product wash and purification, boiler blow downs, cooling water, steam fed vacuum pumps, pumps and compressors, etc. Potential sources of solid waste from petrochemical industries are: Cracker, reformer spent catalyst; sludge from process and waste treatment facilities, catalyst regeneration units.

Sour water generation is also a problem in refinery and petrochemical complexes. Sour water is produced at a number of distinct pints within a refinery and petrochemical complex which is drained in sour water collection system. Sour water results in different sections when steam and/or water used in various processes pickup H_2S and NH_3 which if not properly treated, affect the waste water treatment system.

Oily sludge, spent catalyst, spent caustic are some of the major sources of waste generation in petroleum and petrochemical complexes. Oily sludges are formed through emulsification of oil with water usually in presence of suspended solids. Various sources of oily sludge are sludge from API separators, slop oil emulsion, heat exchanger bundle cleaning sludge, cooling tower sludge and tank bottoms. Spent caustic is discharged from gas purification processes where caustic soda is used for scrubbing the dissolved sulfide, H_2S, acidic gases, mercaptans and acid

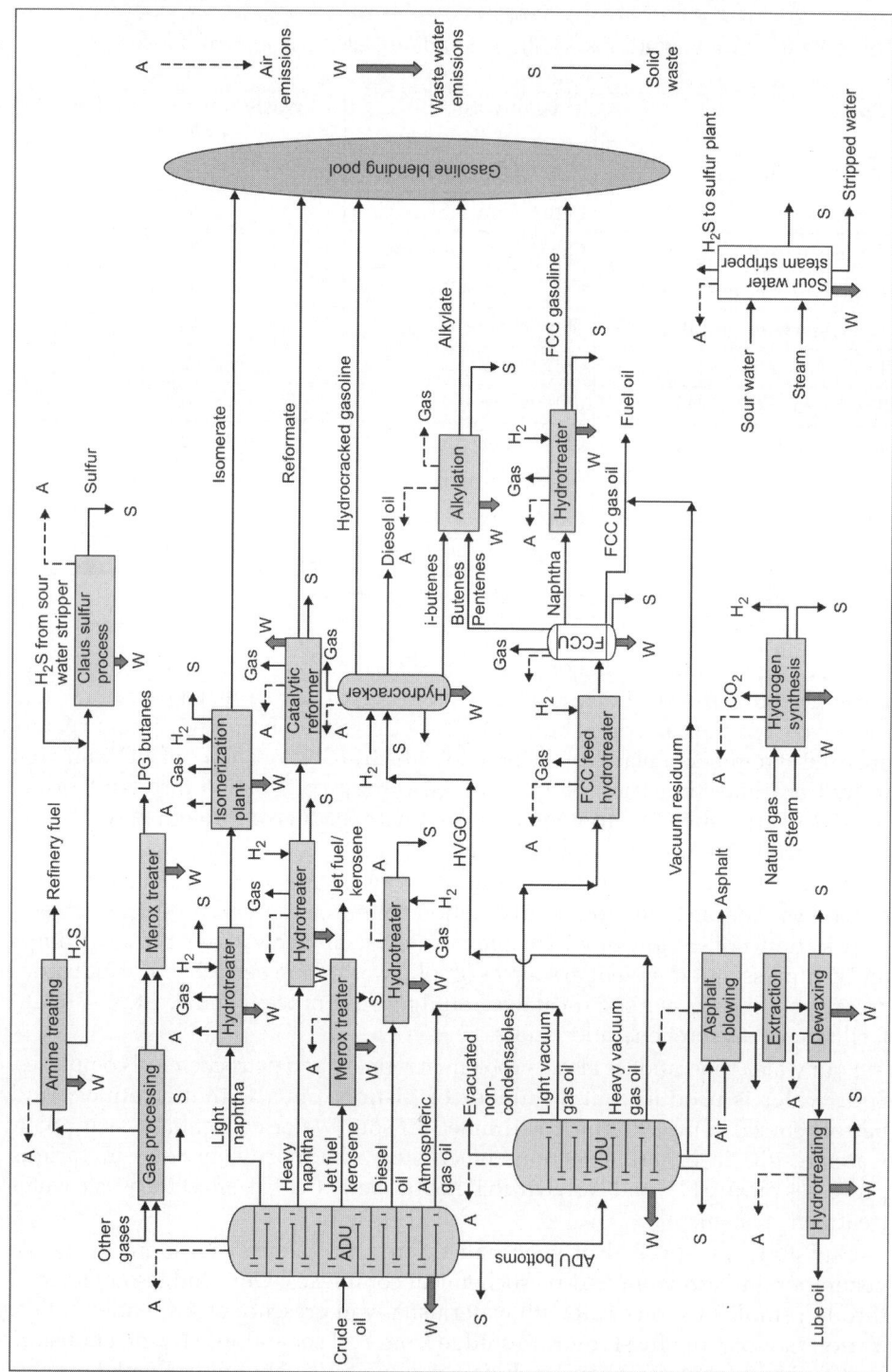

Fig. 24.1 Various pollutants emitted from different parts of petroleum refinery.

gases. Catalyst deactivation through sintering, poisoning or build up of surface deposits and loss due to attrition is the cause of generation of spent catalyst which is unavoidable. However, use of catalyst having high reactivity, activity, stability regenerability, resistance to attrition may result in less generation of spent catalyst.

24.3 ENVIRONMENTAL IMPACT ASSESSMENT

Environmental impact assessment is a pointer to the environmental compatibility of projects in terms of their location, suitability of technology, efficiency in resource utilization, recycling and so on. Environmental impact assessment has been made mandatory for 29 categories of development projects under various sectors such as industrial, mining, irrigation, power, transport, tourism, communication, etc. Petroleum industries are also amongst the top polluting industries. Detail of pollutants generated from petroleum refining is given in Table 24.9.

Some of the basic steps in EIA involve base line study, scoping or impact identification, impact measurement, impact analysis, mitigating measure, monitoring of scheme, disaster management plan, environmental management plant and green belt design. Environmental apprehension from various activities of petroleum refinery are: (1) Deterioration in air quality due to emission of hydrocarbon, oxides of sulfur and nitrogen, hydrogen sulfide, particulates, etc.; (2) deterioration in water quality due to presence of high BOD, COD, turbidity, oil and grease, dissolved solid, toxic organic compounds, cyanide, heavy metals,

Table 24.9   Water and air pollutants discharge/emission from various processes in petroleum refining

Sl. No.	Source	Pollutants
1.	Crude oil storage	Emulsified oil, oily sludge, VOC
2.	Desalting, atmospheric vacuum distillation	Inorganic salts, oily sludge, oil and grease, phenols naphthenic acid, ammonia, VOCs, H_2S, SO_x, NO_x, CO
3.	Thermal cracking, visbreaking, coking	Oil and grease, VOCs, H_2S, SO_x, NO_x, CO, particulates
4.	Fluidized bed catalytic cracking, hydrocracking	Oil and grease, VOCs, H_2S, SO_x, NO_x, CO, spent catalyst, particulates
5.	Catalytic reforming and aromatic separation	Oil and gases, VOCs, aromatics, H_2S, SO_x, NO_x, CO, spent catalyst, particulates, solvents
6.	Hydrotreating and hydrodesulfurization	Oil and grease, VOCs, H_2S, SO_x, NO_x, CO, spent catalyst
7.	Lube oil and wax processing	Oil and grease, solvents, clay, bauxite, SO_2
8.	Sweetening and sulfur recovery plant	VOCs, H_2S, SO_x, NO_x, CO, spent catalyst, spent caustic
9.	Process heaters and flares	VOCs, H_2S, SO_x, NO_x, CO
10.	Power plant cooling tower blow downs	Chromium, calcium and magnesium salts, SO_x, NO_x, CO, particulates, oil and grease
11.	Water treatment plant	Suspended solids, oily sludge, nitrogen and phosphorous compounds, chlorides, heavy metals

carbon, etc.; (3) environmental deterioration due to noise pollution; (4) adverse impact on biological environment due to discharge of various toxic compounds, heavy metal, oil grease, phenolic compounds, etc.; (5) environmental impact due to various solid wastes, and (6) socioeconomic destruction due to influx of labor force, migration from outside, movement of heavy machinery, additional traffic [Mall, 2007]. Characterization of effluent discharge is given in Table 24.10.

Resources, which are likely to be affected due to location/expansion/ modernization of the petroleum and petrochemical plants are [Mall, 2007]:

- *Physical component:* Metrology, air quality, surface water, hydrology, ground water, topography, geology, soil and material.
- *Ecological environment:* Fresh water ecology, terrestrial, forest, fauna, sanctuary, natural vegetation, species diversity, fisheries, animals, etc.
- *Socioeconomic and cultural aspects:* Impact on economic and cultural aspects, economic yield, etc.

24.4 ENVIRONMENTAL POLLUTION CONTROL STRATEGIES IN PETROLEUM REFINERY [MALL, 2007]

At the planning and design stage of each project, therefore, specific provision must be made which clearly indicates control measures to combat the adverse impact on the marine environment. Oil discharges from offshore drilling and production can be controlled by installing equipment and treatment facilities and observance of prescribed procedures. Safety and pollution control on platforms are inter-independent. Advanced standards and practices are needed for prevention and control of oil spills from petroleum marine transportation. Methods to ensure proper handling of oily wastes should be established for all transportation vessels. Coastal and estuarine areas are subject to serious pollution due to discharge of effluents and wastes from many sources as far as possible, methods of discharge through conduits beyond the lowest tide mark should be adopted. Water produced at onshore drilling sites should be treated at effluent treatment plants and disposed off by subsurface injection at appropriate depths, or used for pressure maintenance or into abandoned wells. The main air pollutants from oil and gas processing units

Table 24.10 Typical characteristics of combine effluent from an integrated refinery and petrochemical complexes

Parameter	Unit	Refinery	Petrochemical
pH			6.0–6.5
Oil and Grease	mg/L	10	–
BOD	mg/L	15	50
COD	mg/L	–	250
TSS	mg/L	20	100
Sulfides	mg/L	0.5	2
Phenols	mg/L	1	5

are VOCs, H_2S, SO_x, NO_x, CO, and particulates. Typical sources of emissions in exploration of oil and production and control strategies are given in Tables 24.11 to 24.13.

Waste management comprises quick identification of the waste generated/caused economic reduction, efficient collection and handling, optimal reuse and

Table 24.11 Typical sources of VOCs from process plants

Sources	Relative (%)
Fugitive equipment leaks	50–60
Loading	20–30
Waste water treating	10–15
Storage tanks	10–15

Source: Hydrocarbon Processing, Aug 1995, p77.

Table 24.12 Fugitive emissions control

Initiate a leak detection and repair program
Install new packing sets in block and control valves
Upgrade pump seals to multiseal designs

Source: Hydrocarbon Processing, Aug 1995, p77.

Table 24.13 Tank-emissions control [tijst.net]

Fixed-roof tanks
Install vapor balance systems
Install vapor recovery/destruction
Install internal floating roof
External floating roof tanks
Check condition of existing seals
Replace vapor mounted primary with liquid mounted primary seal
Control losses from roof fittings
Install secondary rim seal
Convert tank to internal floating roof design
Install vapor recovery/destruction
Internal floating roof
Check condition of exiting seals
Replace vapor mounted primary seal with liquid mounted primary seal

Contd.

Table 24.13 Tank-emissions control [tijst.net] *(Contd.)*

Control losses from roof fittings
Install secondary rim seal
Install vapor recovery/destruction

Source: Hydrocarbon Processing, Aug 1995, p77.

recycling, effective disposal leaving no environmental problems [Sushil, 1990]. Industrial work management has now evolved from an end pipe treatment mentality to holistic environmental waste management with source reduction as a preferred option [Wang, 1998]. The best way to manage waste is not to generate it, however, practically it is not feasible and waste generation is always inevitable. The basic steps for waste reduction are: Recognize the waste, determine the cause, plan corrective action, eliminate the cause, and establish controls to prevent reoccurrence.

Due to ever increasing and restrictive environmental regulations the petroleum industries will have to renovate and revamp the waste management facilities by implementing pollutant prevention methods and improved operating and maintenance procedures. The seemingly simple process of removing oil/grease, solids and COD from wastewater is now more challenging due to new requirements for stricter control of hydrocarbon vapors to atmosphere and elimination of the potential for contaminated water into ground and surface water streams [Guida and Fruge, 1995].

Environmental strategic planning begins during conceptual design, construction, start up operations and final shut down facility using a life cycle approach which is a cradle-to-grave approach. It will be possible to improve its environmental regulatory compliance, reduce environmental-equipment investment and facilitate a speedy closure at shut down [Breman and Hattway, 1995].

Housekeeping procedures, visual inspections, spill prevention program, implementing pollution/waste consciousness training program, supporting plant-wide waste recycling program, conducting leak detection and repair program play key role in minimizing and controlling the generation of waste water, reducing emissions to air water and soil.

Some of the major loss areas in a hydrocarbon industries are: Measurement and accounting of raw materials and intermediate/finished products, impurities in the feedstocks, effluent streams from individual equipment/battery limit/entire complex, vents, drains, overflows and equipment leaks, losses due to plant upsets and shut down and start up, climatic condition such as high temperature in summer and washing effects in rainy season, effects of varying throughputs and storage loss [Goyal, 1999].

Waste water generated from petroleum is quite complex and contains wide variety of pollutants making the waste water treatment system more complicated. It is going to be more and more complicated with increasing trend of integration of refinery and petrochemicals for value added products. No longer we are only concerned with reducing the level of BOD/COD, but also the specific pollutants.

Efficient water management in any large complex is crucial to controlling costs and satisfying environmental obligations. Basic approach for waste water management is—minimize generation, segregation, reuse/recycle, treatment and disposal. Reduction in waste water generation can be achieved by careful planning and selection of the process method, raw materials, operating conditions, equipments, product substitution, monitoring of water used and waste water discharge by taking corrective measures. By segregation of waste water streams, more recycling of water in the process can also help in reduction of waste. Prior to treatment of combined effluent from petrochemical complex source treatment of the individual effluent from various sections is provided. To be cost effective multilateral approach that characteristics process streams, identifies contaminant sources, quantifies pollutant level, etc. must be used. Hydrocarbon losses in refinery and petrochemical complexes can be substantially reduced by taking preventive actions, which vary from monitoring operating conditions to implementing justified cost effective capital projects. In many refineries and petrochemical plants, loss reduction programs are very successful.

The objective of a effective management plan should concentrate on—review of current and future projects, determination of impacts on waste water generation, reuse and secondary treatment, development; recommendation to minimize waste water generation and increase recycle, examination of solids and sludge generation/disposal and recommendation for reduction and treatment options, review of current and future air emissions and proposal for reduction options, determination of effluent quality criteria to meet current and future regulations.

Spent caustic from naphtha cracker/gas crackers and various sweetening processes are one of the major sources of water pollution in petroleum and petrochemical complexes and is contaminated with sulfides, mercaptans and phenols. Spent caustic is considered to be hazardous and cannot be directly managed by biological waste water treatment processes and must be treated separately before final biological treatment. Wet air oxidation technology is gaining popularity for treating spent caustic which detoxifies the spent caustic by oxidizing the sulfide and mercaptans to sulphate and breaks down the toxic naphthenic and cresylic compounds [Carlos, 2002]. After flow equalization spent caustic is pumped to charcoal filters for removal of tarry ad polymers and is fed to oxidation system. Both low temperature and high temperature have been tried. In low temperature oxidation, the sulfide is converted to thiosulphate while in high temperature it is converted to sulphate.

Increasingly stringent discharge requirement has necessitated the treatment of the waste water so that it can be recycled or discharged safely. Water use and waste water monitoring system is the first step in controlling and regulating the discharge of waste water. Various waste water treatment technologies are given in Fig. 24.2. Typical treatment system for petroleum industry may involve primary screening, flow equalization, oil/water separation, biological treatment, clarification, tertiary treatment and solids handling. Prior to collection of effluent for equalization and biological treatment, primary source treatment of the effluent from the individual effluent is carried out as per the requirement. Primary and secondary oil separation systems using API oil separators or parallel plate separators. Secondary separation may be required when oil content is high. This

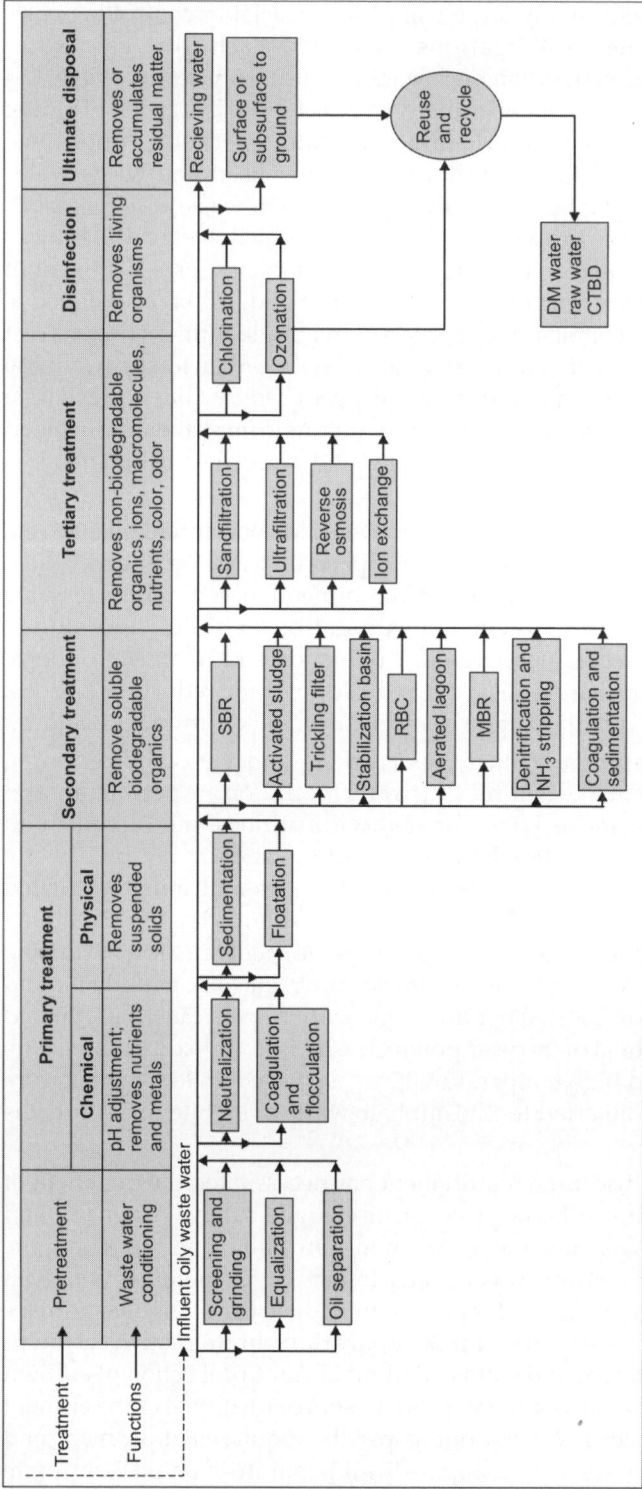

Fig. 24.2 General layout of effluent treatment plant of an oil refinery

involves use of coagulating agents like ferrous sulfate, ferric sulfate lime, poly-aluminum chloride have been used for de-emulsify an oil separator effluent. Sulfide precipitation is suited to waste water containing sulfide and mercaptans. Biological system includes trickling filter, activated sludge process, oxidation ponds, aerated lagoons are commonly used. Depending on the nature of waste water, a single or a combination of the units is used.

Some of the generic strategies for minimizing air toxic emission are process chemistry modification, changes in the specific constants responsible for emission, operational modification and preventive maintenance practice. Although emission from various sources can be minimized by these methods, however, some external treatment methods may be required in order to meet the environmental standards. In addition to this specific treatment, technologies may be required for the removal of some of the target toxic pollutants depending on the type of processes and products.

For reducing the CO emission from FCC recovery regenerator many refineries have installed CO converter which convert the CO to CO_2.

In petroleum complex, many plant components are potential sources for emitting volatile organic compounds. Monitoring and maintenance use of lower leak equipment collecting emissions and use of leak less technology can control fugitive emission [Siegell, 1996]. Some of the volatile organic compound control options are recovery vapor balancing, absorption, adsorption, refrigeration and destruction, thermal oxidation, catalytic oxidation and biofilters [Siegell, 1996]. Fugitive emissions can be controlled by elaborate leak detection and repair and equipment modifications. Many plants have implemented leak detection and repair (LDAR) programs in which various sources of fugitive emission like leaks from valves, flanges, pump seals, etc. are routinely monitored for leaks and maintained on regular basis (Venkatesh Moores, 1998).

Compiling an emission inventory is the first step in reducing or controlling emissions. It can also help in achieving waste minimization, process safety and quality goals. Catalytic reforming unit which is one of the major refinery and petrochemical process units is a potential emissions source of a number of compounds and has been identified as potential source of hazardous air pollutants by the EPA under the title III of the Clean Air Act (CAA) Amendments of 1990 [Beryrand and Siegell, 2003].

Fired heaters which are used in refineries and petrochemical complexes may be the major sources of sulfur emission as they use fuel oil. Fuel oil desulfurization, flue gas desulfurization and use of low sulfur fuel may be some of the methods may be an option for reducing the sulfur emission. Technological option to control NO_x can be controlled either by adjusting the combustion process by adjusting the parameters responsible for high NO_x emission or by flue gas treatment. Low NO_x burner technology for ethylene cracking furnaces has been developed in order to meet for reduction of NO_x emission from combustion processes to less than 10 ppm [Bussiman et al., Env progress, 2003, 1]. Feed desulfurization, use of additives, and flue gas desulfurization are some of the measures for reducing SO_x emission. Emission from sulfur recovery units which is a major source of SO_x emission can be increased by improving sulfur efficiency by integrating Tail gas clean up unit. For reducing the SO_x emission, IFP Clauspol 99.9+ process can be used. With the addition of a solvent desaturation loop, it is now possible to reach highest sulfur conversion rates.

Concept of employing biodegradation to destroy environmental pollutants in different kinds of wastes has increasingly applied [Sorial, 1997]. Bioscrubbing and infiltration are being used commonly for removal of air pollutants. Bioscrubbing is primarily used for contaminants strongly soluble in water. The contaminants are scrubbed from the waste gas in absorption column and then passed to a separate oxidation reactor employing a standard water treatment plant to aerobically degrade the contaminants [Sorial et al., 1997]. The biofiltration process utilizes a biological microbial film fixed on support media within a single process where the contaminants are both absorb from the waste gas and converted to benign end products such as water and CO_2. Over the last two decades biofilters have developed from systems abating odors to technically sophisticated and control units removing specific chemicals from industrial sources. With proper design and operation, VOC removal efifciencies of 95–99% have been achieved [Swanson et al., 1997]. The recent emergence of biofiltration as a cost effective air pollution control alternative has stimulated interest in its applicability among various industries [Leson and Smith, 1997]. Biofiltration technology has been found effective for the removal of major volatile organic compounds [Sorial et al., 1997; Leson and Smith, 1997; Zhu et al., 1998; Swanson and Loehr, 1997].

VOC recovery processes are adsorption, absorption, vapor balance and refrigeration. VOC destruction processes are: Use of flares, incinerators, catalytic oxidation and biofilters. The choice of the control method depends on many factors such as value of vapors, vapor quantity, number and location of emission points and viability of appropriate abatement technologies for the vapor under consideration [Ranade and Braswell, 1995]. Waste water plants in refinery and petrochemical complexes are the sources of VOC emission include stripping, adsorption, chemical oxidation, membrane separation [Sarvan and Pimparker, 2001].

Process vents emission can be controlled by various recovery and control devices like absorbers, chillers, catalytic incinerators, thermal oxidizer, flares, routine the vents to a boiler or process heaters are used. Use of floating roof tanks, proper design of storage tank piping to prevent the tank from being over- or under-pressurized and to prevent plugging are some critical issues for controlling storage tank emission. For controlling the emissions from equipment decommissioning and maintenance various piping designs, equipment designs, and equipment decommissioning procedures have been developed by the facilities that process extremely toxic or lethal materials, such as acrylonitrile and vinyl chloride [Venkatesh and Moorers, 1998]. The most cost effective control measure to reduce total fugitive emissions is to initiate leak detection and repair (LDAR) or monitoring and maintenance. The LDAR program uses a sensitive gas detection instrument and samples each piping component individually to determine the concentration of hydrocarbon adjacent to a potential leak site. Other methods for fugitive emission control are installation of new packing sets in block and control valves, upgrading of pumps seals to multiseal designs.

Storage tanks are also important sources of VOCs emission and account for 10–15% of total VOC emission from process plants. Three types of tank designs are used for the atmospheric storage of petroleum products: Fixed roof, external roof and internal floating roof. Some of the methods used for controlling VOC emission from fixed roof tanks are: Vapor balancing, vapor recovery and destruction and installation of internal floating head. Some of the tank emission controls in external floating roof tanks are checking of the condition of existing seals, replacement of vapor mounted primary with liquid mounted primary seal, controlling losses from roof fittings, installation of vapor recovery/destruction [Siegel, 1995].

Emission during loading can be controlled by use of submerged loading in place of splash loading, vapor balancing during loading and installation of vapor recovery system. Emissions from waste water treating system can be controlled by good housekeeping, by reducing waste water volume and concentrations, optimizing stripper operation, installation of sewer system suppression, reducing air/water contact area (covering, nitrogen blanketing) and replacing API separator with covered corrugated plate interceptor [Siegel, 1995].

Solid waste from petroleum and petrochemical complexes include oily sludge, spent catalyst, coke from cracking units, effluent treatment sludge, spent carbon and resins, contaminated soil, tank sediments, etc. and disposal techniques for solid waste include pretreatment and dewatering, detoxification, incineration/deep well injection, chemical and biological treatment, salvage of recoverable, materials and disposal by controlled sanitary and landfills.

REFERENCES

1. Minimal National Standard, Petrochemicals Industry (Basic and Intermediates). Central Pollution Control Board, Delhi.

2. Pollution Control Acts, Rules, Notifications Issued Thereunder. Central Pollution Control Board, Ministry of the Environment and Forest, Govt. of India, 3rd edition, 2000.

3. Allen DT and Rosselot KS. Pollution Prevention for Chemical Processes, John Wiley & Sons Inc (1997).

4. Bertrand RR and Spiegell JH. Emission of Trace Compounds from Catalytic Reforming Units. Environmental Progress, Vol 22, No. 1, 74, April 2003.

5. Brenneman DR and Hattaway DR. Incorporate Pollution Prevention into Pour Business Plan. Hydrocarbon Processing, Vol 4, No. 8, 84, 1995.

6. Bussman W, Poe R, Hayes B, Mcadams J and Karan J. Environmental Progress, Vol 21, No. 1, 1, April 2002.

7. Carlos TMS. Manage Refinery Spent Caustic Efficiently. Hydrocarbon Processing, Feb 2002, p89.

8. Guida JW and Fruge DE. Improve Waste Water Treatment. Hydrocarbon Processing, Vol 74, No. 8, 55, 1995.

9. Leson G and Smith BJ. Petroleum Environmental Research form Field Study on Biofilters for Control of Volatile Hydrocarbons. Jl of Env. Engg, 556, June 1997.

10. Ranade SM and Braswell JB. How to Design a Styrene Vapor Collection and Disposal System and Braswell. Hydrocarbon Processing, Vol 74, No. 8, 1995, p89.

12. Sarvanan R and Pimparkar PM. Sources and Control of Air Emissions from Refinery Complex. Hydrocarbon Technology, Issue No. 54, 51, March 2001.

13. Siegell JH. Exploring VOC Control Options. Chemical Engg, June 1996, p92.

14. Siegell JH. Control VOC Emissions. Hydrocarbon Processing, Vol 74, No. 8, 1995, p77.

15. Siegell JH. Control VOC Emissions. Hydrocarbon Processing, Vol 76, No. 4, 1997 p119.

16. Sorial GA, Smith FL, Suidan MT, Pandit A, Biswas P and Brenner RC. Evaluation of Trickle Bed Filter Air Biofilter Performance for BTEX Removal. Jl of Env. Engg, 530, June 1997.

17. Swanson WJ and Loehr RC. Biofiltration: Fundamental, Design and Operations, Principle and Applications. Jl of Env. Engg, 538, June 1997.

18. Venkatesh PE and Moores CW. Control Air Toxics from Difficult Process Sources. Chemical Engg. Progress, 26, Nov 1998.

19. Wang JM. Petrochemicals. Water Environment Research, Vol 70, No. 4, 658 (1998).

20. Zhu L, Abumaizar RJ and Kocher WM. Biofiltration of Benzene Contaminated Air Streams Using Compost–Activated Carbon Filter Media. Env. Progress, Vol 17, No. 3, 168, 1998.

25
Safety in Hydrocarbon Processing

Oil and gas industry carries out complex activities ranging from crude oil exploration/production, refining to distribution. Petroleum refinery has been playing a dominant role in fueling our socioeconomic growth by providing source of energy, feedstocks for petrochemical industry. With recent trends of integration of refinery with petrochemical, now the new concept of petrochemical refinery operates in a high risk environment. Now the hydrocarbon industry is subjected to high safety risk on account of potential hazard from fire explosion due to handling, processing and storage of highly flammable liquid with low flash point, [local.backpain.org] as well as gas and vapor and operation of the activities at elevated pressure [http://osid.govt.in/refining1.htm]. Process safety is always an essential part of oil and gas industry and a core value requires continual improvement [Mannanan, 2010].

In all associated activities starting from oil and gas exploration, petroleum refining, and petrochemical production are prone to potential hazard on account of explosions due to leakage, blow downs, process upsets, corrosives gases, extreme physical conditions in addition to potential fire and explosion from accidental release of flammable hydrocarbon. Severity of consequences may be human damage due to ill-health and injury, equipment damage, environmental damage, financial loss on equipment or production and increased liabilities. The recent developments in refining technology like catalytic hydrotreating, isomerization, hydrocracking and alkylation are increasing the complexity of the process and demanding added responsibilities towards safer environment. Due to rapid development in information technology and communication, safety awareness has grown so much that, today without safety every existence and growth in industry not imaginable. Various types of hazards associated in petroleum industry may be fire and explosion hazards, health hazards due to exposure of toxic/carcinogenic, corrosive and reactive chemical, mechanical and electrical hazards. Innovative offshore technology is being developed to carry out deep water production and operations. However, these hazardous operations are creating new and unique hazards [Mannanan, 2010].

Safety in the oil gas industry has become subject of increasing importance in recent years due to increasing stringent regulation in the area of safety, health and environment. Oil and gas organization has shifted from a prescriptive approach to a goal setting approach [Santos-Reyes and Beard, 2001]. Some of the factors influencing process hazard are:

- Reaction types (hydrolysis, oxidation, reduction, polymerization, isomerization, alkylation)
- Reaction parameters (stability, reactivity, exothermically pressure, temperature)
- Physical and chemical properties of substances, their toxicity
- Quantities of substances used and stored
- Storage characteristics (pressure, temperature) [Tixier, 2002].

It has been observed that human error is one of the major causes of incidents. Other causes are equipment failure, static electricity, design deficiency and others. Five principles of safety are [Gupta, 2010, Reference OSHA]:

- All accidents are preventable
- All levels of management are responsible for safety
- Every employee has the responsibility for himself/herself, their co-workers and their families to work safely
- To eliminate accidents, management must ensure that all employees are properly trained on how to safely and efficiently perform every job
- Every employee must be involved in every area of safety and production process.

25.1 SAFETY AND HAZARD IN OFFSHORE AND ONSHORE OIL AND GAS PRODUCTION

Oil and gas production activity in offshore and onshore has considerably increased during recent years due rising demand of petroleum products. An offshore rig is a platform which is exposed to a unique combination of hazards. An offshore platform is uniquely hazardous in that persons are miles out to sea and surrounded by huge quantities of powerfully combustible material [cedengineering.com]. Offshore operations has a very special environment, involving drilling, production and transport as well as emergency response to incidents [Mannanan, 2010]. A number of accidents has occurred in offshore drilling activity. Some of the principal hazards in offshore gas production are: Structural failure; ship collisions, severe weather, earthquake; falling objects, blowout, fire/explosion [Lees, 1996]. Some key hydrocarbon bearing facilities at an offshore are; well head, separators, gas compressor, gas or oil riser, oil and gas pipelines [Jones, 2003]. The whole platform structure is also prone to corrosive atmosphere. Some of the major hazards associated are corrosion of structures, leakages of hydrocarbons from pipelines and storage followed by ignition some of the major taken in event of leak are emergency shutdown and engagement of fire water protection systems. Some of the sources of leaks are reactor, tanks, pipe, flange, hose, valves, sight glass, pumps, flares, valve opened, venting, and nipples.

Some of the major offshore accidents are:

- Failure of oil riser in Norwegian Sector of North sea in 1975 resulting six deaths
- Piper alpha disaster in British sector of north sea in 1988 resulting in death of 167 persons

- ONGC helicopter crash in August 2003
- Major fire in ONGC's offshore platform in 2005 due to collision as vessel docked nearby in high tide

Government of India promulgated Petroleum and Natural Gas (Safety in Offshore Operations) Rules, 2008 in June this year. Special features of Offshore Safety Rules are:

- Cover all phases of offshore petroleum activities
- Based on goal setting approach
- Do not supplant rules/guidelines/clearance requirements of other Government departments and ministries
- Operator to identify risks associated with his operations and work out solutions
- Operator to identify requirements arising out of these rules and specify standards/codes/performance level for his activities, in order to comply with the rules.

Some of the major approaches to offshore hazard management are: Inherent safety design of platform, proper plant layout, electric power system, fire protection system, emergency shut down system, communication system, evacuation, escape and rescue system, fire protection system and explosion protection system. An increasingly common method of explosion protection is the use of blast walls [Lees, 1996]. Evacuation, escape and rescue system in offshore activity is built around three main means of leaving the platform, which are helicopter, life boat and life raft [Lees, 1996].

Fire and gas detection: There are two types of detectors which are commonly used in offshore installation: Heat, flame and smoke and flammable gas instruments. The most significant for risk detection are gas detection systems since they give earliest warning of hazardous situations. Infrared (IR), line-of-sight or point type of detectors are also used.

The fire protection system comprises a number of complementary elements which include: The hazardous are classification, the fire walls, fire and gas detection systems, the fire pumps, the foam system, the halon systems and fire fighting arrangements.

25.2 SAFETY AND HAZARDS IN CRUDE OIL AND NATURAL GAS PROCESSING

Separation of oil and gas from offshore and onshore facilities is done in oil and gas processing units at different locations and crude oil is transported to the refinery through pipeline tankers. Crude oil and natural gas processing units are associated with handling, storage and pressing of natural gas and crude oil.

25.3 SAFETY AND HAZARDS IN PETROLEUM REFINERY

Modern petroleum refinery is a complex operation consisting of large number of process units and handle wide variety of feed and products consisting of gas, liquids and solids which are hazardous in nature. The complexity has further increased with coming of newer processes and integration of refinery with petrochemical complexes. In recent past, the industry throughout its value chain has seen major upgradations and capacity built up in petroleum refining with introduction of newer technologies like catalytic hydrotreating, isomerization, alkylation, hydrocracking, FCC, delayed coking to extract maximum bottom and barrel of oil. The principal hazards in refineries are due to fire and explosion as refineries process large number of products with low flash points. Many hazardous chemicals are present in the refineries which result in fire, explosion, toxicity, corrosiveness and asphyxiation. Hydrogen sulfide is a potential problem in the transport of crude oil and storage of crude oil. Alkylation process in the refinery uses hydrofluoric acid or sulfuric acid which are highly toxic and hazardous in nature.

Safety health and environmental management has become one of the major activities in refinery to have a safer environment and sustainable development.

Petroleum refinery inherently hazard prone because of [Gupta, 2010]:

- Large inventory of petroleum raw material and product which are highly inflammable
- Processing at high pressure, temperature
- Using hazardous chemicals having high corrosivity and toxicity
- Complexity and process integration
- Uncontrolled process reactions
- Selfignition on leakage from system.

25.3.1 Explosion and Fire in Isomerization Unit

Explosion and fire incident during start up of isomerization unit due to flooding of splitter tower with hydrocarbon resulting in release of HC to atmosphere and ignition from nearby source. Plant was shattered and there were fatality resulting in death of 15 workers. Some of the reasons for above hazard were

- Pressure relieve system was not connected to flare
- Malfunction of level indicator, alarm and control valve
- Failure to follow procedure leading to overflow of tower
- Process safety was neglected.

25.3.2 Explosion and Fire Incidents During Commissioning of Slurry Settler at FCC Unit

Explosion and fire incident during commissioning of slurry settler at FCC unit where slurry settler got overpressurized due to presence of steam condensate and reactor and slurry settler was damaged due to pressure effect. Accident causes in oil and gas industry are given in Table 25.1.

25.4 OIL SPILL HAZARD AND MANAGEMENT

Demand of crude is increasing faster than to the proven reserves and major portion of the demand is met through import. Ocean remains the frontier of international trade and shipping is the main route for import of crude. India, South East Asia, East Asia, Japan and China are heavily dependent on import of crude from West Asia via Indian Ocean through Arabian Sea. However, narrowness of lanes, routes is accident prone. India's 7500 km long shoreline is, therefore, at risk of a serious ecological disaster from oil spills. The major source of oil spills, next to transportation

Table 25.1   Some of the major accidents in oil and gas industry

Incidents	Description
Feyzin Refinery, 1966	Leakage from propane storage sphere ignited, caused a fire which burned fiercely around the vessel and led to BLEVE
Linden, New Jersy, 1970	Local overheating in a hydrocracker unit caused an explosive failure.
Hearne, Texas, 1972	Ignition and explosion occurred in crude oil pipeline
Texas City, Texas, 1978	A series of fires and explosion occurred at LPG storage
Milford Haven, 1983	Fire broke out on the roof of a large crude oil tank a the refinery
Romeovillew, Illinois, 1984	Spontaneous failure occurred due to vapor cloud explosion in an absorption column where propane/butane feedstock was contacted with methanol amine
Grangemouth Refinery, 1987	Explosion occurred in hydrocracker plant resulting in severe fire.
Norco, Louisiana, 1988	Explosion occurred on catalytic cracker at the refinery. In depropanizer overhead piping due to internal corrosion
Texas Refinery in March'05	Explosion and fire incident during start up of isomerisation unit due to flooding of splitter tower with hydrocarbon resulting in release of HC to atmosphere and ignition from nearby source
Incident at Indian Refinery in Oct'04	Explosion and fire incident during commissioning of slurry settler at FCC unit
Offshore Platform in July '05	A multipurpose support vessel (MSV) hits the riser pipes carrying oil and gas to the offshore platform. Impact resulted in leak of HC which ignited engulfing the MHN platform. Platform completely collapsed into sea.
Oil Storage Depot at UK in Dec '05	Motor sprit escaped from storage due to overflow. Vapor cloud got ignited and large fire engulfed 20 storage tanks. Significant damage occurred in the vicinity around the site.

Source: Verma, 2009.

is the hydrocarbon exploration arena. With the increase in exploration and production, the probability of oil spills increased during recent years. Oil spill also occurs due to natural disaster.

Road map for spill management for India has been prepared by Project Review and Monitoring Committee (PRMC) for Oil Spill Management. Standardization of practice and policies and a single point approach to deal with all aspects of oil spillage management by a centralized independence agency was recommended by PRMC. Some of the entities are Oil Spill Management and Regulatory Authority (ORSA), Oil Spill Authority of India (OSA), Oil Spill Management Authority of India (OSMA) [PRMC, 2003].

25.5 HAZARDS IN FLARING

Flares are commonly used in crude oil and natural gas processing and refinery for safe disposal of unwanted hydrocarbons. However, malfunctioning of flares can have serious consequence and may result to explosion due to entry of air into the structure of pipe and valves that convey the gas for burning [Jones, 2003].

25.6 SAFETY IN TRANSPORTATION OF HYDROCARBON AND COMPRESSED GASES

Due to increasing industrialization, population growth and revolution in automobile and domestic and industrial application of hydrocarbon—liquids and gases in the form of compressed natural gas (CNG), LPG, petrol, diesel, kerosene fuel oil transportation of hydrocarbon gases and liquid has become one of the major activities. Various ways of transportation are: Through shipment, mobile tankers, gas cylinders and pipeline. Safety associated with transportation of hydrocarbons has become one of the major causes of during recent years.

Some of the general rules for transportation of hydrocarbons are [Nigam, 2007]:

- Liquid hydrocarbon, i.e. petroleum: Petroleum Rules, 2002 framed under the Petroleum Act, 1934.
- The transportation of compressed gases in bulk by mobile transport vehicles is regulated under Static and Mobile Pressure Vessels (unfired) Rules, 1981.
- Transportation of compressed gasses
- By cylinders (under Gas Cylinders Rules, 2004).
- By tankers (under SMPV(U) Rules, 1981 [nidm.gov.in])

Implementation of effective corrosion prevention program can help in achieving the safe, economical and reliable operations of pipelines. Approach is based on the following five components [Sinha, 2009].

- Identification of possible causes of corrosion
- Evaluating approaches and methods to deal with corrosion.
- Selecting corrosion monitoring and mitigation methods with CP and diverse coating solutions.
- Dealing with challenges in achieving effective corrosion mitigation measures
- Instituting an effective corrosion mitigation program and experience sharing.

25.7 PIPELINE SAFETY

Pipelines are considered as the safest, most economical and environment friendly mode of transportation of crude oil and gas and petroleum products, and provide important link of petroleum supply chain management and cutting edge to petroleum industry. The largest beneficiary of this infrastructure is the customer who has to pay much less towards transportation cost from port/refineries to consumption center. There have been significant losses due to failure of piping and piping leaks. Causes of failure are mentioned in Table 25.2. Major hazards associated in pipeline safety is due to [Gail India, 2007]:

- Corrosion—internal or external
- Wrong metallurgy
- Human errors during pigging hot tapping, valve operation
- System procedure failure—inspection, operation, start up/shut down, material specification and testing
- External reason—accidental excavation, earthquake, flood, fire, lightening, rail/road accident.

Major causes of pipeline external corrosion are poor/defective coatings, inadequate cathodic protection (CP), and coating defects combination with inadequate CP, interference due to external agencies, stress and bacterial corrosion. Major causes of internal corrosion is corrosive nature of fluid transported through pipeline, erosion—corrosion, localized chemical attack/bacterial corrosion.

Table 25.2   IOC pipeline static of last decades

Sl. No.	Cause of failure	Reason	Percentages
1.	Mechanical	Due to material or weld defects and construction faults	9%
2.	Corrosion	Due to internal or external corrosion	38%
3.	Natural hazard	Due to land slide, flood, storm, soil instability, etc.	1%
4.	Pilferage/sabotage	Damage caused by third party with malicious intention	50% Pilferage (46%) Sabotage (4%)
5.	Others	Unclassified reasons	2%

Source: Sinha, 2009.

The salient features ensuring the safety in petroleum transportation are listed as under [Nigam, 2007]:

- No leaky tank or container contain petroleum shall be tendered for transport. Barrels, drums and other container filled with petroleum shall be loaded with their bung upwards.
- No ship, vessel or vehicle shall carry petroleum (class A or B or C) in bulk, if it is carrying passengers or any other combustible cargo other than petroleum.
- No person while engaged in loading/unloading or transporting petroleum shall smoke or carry matches, lighters or any other appliances capable of producing ignition or explosion.
- Petroleum shall not be loaded into or unloaded from any ship vessel or vehicle between the hours of sunset or sunrise unless adequate electric lighting is provided at the place of loading or unloading and adequate fire fighting facilities with trained personnel are kept in place for immediate action in the event of fire [explosives.nic.in].

Mitigation of Pipeline Hazards

- Protection against external corrosion
- Protection against internal corrosion
- Protection against third party damages
- Protection of pipeline supports
- Leak detection system
- Pig based monitoring system
- Protection against overpressures
- Protection against detonation hazards.

25.8 VAPOR CLOUD EXPLOSION

Vapor cloud explosion has been a cause of major concern in oil and gas sector. Vapor cloud explosions incidents cause damage to life, environment and property. Explosion results instant and wide spread damage to life and property. Some examples of vapor cloud explosion are Feysin, 1966; Fixbourgh, 1974; Mexici, 1984; Piper Alpha, 1988; Pasadena, 1989; Esso Longford, 1998; Buncefield, 2005; BP Texas Refinery, 2005; Jaipur (India). In almost VCE incidents, there was a huge and instant release of hydrocarbons/gas by ignition of vapor cloud mostly without much time delay. Efficient mitigation to prevent vapor cloud explosions is the requirement [Tolmare, 2011].

25.9 SAFETY, HEALTH AND ENVIRONMENTAL (SHE) MANAGEMENT

Process safety, occupational health and environmental issues are ever increasing in importance in response to heightening public concerns, and the resultant tightening of regulations [Venkatasubramanian et al., 2000]. SHE management in hydrocarbon industry is a multidisciplinary function and every person needs to keep vigil against potential cause of fire and accidents and draw lessons from past incidents. Safety involves:

- Protection of personnel
- Protection of plant and property
- Production uninterrupted
- Legal requirements
- Welfare of community
- Good reputation.

Some of the commonly used approaches to prevent accidents and ensure safety are enforcement of safety rules and human resource development approach are: Psychological approach, safety management approach, engineering analytical approach, total quality management approach for ultimate safety [Sood, 2005]. Gas Authority of India Limited (GAIL) has taken following initiative to implement SHE management [GAIL, 2007b].

- SH and E Leadership and Commitment
- Employees Participation
- Facility, Design Construction and Prestart up Review
- Process Safety Information
- Risk Analysis and Management

- Third Party Services
- Personnel Safety
- Control of Defeat and Reliability of Critical System and Devices
- Work Permit System
- Operation and Maintenance
- Inspection and Maintenance
- Management of Change
- Training
- Incident Investigation and Analysis
- Occupational Health
- Environment Management
- Emergency Planning and Response
- Compliance Audit [www.archfin.com].

25.10 ACCIDENTS

Accidents can be classified into five groups [Venkateshwar Rao et al., 2002]:
- Release of flammable gases, flash fire and VCE
- BLEV
- Toxic vapor release
- Pool fires
- Fragmentation.

Accidents can be instantaneous or they can be consequence of minor events leading to a major one. Outcomes of accidents are injury to the person: Suffering from pain, worry, absences, incapability to perform, loss of wage, medical expense, loss of limb or life, organization loss: Good man is lost, new person for the job to be trained, more supervision, loss of output, medical expenses, loss of morale, loss of prestige; damage to equipment and property, litigation costs and lost productivity.

25.11 SAFETY MANAGEMENT

Safety management in an organization can be reactive or proactive. Safety management will be reactive, if safety measures are implemented after an accident has occurred. However, it will be proactive, if safety management implies realization of actions ahead of accidents and incidents.

Proactive actions to prevent incidents are [Verma, 2009]:
- Process safety through hazard identification
- Develop safe operating procedure to minimize operational risk.
- Ensure mechanical integrity of process equipment and facilities including safety instrumentations
- Minimize human error through training of workforce
- Develop core competency with continuous focus on safety
- HSE aspects to be included in the curriculum of engineering institute to minimize the knowledge gap of fresh engineers.

The 3 Ps of safety management [Raichur, 2009]:

- **Safety policy:** Specifying how safety will be achieved
- **Safety procedures:** What management wants people to do to execute the policy
- **Safety practices:** What really happens on the job.

25.11.1 Hazard Identification and Assessment

Industrial accidents are usually accompanied by negative consequences for human life, property and environment. These accidents provide new knowledge of chemical processes and their properties [Laskova and Tbas, 2008].

Hazard identification and control techniques: [Bhardwaj, 2005] Hazard is a latent factor of the unforeseen occurrences that represent a potential for an accident.

The techniques most frequently used to identify and assess hazards are HAZOP, What-if method, Check Lists, Failure mode and Effect Analysis, Fault Tree Analysis, Event Trees and Task Analysis.

Objectives of a HAZOP study to review the process design of the unit, identify all deviations from the way the design is intended to work, their causes and all potential hazards or operability problems associated with these deviations. Causes and consequences of hazards are mentioned in Table 25.3.

Table 25.3   Hazards are identified with possible risks

Hazards	Activities	Consequences
Chemical hazard	Chemical handling/storage	Fire, acute/chronic toxicity, ecotoxic
Electrical hazard	Plant operation	Fire, shock, burns
Mechanical hazard	Plant operation	Injury
Physical hazard	DG operation	Accident

25.11.2 Risk Assessment and HAZOP Study

A hazard is anything that may cause harm, ill-health or injury, damage to property, plants, products or environment, production losses or increased liabilities. Examples of hazards are hydrocarbons under pressure, objects at height, toxic chemicals, electricity, etc. The risk is the chance, high or low, that somebody could be harmed by these or other hazards, together with an indication of how serious the harm could be [Mukherjee, 2005]. There are different stages of risk assessment process that have been mentioned in Fig. 25.1.

Risk assessment: A risk assessment is simply a careful examination of what, in your work, could cause harm to people, so that you can weigh up whether you have taken enough precautions or should do more to prevent harm. Risk management process involve [Shah and Moosemiller, 2012]:

- Risk identification : Determine what events are possible
- Risk significance   : Evaluate and prioritize risks
- Risk options   : Develop and evaluate risk-reducing options
- Risk decisions   : Decide what to do

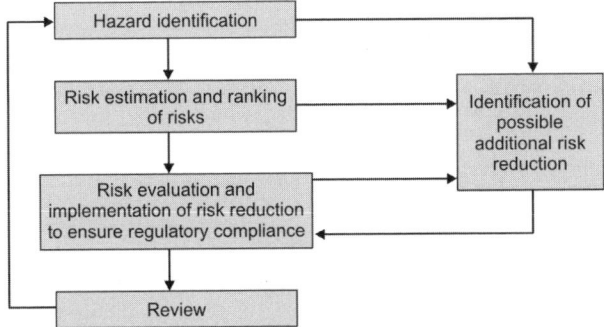

Fig. 25.1 Main stages of risk assessment process

Steps in a risk assessment are given below [local.backpain.org]:
- Identify the hazard (hazard identification)
- Assess the hazard (hazard assessment)
- Assess the consequences, whether acceptable or otherwise (consequence modelling)
- If the consequences are serious, quantitative evaluation of event probability must be made.
- If consequences are not acceptable and magnitude cannot be reduced, control measures will need to be implemented for incident prevention.
- The resultant estimation of risk is then compared with agreed criteria (risk estimation).
- If the criteria is met, then it is classified as acceptable.

The severity of the risks could be classified as follows [Dhar and Katoch, 2007]:
- High Risk: High probability and severe consequences.
- Moderate risk: High probability with mild consequences or low probability with severe consequences.
- Low risk: Low probability with mild consequences.

The techniques used to evaluate are [Dhar and Katoch, 2007]:
- Fault tree analysis: Evaluation of probability of occurrence
- Risk ranking matrix: Helps in ranking the severity of the identified hazards/risks.
- Consequence analysis: Evaluation of consequences (generally by the use of risk analysis software under different simulated situations).

Risk Ranking

The best strategy for controlling risks is to first draw up an inventory of hazards/risks and then identify the more severe ones using risk ranking matrix given in Table 25.4 and then formulate an action plan for either of the options in the order of priority as follows [Dar and Katoch, 2007]:

HAZOP study

HAZOP is a structured technique which is used to examine potential deviations that coil occur for each part of the plant. The consequence (including knock on

Table 25.4   Risk ranking matrix

Probability→ Severity↓	Likely	Unlikely	Highly unlikely
Serious	Intolerable	Significant	Moderate
Moderate	Significant	Moderate	Tolerable
Minor	Moderate	Tolerable	Insignificant

effects) of each deviation judged to have a credible cause is considered [Anand and Row, 2008]. HAZOP identifies potential hazards and operability problems in new designs and continuous processes [EHSC Royal society of Chemistry www.rscx.org/lap/rsccom/ehsc/ehscnotes.htm.].

Emergency preparedness and emergency plans

Disaster is a reality which co-exist with mining, however, much we may like to get rid of it. Hence emergency management plan is an essential component of a hazardous industry. Preparation of emergency plan is a statutory requirement. Relevant portion of statute in brief is given below:

Regulation 51A of OMR, 1984 provides for preparation of emergency plan for group gathering station and submission to the regional inspector and district magistrate specifying:

a. Duties and responsibilities of each key personnel
b. Alarm and communication system
c. Equipment plan
d. Plan for training of personnel and for mock drills.

25.12 PROCESS SAFETY MANAGEMENT (PSM) SYSTEM

Process safety management system in India is still at development stages, although it came into picture in the late 1980s and early 1990s [Gopal, 2013]. PSM system has implemented in ESSAR. PSM provides a systematic approach towards achieving these ends, resting on the three pillars of integrity such as operational integrity, plant integrity and design integrity (Fig. 25.2). Looking to the importance of safety ESSAR has implemented refinery integrated management system.

25.13 EMERGENCY RESPONSE AND DISASTER MANAGEMENT PLAN

Emergency response and disaster management plan is shown in Fig. 25.3, has an important aspect of sound safety management to reduce the probability of serious loss to people, equipment, material, environment, process, reservoir, etc. A disaster management plan should include hazard, assessment, loss prevention methodology, emergency response programs and overall disaster management system [Venkateshwar Rao et al., 2002].

Four elements [PRRP] of disaster management recognized by Commonwealth Government [Gupta, 2010]:

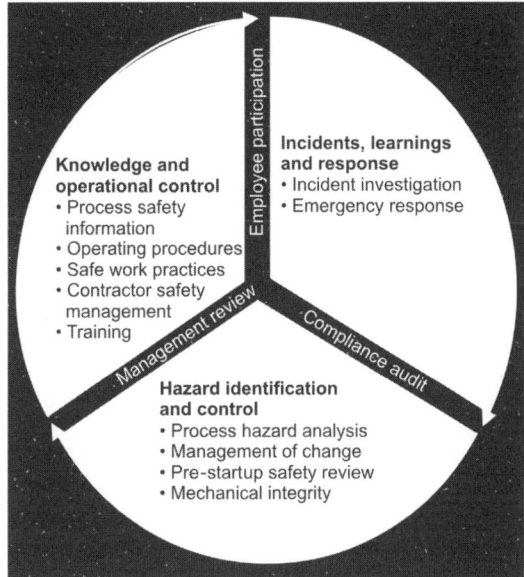

Fig. 25.2 Operational, plant and design integrities

Source: Gopal, 2013.

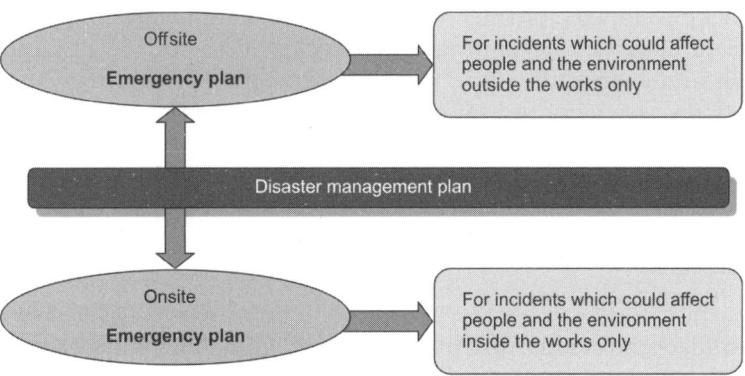

Fig. 25.3 Disaster management plan

- Preparedness
- Response
- Recovery
- Prevention.

Effect of occupational exposure of hazardous airborne chemicals in HC industry [Brij Mohan, 2009]:

- Petroleum and natural gases are important sources of alkanes, alkenes, cycloalkenes and arenes.
- On exposure to hydrocabon irritation, asphyxiation, narcosis and systemic damages are observed among workmen whereas chemicals like benzenes are carcinogenic.

- Cases of dermatitis, folliculitis and skin cancer are reported on repeat skin contact with some chemicals like pitch, asphalt and tar in HC industry.

Degree of hazards depends upon nature of chemical, concentration and length of exposure, method of handling and individual susceptibility.

25.14 OIL INDUSTRY SAFETY DIRECTORATE (OISD)

OISD is a technical directorate under the Ministry of Petroleum and Natural Gas, Govt. of India, was setup with the aim at enhancing SHE performance in oil gas sector in India, formulates and develops safety standards/guidelines for oil and gas industry with the jurisdiction in exploration and production, petroleum refining and gas processing, cross country pipelines, rail/road transportation and marketing installations. Oil industry safety directorate has developed standards and guidelines covering [Verma, 2009]:

- Design and layout
- Operation and inspection practices
- Safety audit procedures/guidelines
- Safety management system
- Emergency preparedness
- Training in safety.

OISD has developed more than 100 standards and guidelines covering the entire spectrum of activities pertaining to hydrocarbon sector: design and layout, operation and inspection practices, safety audits procedures/guidelines, emergency preparedness and safety management training.

- Exploration and production: On land/offshore
- Refining and gas processing
- Transportation: Pipeline, rail, road
- Storage and distribution
- Environment protection.

To formulate and standardize procedures and guidelines in the areas of design, operations and maintenance as also the creation of new assets with a view to achieve the highest safety standards in a cost effective manner [worldlpgas.com].

- Design and layout facilities
- Construction, operation, maintenance and inspection procedures
- Work permit system
- Safety audit procedures/guidelines
- Training
- Personal protection equipment
- Occupational health monitoring
- Fire protection and fighting facilities
- Environment protection, oil spill
- Emergency preparedness
- Safety management system.

25.14.1 Importance of Safety

Protection of personnel, protection of plant and property, production uninterrupted, legal requirements, welfare of community and good reputation.

Safety management system based on OISD STD 206 and ISRS

- Safety policy
- Safety committees
- Employees participation
- Process safety information
- Process hazard analysis
- Operating procedures
- Training
- Contractors
- Pre-startup safety review
- Mechanical integrity.

OISD

- Exploration and production: On land/offshore
- Refining and gas processing
- Transportation: Pipeline, rail, road
- Storage and distribution
- Environment protection.

To formulate and standardize procedures and guidelines in the areas of design, operations and maintenance as also the creation of new assets with a view to achieve the highest safety standards in a cost effective manner.

- Design and layout facilities
- Construction, operation, maintenance and inspection procedures
- Work permit system
- Safety audit procedures/guidelines
- Training
- Personal protection equipment
- Occupational health monitoring
- Fire protection and fighting facilities
- Environment protection, oil spill
- Emergency preparedness
- Safety management system.

25.14.2 Safety Culture

Safety does not happen by itself or by external enforcement. It has to be inbred, developed, nurtured and encouraged by management and every member of the organization [Bhular, 2007]. Five important key organization cultures that need to be taken into account are [McKay and Lacourseiere, 2008]:

- Maintain sense of vulnerability
- Establish and imperative for safety
- Perform valid/timely hazard/risk assessments

- Ensure open and frank communications
- Learn and advance the culture [Mckay, 2008].

Kao et al., 2008 have reported eight global dimensions of safety culture such as safety commitment and support, safety, attitude and behavior, safety communication , and involvement, safety training and competence, safety supervision, and audit, safety management organization, accident investigation, and emergency planning and reward benefit [homepages.wmich.edu].

25.14.3 Safety Report [Kerada, 2009]

Safety Report is a comprehensive documentation of a particular organization and includes:

- The name and address of mine, owner and GGS
- Description of installation namely
  - o Site
  - o Construction design
  - o Identification of hazardous area and safety distances
  - o Maximum number of persons working on the site and those exposed to hazard
- Description of process
- Description of the hazardous substances
- Information on the preliminary hazard analysis

  Description of safety relevant units, amongst others:
- Special design criteria
- Control and alarms
- Special relief systems
- Quick acting valves
- Sprinkler system
- Fire fighting, etc.

  Information on the hazard assessment namely:
- Identification of hazards
- The causes of major accidents
- Assessment of accident consequences
- Safety system
- Known accident history

  Organization to carry on the activity safely namely:
- Maintenance and inspection schedule
- Guidelines for the training of personnel
- Allocation and delegation of duties for the safety of installation
- Implementation of safety procedures [www.labourandemployment.gov.in].

IS 18001: 2000 Indian Standards on Occupational Health and Safety (OH & S) Management System [Diundi, 2005]

IS 18001 provides the industry a systematic, structured and documented framework to address OH & S issues effectively and in an auditable manner. The element of OH & S management systems are as follows:

- OH & S policy
- Planning
- Implementation and operation
- Measurement and evaluation
- Management review.

Some of the benefits of OH & S management system are [Diundi, 2005]:
- Reducing the number of personnel injuries through preventive and control of workplace
- Reducing the risk of major accident
- Reducing the material loss caused by accident and in production interruption
- Ensuring the appropriate legislation is addressed
- Meeting increasing importance of OH & S for public image
- Serving the possibility for an integrated management system including quality, environment and OH & S [www.goldenpeacockawards.com].

25.15 FIRE PROTECTION AND FIRE FITTING FACILITIES

Fire is the greatest hazard in oil and gas sector. Along with an effective safety management, fire protection and fire effective fire fighting facilities is very important. An effective fire protection system consist of [Gopal, 2012]:

Fixed fire protection system	Fire water pump, hydrate/monitor/risers, medium velocity spray system, semi-fixed foam pouter system, deluge valves, F & G detection/alarm system, Jetty tower monitor system, Jetty jumbo water curtain, gas detectors, fire and smoke alarm
Portable fire protection system	Fire extinguishers, portable monitors, hoses, foamy trolly, fire buckets, foam fire retarders
Mobile fire protection system	Fire cum water tender, foam nurser, DCP tender, water tankers, fire jeeps
Fire hydrant system	Fired water line, fire hydrants, water monitors, water cum foam monitors, pressurized hydrant system, automatic water spray system and remote operated valves at LPG horton spheres

REFERENCES

1. Nigam A. Safety in Transportation of Hydrocarbon and Compresses Gases/Safety Issues and Handling of Emergency. Industrial Safety and Socioeconomic Growth. The Institution of Engineers (India) 2007, p8.

2. Bhardwaj. Moser Baer Corporation Initiatives Developing Environment, Health and Safety Culture. Proceedings of Safety Convention 2005, organized by Safety and Quality Forum, The Institution of Engineers (India) and Quality Council of India, Hotel Le Meridian, New Delhi, INDIA, Feb 14–16, 2005, p346.

3. Brij Mohan. Industrial Hygiene and Occupational Health in Hydrocarbon Industry.

4. Bhular RS. What can You do to Enhance the Safety of Operations. Hydrocarbon Processing. Oct 2007.

5. Dhar AK and Katoch RC. Industrial Risk Analysis. Proceedings of Safety Convention 2005, organized by Safety and Quality Forum, The Institution of Engineers (India) and Quality Council of India, Hotel Le Meridian, New Delhi, India, Feb 14–16, 2005, p263.

6. Diundi IS. IS 18001: 2000 Indian Standards on Occupational Health and Safety (OH & S) Management System—Specification with Guidance for Use. Proceedings of Safety Convention 2005, organized by Safety and Quality Forum, The Institution of Engineers (India) and Quality Council of India, Hotel Le Meridian, New Delhi, India, Feb 14–16, 2005, p357.

7. GAIL, 2007a GAIL (India) Limited, Agartala Proceedings of Safety Convention 2007, organized by Safety and Quality Forum, The Institution of Engineers (India) and Quality Council of India, Hotel Le Meridien, New Delhi, India. Sep 5–7, 2007, p113.

8. Gopal J. Process Safety Management System. Implementation at ESSAR Oil. Chemical News, March 2013, p30.

9. Gopal J. Best HSE Practices at ESSAR Oil. Chemical News, March 2012, p16.

10. Gupta KK. Health and Environment: Issues and Challenges. 6th Summer School on Petroleum Refining and Petrochemicals, IIPM Gurgaon, June 7–11, 2010, organized by IOCL and Petrotech 2010.

11. GAIL, 2007b GAIL (India) Limited, Vaghodia. Proceedings of Safety Convention 2007, organized by Safety and Quality Forum, The Institution of Engineers (India) and Quality Council of India, Hotel Le Meridien, New Delhi, India, Sep 5–7, 2007, p131–132.

12. Jones JC. Hydrocarbon Process Safety. PennWell, 2003.

13. Kao, Chen Shan, Lai WH, Chuang TF and Lee J. Safety Culture Factors, Group Differences and Risk Perception in Five Petrochemical Plants. Process Safety Progress, Vol 27, No. 2, June 2008, p145.

14. Kerada NK. Workshop Safety, Health and Environment Management In Hydrocarbon Industry, Jan 22–24, 2009 at IIT Roorkee, sponsored by Lovraj Kumar Memorial Trust and Petrofed, Petrotech.

15. Laskova A and Tbas M. Method for the Systemical Hazard Identification. Process Safety Progress, Vol 27, No. 4, Dec 2008, p289.

16. Lee FP. Loss, Prevention in the Process Industries, Vol 1, Butterworth Heinemann, 1996.

17. Mannan MS. Stretch in Technology and Gaps in Process Safety for the Hydrocarbon Industry. Hydrocarbon Processing, Nov 2010, p31.

18. Mckay M and Lacoursiere. Development of a Process Safety Culture of Chemical Engineers. Process Safety Progress, Vol 27, No. 2, June 2008, p153.

19. Mukherjee. Safety, Health and Environmental Management in Hydrocarbon Industry, Jan 22–24, 2009, Indian Institute of Technology, Roorkee.

20. Nigam A. Safety in Transportation of Hydrocarbon and Compressed Gasses/Safety Issues and Handling of Emergency. Proceedings of Safety Convention 2007, organized by Safety and Quality Forum, The Institution of Engineers (India) and

Quality Council of India, Hotel Le Meridien, New Delhi, India, Sep 5–7, 2007, p11–16.

21. PRMC, 2003. Road Map for Oil Spill Management for India. Project Review and Monitoring Committee for Oil Spill Management, May 2003.

22. Raichur AA. Safety Management System Workshop. Safety, Health and Environment Management in Hydrocarbon Industry, Jan 22–24, 2009 at IIT Roorkee, sponsored by Lovraj Kumar Memorial Trust and Petrofed, Petrotech.

23. Santos-Reyes J and Beard AN. A Systemic Approach to Fire Safety Management. Fire Safety Journal, Vol 36, p359–390 (2001).

24. Shah JN and Moosemiller MD. Dynamic Risk Management. Chemical News, Sep 2012, p27.

25. Sinha JP. Integrity of Cross Country Pipelines Workshop. Safety, Health and Environment Management In Hydrocarbon Industry, Jan 22–24, 2009 at IIT Roorkee, sponsored by Lovraj Kumar Memorial Trust and Petrofed, Petrotech.

26. Sood VR. Managing Safety: Fertilizer Industry and Example. Managing Safety Challenges Ahead, edited by Chaturvedi V, organized by Safety and Quality Forum Council, Feb 14–16, 2005.

27. Tixier J, Dusserre G, Salvi O and Gaston D. Review of 62 Risk Analysis Methodologies of Industrial Plants. Journal of Loss Prevention, Vol 15, 2002, p291.

28. Tolmare GB. Vapor Cloud Explosion—Prevention or Mitigation Compendium 16th Refinery Meet, organized by Center for High Technology and Indian Oil Corporation, Feb 17–19, 2011, p357.

29. Venkatassubramanian V, Zhao J and Viswanathan S. Intelligent Systems for HAZOP Analysis of Complex Process Plants. Computers and Chemical Engineering, 24 (200), 2391–2302.

30. Venkateshwar Rao S, Ramanayya KV, Charey V and Khan AA. Disaster Management Plan. Chemical Industry Digest, Jan–Feb 2002.

31. Verma JB. Invited Lecture Workshop. Safety, Health and Environment Management in Hydrocarbon Industry, Jan 22–24, 2009 at IIT Roorkee, sponsored by Lovraj Kumar Memorial Trust and Petrofed, Petrotech, India.

26

Oil Movement, Storage, Crude and Product Blending and Handling

Oil movement, storage, blending and product blending and handling is an important function in a refinery. Refinery tank farm provides linkages with all process units as well as dispatch terminals for movement of crude oil, products from different sections and blending of finished product and their storage. Entire tank farm in a refinery consists of a number of tanks, extensive piping network and various pump houses. Crude receipt, storage and processing is very important operation as refineries are processing both indigenous and imported crude, receipt, storage and processing of crude through crude pipeline and storage in tanks and finally blending of crude oil. With increasing use of different types of crude and use of a number of secondary conversions processes to improve the quality of fuel, product of same quality from various streams are also blended. Natural gas storage and transportation infrastructre has also received much importance in refinery, petrochemical, energy and transportation sectors. Increasing production and use of the shale gas, requires additional storage and transportaion infrastructure. Global natural gas demand is expected to double by 2020 which has increased activity in natural gas exploration. Globally 75% of the natural gas is transported through pipeline and remaining 25% as liquefied natural gas (LNG) [Saraf, 2007]. In recent years, LNG and CNG have emerged as primary source for natural gas.

Major types of oil and gas and products requiring transportation and storage are given below:

- *Natural gas:* Natural gas associated and non-associated, condensate, LNG and CNG, shale gas requiring natural gas gathering, processing, storage and transportation infrastructure facilities.
- *Crude oil:* Crude oil from oil field, imported crude, opportunity crude gathering, processing, storage and transportation infrastructure facilities.
- *Petroleum products and LPG:* Storage, blending, transportation of petroleum products including LPG. Transportation may be by pipe, marine, rail/truck tankers.

Increasing use of variety of crudes, natural gas and petroleum products has increased infrastructure investment.

26.1 REFINERY TANK FARM

Different types of storage tanks are located in the refineries for storage of crudes, blending stock for products and products. Tank farms are grouping of storage tanks at producing fields, refineries, marine, pipeline and distribution terminals and bulk plants which store crude oil and petroleum products [Stellman]. Tank farm is considered as focal point of refiner with increasing use of various types of crudes as well as different types of products requiring more storage tanks. The problem has become more complex with increasing trend of integrating refinery with petrochemicals with number of petrochemical intermediates. As tanks are expensive to build or maintain tanks it has become very important to reduce the number of storage tanks to allow significant saving on tank inspection and maintenance activities [Painter, 2011]. Optimizing the storage within a refinery is a complex task. Conventional LP model is not some of the models use by refineries for planning with optimal level of storage are conventional LP model, traditional RAM (reliabilty, availability and maintainability). Conventional LP model assumes that palnt operations are predictable and steady state in nature. Model, total review and optimization (TARO) simulation model is a discrete event simulation tool that is capable of quantifying the expected performance of an entire refinery complex or of individual equipment items [Painter, 2011].

For ensuring optimized operation and operational safety some of the automation packages used are:
- Tank information system
- Oil movement information system
- Digital blender system
- Blend optimization and supervisory system.

26.2 STORAGE TANKS

Different types of storage tanks used are fixed roof tank, floating roof tank, fixed cum floating roof tank and doom type of tanks. The tanks are provided with manholes, product inlet and outlet nozzle, drain, staircase and lader, level gauge, open vane with wire mesh or breather valve or vents with flame arrester. Depending on the service the tanks are provided with sampling device, temperature gauge, inert gas blanketing, steam heating coils and jet mixing nozzle. LPG is stored in bullets and horton sphere, mounded bullets. Lighter products are in floating roof tanks while heavier crudes are stored in cone roof tanks. Product tanks with heavier products are fitted with steam coils and insulated. Storage of special products are done under inert gas blanketing. Refrigerated tanks are used for LPG. Underground and mounted storage tanks are used for storage of gasoline, diesel and other fuels. Underground storage tanks are also used in case of shale gas.

26.3 CRUDE BLENDING

Now refineries are using different types of crudes both lighter and heavier crudes in order to meet the changing requirement of distillates and to reduce cost of distillate. Opportunity crude oils and their blends play an important role in increasing profitability of refineries [Rathore et al., 2011]. Now refineries are blending high value light and heavier crudes or using blended crudes. Now low quality opportunity crudes are available at lower cost. Blending the high quality lighter crude with low cost heavier crudes to produce crude blend to have optimal properties at minimum cost is very important activity in refineries. However, if proper precaution is not taken during proper blending, the presence of impurities like destabilized asphaltenes, waxes and metal contents. Now crude blending is performed either by blenders or by refineries. Crude blending is applied directly by refiners to prepare low cost and compatiable blends for internal consumption or for trading in the market.

Some of the problems associated with the opportunity crude are stable oil-water emulsion problems, heat exchanger fouling and catastrophic coking in furnace tubes, flocculation and deposition of asphaltenes in crude oil blends [Rathore et al., 2011]. Prediction of insolubility number and solubility number are the key parameters in the prediction of flocculation and crude oil blend compability [Rathore et al., 2011].

Online analyzers are available which measure instantaneously the downstream blend. Simulation software is widely used to predict the ratio of individual components in the blend to produce crude blend with desired distillates at optimal yields. Nuclear magnetic resosance (NMR) process analyzers are now available which provide an effective tool for efficient blending of the crudes [Shahnovsky, 2014].

26.3.1 In-tank Blending

Different types of crudes are stored separately in tanks and loaded in blending tanks and mixed to achieve homogeneous mixture. This process is time consuming and expensive.

26.3.2 In-line blending

In in-line blendind, different crudes are simultaneously transferred through a static mixing device to the blend tank. The predetermined flow ratio between different crudes provides a blend of required quality.

26.4 PRODUCT HANDLING AND BLENDING

In refinery number of intermediate products produced from primary separation processes and secondary and treating processes which are blended in different proportions to produce desired product of required specification. First the products from various steps are allowed for water settling followed by product blending and dosing of additives and chemicals. Product blending has received much attention during recent years as number of secondary conversion processes have increased. Traditional blending used to be batch blending requiring large number of tanks and manpower depending on type of blending component. Now fuel blend analyzers are available.

Fuel blend analyzer: In refinery gasoline and diesel are obtained from various processes which are blended to meet the specification of final productrs. Near infrared (NIR) analyzer is being increasingly used for optimizing fuels blending. Barsamian, 2008; reported on the basis of survey that shows the percentage of NIR users compared to conventional analyzers. Fifty-nine percent used NIR blending analyzer. Some of the properties of gasoline and diesel are given in Table 26.1 [Barsamian, 2008]. Typical blend of various products are given in Table 26.2.

26.5 TRANSPORTATION OF CNG, LNG, CRUDE AND REFINERY PRODUCTS

Both imported and indigenous crude is used in different parts of the world. Indigenous crude oils from offshore/onshore are supplied by pipelines. Some of the important characteristics for effective movement of crude oil are density, temperature and viscosity. Road is the most significant mode of transport of finished products within in the country.

- *Marine transportation:* As most of the countries are dependent on crude oil from Middle East and Africa, crude is being supplied is transported in cargo ships. Although earlier tankers were of smaller capacity now larger capacity tanks are available.
- *Rail/road transportation:* Crude oil and different grades of products like gasoline, kerosene, diesel, fuel oil, LPG, etc. are transported via this mode. Tankers may be pressurized or unpressurized.
- *Pipeline transportation:* Crude, natural gas and various products are transported by pipeline transportation.
- *CNG, LNG shipment:* CNG is transferred via ships in cylinders or pipes capable of handling high pressure. Although CNG plant is less capital intensive than

Table 26.1   Properties of gasoline and diesel measured by NIR

Gasoline properties measured by NIR	RON, MON, Rvp, T10, T50, T90, FBP, E70C, E100C, E150C, aromatics, olefins, benzene, oxygenates/oxygen/ethers, ethanol, density
Diesel properties	Cetane number, cetane index, viscosity @ 40 °C, flash point, cloud point, CFPP, pour point, TBP, FBP, acidity, density

Table 26.2   Blends of products from various process streams

Product	Sources
LPG	Straight run LPG from crude distillation unit, cracked LPG, butane
Naphtha	Straight run naphtha and cracked naphtha, heavy naphtha
Motor sprit (gasoline)	Straight gasoline of high octane, reformate, light FCC gasoline, heart cut FCC gasoline, isomerate, alkylate, oxygenates (MTBE.TAME, ETB0 DME), alcohol, butanol, methanol
High speed diesel	Light gas oil, heavy gas oil, light cycle oil, low sulfur heavy gas oil, heavy naphtha, superior kerosene, biodiesel
Lubricating oil	Various cuts of vacuum distillation unit

LNG plant, however, the transportation of CNG is higher than LNG. LNG is supplied as a cryogenic in specialized marine vessels and at delivery port it is unloaded to storage facilities or regasification plants.

The petroleum products are dispatched to various places through railway ship/ barge, wagon, truck tanker and pipeline. Tanker's ranges in size depending on the amount of product to be transported. Large range and very large range carriers are used to transport crude oil. Smaller tankers are used for products. LPG is dispatched in cylinders and by bulk in LPG tankers. Bulk dispatch of products like LPG, gasoline, aviation turbine fuel (ATF), diesel, kerosene, fuel oil, lubricating oil blend, bitumen are normally done by road, by rail. Automatic loading is done using DCS and PLC control. Some of the products like naphtha, gasoline, kerosene, diesel are also dispatched through pipeline. Single pipeline is used for multiple products. Globally cross country pipelines are recognized as the safest, cost effective and environment friendly mode of transportation of crude oil and petroleum products. Indian Oil Corporation operates network of 11,214 km crude oil, product and gas pipeline with a capacity of 77.258 million tons per annum oil and 10 million cubic meter per day of gas [http://www.iocl.com/aboutus/Pipelines.aspx]. Both single and multiple product lines are used which are more complex in operation than a single pipeline [Richardson and Ward, 2001].

- *LPG storage and transportation:* Bulk storage of LPG is done in Horton sphere and Bullets LPG obtained from various units is stored in bullets and transferred to Horton sphere. Bulk loading of LPG in truck cylinder is done from Horton sphere and bullets. In LPG botteling plant LPG is filled in cylinders by guns and the cylinders are checked for the bung leaks and body leaks. The faulty cylinders are taken aside and the leaky valve is replaced after emptying. Mercaptan is added to LPG @ 20 ppm.

26.6 TYPES OF PIPELINE

Various types of pipelines are:
- Flow lines
- Gathering and feeder lines
- Crude tank lines
- Petroleum product trunk line.

Field gathering system consists of pipeline that moves oil from the well side storage tanks and treatment facilities to pump station where the oil is delivered to main transporting pipeline. Natural gas gathering system is used for natural gas collection and supply.

26.7 TERMINAL LOADING

Although both top loading and bottom loading have been in use, however, bottom loading is being commonly used because of increased productivity, improved safety, cleaner environment and reduced costs. As the terminals now operate in a new environment and rapidly changing market conditions, new automation system has been introduced in the terminals. Terminal automation system offers opportunities to reduce unit costs investment and provides reduced storage and handling cost, delivery expenses, improved customer service, reduced capital requirements, improved security and control procedures and increased terminal safety. Some of the main functions of terminal are product receipt, tank farm operation, loading rack operations, accounting and support [Richardson and Ward, 2001].

REFERENCES

1. Barsamian JA. Optimize Fuels Blending with Advanced Online Analyzers, Sep 2008, p121.

2. Panter A. Evaluating Tank Farm Capacity, Petroleum Technology, Quarterly, 2011, Q4, p63.

3. Rathore V, Brahma R, Throat TR, Rao PVC and Choudhary NV. Assessment of Crude Oil Blends Petroleum Technology, Quarterly, 2011, p111.

4. Shahnovsky G, Cohen T and Mcmurray R. Advanced Solutions for Efficient Crude Blending Petroleum Technology, Quarterly, 2014, p103.

5. Stellman JM. Storage and Transportation of Crude Oil, Natural Gases, Liquid Petroleum Products and Other Chemicals. Encyclopedia of Occupational Health and Safety.

6. Maity LN. Product Blending Handling and Dispatches. Petrotech Society's Summer School Program on Advances in Petroleum Refining Industry, July 3–8, 2006.

7. Saraf S. Shipping Natural Gas—New Frontiers. Hydrocarbon Processing, Jan 2007, p15.

8. Reichardson J and Ward CJM. Marketing Operations: Storage and Distribution. Modern Petroleum Technology, Vol 2, Downstream, edited by Lucas AG, 6th edition, John Wiley & Sons Ltd, 2001.

Glossary of Terms

GLOSSARY OF TERMS

Absorption	It involves bringing of the contaminant effluent gas in contact with liquid absorbent so that the liquid removes one or more constituents of the effluent gas.
Additive	Chemicals used in small amount to improve the quality of products.
Activated sludge	It is a suspended culture system where part of the settled sludge containing living or active microorganism is returned to the reactor to increase the available biomass.
Adsorption	Process of selective collection of molecules by external or internal surface of solids or by the surface of liquids.
ADIP process	Regenerative amine process for removal of H_2S and CO_2 using aqueous solution of the secondary amine, disopropanol amine or the tertiary amine, methyl diethanol amine.
Aerobic digestion	Stabilization of organic matter by aerobic digestion.
Alkenes (olefins)	Characterized by double bond. General formula, $C_n H_{2n}$.
Aliphatic hydrocarbons	Hydrocarbons characterized by open chain structures.
Alkylation	The addition of alkyl group to any compound is called alkylation. It is used for producing high octane number in petroleum refinery by introducing high octane motor fuel by combining light olefins with isobutene to form higher molecular weight branched chain paraffinic hydrocarbons with exceptional antiknock and clean burning properties. This is also used for production of petrochemicals like ethyl benzene, linear alkyl benzene, etc.
Alkynes (acetylenes)	Characterized by triple bond. General formula, $C_n H_{2n-2}$.
Amine treatment	For removal of sour gases using ethanol amines. The process is also used in petrochemical and fertilizer industries.

Aniline point	Lowest temperature at which the oil is completely miscible with an equal volume of aniline. High aniline point fuel is highly paraffinic.
API gravity	Degree API = 141.5/sp. gravity at 15.6/15.6 °C – 131.5.
Aromatics	Cyclic compounds containing at least six carbon atoms contain benzene ring, e.g. benzene, toluene, xylenes, cumene, etc.
Asphalt/asphaltene	Asphalts are the heaviest products obtained during distillation of crude oil and are soluble in carbon disulfide. It indicates the presence of heavier hydrocarbons.
ASTM distillation	It is the test procedure used by American Society for Testing and Materials to measure the volume percent distillated at various temperatures. An atmospheric batch type of distillation process.
Atmospheric crude distillation	Process of crude distillation at below 385 °C to gasoline, kerosene, jet fuel, diesel and atmospheric residue.
Autoignition temperature	It is the lowest temperature at which self-sustained combustion of solid, liquid and gases is initiated or caused in the absence of spark or flame.
Aviation fuel	Aviation gasoline and jet fuel.
Azeotropic distillation	It is the process of separation of azeotropic liquid mixtures by addition of third component to form ternary mixture, which is then separated by distillation. The third component is found in appreciable quantity in distillate product.
Biochemical oxygen demand	The amount of oxygen consumed during microbial decomposition of organics.
Biofuel	Gasoline and diesel derived from renewable material.
Biodiesel	Produced by the reaction of vegetable oil or animal fat with methanol to produce methyl esters known as biodiesel.
Bitumen	A semisolid or solid hydrocabonaceous material from the vacuum residue of the crude oil distillation.
Bitumen blowing	A special grade of bitumen obtained by blowing air through liquid bitumen to obtain harder and more brittle bitumen having lower penetration and higher softening point.
Blending	Process of mixing two or more oils.
Bureau of mines correlation index (BMCI)	$\text{BMCI} = \dfrac{}{} + 473.5 \,(\text{sp. gr.}) - 456.8$ where VABP °R is volume average boiling point in degrees rankine.
Bromine number	Bromine number is a measure of the degree of unsaturation of the material to which the test is applied.

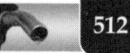

Bromine index	Bromine index is 1,000 times of the bromine number.
BTX	Benzene, toluene, xylene.
Bubble column	Cup-like element of bubble cap tray placed over risers.
Butamer process	Fixed bed, vapor phase butane isomerization process developed by UOP.
Carbon residue	Carbon residue of a fuel is the tendency to form carbon deposits under high temperature conditions.
Catalyst	Catalyst is a substance which alters or changes (increases/decreases) chemical reaction rate without undergoing a change in mass and composition at the end of reaction. In homogeneous reactions, both catalyst and reacting substances are in the same physical phase while in heterogeneous reactions catalyst and reactants are in different physical phases.
Catalyst activity	Catalyst activity at any time is defined as the rate at which the catalyst converts the reactant divided by rate of reaction of the reactant with a fresh catalyst.
Catalyst poisons	Substances which exert an appreciable inhibitive effect on catalyst even when present in very small amounts.
Catalytic reforming	A process in which naphthenes and paraffins are converted to aromatics and isoparaffins under selected operating conditions and catalyst for improving octane number and producing aromatics (in petrochemical production).
Catalyst regeneration	It is the process by which coke deposited on catalyst is burnt off and the catalyst thereby is reactivated.
Catalyst selectivity	Percentage of desired product yield from the feedstock.
Catalyst stability	Catalyst stability is the ability of a catalyst to maintain its activity and selectivity over a reasonable period.
Characterization factor	A parameter used to pressurizing petroleum feedstock. It is given by $\sqrt{}/S$, where t_B is the boiling point and S is the specific gravity at 15.6 °C/15.6 °C.
Chemical oxygen demand	Amount of oxygen required for chemical oxidation of organic matter in a sample.
Chloride corrosion	It is caused by hydrogen chloride which is formed from hydrolysis of chloride contained in crude oil.
Claus process	It involves the combustion of H_2S in carefully controlled stream of air to convert H_2S to SO_2 and subsequently to elemental sulfur.
Cloud point	It is the temperature at which a cloud or haze of wax crystals appears at the bottom of the test pan when oil is cooled under prescribed conditions. It gives a rough idea of the temperature above which the oils can be safely handled without any fear of congealing or filter clogging.
Coalbed methane	Natural gas entrapped in coalbed.

Condensate	Liquid hydrocarbons obtained by condensing heavier hydrocarbons.
Coking	Coking in steam cracker is the formation of coke in the furnace tube. Coking in the petroleum refinery is the process of conversion of heavy low grade oils into lighter products.
Coke factor	A relative number proportional to specific coke. Used in FCC catalyst analysis.
Corrosion	Corrosion is the destruction or deterioration of material because of its interaction with its environment.
Corrosion inhibitor	Any substance, the presence of which even in small amounts greatly diminishes the corrosion rate.
Cracking	The process of cracking of heavy molecule hydrocarbon to lighter fraction.
Cracking (catalytic)	Cracking of heavy crude oil fraction under selected operating conditions and catalyst to more valuable gasoline and lighter products.
Cracking (hydro)	Cracking of cyclic oil and coker distillate in presence of hydrogen and zeolite catalyst to improve the gasoline yield.
Cracking (steam)	Cracking of naphtha and natural gas to yield olefins and other by-products.
Cracking (thermal)	Cracking of atmospheric and vacuum distillation column heavy residue products under high temperature without use of any catalyst.
Crude oil	It is a mixture of hydrocarbons with some sulfur, nitrogen and oxygen compounds.
Crude oil assay/evaluation	Assessing the value of various properties of crude oil. Evaluatiuon may be preliminary assay, short evaluation or detailed evaluation.
Cycle oil	Heavier residue obtained after fraction of cracked product of FCC.
Cyclic reformer	Cyclic reformer is used for catalytic reforming.
Cryogenic CNG	It refers to condition below a temperature of –100 °C.
Dealkylation	It is a process of removal of alkyl group.
Decoking	Process of removal of coke from naphtha cracker tubes is called decoking.
Deflection temperature	It is the highest continuous temperature that a material will withstand.
Delayed coking	It is a semi-continuous process used for conversion of heavy low grade oils to light fraction.
Demulsifier	Equipment or chemical used for separation of emulsions into separate layers.

Desalting	Desalting is a process of removal of salt from crude oil for minimizing fouling and corrosion.
Deasphalting	Process of removal of asphalt.
Deasphalted oil	Oil left after removal of asphalt.
Desulfurization	Process of removal of sulfur compounds from hydrocarbon.
Dehydrogenation	Catalytic removal of hydrogen.
Dewaxing	Removal of wax from lubricating oil. Dewaxing may be solvent dewaxing (propane dewaxing) or catalytic dewaxing.
Diesel	Hydrocarbon fraction in the boiling range of 250–320 °C obtained from crude oil distillation and other conversion/cracking processes. Diesel may be low speed, medium speed or high speed.
Diesel index	Measure of diesel fuel quality ignition quality.
Diethanol amine	Used for absorption of acid gases.
Diolefins	These compounds contain two double bonds and are even more reactive than olefins, e.g. butadiene.
Disproportionation	It is the conversion of two moles of a single aromatic to one mole each of two different aromatic compounds.
Distillation	Distillation is the process used to separate two or more liquid compounds of a liquid into more or less pure fraction and involves pressurizing and subsequent condensation.
Doctor test	Test used to test the sweet stock.
Ductility	An important characteristic of bitumen.
Edeleanu process	It uses liquefied sulfur dioxide and is used to dearomatize kerosene, gas oil and separation of aromatics.
Electric desalting	Process of desalting using high voltage DC current for removal of salt from crude oil.
Environmental impact assessment (EIA)	It is the systematic identification and evaluation of the potential impacts of proposed projects, plans, programs or legislative actions relative to the physicochemical, biological, cultural and socioeconomic components of the total environment.
Esterification	Reaction of alcohol with organic and inorganic acids (except halogen acid).
Etherification	Reaction between isoolefin and methanol/ethanol.
Ether	A class of organic compounds in which oxygen is interposed between two carbon atoms (organic group). Generic formula: ROR.
Explosive limits	Lower explosive limit: Minimum concentration of vapor, gas or dust in air below which the propagation of flame does not occur in contact with the source of ignition.Upper

explosive limit: Maximum concentration of vapor, gas or dust in air above which propagation of flame does not occur in contact with the source of ignition.

Extractive distillation

It is the process of separation of azeotropic liquid mixtures by addition of third component to form ternary mixture, which is then separated by distillation. The third component is found in appreciable quantity in bottom product.

FAME

Fatty acid methyl ester.

Fluidized bed cracking (FCC)

Catalytic cracking of heavy residues using fluidized bed reactor.

Flammability

It is the measure of the ability of a material to support combustion.

Flash point

It is the lowest temperature at which a material gives off enough vapor to form flammable mixture with air near the surface of the liquid or within the container. It is determined by closed cup method or open cup method.

Fluid coking

Continuous thermal cracking of heavy low grade oils to low fraction using fluidized bed reactor for cracking.

Fluidization

Fluidization is the condition of fully suspended solid when liquid or gas is passed through a bed of solid. The motion of solid is created at superficial velocities far below the terminal settling velocity of the solid particle.

Isomerization

Conversion of hydrocarbons to their isomers (rearrangement of a molecule to a different chemical configuration which maintain same atomic number). Light paraffin isomerization is commonly used for improving octane.

Kerosene

Fraction of hydrocarbon obtained from atmospheric column having boiling in the range of 150–250 °C used fuel and illuminant. Paraffin derived from kerosene is used in linear alkyl benzene manufacture.

Kerox

Fixed bed sweetening process.

Knockout drum

A vessel for separating liquid from gas or vapor.

Lean oil

Absorbent oil used for stripping gas.

Liquid-liquid extraction

It is the process by means of which one or more components of a liquid are transferred to another liquid.

LD$_{50}$

LD$_{50}$ is a calculated dose capable of killing 50% of a population of experimental animals exposed through a route other than respiration.

Liquefied natural gas (LNG)

Natural gas is liquefied either by cooling and pressurizing or by refrigerating down to the boiling point of –162 °C.

Logging

Well-logging is an operation of continuous recording of characteristics of the formation penetrated by a drill hole, as a function of depth.

Liquefied petroleum gas (LPG)	Liquefied petroleum gas containing mainly propane and butane.
Lubricants	Heavier hydrocarbons derived from petroleum used to reduce frictions.
Melt index	It is a measure of the amount of material that is exposed from a small orifice for 10 minutes at 300 kPa.
Membrane processes	Membrane process is considered like a unit operation in chemical engineering to separate mixtures. According to driving force of the operation it is possible to be distinguished.
Merox extraction	Chemical treatment of sulfur compounds present in hydrocarbon fractions using extraction developed by UOP.
Merox sweetening	Process of converting mercaptans to less objectionable disulfides using merox catalyst developed by UOP.
Microactivity test	Test used for determining the activity of deactivated FCC catalysts.
Minalk process	Dual stage sweetening process using two oxidation zones developed by UOP.
Molex process	Adsorption process using molecular sieve to separate paraffins.
Methyl tertiary butyl ether (MTBE)	It is used as an oxygenate and is made by reaction of methanol with isobutene.
N + 2A	N = %wt of naphthenes, A = %wt of aromatics.
Naphtha	Naphtha (light, medium, heavy) is the low boiling fraction of crude oil having boiling point in the range of 37–200 °C. Light naphtha is a major component of gasoline.
Naphthenes	Saturated hydrocarbons (alicyclic paraffins) containing one or more rings which may have one or more paraffins side chain having formula C_nH_{2n}.
Naphthenic acid	Naphthenic acids are low carbon number organic acids like formic acid, acetic acid, propionic acid and butyric acid.
Natural gas	Natural gases are those containing methane along with other alkanes like ethane, propane, butane and other hydrocarbons, etc., along with sulfur compounds may be present as associated gas or as non-associated.
N-methylpyrrolidone (NMP)	It is used as a solvent for separation of aromatics from non-aromatics.
Nucleophilic addition	Reactions in which the addition is initiated by a negative group or nucleophile.
Octane number	It is a measure of the tendency of a motor fuel to knock in an internal combustion engine and is defined as the percentage of iso-octane in a mixture of iso-octane (octane number = 100) and n-heptane (octane number = 0). Motor octane number (MON) is the measure of antiknock characteristics of gasoline and is a laboratory simulation of engine performed at high speed. Research octane

	number (RON) is also a measure of antiknock characteristics of gasoline and is tested under less severe conditions, i.e. at lower speed.
Olefins	Unsaturated hydrocarbon having formula C_nH_{2n}.
OLEX	Process for separation of olefin using adsorption.
Oligomerization	The polyaddition of monomer to its dimer, trimer, or tetramer form.
Parisom process	UOP process of catalytic isomerization of light paraffins, e.g. pentanes, hexanes and mixture to improve octane number.
Paraffins	Saturated hydrocarbon having general formula C_nH_{2n+2}. Paraffin hydrocarbons may be straight chains or branched chains.
Paraffin wax	Normal paraffins above nC_{16}. Presence of wax affect the flow behaviors.
Penetration test	Used for classification of different grades of bitumen.
Platforming	It is the catalytic reforming process developed by UOP using continuous/semi-regenerative/cyclic type of reactor.
Polymerization	Catalytic process used for polymerization of light olefins to high octane gasoline component in petroleum refinery.
Pour point	Used for measuring the low temperature pumpability characteristics of fuel oil.
Powerforming	It is the catalytic reforming process developed by Exx Engineering using semi regenerative/cyclic type of reactor.
Pressure swing adsorption (PSA)	PSA involves adsorption at high pressure followed by desorption by reducing pressure.
Prism membrane	Membrane process to recover and purify hydrogen or to reject hydrogen from refinery, petrochemical or gas processing streams.
Propane deasphalting	Process of removal of asphalt using propane as solvent.
Propane dewaxing	Process of removal of wax from oil using propane solvent.
Proven reserves	See reserves.
Probable reserves	See reserves.
Possible reserves	See reserves.
Reactive distillation	Simultaneous implementation of reaction and distillation.
Reformate	The main liquid product obtained from the catalytic reforming process used for production of high octane gasoline and aromatics.
Reid vapor pressure	It indicates the relative percentage of gaseous and lighter hydrocarbons. RVP is the pressure exerted by vapor when it is in equilibrium with the liquid.
R2R technology	Residue cracking technology using one stage of reaction and two stages of catalyst regeneration.

Regenerativity	Regenerativity of catalyst means recovery of its activity, selectivity and stability.
Reserves proven	Proven reserves of oil or gas are considered as certain with high probability of production of oil and gas.
Reserves probable	Probable reserves are reserves which have not been proved based on limited evidence.
Reserves possible	Possible reserves are reserves which may exist.
Risk assessment	A risk assessment is simply a careful examination of what in your work, could cause harm to people, so that you can weigh up whether you have taken enough precautions or should do more to prevent harm.
Selectivity of catalyst	Percent of desired product yield from the feedstock.
Semi-regenerative reformer	Reformer that processes feedstock for a time and then shuts down for regeneration.
Severity	Severity in cracking process is the measure of the extent or depth of cracking. Severity function is defined as the logarithmic function of conversion X_f of reaction hydrocarbon present.
Short path distillation	Short path distillation is a continuous fractionating process without steam injection, using wiped film evaporator with an internal condenser. Short path distillation covers a range of applications not possible in traditional evaporation units. Specifically short path distillation requires high temperature and high vacuum operation.
Smoke point	This is a measure of the burning cleanliness of kerosene/jet fuel. Low smoke point is desirable.
Soaker drum	Used in visbreaking process as an additional reactor where thermal cracking started in the furnace is completed.
Softening point	It is a temperature at which a substance attains a particular degree of softness.
Solutizer process	A process of mercaptans extraction using mixture of KOH and tricresols.
Solvent deawxing	Process of removal of wax from oil using solvent by means of chilling of waxy oil and solvent followed by filtration/centrifuging for removal of precipitated wax.
Solvent deasphalting	See deasphalting.
Space velocity	The volume of feed treated per unit time per unit volume of effective reactor or catalyst bed. Liquid hourly space velocity (LHSV) is the volume per hour of reactor charge per unit weight of catalyst. Weighted hourly space velocity (WHSV) is the weight per hour of reactor charge per unit weight of catalyst.
Stabilization	Removal of low boiling hydrocarbon fraction (volatile matter) from oil.

Stabilizer	Fractionating column used for removal of low boiling hydrocarbon fractions from gasoline/naphtha and other hydrocarbon fractions.
Stability of catalyst	Stability of catalyst is the ability of the catalyst to maintain its activity, selectivity over a reasonable period.
Sulfreen	Catalytic purification of Claus gas or lean H_2S waste gas.
Sulfolane	It is produced by hydrogenation of sulfolane, which is made by reaction of butadiene with sulfur dioxide. It is used as solvent in extraction of aromatics from non-aromatics.
Suspension polymerization	Polymerization in which the monomer is maintained in suspension by proper agitation without emulsifying it.
Sweetening	The process of treating a hydrocarbon for removing or converting to a less undesirable form of sulfur compounds present.
Tertiary amyl methyl ether (TAME)	It is made by esterification of 2-methyl butane and is used as oxygenate.
Tank farm	Tank farms are grouping of storage tanks at various places.
Thermal stability	It is used for measuring high temperature stability of fuels.
Trace metals	Those metals present at very low level and harmful to the catalyst/human beings.
Threshold limit value (TLV)	It is a concentration of air contaminants to which nearly all persons can be exposed day after day without adverse effect. TLV-TWA (time weighted average): TLV-TWA is a time weighted average for a normal 8 hours work day or 40 hours work week.
Thiolex/regen	Fibre film contractor technology for removal of H_2S, COS and mercaptans from gases and liquid stream.
Trap	Trap is an area bounded by barrier lying upwards from flow and is present in reservoirs.
Trickling filter	It is an attached growth aerobic process with randomly packed solid, which provides surface for biofilm growth.
Ultraforming	It is a catalytic reforming process licensed by Standard Oil Co.
Unicracking	UOP hydrocracking process using moderate temperature and pressure.
Unionfining	Hydrotreating process by UOP and Union Oil Co.
Visbreaking	It is a mild form of thermal cracking of heavy fraction of crude oil.
Vacuum distillation of crude oil	Fractionating column used for fractionation of heavier residue from atmospheric distillation of crude oil under vacuum to avoid cracking.
Viscosity index (VI)	An arbitrary

Weighted average inlet temperature (WAIT)	[Weight of catalyst in reactor 1 × inlet temperature in reactor 1 + weight of catalyst in reactor 2 × inlet temperature in reactor 2 + weight of catalyst in reactor 3 × inlet temperature in reactor 3]/total weight of catalyst, i.e. $[WCR_1 \times R_{1IT} + WCR_2 \times R_{2IT} + WCR_3 \times R_{3I}]/(WCR_1 + WCR_2 + WCR_3)$, where WCR_1, WCR_2, WCR_3 are the weights of catalysts in reactors 1, 2, 3 and R_{1IT}, R_{2IT}, R_{3IT} are the inlet temperatures for reactors 1, 2, 3.
Weighted average bed temperature (WABT)	$[WCR_1 (R_{1IT} + R_{1OT})/2 + WCR_2 (R_{2IT} + R_{2OT})/2 + WCR_3 (R_{3IT} + R_{3OT})/2]$ / total weight of catalyst, where WCR_1, WCR_2, WCR_3 are the weights of catalysts in reactors 1, 2, 3; R_{1IT}, R_{2IT}, R_{3IT} are the inlet temperatures for reactors 1, 2, 3 and R_{1OT}, R_{2OT}, R_{3OT} are the outlet temperatures for reactors 1, 2, 3.
Weathering test	This test shows the volatility of LPG.
Well-logging	See logging.

Appendices

APPENDIX A

Commonly Used Abbreviations in Petroleum and Petrochemical Industries

ACGIH	American Congress of Governmental Industrial Hygienists
ADU	Atmospheric Distillation Unit
AIHA	American Industrial Hygiene Association
ANGM	Association of Natural Gasoline Manufacturers
API	American Petroleum Institute
Aq.	Aqueous
ARCO	Atlantic Richfield Co.
ASTM	American Society For Testing And Materials
Atm.	Atmospheric Pressure
BMCI	Bureau of Mines Correlation Index
BP	Boiling Point
BP	British Petroleum
BRPL	Bongaigaon Refinery and Petrochemicals Limited
BTX	Benzene, Toluene, Xylene
CCR	Continuous Catalyst Regeneration
CCRU	Continuous Catalytic Reforming Unit
CEL	Shell Corrected Energy and Loss Index
CF	Combustion Factor
CFC	Continuous Film Contactor
CFR	Compagnie Fransaus Deraffinage
CLO	Clarified Light Oil
CNG	Compressed Natural Gas
Co.	Corporation

COC	Cleve Land Open Cup
Conc.	Concentration
CPCL	Chennai Petroleum Corporation Limited
CPW	Chlorinated Paraffins Wax
CR	Catalytic Reforming
CSD	Critical Solvent Dewaxing
CTC	Coefficient of Thermal Expansion
DAO	Deasphalted Oil
DCC	Deep Catalytic Cracking
DEA	Diethanol Amine
DEG	Diethylene Glycol
DGA	Diglycol Amine
Dil.	Dilute
DIPE	Diisopropyl Ether
DMF	Dimethyl Formamide
DMSO	Dimethyl Sulfoxide
EIA	Environmental Impact Assessment
EII	Energy Intensity Index
EOR	End-of-Run
EPA	Environmental Protection Agency
EPDM	Ethylene-propylene Diene Rubber
ETBA	Ethyl Tertiary Butyl Alcohol
FCC	Fluid Catalytic Cracking
FEP	Fluorinated Ethylene Propylene
FO	Fuel Oil
FP	Freezing Point
FTT	Furnace Transfer Temperature
GAIL	Gas Authority of India Limited
GAZOP	Hazard Operability Study
GTL	Gas to Liquids
GTBA	Gasoline Grade Tertiary Butyl Alcohol
HCO	Heavy Cycle Oil
H:HC	Hydrogen to Hydrocarbon Ratio
HSD	High Speed Diesel Oil
HTSD	High Temperature Simulated Distillation
IBP	Initial Boiling Point
ICI	Imperial Chemical Industries
IFP	Institut Francais Due Petrole

IPA	Isopropyl Alcohol
LAB	Linear Alkyl Benzene
LCO	Light Cycle Oil
LD_{50}	Lethal Dose
LDO	Light Diesel Oil
LEL	Lower Explosive Limit
LFL	Lower Flammability Limit
LHSV	Liquid Hourly Space Velocity
LNG	Liquefied Natural Gas
LPG	Liquefied Petroleum Gas
Ltd.	Limited
LWD	Logging While-Drilling Technologies
MAT	Micro Activity Test
MBK	Methyl Butylketone
MCP	Molecular Collision Parameter
MDD	Maximum Drilling Depth
MDDW	Mobil Distillate Dewaxing
MDEA	Methyldiethanol Amine
MEA	Monoethanol Amine
MEK	Methyl Ethyl Ketone
MHC	Mild Hydrocracking
MLTD	Mobil Low Temperature Disproportionation
MON	Motor Octane Number
MOGD	Mobil Olefin To Gasoline And Distillate
MP	Melting Point
MSCC	Millisecond Catalytic Cracking[sm]
MTBE	Methyl Tertiary Butyl Ether
MVPI	Mobil Vapor Phase Isomerization
MWD	Measurement While-Drilling
NA	Naphthenic Acid
NAC	Naphthenic Acid Corrosion
N + 2A	Naphthene + Two Times Aromatic Content
NGHs	Natural Gas Hydrates
NIOSH	National Institute for Occupational Safety and Health Services
NMP	N-Methyl Pyrrolidone
NRCC	Non-regenerative Catalytic Cracking
OH & S	Occupational Health and Safety
OHCN	Once Through Hydrocaracker Unit

OISD	Oil Industry Safety Directorate
ON	Octane Number
ONGC	Oil And Natural Gas Corporation
OSA	Oil Spill Authority of India
OSHA	Occupational Safety and Health Administration
OSMA	Oil Spill Management Authority of India
OSW	Österreichische Stickstoff Werke
PDM	Positive Displacement Motor
PRMC	Project Review and Monitoring Committee for Oil Spill Management
PSA	Pressure Swing Adsorption
QRA	Quantified Risk Assessment
RFBD	Residue Fluidized Bed Cracking
RFCC	Residue Fluid Catalytic Cracking
RFCC	Reduced Feed Fluid Catalytic Cracking Unit
RFG	Reformulated Gasoline
(R + M)/2	(RON + MON)/2
RON	Research Octane Number
RVP	Reid Vapor Pressure
SAC	Solid Acid Alkylation Catalyst
SOP	Super Oil Cracking
SOR	Start-of-Run
SCC	Stress Corrosion Cracking
SEC	Specific Energy Consumption
SHE	Safety, Health and Environmental Management
SYDC	Selective Yield Delayed Coking
TAA	Tertiary Amyl Alcohol
TAEE	Tertiary Amyl Ethyl Ether
TAN	Total Acid Number
TAME	Tertiary Amyl Methyl Ether
TBA	Tertiary Butyl Alcohol
TBP	True Boiling Point
TEA	Triethanol Amine
TEG	Triethylene Glycol
TEL	Tetraethyl Lead
TFE	Tetrafluoroethylene
THEME	Tertiary Hexyl Methyl Ether
THD	Thermal Hydrodealkylation

T_m	Melting Point Temperature
TPR	Throughput Ratio
TOC	Tag Open Cup
TSA	Temperature Swing Adsorption
UCC	Union Carbide Co.
UDEX	Universal Dow Extraction
UEL	Upper Explosive Limit
UOP	Universal Oil Product
VDS	Vortex Disengager Stripper[sm]
VGC	Viscosity Gravity Correlation
VGO	Vacuum Gas Oil
VI	Viscosity Index
VLI	Vapor Lock Index
VSS	Vortex Separation System
WABT	Weighted Average Bed Temperature
WAIT	Weighted Average Inlet Temperature
WHSV	Weight Hourly Space Velocity
ZMS	High Silica/Alumina Ratio Zeolite

APPENDIX B

Physical and Chemical Characteristics of Major Hydrocarbons

Compound	Appearance/odor	BP (°C)	FP (°C)	MP (°C)	Specific gravity (20/20 °C)	Flash point (°C)	Flammability limit lower (°C)	Flammability limit upper (°C)	Auto-ignition temperature (°C)
Ammonia, NH_3	Colorless gas liquid with sharp, intensely irritating odor; lighter than air; toxic and irritant; tolerance 25 ppm in air.	−33.5	Gas	−77.7	0.77 at °C		16.0	27.0	650
Aniline, $C_6H_5NH_2$	Colorless oily liquid; characteristic odor and taste; toxic; an allergen; tolerance 2 ppm in air.	184.4	−6.2		1.0235	70 (closed cup)	1.3	20.25	615
Asphalt	Black solid or viscous liquid; moderately toxic by inhalation of fumes.	371	> 204	93	1.0	132	–	–	485
Benzene, C_6H_6	Colorless to light yellow liquid with aromatic odor; flammable, dangerous fire risks; tolerance limit in air 10 ppm, carcinogenic.	80.1	5.5	5.0	0.879 (at 20/4 °C)	−11 (closed cup)	1.3	7.1	562.2
1,3-butadiene, $CH_2 = CH-CH = CH_2$	Colorless gas, non-corrosive with mildly aromatic odor; highly flammable gas or liquid.	−4.4	−108.9	−135	0.6211 (liquid) (at 20 °C)	76.0	2	11.5 (%vol)	417.8
Butane, C_4H_{10}	Colorless gas with natural gas odor; highly flammable; anasphyxiant gas.	0.6	−138.3	−135	0.6	−60	1.6	8.4	405

Contd.

Physical and Chemical Characteristics of Major Hydrocarbons (Contd.)

Compound	Appearance/odor	BP (°C)	FP (°C)	MP (°C)	Specific gravity (20/20 °C)	Flash point (°C)	Flammability limit lower (°C)	Auto-ignition temperature (°C) upper (°C)
n-butanol, C_4H_9OH	Colorless liquid; toxic on prolong inhalation; irritant to eyes; tolerance 50 ppm in air.	117.7	–89.0		0.8109	28.85 (closed cup)		365
1-butene, $CH_3CH_2CH=CH_2$	Colorless gas; highly flammable; dangerous fire and explosion hazard.	–6.3	–185	–130	0.5951 (at 20/4 °C)	–79	1.6	9.3 371
Carbon disulfide, CS_2	Colorless or faintly yellow liquid; highly flammable; dangerous fire and explosion hazard.	46.3	–111		1.260 (at 25/25 °C)	–30		100
Carbon monoxide, CO	Colorless; highly toxic by inhalation; solidification point: –207°C.	–190	Gas	–206	0.967	Gas	12.5	74.2 609
Carbonyl sulfide, COS	Colorless gas with typical sulfide odor; flammable.	50.2 (1 atm)	–138.8	–138	2.10		11.9	28.5 –
Carbon tetrachloride, CCl_4	Colorless liquid; toxic by ingestion inhalation; tolerance 5 ppm in air.	76.8	–22.9	–22.6	1.595	None		
Chloroform, $CHCl_3$	Colorless, highly refractive, heavy liquid with characteristic odor; toxic by inhalation; narcotic; prolong inhalation or ingestion may be fatal; tolerance 10 ppm in air; carcinogenic.	61.2	–63.5		1.485			

Contd.

Physical and Chemical Characteristics of Major Hydrocarbons (Contd.)

Compound	Appearance/odor	BP (°C)	FP (°C)	MP (°C)	Specific gravity (20/20 °C)	Flash point (°C)	Flammability limit lower (°C)	upper (°C)	Auto-ignition temperature (°C)
Chloromethane, CH_3Cl	Colorless gas, etheral odor, sweet taste; flammable; dangerous fire risk	−23.7	−97.6		0.92	0	8.1	17	632
Chloroprene, $H_2C{=}CH{-}C(Cl){=}CH_2$	Colorless liquid; flammable; dangerous fire risk; toxic by ingestion and inhalation.	59.4			0.9583	−20	4	20	
Cumene (isopropyl benzene), $C_6H_5CH(CH_3)_2$	Colorless liquid, aromatic odor; moderately toxic by ingestion; inhalation; tolerance 50 ppm in air.	152.39	−96.03		0.8619 (at 20 °C)	44	0.9 (%vol)	6.5	424
Cyclobutane, C_4H_8	Colorless gas	13	−80			10			
Cyclohexane, C_6H_{12}	Colorless mobile liquid with pungent odor.	80.7	6.3		0.779 (at 20/4 °C)	−18.3	1.3	8.4	245
Cyclohexanol, $C_6H_{11}OH$	Colorless viscous liquid, hygroscopic.	160.9		23	0.9429 (at 25/4 °C)	67.7			299.7
Cyclopantane, C_5H_8	Colorless liquid	44	−135.2		0.772	−29			
Cyclopentadiene, C_5H_6	Colorless liquid; decomposes violently at high temperature; tolerance 75 ppm in air.	42.5			0.805				
n-decane, $C_{10}H_{22}$	Colorless liquid; moderate fire risk.	173	−30	−32	0.729	44 (closed cup)	0.67	2.6	250

Contd.

Physical and Chemical Characteristics of Major Hydrocarbons (*Contd.*)

Compound	Appearance/odor	BP (°C)	FP (°C)	MP (°C)	Specific gravity (20/20 °C)	Flash point (°C)	Flammability limit lower (°C)	upper (°C)	Auto-ignition temperature (°C)
Diacetone alcohol, $CH_3COCH_2C(CH_3)_2OH$	Colorless liquid with pleasant odor; flammable; dangerous fire risk; toxic and irritant; tolerance 50 ppm in air.	169.1	–42.8	140	0.9406 (at 20/20 °C)	23–38 (depending upon grade)	1.8	6.9	603
Di-n-butyl amine, $(C_4H_9)_2NH$	Colorless liquid with amine odor	159.6	–62		0.7613	51.6			
Di-butyl maleate, (DBM)	Colorless oily liquid	280.6	–85		0.9964	140.4			
Dicyclopentadiene, $C_{10}H_{12}$	Chemical intermediate for pesticide; flammable; moderate fire risk; toxic by ingestion, inhalation; tolerance 5 ppm in air.	172		33.6	0.979	32.2			
Ethane, C_2H_6	Colorless, odorless gas; severe fire risk, if exposed to sparks or open flame.	–88.63	–183.23	–172	0.446 (liquid, 0°C)	–135	3.12	15.0	515
Ethanolamine									
Monoethanol amine (MEA), $(HOCH_2CH_2)NH_2$	Colorless, hygroscopic, viscous liquid; moderately toxic.	170.5 (decomp)	10.56	1.029 (open cup)	93.3				
Diethanolamine (DEA), $(HOCH_2CH_2)_2N$	Colorless crystal or liquid, toxic; tolerance in air 3 ppm.	270	28	1.092 (at 30/20 °C)	152 (open cup)				228.9

Contd.

Physical and Chemical Characteristics of Major Hydrocarbons (*Contd.*)

Compound	Appearance/odor	BP (°C)	FP (°C)	MP (°C)	Specific gravity (20/20 °C)	Flash point (°C)	Flammability limit lower (°C)	upper (°C)	Auto-ignition temperature (°C)
Triethanol amine (TEA), $(HOCH_2CH_2)_3N$	Colorless viscous, hygroscopic liquid; combustible; low toxicity	360 (decomp)	21.2	1.13 (open cup)	19.5				371
Ethyl benzene, $C_6H_5C_2H_5$	Colorless liquid with aromatic odor; flammable, dangerous fire risk; moderately toxic; tolerance 10 ppm in air.	136	−95		0.867 (20 °C)	15			432
Ethanol, C_2H_5OH	Clear, colorless liquid with pleasant odor; flammable, dangerous fire risks; tolerance in air 1000 ppm in air.	78.32	−114.1	−117	0.7893 (25 °C)	14.0 (closed cup)	3.3 (%vol)	19.0	423
Ethylene oxide, C_2H_4O	Colorless gas; highly flammable; dangerous fire and explosion risk; tolerance 10 ppm in air.	10.4	−111.7		0.8711	<−18 (TOC)	3	100	429
Ethylene, C_2H_4	Colorless, slightly sweet odor; highly flammable; dangerous fire and explosion risk.	−103.71	−169.15	−169	0.610	−135	3	36	543
Ethylene, C_2H_4	Colorless, slightly sweet odor; highly flammable; dangerous fire and explosion risk.	−103.71	−169.15	−169	0.610	−135	3	36	543

Contd.

Physical and Chemical Characteristics of Major Hydrocarbons (*Contd.*)

Compound	Appearance/odor	BP (°C)	FP (°C)	MP (°C)	Specific gravity (20/20 °C)	Flash point (°C)	Flammability limit lower (°C)	Auto-ignition temperature (°C)	upper (°C)
Ethylene dichloride, $C_2H_4Cl_2$	Colorless oily liquid; flammable, dangerous fire risk; toxic by inhalation, ingestion; tolerance limit in air 10 ppm.	83.5	–35.5		1.2554	13.3	6		16
Ethylene glycol									
MEG, CH_2OHCH_2OH	Colorless liquid with sweet taste; combustible; toxic by ingestion and inhalation.	197.6	–13.5		1.1155	116		412	
DEG, $(CH_2CH_2OH)_2O$	Colorless liquid.	245	–8.0		1.1184	123.9		228.9	
TEG, $HO(C_2H_4O)_3H$	Colorless, hygroscopic liquid.	287.4	–7.2		1.2544	176.6		371	
Ethylene glycol monobutyl ether, $HO.CH_2.CH_2.O. C_4H_9$	Colorless liquid with mild odor; toxic; tolerance limit in air 25 ppm.	171.2			0.9019	61		244	
Ethylene glycol monoethyl ether, $CH_3OCH_2CH_2OH$	Colorless liquid; moderately toxic by ingestion and inhalation; tolerance limit 5 ppm in air.	135.6	–85.1		0.9311	48.9		237	
Formaldehyde, HCHO	Colorless gas, pungent suffocating odor.	–19	–118		0.8153 (at –20 °C)		7.0 (mol. %)	430	7.3
Gas oil		232–426		–	–	65.5		338	–

Contd.

Physical and Chemical Characteristics of Major Hydrocarbons (Contd.)

Compound	Appearance/odor	BP (°C)	FP (°C)	MP (°C)	Specific gravity (20/20 °C)	Flash point (°C)	Flammability limit lower (°C)	upper (°C)	Auto-ignition temperature (°C)
Furfural	Colorless liquid when pure, toxic, absorbed by skin, tolerance 2 ppm in air, irritant to eyes, skin and mucous membrane.	161.7	−36.5	−38	1.1598 (20/4 °C)	60	2.0	–	393
Gasoline	Highly flammable, dangerous fire and explosion risk.	38–204	−45.6	< −45.6			1.3	6.0	257
n-heptane, C_7H_{16}	Volatile, colorless liquid, flammable, dangerous fire risk, tolerance 400 ppm in air.	98.428	−90.595	−90	0.684	−3.89	1.0	6.0	222
Hexane, C_6H_{14}	Colorless volatile liquid with faint odor; flammable dangerous fire risk; tolerance 500 ppm in air.	69	−95.6	−94	0.659	−22.7	1.2	7.5	260
Hydrazine, $H_2N.NH_2$	Colorless, fuming, hygroscopic liquid; highly toxic by ingestion, inhalation and skin absorption; tolerance 0.1 ppm in air, a known carcinogen.	113.5		1.4	1.004 (at 25/4 °C)	52.2 (open cup)			270
Hydrogen, H_2	Highly flammable and explosive, dangerous when exposed to heat or flame.	−252	−259	−260	0.069		4.1	74.2	580
Hydrogen chloride, HCl	Colorless, fuming gas with suffocating odor; toxic by inhalation, strong irritant to eyes and skin. Tolerance in air 5 ppm.	−85		−114	1.268				

Contd.

Physical and Chemical Characteristics of Major Hydrocarbons (*Contd.*)

Compound	Appearance/odor	BP (°C)	FP (°C)	MP (°C)	Specific gravity (20/20 °C)	Flash point (°C)	Flammability limit lower (°C)	upper (°C)	Auto-ignition temperature (°C)
Hydrogen cyanide, HCN	Colorless gas or liquid with characteristic odor; highly toxic by ingestion, inhalation and skin absorption.	26	–13.3		0.688 (liquid) (at 20/4 °C)	–17.7 (closed cup)	6	41	538
Hydrogen sulfide, H_2S	Colorless gas with offensive odor; toxic by inhalation; highly flammable, dangerous fire risk; tolerance 10 ppm in air.	–60		–83	1.189		4.3	46	260
Isobutene, $(CH_3)_2C{=}CH_2$	Colorless volatile liquid.	–6.9	–139		0.6		1.8	8.8	465
1, 3-isobutadiene, C_4H_6	Colorless gas	–4.4		–108.9	0.64	–85			
Iso-octane									
Isopentane, $CH_3CH(CH_3)CH_3$	Colorless liquid, pleasant odor; highly flammable.	27.85	–159.89		0.6197	–57			420
Isophorone, $C(O)CH{-}C(CH_3)$ $CH_2C(CH_3)_2CH_2$	Water white liquid; toxic, irritant to skin and eyes.	215.2	–8.1		0.9229 (at 20/20 °C)	96			462
Isopropyl alcohol, C_3H_7OH	Colorless liquid, pleasant odor; flammable, dangerous fire risk; toxic by ingestion and inhalation; tolerance 400 ppm in air.	82.3	–86		0.78634 (at 20/20 °C)	11.7 (closed cup)	2.0 (% vol)	12	453

Contd.

Physical and Chemical Characteristics of Major Hydrocarbons (*Contd.*)

Compound	Appearance/odor	BP (°C)	FP (°C)	MP (°C)	Specific gravity (20/20 °C)	Flash point (°C)	Flammability limit lower (°C)	Auto-ignition temperature (°C) upper (°C)	
Kerosene	Moderate fire risk, moderately toxic by inhalation.	–	180–300	–	0.81	37.7–65.5	0.7	0.5	228
Methane, CH_4	Colorless, odorless, tasteless gas.	–161	–184		0.554				537
MIBK	Colorless, stable liquid, pleasant odor.	115.8	–85.5		0.8042	22.7	–	–	460
MEK, $CH_3COCH_2CH_3$	Colorless liquid; flammable, dangerous fire risk.	80	–86.4		0.8256 (at 0/4 °C)	–4.4 (TOC)	1.81	11.5	515
Methanol, CH_3OH	Clear, colorless mobile liquid.	64.7	–97.8		0.7924 (at 25 °C)	463	6 (%vol)	36	464
Methyl chloride, CH_3Cl	Colorless compressed gas or liquid.	–23.7	–97.6		0.92	0			632
Methyle cyclo-hexane, $CH_3C_6H_{11}$	Colorless liquid.	100.8	–126.9		0.769	–3.89			285
Methyl cyclopentane									
Methyl tert-butyl ether (MTBE)	Colorless pungent smelling.	55	–100	8.4	0.74				
Methyl ethyl ketone, $CH_3COCH_2CH_3$	Colorless liquid, acetone like odor.	79.57	–85.9		0.80620 °C	–4.4 (TOC)	1.8	10	515

Contd.

Physical and Chemical Characteristics of Major Hydrocarbons (Contd.)

Compound	Appearance/odor	BP (°C)	FP (°C)	MP (°C)	Specific gravity (20/20 °C)	Flash point (°C)	Flammability limit lower (°C)	Auto-ignition temperature (°C) upper (°C)
2-methyl pentane N-methyl pyrroliodone, (NMP)	Colorless liquid.	202	–24		1.027			
Naphthalene, $C_{10}H_8$	White soild crystals, odor of coal tar, highly flammable	218		80.5	1.14	79.87		525
Nitroparaffin $C_nH_{2n+1}NO_2$	Colorless mobil liquid, pleasant odor	101–131	–28 to –104		0.983–1.131	28–38		
Nitrobenzene, $C_6H_5NO_2$	Oily liquid, greenish yellow crystal.	210.5		5.7	1.19867	87.7		482
Nonane, C_9H_{20}	Colorless liquid	151		–53	0.728	31		205
n-octane, C_8H_{18}	Colorless liquid.	125	13.3	–56	0.7026 (at 20/4 °C)	13.3		220
n-pentane, C_5H_{12}	Colorless liquid with pleasant odor; highly flammable.	36	–129.7		0.62624	–40	1.4	8.0
Phenol, C_6H_5OH	White, crystalline mass, burning taste, distinctive odor.	182	79.4	42	1.07	78 (closed cup)	–	715
Propane, C_3H_8	Colorless gas, highly flammable, dangerous fire risk gas.	–42.2	–187.1		0.531 (at 0 °C)	–105	2.3	468.7 9.5
Sulfur		444		119	1.96	207	–	232 –

Contd.

Physical and Chemical Characteristics of Major Hydrocarbons (Contd.)

Compound	Appearance/odor	BP (°C)	FP (°C)	MP (°C)	Specific gravity (20/20 °C)	Flash point (°C)	Flammability limit lower (°C)	upper (°C)	Auto-ignition temperature (°C)
Terephthalic acid, $C_6H_4(COOH)_2$	White crystal or powder, insoluble in water and many solvents.		sublimes > 300		1.51				
Toluene, $C_6H_5 \cdot CH_3$	Colorless, distinctive aromatic odor.	110.7	−94.5	−95	0.860 (at 20/4°C)	5	1.2 (%vol)	7.1	536
Trinitrotoluene	Pale yellow needle-shaped crystal poisonous and allergic.	295		80.35		1.654			
Urea, NH_2CONH_2	White crystalline solid, readily soluble in water and alcohol.			132	1.335				
Vinyl acetate, $CH_3COOCH=CH_2$	Colorless, flammable liquid.	72.7	−100.2		0.9345	−8 (closed cup) −1.1 (open cup)	Exp. 2.6 (%vol)	13.4	426.9
Vinyl chloride or Chloroethylene, $CH_2=CH-Cl$	Colorless liquid or gas, faintly sweet odour.	−13.4	−159.7		0.9195 (at 15/4 °C)	77 (COC)	4	22	472
m-xylene, $C_6H_4(CH_3)_2$	Colorless liquid.	139		−47.9	0.8684 (at 20/4 °C)	28.9	1.1	7.0	527.7
o-xylene, $C_6H_4(CH_3)_2$	Colorless liquid.	144	−25.5	−	0.88 (at 20/4 °C)	46.1			463.8
p-xylene, $C_6H_4(CH_3)_2$	Clear liquid.	138.3		13.2	0.861 (at 20/4 °C)	27.2 (TOC)	1.1	7.0	

APPENDIX C

SI Units and Conversion Tables

Quantities and SI Units		
Quantities	SI units	Symbol
Length	Micrometer	μm
	Millimeter	mm
	Meter	m
	Kilometer	km
Area	Square meter	m^2
	Square kilometer	Km^2
	Hectare	Ha
Volume	Liter	L
	Cubic meter	m^3
	Mega liter	mL
Time	Second	sec or s
	Minute	min
	Hour	hr
Velocity	Meter/second	m/s
Flow rate	Cubic meters/second	m^3/s
	Liters/second	L/s
	Mega liters/day	mL/d
Mass	Microgram	μg
	Milligram	mg
	Kilogram	kg
Force	Newton	N
	Kilonewton	KN
Pressure	Pascal	Pa
Energy	Joule	J
Power	Watt	W
	Kilowatt	KW
Temperature	Kelvin	K
	Degree celsius	°C
Acceleration	Meter per second square	m/s^2
Concentration	Mole per cubic meter	mol/m^3
Current density	Ampere per square meter	A/m^2
Density	Kilogram per cubic meter	kg/m^3
	Gram per cubic centimeter	g/cm^3
Kinematic viscosity	Square meter per second	m^2/s

Conversion to Metric System

Length	
1 inch	2.54 cm
1 foot	0.3048 m
1 yard	0.914 m
1 mile	1.609 km
Area	
1 ft^2	0.0929 m^2
1 acre	0.4047 hectares
1 sq mile	2.59 km^2
Volume	
1 cu inch	16.39 cc
1 US gallon	3.785 L
1 cu ft	0.0283 m^3
1 US barrel	0.159 m^3
Weight	
1 grain	0.0648 g
1 ounce	28.35 g
1 pound (lb)	454 g
1 short ton	907.2 kg
1 long ton	1,016 kg
1 ton	1,000 kg
Velocity	
1 ft/min	0.00508 m/s
1 mile/hour	0.447 m/s
Temperature	
°F	$1.8 \times °C + 32$
°C	$(°F - 32)/1.8$
Viscosity	
1 cp	10^{-3} kg/m s
Thermal conductivity	
1 Btu/hour ft °F	0.5778 W/m K
1 Btu/hour ft °F	241.75 cal/s cm K
Heat transfer coefficients	
1 Btu/hour ft^2 °F	0.1761 W/m^2 K
1 Btu/hour ft^2 °F	7,368.6 cal/s cm^2 K

Conversion of Energy Units

	kg-m	m-cal	Joule	KW-hour	HP-hour
kg-m	1	0.00235	9.81	0.27×10^{-5}	0.365×10^{-5}
m-cal	427	1	4,179	0.001161	0.001556
Joule	0.102	0.000239	1	27.77×10^{-8}	37.23×10^{-8}
KW-hour	366.973	860	3.6×10^{-10}	1	1.3411
HP-hour	273.745	642.5	2,685.5	0.74565	1

Conversion of Power Units

	kg m/s	Kcal/s	KW	HP
kg-m/s	1	0.00234	0.00981	0.0133
Kcal/s	427	1	4.17	5.59
KW	101.97	0.239	1	1.341
HP	76	0.178	0.746	1

Conversion of Pressure Units

	psi	Kg/cm^2	Atm	Bar	mm WG	Pascal	Pieze
psi	1	0.07031	0.06804	0.069	703.1	6,896	6.895
kg/cm^2	14.233	1	0.9678	0.981	10,000	98,087	98.08
Atm	14,069	1.033	1	1.0133	10,331	1,01,325	101.32
Bar	14.5	1.019	0.986	1	10,200	1,00,000	100
mm WG	14.233×10^{-4}	1×10^{-4}	96.78×10^{-6}	9.81×10^{-5}	1	9.808	98.04×10^{-4}
Pascal	145×10^{-6}	10.19×10^{-6}	9.869×10^{-6}	10^{-5}	0.1019	1	0.001
Pieze	0.145	0.01019	0.00986	0.01	102	1000	1

Conversion of Weight Units

	Grams	Kilograms	Ounces	Pounds	Tons
Grams (g)	1	1,000	28.35	453.6	1.016×10^6
Kilograms (kg)	0.001	1	2.835×10^{-2}	0.4536	1,016
Ounces (oz)	3.527×10^{-2}	35.27	1	16	3.584×10^4
Pounds (lb)	2.205×10^{-3}	2.205	6.250×10^{-2}	1	2,240
Tons (t)	9.842×10^{-7}	9.842×10^{-4}	2.790×10^{-5}	4.464×10^{-4}	1

Conversion of Speed Units

	m/s	m/min	km/hour	ft/s	ft/min	mph
m/s	1	1.667×10^{-2}	0.2778	0.3048	5.080×10^{-3}	4,770
m/min	60	1	16.67	18.29	0.3048	26.82
km/hour	0.278	0.06	1	1.097	1.829×10^{-2}	1.609
ft/s	3.281	5.468×10^{-2}	0.9113	1	1.667×10^{-2}	1.467
ft/min	196.8	3.281	54.68	60	1	88
mph	2.237	3.728×10^{-2}	0.6214	6,818	1.136×10^{-2}	1

Main Conversions Used in Petroleum Industry

Item		Conversion factor
Crude oil	1 ton	7.33 barrel 1.165 cubic meters (kiloliters)
	1 barrel	0.136 ton 0.159 cubic meters (kiloliters)
	1 cubic meter	0.858 ton 6.289 barrels
	1 million ton	1.111 billion cubic meters natural gas 39.2 billion cubic feet natural gas 0.805 million ton LNG 40.4 trillion British thermal units
Natural gas	1 million ton	35.3 billion cubic feet natural gas 0.90 million ton crude oil 0.73 million ton LNG 36 trillion British thermal units 6.29 million barrels of oil equivalent
LNG	1 million ton	1.38 billion cubic meters natural gas 48.7 billion cubic feet natural gas 1.23 million tons crude oil 52 trillion British thermal units 8.68 million barrels of oil equivalent
CNG	1 kilogram	1.244 standard cubic meters natural gas 1.391 liters of petrol 1.399 liters of HSDO

1 cubic feet of natural gas = 1,000 Btu; 1 m^3 of natural gas = 9,000 Kcal = 37,656 KJ
1 kWh = 3,412 Btu = 860 Kcal.

Source: 1. BP Amoco Alive Statistical Review of World Energy.
         2. OPEC Annual Statistical Bulletin.

Conversion Factors for Petroleum Products

Products	TOE/ton	Barrel/ton
Refinery gas	1.150	8.00
Ethane	1.130	16.85
LPG	1.130	11.60
Aviation gasoline	1.070	8.90
Motor gasoline	1.070	8.53
Jet gasoline	1.070	7.93
Jet kerosene	1.065	7.93
Other kerosene	1.045	7.74
Naphtha	1.075	8.50
Gas/diesel oil	1.035	7.46
Heavy fuel oil	0.960	6.66
Petroleum coke	0.740	5.50
White spirit	0.960	7.00
Lubricants	0.960	7.09
Bitumen	0.960	6.08
Paraffin waxes	0.960	7.00
Non-specified products	0.960	7.00

TOE: Tons of oil equivalent

Source: International Energy Agency Statistics.

APPENDIX D

Major Petroleum and Petrochemical Complexes in India

Name of unit	Address	Product
The Andhra Petrochemicals Ltd.	202A, My Home Sarovar Plaza Secretariat Road, Saifaba Hyderabad 500 063 Tel.: 040-23420666–67 Fax: 040-23420665 Factory: Opp. Naval Dock Yard, P.B. No. 1401, Visakhapatnam, AP 530014 Tel.: 0891-2891500; Fax: 0891-2577751	Petrochemicals
Assam Petrochemicals Ltd.	4th Floor, Orion Place Bhangagarh, G.S. Road, Guwahati 781005 Tel: +91-361-2461470/2461471/ 2461594 Factory: Namrup, PO Parbatpur, Dist Dibrugarh, Assam 786623 Tel.: +91-374-2500331/212/518	Formaldehyde, methanol, carbon dioxide
Bayer (India) Ltd.	Express Towers, Nariman Point, Bombay 400 021 Grams: Bayerind Factory: Kolshet Road, Thane 40060766/1 to 75/2, GIDC Estate, Himmatnagar 383001, Gujarat Tel.: +91 22 2531 1234	Agrochemicals, rubber chemicals and health care products
Bharat Petroleum Corporation Ltd.	P.O Box No 1725, Mumbai 400 001 Mahul, Mumbai 400074 Phone: 022-25533888 Fax: 022-25542970 Gram: Bharefin www.bharatpetroleum.com	Petroleum products, propylene, greases, refineries, toulene, bitumen, carbon black, diesel, furnace oil, hexane, kerosene, LPG, naphtha, petroleum ether, propylene, rubber processing oil, sulfur, transformer oil, turpentine
Bharat Oman Refineries Ltd.	Post BORL Residential Complex Bina 470 124, Dist. Sagar, Madhya Pradesh, India. Tel.: +91-7580-22 6000, 27 6000 Fax: +91-7580-22 6903	Liquefied petroleum gas (LPG), naphtha, kerosene, aviation turbine fuel (ATF), petcoke and sulfur
Bongaigaon Refinery and Petrochemicals Ltd.	P.O. Dhaligaon, Dist: Bongaigaon, Assam 783385 Tel.: 03664-241231 Fax: 03664-241230 http://biz.yahoo.com	Petroleum products, gasoline, kerosene, diesel, lubricating oil, bitumen, p-xylene, DMT, polyester

Contd.

Major Petroleum and Petrochemical Complexes in India *(Contd.)*

Name of unit	Address	Product
Bharat Petroleum Corporation Ltd., Kochin Refinery	Ambala mugal, Ernakulam Kerella 682302 Ph: 0484-2822644	Gasoline, kerosene, diesel, lubricating oil.
Gas Authority of India Ltd.	U.P. Petrochemical Complex, Pata, Dist. Etawah (UP) 206241 Ph: 05683-282356, 282049, 283403-5;	Ethylene propylene, polyethylene (HDPE, LDPE), butene, LPG, propane, pyrolysis gasoline.
Gujarat Petro-synthese Ltd.	Plot No.1, Marol Co-Oprative Indl. Estate, M.V. Road, J.B. Nagar Post, Andheri East, Mumbai India 400059 Tel.: 022-28509396 Fax: 022-28509394 Factory: Petrochemical Complex, PO Petrofil, Dist Baroda, Gujarat 391347 Tel.: 0265-372994	Polybutene
Haldia Petrochemicals Ltd.	1 Auckland Place, Calcutta W. Bengal 700017 Tel.: 033-2283 1640/1643 Fax: 033-2280 2390 www.haldiapetrochemicals.com	Organic chemicals, petrochemicals, polymers and plastics, speciality oils, synthetic rubber, benzene, butadiene, carbon black, ethylbenzene, isobutylene, gasoline, HDPE, ethylene, propylene isobutylene, LLDPE/HDPE, polypropylene, pyrolysis gasoline, styrene butadiene, rubber.
Herdillia Chemicals Ltd.	Air India Bldg., Nariman Point, Bombay 400021 Ph.: 9122-22024224 Grams.: Herdillia; Telex: 1183754 HCL IN Fax: 22042379 Works: Plot No. 2-1, TTC Industrial Area, Thane, Belapur Road, KV Bazar Post Vashi, New Bombay 400705 Ph.: 0215-681153 www.herdillia.com	Phenol, acetone, diacetone, alcohol, pthalic anhydride, cumene, a-methyl styrene and its dimers, fumaric acid, capacitor fluids and other speciality fluids, dodecyl and other alkyl phenols, isobutyl benzene, diphenyl oxide, cumyl phenol, etc.
Hindustan Petroleum Corp. Ltd.	Petroleum House, 17, J Tata Rd, Churchgate, Mumbai 400001 Tel.: 022-22863900 Fax: 022-2872992	

Contd.

Major Petroleum and Petrochemical Complexes in India *(Contd.)*

Name of unit	Address	Product
	Factory 1: Corridor Rd, Mahul, Mumbai P.O.Box 18820, B.D.Patil Matg Mahul, Mumbai 400 074 Tel.: 022-5562830 Fax: 022-5562008 Factory 2: P.B. No. 15, Visakhapatnam, AP 530001 Ph: 2895000	Petroleum products, coke, LPG, lube oil base, naphtha, propane, propylene, sulfur
HPCL-Mittal Energy Ltd.	Corporate office: INOX Towers, Sector 16 A, Plot No. 17 Noida 201301 (U.P.) India. Ph: +91 120-4634500 Fax: +91 120-4271940 Refinery: Phullokari Village Talwandi Saboo Taluka District Bathinda 151301, Punjab, India.	LPG, petcoke, aviation fuel, sulfur, polypropylene
Indian Oil Corporation Ltd. Barauni refinery	P.O. Oil Refinery P.O. Barauni Oil refinerry Dist. Begusarai (Bihar) Ph: 06243 23257, 75271 Fax: 06243 24005, 23258, 24038 www.iocl.com	Petroleum products
Indian Oil Corporation Ltd., Gujarat Refinery	P.O. Jawahar Nagar, Dist. Vadodara, Gujarat 391320 Grams: Guajrat refinery	Petroleum products, MTBE
Indian Oil Corporation Ltd., Haldia refinery	P.O. Haldia Oil Refinery, Dist. Midnapore 721606 (W.B.) Phone: 03224-52147, 52151 Grams: Oil Refinery Fax 91-3224-52204, 52141, 52284	Petroleum products
Indian Oil Corporation Ltd., Mathura refinery	Mathura 261 005 (U.P.) India Ph: 845315 Fax: 0565 842167, 842169, 430019, 430298 Telegram: Oil Refinery Internet sit www.Indianoilcorp.com	Petroleum products
Indian Oil Corporation Ltd., Panipat refinery	P.O. Panipat refinery Panipat 132 140 (Haryana) Fax: 01742 78833/43,4478825/26	Petroleum products

Contd.

Major Petroleum and Petrochemical Complexes in India *(Contd.)*

Name of unit	Address	Product
Indian Petrochemical Corporation Ltd.	PO Petrochemicals 391346, Dist. Vadodara, Gujarat Ph: 72011, 72031, Telex: 0175-365 Grams: Petcomplex	Petrochemicals, viz. aromatic chemicals (DMT and xylenes) and olefins (ethylene, propylene, butadiene, benzene and pyrolysis gasoline), other chemicals such as butadiene, ethylene, propylene, acetonitrile, benzene, heavy paraffins and carbon black feedstock, trading in polymer, fiber and fiber intermediate and chemicals
IG Petrochemicals Ltd.	401/404, Raheja Centre, Free Press Journal Marg, 214, Nariman Point, Mumbai 400021 Tel.: 022-30286100 Factory: T-2, MIDC Area, Taloja, Raigad, Maharashtra Tel.: 022-27410230 Fax: 022-27410192	Organic chemicals, phthalic anhydride
Indian Oil Corporation Ltd.	Indian Oil Corporation Ltd, Corporate Office, 3079/3, JB Tito Marg, Sadiq Nagar, New Delhi 110049 Tel: 011-26260000	Petrochemicals, refineries, benzene, diesel, gasoline, hydrogen, LPG, methyl tert-butyl ether, paraffin wax, phenol
Chennai Petroleum Corporation Ltd. Subsidiary: M/s Indian Additives Ltd.	New No: 536, Anna Salai, Teynampet Chennai 600 018 Ph: 044-24349519, 24349542 Fax: 044-24341753	Refining of crude oil and manufacture of petroleum and by-products
Manali Petrochemicals Ltd.	Manali Petrochemicals Limited Ponneri High Road Manali Chennai 600 068 Ph: 044-2594 1025	Propylene oxide, propylene glycol and Polyol
Mysore Petrochemicals Ltd.	D/4 Jyothi Complex, 134/I, Infantry Rd, Bangalore 560001 Ph: (080) 22868372 Factory: Phthalic Anhydride Plant: Near Chicksugar Railway Station Raichur, Karnataka Ph: (08532) 246425 Maleic Anhydride Plant: T-1, MIDS Indl Area, Taloja, MS Synthetic Diamonds Division: Shed No. 1102-1103 GIDC Panoli, Dist. Bharuch, Guj	Phthalic anhydride and plasticizers.

Contd.

Major Petroleum and Petrochemical Complexes in India *(Contd.)*

Name of unit	Address	Product
Numaligarh Refinery Ltd., Numaligarh Assam	P.O.: Numaligarh Refinery Project Dist.: Golaghat, Assam, India Pin 785699 Phone: +91-03776-265493 Fax: +91-3776-265800	Petroleum products including LPG.
Reliance Industries Ltd.	Reliance Industries Limited Makers Chambers-IV, Nariman Point, Mumbai 400 021. India. Tel: 91-22-2278 5000 Factory: B-4, MIDC indl Area, Patalganga Off Mumbai-Pune Rd, Nr. Panvel, Raigad, MS 410207 Tel.: 02192-502015/50601-5 www.ril.com	Organic intermediates, petrochemicals, benzene, carbon black, diethylene glycol, ethylene, ethylene glycol, ethylene oxide, formalin, linear alkyl benzene, LLDPE, HDPE, xylene, vinyl chloride, etc.
Reliance Industries Ltd.	Vill. Mora, Bhatha, PO Surat Hazir Road Dist. Surat, Gujarat 394510 Ph.: 96307, 96211, 96203, 96205 Telex: 188-447 RIL IN Fax: 0261-42332	Monoethylene glycol, higher ethylene glycols, high density polyethylene, polyvinyl chloride, chlorine and caustic soda.
Reliance Industries Ltd.	15, Walchand Hirachand Marg, Ballard Est, Mumbai 400028 Tel.: 022-30327000; Fax: 022-22870072 Factory: Village Motikhavadi PO Digvijaygram, Jamnagar, Gujarat 361140 Tel.: 0288-510000	Petroleum products.
Rock Hard Petrochemical Industries Ltd.	16 Manishpuri, Saket, Indore Madhya Pradesh 452001 Tel: 0731-2560945 Fax: 0731-2566568 Factory: Vill Kundla, Tehsil Agar (Malwa), Dist. Shajapur, MP Tel.: 07362-52176/52880	Fine chemicals, organic chemicals, formaldehyde, hexamine.
Shri Ambuja Petrochemicals Ltd.	7-C, Surya Towers, 105, Sardar Patel Road, Secunderabad 500 003 Ph: 040-27841717 Works: Nos. 17, 18 and 20 IDA, Phase-I, Patancheru 502319, Medak Dist. AP.	Phthalic anhydride, fumaric acid and maleic anhydride.

Contd.

Major Petroleum and Petrochemical Complexes in India *(Contd.)*

Name of unit	Address	Product
Sriman Petrochemicals Ltd.	5th Floor,Ghanshyam Chambers Link Road, Andheri (W), Mumbai 400053. Maharashtra. India. Tel: +91-22-26732894 Fax:+91-22-66779569 Factory 1: A-171, Phase 1, MIDC, Dombivli (E), Thane, MS Factory 2: C-366/376, TTC Indl Area, Turbhe, Navi Mumbai	Organic intermediates, meta-chloro aniline, o-anisidine, p-anisidine, o-chloroaniline
Supreme Petrochem Ltd.	Building No. 11, 5th Floor, Solitaire Corporate Park,167, Guru Hargovindji Marg, (Andheri-Ghatkopar Link Road) Chakala, Andheri (East), Mumbai 400093 Tel: +91-22-6709 1900 Factory:Vill Amdoshi, Wakan-Roha Rd, Tal. Roha, Dist. Raigad, MS 402106 Tel.: 0211442-2540-47 Fax: 2537,2617	Petrochemicals, polystyrene.
Tamil Nadu Petroproducts Ltd. (A joint sector company of TIDCO and SPIC)	68, Manali Express Highway, Manali, Madras 600008 Ph.: +91-44-25941501-10 Telex: +91-44-25941139	Linear alkyl benzene and heavy normal paraffin.
UB Petroproducts Ltd.	McDowell House, 3-Second Line Beach, Madras 600001 Ph.: 520025-26-28 Telex: 0418787 UBPL IN Factory: Sathangadu Village, Manali, Madras 600 066	Petrochemical products including propylene oxides, polyols, propylene glycol.
United Carbon India Ltd.	NKM International House, Babubhai M., Chinar Marg, 178, Backbay Reclamation, Bombay 400020 Gram: Unicarbon; Telex: 011-83107 Ph.: 202-1914, 202-7962, 202-7846, 202-8285, 202-0767; Fax: 2850406 Factory: MIDC Plot No. 3, Trans Thane Creek Area, Thane, Belapur Road, Post Ghansoli, Thane 400701, Maharashtra	Furnace black and various grades of carbon used in tyre production.

APPENDIX E

Major Research and Consultancy Organizations

Institution	Areas of research/consultancy	Contact
Council of Scientific and Industrial Research (CSIR)	Established in 1942, CSIR has evolved more than 3,000 technologies; with about 6,000 clients using them. Based on CSIR technologies, the annual industrial production today has risen to about ₹ 5,000 crore. CSIR had then defined its functions—promotion, guidance and coordination of scientific and industrial research, and setting up laboratories to spur research and development for growth of industry. Chemicals and petrochemicals are the thrust areas, CSIR made major strides in this sector, making India among the top five countries possessing world class capabilities for formulating new catalysis. Today, many of the CSIR technologies have made their ways to world markets. CSIR developed its own brand of zeolite catalyst called encilites, which is exported to several countries and used for diverse industrial processes including the production of petrochemicals such as p-xylene, ethyl benzene and olefin, which in turn are used as raw materials for almost all entire plastic, rubber and dye industries. **Corporate affairs:** • R and D planning and Business Development Division (RPBDD). • International Science and Technology Affairs Directorate (ISTAD). • Human Resource Development Group (HRDG). • Intellectual Property Management Division (IPMD). • Societal and Technology Mission and Societal Program Division (STMD). • Unit for Science Dissemination (USD).	Director General CSIR, Rafi Marg, New Delhi 110001 Tel: 011-3730681 Fax: 011-3710340/3710618 E-mail: csirhq@sirnetd.ernet.in

Contd.

Major Research and Consultancy Organizations *(Contd.)*

Institution	Areas of research/consultancy	Contact
Central Building Research Institute (CBRI)	Architecture/Building/Construction Engineering/Engineering/ Geotechnical Engineering/Structural Design.	Roorkee 247667 Tel: +91-1332-72243 Fax: +91-1332-72272/72543 Email: director@cbrimail.com
Center for Biochemical Technology (CBT)	Allergy/Biochemicals/ Biochemistry/ Bioscience/ Diagnostics/Enzyme/Immunology/ Space Medicine.	Director Near Jubilee Hall, Univ. Campus, Mall Road, Delhi 110007 Tel: (+91) 11 7257-578, 298, 7416489 Fax: (+91) 11 7257471, 7416489 E-mail: csircbt@del2.vsnl.net.in
Center for Cellular and Molecular Biology (CCMB)	Biochemistry/Biomolecule/ Bioscience/Cancer/Cell Biology/ DNA/Embryo/Fertilization/ Genetic Engineering/Molecular Biology/Oncogenes/Reproduction/ Transgenetics/Tumor/Virus.	RRL Campus, Uppal Road, Hyderabad 500007 Tel: (+91) 40 7172241-50, 7170130-39 Fax: (+91) 40 7171195 E-mail: lalji@ccmb.apnic.in
Central Drug Research Institute (CDRI)	Bioscience/Drugs/Embryo Transfer/ Hormone/Mutagenicity/Natural Products/Pharmacology/ Teratogenicity/Toxicity/Vaccine/ Virus.	Chattar Manzil Palace P.B. No. 173 Lucknow 226001 Tel: (+91) 522 2223286, 2210932, 2212411-18 Fax: 091-(522)-2223405/ 2223938/2229504 E-mail: drcmg@satyam.net.in
Central Electrochemical Research Institute (CECRI)	Battery/Chemical Science/Corrosion/ Electrobiology/Electrochemistry/ Electronics/Electrometallurgy/ Fuel Cells/Instrumentation/Metal Finishing/Pollution Control.	Karaikudi 630006 Tel: (+914565 22064, 22065 Fax: (+91) 4565 22088 E-mail: ragha@cscecri.ren.ni.in
Central Electronics Engineering Research Institute (CEERI)	Audio Visual System/Design Engineering/Electronics/ Instrumentation/Microwave Tube/ Physical Science/Semiconductors/ Telematics.	Pilani 333031, Rajasthan Tel: (+91) 1596 42111, 42133 Fax: (+91) 1596 42294 E-mail: rnb@cecri.ernet.in
Central Fuel Research Institute (CFRI)	Briquettes/Carbonization/Coal/ Coal Ash/Combustion/Engineering/ Fuel/Gasification/Industrial Carbon/ Petrochemistry/Pilot Plant/Quality Assessment/Waste Utilization.	POFRI Dist. Dhanbad Dhanbad 828108 Tel: (+91) 326 460141, 461710 Fax: (+91) 326 464350 E-mail: cfri@sirnetd.ernet.in

Contd.

Major Research and Consultancy Organizations *(Contd.)*

Institution	Areas of research/consultancy	Contact
Central Food Technological Research Institute (CFTRI)	Animal Products/Bioscience/Crops/ DNA/Energy Conversion/Food/ Fruit/Microbial Genetics/Molecular Biology/Nutrition/Packaging/ Process Optimization/Vegetable.	Director Mysore 570013 Tel: (+91) 821 517760 Fax: (+91) 821 516308, 517233 E-mail: director@nicfos.ernet.in
Central Glass and Ceramic Research Institute (CGCRI)	Bioceramics/Ceramics/Chemistry/ Engineering/Glass/Glass Electrode/ Laser Glass/Optical Coating/Optical Communication Fiber/Refractory/ Silicate.	Dr HS Maiti, Director, 196 Raja SC Mullick Road Kolkata 700032 (India) Ph: 091-33-2473-3496/69/ 76/77, 2483-8079/82 Fax: 2473-0957 E-mail: liaison@cgcri.res.in
Central Institute of Medical and Aromatic Plants (CIMAP)	Alkaloids/Aromatic Plants/ Bioscience/Disease Control/Drugs/ Fermentation/Genetic Transformation/ Medical Plants/Oil/Pest Control/ Pharmacology/Virus.	Director P.O. CIMAP, Near Kukrail Picnic Spot, Lucknow 226015, India Tel: +91-522-2359623 Fax: +91-522-2342666 E-mail: director@cimap.res.in
Central Leather Research Institute (CLRI)	Bioinformatics/Chemical Science/ Collagen/DNA/Immunology/ Leather/Microbial Enzymes/ Molecular Biology/Peptide Chemistry/Skin/Tanning/Virus/ Waste.	Director Adyar, Madras 600020 Tel: (+91) 44 24910897, 24910846, 24912150 Fax: (+91) 44 4911589 E-mail: clrim@giasmd01.vsnl.net.in
Central Mechanical Engineering Research Institute (CMERI)	Cryogenics/Design Engineering/ Energy System/Engineering/ Environmental Engineering/ Machines/Material Evaluation/ Mechanical Engineering/ Refrigeration/Robotics.	Director Mahatma Gandhi Avenue, Durgapur 713209 Gram: Mechsearch, Durgapur Ph: (0343) 546749 Fax: (0343) 546745 General: (0343) 546818, 546826, 546828 E-mail: director@cmeri.res.in
Central Mining Research Institute (CMRI)	Cast Mining/Coal Conservation/ Coal Mining/Engineering/ Excavation Engineering/Explosives/ Geomechanics/Hazard/Mining.	Director Barwa Road, Dhanbad 826001, India. Tel: 91-0326-2202326, 2203043. Fax: 91-0326-2202429, 2205028 E-mail: dcmri@csir.res.in

Contd.

Major Research and Consultancy Organizations *(Contd.)*

Institution	Areas of research/consultancy	Contact
Central Road Research Institute (CRRI)	Air Pollution/Bridge Constructions/ Bitumen/Cement/Concrete/ Engineering/Geotechnical Engineering/Highway Engineering/ Road Pavement/Road Policy/Road Safety/Traffic/Transport.	Prof PK Sikdar, Director P.O. CRRI New Delhi 110020 Tel: (+91) 11 6848917 Fax: (+91) 11 6845943 Email: director@cscrri.ren.nic.in
Central Scientific Instruments Organisation (CSIO)	Agroelectronics/Instrumentation/ Medical Electronics/Microelectronics/ Optical Systems/Physical Science/ Pollution Monitoring/Process Control/Space Application/Testing.	Director Sector 30, Chandigarh 160020 Tel: (+91) 172 657190 Fax: (+91) 172 657267, 657082 E-mail: root@cscsio.ren.nic.in
Central Salt and Marine Chemicals Research Institute (CSMCRI)	Algae/Biosalinity/Chemical Science/Desalination/Effluent Treatment/Fertilisers/Marine Chemistry/Nitrogen/Photocatalysis/ Phytosalinity/Solar Pond/Solar Thermal.	Dr PK Ghosh, Director Gijubhai Badheka Marg Bhavnagar 364002 Tel: (+91) 278 566970/567562 Fax: (+91) 278 566970 E-mail: csmcri@csmcri.org
Institute of Himalayan Bioresource Technology (IHBT)	Floriculture/Tea Science/ Biotechnology/Natural Plant Products/Biodiversity.	Director Palampur, HP 176061 Tel: 01894-30411 Fax: 30433, E-mail: director@csihbt.ren.nic.in
Indian Institute of Chemical Biology (IICB)	Bacteria/Biomolecule/Bioscience/ Drug Delivery/Drugs/Enzyme/ Fermentation/Fertility/Hormone/ Immunology/Leishmania/ Microbiology/Molecular Biology/ Neurobiology/Parasite/ Reproduction/Vibrio.	Director 4, Raja SC Mullick Road Kolkata 700032 Tel: (+91) 33 4733492/0492 Fax: (+91) 33 4730286
Indian Institute of Chemical Technology (IICT)	Agrochemicals/Catalyst/Chemical Engineering/Chemical Science/Coal/ Coating/Drugs/Fat/Information/ International Cooperation/Oil/ Organic Intermediates/Pesticides/ Polymers.	Director Uppal Road, Hyderabad 500007 Tel: (+91) 40 27193943, 27193030, 27193234 Fax: (+91) 40 27160386, 27193626 E-mail: kvr@iict.ap.nic.in
Indian Institute of Petroleum (IIP)	R & D activity of IIP covers the fields of petroleum refining, chemical science. Product applications, combustion and biotechnology/ Biomass/Bitumen/Catalyst/	Director P.O. IIP, Mohkampur Dehradun 248005 Tel: (+91) 135 624508 Fax: (+91) 135 671986

Contd.

Major Research and Consultancy Organizations *(Contd.)*

Institution	Areas of research/consultancy	Contact
	Chemical Science/Emission Control/ Energy Conservation/Fuel Oil/ Hydrocarbon/Hydrogen/Lubricants/ Microbial Dewaxing/Petrochemistry/ Refining/Solvents/Surfactants. **Major achievement:** *Processes:* Benzene and toluene through solvent extraction, solvent dearomatizing of naphtha, food grade hexane through solvent extraction, superior kerosene/ATF through solvent extraction, solvent dewaxing and deoiling, visbreaking technology, delayed coking technology, catalytic reforming, Pt-Re bimetallic reforming catalyst, hydrodesulfurization of naphtha, kerosene, and gas oil, pyrolysis gasoline hydrogenation, additives for petroleum industry, speciality chemicals, sulfonate dehydration catalyst for ethylene from ethanol, adipic acid, high temperature antioxidants.	Email: iipddn@del2.vsnl.net.in
Institute of Microbial Technology (IMT)	Bacteria/Biochemical Engineering/ Bioscience/DNA/Drug Delivery/ Drugs/Fungi/Genetic Engineering/ Immunology/Microbial Genetics/ Molecular Biology/Tissue Culture/ Vaccine/Yeast.	Director P.B. 1304, Sector 39-A Chandigarh 160036 Tel: (+91) 172 690908, 690713, 690785, 690263 Fax: (+91) 172 690585, 690632 E-mail: root@koel.imtech.ernet.in
Indian National Scientific Documentation Centre (INSDOC)	Bibliography/Bibliometrics/ Database/Documentation/ Information/Translation.	Director 14, Satsang Vihar Marg, New Delhi 110067 Tel: (+91) 11 6515837, 665072 Fax: (+91) 11 6862228 E-mail: teevee@csird.ernet.in
Industrial Toxicology Research Centre (ITRC)	Agrochemicals/Bioscience/Effluent Waste/Environment/Environmental Biotechnology/Genetic Engineering/ Immunology/Industrial Chemicals/ Pollution Control/Toxicology/ Waste Water.	Director P.O. 80, Mahatma Gandhi Marg, Lucknow 226001, India Ph: +91-522-2621856, 2228227, 2213357 Fax: +91-522 2228227, 2211547 E-mail: itrc@itrcindia.org

Contd.

Major Research and Consultancy Organizations *(Contd.)*

Institution	Areas of research/consultancy	Contact
National Aerospace Laboratories (NAL)	Aerodynamics/Aerospace Research/ Biogas/Coating/Combustion/ Composite/Engineering/Fiber/ Fluid Dynamics/Geothermal/Heat Transfer/Propulsion/Renewable Energy/Resin/Rotor/Wind Energy.	Director P.B. 1779, Bangalore 560 017, India. Tel: 91-80-5270584, 5265579. Fax: 91-80-5260862, 5270670. E-mail: bhogle@css.cmmacs.ernet. in
CSIR Center for Mathematical Modelling and Computer Simulation (C-MMACS)	Modelling of resources/Climate/ Environment/Hazard quantification/ Non-linear dynamical systems/ Design of Engineering Systems.	Dr. Anand Kumar Scientist-Incharge (Acting) NAL Belur Campus, Bangalore 560037 Tel: 080-5274667, 5274649 Fax: 5260392 E-mail: rnsingh@cmmacs.ernet.in
National Botanical Research Institute (NBRI)	Biochemistry/Bioscience/Botany/ Drugs/Herbs/Morphology/ Ornamental Plants/Pharmaceuticals/ Plant Biomass/Plant Breeding/Seeds/ Taxonomy/Trees.	Director P.B. 436, Rana Pratap Marg Lucknow 226001 Tel: (+91) 522 2205839, 2205848, 2207648 Fax: (+91) 522 2205839 E-mail: manager@nbri.sirnetd. ernet.in
National Chemical Laboratory (NCL)	Alloys/Biochemical Engineering/ Biosensors/Catalyst/Ceramics/ Chemical Science/Composite/Drugs/ Energy Conversion/Fiber/Forest Trees/Plantation Crops/Polymers/ Tree Tissue Culture.	Director Dr Homi Bhabha Road Pune 411008 Tel: +91-20-5893030 Fax: +91-20-5893355 E-mail: director@ems.ncl.res.in
National Environmental Engineering Research Institute (NEERI)	Air Quality/Bacteria/Biogas/ Drinking Water/Engineering/ Environmental Biotechnology/ Environmental Monitoring/ Gasification/Methane/Microbial Degradation/Toxic Waste/Waste Management.	Director Nehru Marg, Nagpur 440020 Tel: 0712-2223983 Fax: 0712-2222725 E-mail: dirneeri_ngp@sancharnet.in
National Geophysical Research Institute (NGRI)	Geophysics/Minerals/Mining Mechanics/Ores/Petroleum/Physical Science/Seismology/Soil Science.	Director Uppal Road, Hyderabad Tel: (+91) 40 7171124 Fax: (+91) 40 7170491/ 7171564

Contd.

Major Research and Consultancy Organizations *(Contd.)*

Institution	Areas of research/consultancy	Contact
National Institute of Oceanography (NIO)	Environmental Monitoring/ Fisheries/Food/Marine Biotechnology/Marine Drugs/Marine Science/Minerals/Ocean Climate/ Oceanography/Pharmaceuticals/ Physical Science/Polar Research.	Director Dona Pula, Goa 403004 Tel: (+91) 832 226253, 221322 Fax: (+91) 832 223340, 229102 E-mail: ocean@darya.nio.org
National Institute of Science Communication (NISCOM)	Aromatic Plants/Database/ Environmental Information/Health Information/Information/MAPIS/ Medical Plants/Raw Materials/ Scientific Journal.	Director-In-charge Dr. KS Krishnan Marg New Delhi 110012 Tel: (+91) 11 5786301, 5746024 Fax: (+91) 11 5787062 E-mail: niscom@sirnetd.ernet.in
National Institute of Science Technology and Development Studies (NISTADS)	Database/Forecasting/Geographic Information System/Information/ Research Analysis/Research Policy/ Science Management/Software Development/Technological Transfer/ Technology Assessment.	Director Pusa Gate, KS Krishnan Marg, New Delhi 110 012, India Tel: +91 11 25846064, 25843227 Fax: +91 11 25846640 E-mail: director@nistads.res.in
National Metallurgical Laboratory (NML)	Alloys/Corrosion/Creep Testing/ Engineering/Metallurgy/Metals/ Minerals/Pollution Control.	Director P.O. Burma Mines Jamshedpur 831007 Tel: (+91) 657 2271715/ 2270092 Fax: (+91) 657 426527 E-mail: dnml@nmlindia.org
National Physical Laboratory (NPL)	Atmospheric Physics/Biosensors/ Carbon Fiber/Ceramics/Composite/ Cryogenics/Greenhouse/Physical Science/Polymers/Quality Control/ SQUID/Silicon/Solar Cells/ Standardization/Super Conductor	Director Dr. KS Krishnan Road New Delhi 110012 Tel: (+91) 11 5741440 Fax: (+91) 11 5752678 E-mail: npl@sirnet.ernet.in
Regional Research Laboratory-Bhopal (RRL-B)	Aluminum Alloys/Building Materials/Ceramics/Coal/Composite/ Corrosion/Engineering/Fertilizer Minerals/Material Science/Metallurgy/ Mining Equipment/Natural Fiber/ Ores/Polymers/Sisal.	Director Hoshangabad Road Habibganj Naka Bhopal 462026 Tel: (+91) 755 587105, 580836 Fax: (+91) 755 587042 E-mail: rrlbho@sirnetd.ernet.in

Contd.

Major Research and Consultancy Organizations *(Contd.)*

Institution	Areas of research/consultancy	Contact
Regional Research Laboratory-Bhubaneswar (RRL-BH)	Agrotechnology/Alloys/Bioleaching/Biomass/Biotechnology on Metals/Engineering/Metals/Minerals/Natural Products/Ores/Pollution Monitoring/Recultivation/Slurry Transport.	Director Sachivalya Marg Bhubaneshwar 751013 Tel: (+91) 674 581126 Fax: (+91) 674 581637 E-mail: root@csrrlbhu.ren.nic.in
Regional Research Laboratory-Jammu (RRL-Ja)	Energy/Aromatic Plants/Bioscience/Biotechnology/Drugs/Fermentation/Fruits/Genetic Engineering/Geothermy/Medical Plants/Minerals/Renewable Energy/Solar Energy/Vegetables.	Director Canal Road, Jammu Tawi 180 001 Ph: 0191-2546368, 0191-2549051 Fax: 548607 Email: qazi_gn@yahoo.com
Regional Research Laboratory-Jorhat (RRL-J)	Chemical Plant/Chemical Science/Drugs/Energy Conversion/Hydrocarbons/Microbial Engineering/Microbial Genetics/Microbial Transformation/Paper/Petroleum/Soil Engineering.	Director Jorhat 785 006, Assam, India. Tel: +91-376-370121/370086 Fax: +91-376-370011 E-mail: drrljt@csir.res.in
Regional Research Laboratory-Trivandrum (RRL-T)	Aluminum Alloys/Biochemical Engineering/Energy Conversion/Engineering/Fermentation/Molecular Biology/Photochemistry/Plantation Crops/Process Optimization/Waste Treatment.	Director Trivandrum 695019 Tel: (+91) 471 490324, 490674, 490224 Fax: (+91) 471 490186 E-mail: rrlt@sirnetm.ernet.in
Structural Engineering Research Center-Ghaziabad (SERC-G)	Bridges/Buildings/Chimneys/Engineering/Grid Structures/Structural Design/Structural Dynamics/Structural Engineering/Wind Engineering.	Acting Director PB 10 Kamla Nehru Nagar Ghaziabad 201001 Tel: (+91) 91 721874, 713772, Fax: (+91) 91 721882
Structural Engineering Research Center-Madras (SERC-M)	Concretes/Construction Engineering/Engineering/Expert Systems/Offshore Structures/Ship Construction/Structural Dynamics/Structural Engineering/Transmission Towers/Wind Engineering.	Director CSIR Campus, Taramani Madras (Chennai) 600113 Tel: (+91) 44 2254-2139 Fax: (+91) 44 2254-1508
AMCO	AMCO Corporation Plc is a multi-faceted construction group with global expertize in niche areas of the industry. AMCO has extensive contracting experience in the	Amco Corporation Plc 25, Moorgate Road Rotherham, South Yorkshire S60 2AD United Kingdom

Contd.

Major Research and Consultancy Organizations *(Contd.)*

Institution	Areas of research/consultancy	Contact
	traditional construction markets of design and build, structural steel, power, rail, mining, drilling and mineral exploration both in the public and private sectors.	Tel: +44 (0)1709 828218 Fax: +44 (0)1709 828499
Engineering Project India Limited (EPIL)	With the objective of entering the hitech emerging areas, EPI has entered into strategic alliances with the world renowned companies in the fields of roads, highways and expressways, bridges, ports, airports, environmental engineering, water supply and waste management, bulk material handling and grain handling. Our foreign associates have undertaken large number of projects using the best technology and construction techniques. EPI is keen to join hands with international organizations for taking up hitech projects in India or any other country.	Executive Director Core-3, Scope Complex. 7-Institutional Area, Lodi Road, New Delhi 110003 Tel: 91-011-4363662 Fax: 91-011-4363426 E-mail: epind@nde.vsnl.net.in; info@engineeringprojects.com
Institut Français du Pétrole (IFP)	The Institut Français du Pétrole (IFP) is an independent research and industrial development, education and training, and information center active in the fields of oil, natural gas and the automobile. Its activities cover all aspects of the oil and gas industry: Exploration, production, refining, petrochemicals, engines and the use of petroleum products. IFP's research and development (R & D) is undertaken with a view to sustainable growth—security of supply and protection of the environment and industrial applications; it is organized around the four fundamental areas of the oil and gas sector: Exploration—Reservoir Engineering, Drilling—Production, Refining—Petro-chemicals and Engines—Energy.	1 & 4, avenue de Bois-Préau 92852 Rueil-Malmaison Cedex France Tel: +33 (0) 147 52 6000 Fax: +33 (0) 147 52 7000

Contd.

Major Research and Consultancy Organizations *(Contd.)*

Institution	Areas of research/consultancy	Contact
Mitsubishi Corporation	Mitsubishi Corporation ranks among the few companies in the world with the financial and technical resources to make large scale investments in petrochemicals, fertilizers, methanol, salt and other basic commodity chemicals. As a joint venture partner to many key supplier nations, we help to ensure a steady supply of commodity chemicals at highly competitive rates. Our global logistics networks, including our new Asian distribution hubs, ensure optimal handling and delivery of customer's products.	Mitsubishi Building 2-5-2 Marunouchi, Chiyoda-ku, Tokyo Mail Code 100-8324
Tata Energy Research Institute (TERI)	A unique developing country institution, TERI is deeply committed to every aspect of sustainable development. From providing environment friendly solutions to rural energy problems to helping shape the development of the Indian oil and gas sector, from tackling global climate change issues across many continents to enhancing forest conservation efforts among local communities; from advancing solutions to growing urban transport and air pollution problems for promoting energy efficiency in the Indian industry, the emphasis has always been on finding innovative solutions to make the world a better place to live-in.	TERI, 4th Main, Domlur II Stage Bangalore 560 071 Karnataka Tel: +91 80 535 6590-97/535 4929/555 1566 Fax: +91 80 535 6589 E-mail: terisrc@teri.res.in
Triune Projects Pvt. Ltd.	Triune Project is an engineering and consultancy organization with specific strength in refineries, chemical and petrochemical plants, power plants, oil and gas production technology.	Triune Projects Pvt Ltd 1501, Hemkunt Chambers 89 Nehru Place, New Delhi 110 019 Tel: (91)(11) 6470790 Fax: (91)(11) 6445685
Triune International New Delhi	Incorporated in 1982, the industries addressed by this group include power, oil and gas, petrochemicals, refiners, telecommunications and information Technology.	Triune International, 11th floor, International Trade Tower, Nehru Place, New Delhi 110019

Contd.

Major Research and Consultancy Organizations *(Contd.)*

Institution	Areas of research/consultancy	Contact
Oil and Natural Gas Corporation Limited (ONGC)	Born as a modest Corporate entity within serene Himalayan settings on 14th August 1956, as Commission, Oil and Natural Gas Corporation Limited (ONGC), has grown into a full-fledged horizontally integrated upstream petroleum company. Today, ONGC is a flagship public sector enterprise and India's highest profit making corporate, which has achieved the landmark of registering a net profit of ₹ 6,197.87 crore in the year 2001–02. Since its inception ONGC has produced more than 600 million metric tons of crude oil and supplied more than 200 billion cubic meters of gas, thus fuelling India's economy. Field of activities: Oil and gas (onshore/offshore), chemicals, petrochemicals, refiners, synthetic fiber, pharmaceuticals, power, environmental protection engg. energy audit and optimization, computer software development, technology/sourcing hazop/risk analysis.	Oil & Natural Gas Corporation Ltd., 9th Floor, BS Negi Bhavan, Tel Bhavan, Dehradun 248003 (Uttaranchal), India. Tel (O): 0135-2793478 Fax: 0135-2758159 Hotline: 51229 Email: j_chaturvedi@hotmail.com.
Exxon Mobil	Exxon Mobil Chemical is one of the largest worldwide petrochemical companies. We have an integrated manufacturer and global marketer of olefins, aromatics, fluids, synthetic rubber, polyethylene, polypropylene, oriented polypropylene packaging films, plasticizers, synthetic lubricant basestocks, additives for fuels and lubricants, zeolite catalysts and other petrochemical products. Exxon Mobil provides quality petrochemical products and services in most efficient and responsible manner to generate outstanding customer and shareholder value. Exxon Mobil pioneered the development of breakthrough metallocene technology.	America's: Recruiting and Employment Exxon Mobil P.O. 2180, Houston, Texas 77252-2180, USA Asia Pacific: Recruiting and Employment Exxon Mobil Chemical Asia Pacific 1 Raffles Place, 27th floor OUB Center, Singapore 048616 Europe: Recruiting and Employment Exxon Mobil Chemical Europe Inc., Hermeslaan 21831 Machelen, Belgium

Contd.

Major Research and Consultancy Organizations *(Contd.)*

Institution	Areas of research/consultancy	Contact
Lurgi	Lurgi is a leading group of companies operating worldwide in the field of process engineering and plant contracting. The Lurgi companies engineer, supply and build turnkey plants for applications in gas and hydrocarbon technology, the petrochemical industry and the growing life science market. In line with its customer's wishes, Lurgi offers comprehensive solutions in many fields of application. Chiefly based on proprietary technologies, the companies build plants for products that are in demand worldwide.	Lurgi India Company Ltd. A 30, Mohan Cooperative Industrial Estate, Mathura Road, New Delhi 110 044 Ph: +91 (11) 2695 0035 Fax: +91 (11) 2695 0042, 2695 0072 E-Mail: lurgi_india@lurgi.de
Gas Authority of India Limited (GAIL)	A truly dominant gas major, has been turning out disciplined financial performance with excellence in project management, service quality and customer satisfaction. Building on with business strength, GAIL has set its future vision to be a "Dominant company in natural gas business, with a significant global presence, integrated in energy and petrochemicals". GAIL's foray into petrochemicals, with its state-of-the-art production facilities at Uttar Pradesh Petrochemical Complex (UPPC), Pata, marks its step towards integration of its business, to fulfil its mission to be a global business organization.	6, Bhikaji Cama Place, RK Puram, New Delhi 110066 Tel: (+91)11 617 2580, 618 2955
Engineers India Limited (EIL)	EIL was established in 1965 to provide technical services for petroleum refineries and other industrial projects. In addition to petroleum refineries, with which EIL started initially, it has diversified into and excelled in other fields such as pipelines, petrochemicals, oil and gas processing, offshore structures and platforms, fertilizers, metallurgy and power. EIL now provides a complete range of project services in these fields and has emerged as Asia's leading design and engineering company.	Engineer's India Bhavan 1, Bhikaji Cama Place, R.K. Puram New Delhi 110066 Ph: 011-26102121, 26104132 Fax: 011-26178210, 26194760 Email: eil.mktg @eil.co.in

Contd.

Major Research and Consultancy Organizations *(Contd.)*

Institution	Areas of research/consultancy	Contact
	Engineers India Limited has provided its services for over two dozen projects with a combined refining capacity of 56 million tons/annum (1.1 million bbls/day) and is working on several other projects with a total refining capacity of over 38 million tons/ annum (7,60,000 bbls/day). The projects include grass roots as well as expansion projects of all refining companies in India apart from refinery projects abroad. Barring the process design of a few licensed units, EIL can execute complete petroleum refinery projects on its own. In addition to technologies for the main refinery units, EIL also has technologies for lube refinery complexes.	
Toray Industries Inc.	Toray is a diversified corporate group with operations in nineteen countries and regions. Underlying operations are technological expertize in the three core fields of organic synthetic chemistry, polymer chemistry and biochemistry. Foundation businesses —fibers and textiles and plastics and chemicals—originated in these technologies. Operations also diversify into such business fields as information and telecommunications-related (IT-related) products, housing and engineering, pharmaceuticals and medical products, and advanced composite materials. Consolidated net sales in the fiscal year ended in 31 March 2002, were 1,015.7 billion.	Toray Industries Inc., Toray Bldg., 2-1, Nihonbashi-Muromachi 2-chome, Chuo-ku, Tokyo 103-8666, Japan. Tel: +81-3-3245-5111 Fax: +81-3-3245-5555 Telex: 222-3811 (Toraytok) J22623 (Toray Inc.)
Indian Oil Technology Ltd.	Indian Oil Technology Ltd. is India's flagship National Oil Company with a market participation of over 53%, a total refinery capacity of 41% and 76% of downstream pipeline transportation capacity. It is the country's largest commercial enterprise with an annual sales turnover of US $ 23.57 billion and of profit US $ 592 million in fiscal 2001.	Indian Oil Technologies Ltd. Plot No. 1, Sector-13, Faridabad 121007, Haryana. Tel: 91-129-2283711 Fax: 91-129-2292012 E-mail: ghoshs@iocrnd.stpn.soft.net

Contd.

Major Research and Consultancy Organizations *(Contd.)*

Institution	Areas of research/consultancy	Contact
	Indian Oil is the sole Indian presence in Fortune's Global 500 listing of the world's largest corporations with a ranking of 226. It has also ranked 17th among the world's largest petroleum companies. Indian Oil has also ranked 112th in Forbes International 500 companies outside the USA. Indian Oil's world class R & D center spread over 65 acres of lush green campus on the outskirts of Delhi has state-of-the-art facilities and has carried out pioneering work in lubricants formation, refinery processes and pipeline transportations. The center has 93 patents to its credit of which 42 are international including 13 in the USA. Indian Oil Technology Ltd. has setup an independent marketing company to market the technologies ready for commercialization. Innovation technologies from Indian Oil are INDMAX (a technology for residue upgradation and LPG/light olefin maximization) Needle coke technology, I-MAX (FCC catalyst additive for boosting LPG yield), INDE treat and INDE sweet, Oilivorous-S (technology for disposal of oily sludge bioremediation), material engineering services, Servo-Ds (novel stabilizer for distillate fuels), Innova-li (diesel fuel lubricity improver), Servo-Ao (novel antioxidant for gasoline), technological services in hydroprocessing, Instrumented Pig (IPIG), Engine and Tribotesting and Maximization of LPG yield in FCC unit, etc.	

Contd.

Major Research and Consultancy Organizations *(Contd.)*

Institution	Areas of research/consultancy	Contact
Technip KT India Limited	A specialist in refineries and downstream activities are now a part of the leading European engineering group, TECHNIP, one of the five biggest worldwide services include state-of-the-art technologies, advanced software skills, combustion engineering furnaces, conceptual and detailed process studies, operations optimization consultancy, plant services and quality fabrication.	Technip KT India Ltd., A-4, Sector-1, Institutional Area, Noida 201301, UP Ph: (+91)-120-2443910 Fax: (+91)-120-2532513 E-mail: sales-TPKTI@technip-coflexip.com
Tata Consulting Engineers Ltd. (TCE)	Consultants for: Industrial, chemical, petroleum and petrochemicals, thermal, hydro and nuclear power, water supply, waste water, irrigation, roads, bridges, ports and harbors, IT infrastructure and environmental and safety systems.	Registered office: Matulya center A, 1st floor, 249, Senapati Bapat Marg Lower Parel (West) Mumbai 400013
Kvaerner Power Gas India Ltd.	Working in fields of petrochemicals, polymers, agrochemicals, speciality chemicals, refineries, synthetic fibers, ferrous, non-ferrous, oil and gas and pharmaceuticals.	Powergas House, 177, Vidyanagari Marg, Kalina, Mumbai 400098 Tel: 022-6915901 Fax: 022-6915934 E-mail: kpgi.marketing@kvaerner.com
SUD-CHEMIE India Pvt. Ltd.	Producing catalysts for ammonia fertilizer plants, petroleum refineries, petrochemical industries, automobile industry, steel industry, cracking, environmental protection, deoxidation/dehydrogenation, stem hydrocarbon reforming and miscellaneous chemical industries.	Main Sales and Technical Services 402/403, Mansarovar, 90, Nehru Place, New Delhi 110019 Tel: 26212353, 26416790, 26416792 Fax: 011-26473326 E-mail: delhi@sud-chemie-india.com
Gharda Consultancy	Gharda is a ₹ 400 crore company, with five national R & D/process engineering awards and several hightech products. It has four manufacturing locations with exports of agrochemicals, pharmaceuticals and polymers to Western Europe and the USA.	Gharda Chemicals Ltd., R & D Center, B-29 MIDC, Dombivli (E), 421203 Dist. Thane, Mumbai, Maharashtra.

Contd.

Major Research and Consultancy Organizations *(Contd.)*

Institution	Areas of research/consultancy	Contact
Petroleum India International	It is a consortium of top ten mega corporations of India with an annual turn over exceeding US $ 28 billion. It works in fields of technical backup, training, turn around maintenance, technical consultancy, management consultancy, information technology and procurement services.	C-5, 'Keshawa' Bandra-Kurla Complex, Bandra (E), Mumbai 400051 India Tel: +91-22-6542411/55/13 Fax: +91-22-6542456/047 E-mail: piiindia@bom3vsnl.net.in
Sarla Technologies	Provides a single, integrated, control and graphical view of all process operations, by communicating to various systems like Siemens, Honey-Well, Yokogawa, Allen Bradley, etc.	B-52, Nand Bhuvan Indl. Estate, Mahakali Road, Andheri (E), Mumbai 400093. Tel: +91-22 832 7805/ 834 4326 Fax: +91-22 821 3572 E-mail: intellution@sarlatech.com
UDHE India Ltd.	Working in fields of fertilizers, petrochemicals and organic chemicals, refinery units, caustic soda and chlorine, inorganic chemicals, cryogenic storages and pharmaceuticals.	Uhde India limited, Uhde house, LBS Marg, Vikhroli (W), Mumbai 400083 Tel: +91-022-578 3701 Fax: +91-022-578 4327 E-mail: uhdein@vsnl.com
Stone and Webster	Products and services, design engineering specialists, feasibility study dervices, petrochemical plant, specialists power station, contractors or designers	Stone and Webster Ltd. 500 Elder Gate, Milton Keynes, MK9 1BA (Road Map), Buchinghamshire. Tel: 01908 668844, Fax: 01908 602211
Kellogg Brown and Root	KBR offers technologies in the areas: Fluid catalytic cracking, hydro-processing, residual upgrading, KBR has phenol design experience in over 50 phenol projects. KBR offers Score™ (selective cracking optimum recovery) technology. For more than half a century KBR has forged achievement in FCC.	
Larsen and Toubro Ltd.	Engineering and construction projects, hydrocarbon and related projects: Refinery, petrochemical fertilizer projects, cement projects	Registered Office & Head Office L & T House, Ballard Estate P.O. Box 278, Mumbai 400 001 India Tel: 022-2618181/2 Fax: 022-2620223/4 E-mail: Iv-ccd@lth.ltindia.com www.larsentoubro.com

APPENDIX F

Major Manufacturing Associations

Name of firm	Address
Adhesive Tape Manufactures' Association	1st Floor, Raj Mahal, 84, Veer Nariman Road, Churchgate, Mumbai 400020 Tel:(022) 22048075, 22048878
All India Food Processors' Association	206, Aurobindo Place Market Complex Hauz Khas, New Delhi 110 016 Tel: 011-26510860, 26518848 Fax: 011-26510860 E-mail: aifpa@nda.vsnl.net.in
All India Industrial Gases Manufactures' Association	215, Square One, C-2 District Centre, Saket, New Delhi 110017, India Tel: +91-11-41076159 Fax: +91-11-41076158
All India Instruments Manufactures and Dealers Association	A/32, Navyug Niwas, 167, Dr D Bhadkamkar Road, Opp Minerva Theatre, Mumbai 400 007 Tel: 022-2307 1868 Fax: 022-2307 1868
All India Rubber Industries Association	601, Pramukh Plaza, B Wing, 485, Cardinal Gracious Road, Opp. Proctor and Gamble, Chakala, Andheri (E) Mumbai 400099 Tel: +91-22-28392095/2107 Fax: +91-22-6710 3211 E-mail: info@allindiarubber.net
Alkali Manufactures' Association of India	3rd Floor, Pankaj Chambers, Preet Vihar Commercial Complex, Vikas Marg, Delhi 110 092 Tel: 011-2243 2003/2241 0150 Fax: 011-2246 8249 E-mail: amai.yrsingh@axcess.net.in
Association of Leasing and Financial Services Companies	(Office): Agra Building, 1st Floor, 131, M.G. Road, Opp: Bombay University, Fort, Mumbai 400 023 Tel: 022-267 5400, 5500 Fax: 022-267 5600
Association of Merchants and Manufacturers of Textile Stores and Machinery (India)	Bhogilia Hargovind Das Building, 18/20, Kaikhushru Dubash Marg, Mumbai 400 011 Tel: 022-2284 4350/2284 4401 Fax: 022-2287 4060 E-mail: ammtsmi@bom7.vsnl.net.in

Contd.

Major Manufacturing Associations *(Contd.)*

Name of firm	Address
Association of Synthetic Fiber Industry	125, Uday Park (First Floor), New Delhi 110 049 Tel: 011-26964154 Fax: 011-26515462
Cement Manufactures' Association	Cement Manufacturers' Association (CMA) Tower A-2E, Sector 24, Noida 201 301 (UP) Tel: 0120-2411955, 2411957, 2411958 Fax: 0120-2411956 E-mail: cmand@cmaindia.org
Confederation of India Alcoholic Beverages Companies	Z-27, Hauz Khas, New Delhi 110 016 Tel: 011-26534038 Fax: 011-26967900
Federation of Indian Mineral Industries	FIMI House, B-311, Okhla Industrial Area, Phase-I, New Delhi 110 020 Tel.: +91-11-26814596 Fax: +91-11-26814593/26814594 E-mail: fimi@fedmin.com
Fragrance of Flavor Association of India	Navin Chandra Ram Chhoddas Shah Hall, 2B, Court Chambers, 35, Sir Vithaldas Thakeray Marg, Mumbai 400 020 Tel: 022-2090184 Fax: 022-2005875 E-mail: fafia@bom3.vsnl.net.in
Indian Chemical Manufactures Association	Sir Vithaldas Chambers, 16-Mumbai Samachar Marg, Mumbai 400023 India Ph: +(91)-(22)-4974308/4944624 Fax: +(91)-(22)-4950723
Indian Pump Manufactures' Association	Indian Pump Manufacturers Association 406–408, "ATMA House" Opp. La-Gajjar Chambers, Ashram Road, Ahmedabad 380 009, Gujarat, India Ph: +91-79-26583049 Fax: +91-79-26584194
Indian Refractory Makers' Association	Indian Refractory Makers Association 5 Lala Lajpat Rai Sarani, 4th Floor, Kolkata 700020, West Bengal, India. Ph: 091-33-22810868 Fax: 091-33-22814357
Indian Soap and Toiletries Makers' Association	614, Raheja Centre, 6th Floor, Free Press Journal Marg, Nariman Point, Mumbai 400 021 Tel: 022-2285 3649/2282 4115 Fax: 022-2285 3649 E-mail: istma@bom3.vsnl.net.in

Contd.

Major Manufacturing Associations *(Contd.)*

Name of firm	Address
Indian Sugar Mills Association	Indian Sugar Mills Association (ISMA), Ansal Plaza, 'C' Block, 2nd Floor, August Kranti Marg, Andrews Ganj, New Delhi 110049 (India) Ph: +91-11-2626 2294-98 Fax: +91-11-2626 3231 E-mails: isma@indiansugar.com
Organization of Pharmaceuticals Producers of India	Peninsula Chambers, Ground Floor, Ganpatrao Kadam Marg, Lower Parel, Mumbai 400 013. Ph: +91 22 24918123, 24912486, 66627007 Fax: +91 22 24915168
Organization of Plastics Processors of India	404/5, Golden Chambers, New Link Road, Andheri (W), Mumbai 400 053 Tel: 022-6326958, 6323644, 6692 3131 Fax: 022-6346975
Process Plant and Machinery Association of India	002 Loha Bhavan, 91/93 , P D'Mello Road, Masjid (E), Mumbai 400 009, India Tel: 91 22 2348 0965, 2348 0405 Fax: 91 22 2348 0426
Textile Machinery Manufacturers' Association (India)	53, Mittal Chambers, 5th Floor, Nariman Point, Mumbai 400 021 Tel: 022-2202 3766/2202 4238 Fax: 022-2202 8017 E-mail: mail@tmmaindia.net
The All India Plastic Manufacturers' Association	AIPMA House, A-52, Street No. 1, MIDC Moral, Andheri (East) Mumbai 400 093 Tel: 22 6777 8899 (100 Lines) Fax: 022-2821 6390
The Asbestos Cement Products Manufacturers' Association	The Asbestos Cement Products Manufacturers Association, 502, Mansarovar 90, Nehru Place New Delhi 110 019, India Tel: + 91 11 46521495 Fax: + 91 11 46521496
The Solvent Extraction' Association of India	142, Jolly Maker Chamber No. 2, 14th Floor, 225 Nariman Point, Mumbai 400 021 Tel: 022-2202 1475/2282 2979/22028911 Fax: 022-2202 1692 E-mail: solvant@mtnl.net
The Vanaspati Manufacturers' Association of India	903, Akashdeep Building, 26-A, Barakhamba Road, New Delhi 110 001 Tel: 011-2331 2640/2331 0758 Fax: 011-2331 5698 E-mail: vmai.vanaspati@smy.sprintrpg.ems.vsnl.net.in

Contd.

Major Manufacturing Associations *(Contd.)*

Name of firm	Address
Boiler and Unfired Pressure Vessels Division	Boiler and Unfired Pressure Vessels Division CII, 23, Institutional Area, Lodhi Road, New Delhi 110 003 Tel: 011-2462 9994 (4 lines) Fax: 011-2463 3168/2462 6149
Compressed Air Division	Compressed Air Division, CII, 23, Institutional Area, Lodhi Road, New Delhi 110 003 Tel: 011-2462 9994 (4 lines) Fax: 011-2463 3168/2462 6149 E-mail: asr@co.ci.ernet.in
Industrial Furnace Division	Industrial Furnace Division, CII, 23, Institutional Area, Lodhi Road, New Delhi 110 003 Tel: 011-2462 9994 (4 lines) Fax: 011-2463 3168/2462 6149 E-mail: sgr@co.cii.ernet.in
Industrial Valver Division	Industrial Valves Division, 23, Institutional Area, Lodhi Road, New Delhi 110 003 Tel: 011-2462 9994 (4 lines) Fax: 011-2463 3168/2462 6149
Instrumentation Division	Instrumentation Division, CII, 23, Institutional Area, Lodhi Road, New Delhi 110 003 Tel: 011-2462 9994 (4 lines) Fax: 011-2463 3168/2462 6149 E-mail: srg@co.cii.ernet.in
National Committee on Chemicals and Petrochemicals	National Committee on Chemicals and Petrochemicals, CII, 23, Institutional Area, Lodhi Road, New Delhi 110 003 Tel: 011-2462 9994 (4 lines) Fax: 011-2463 3168/2462 6149
National Committee on Drugs and Pharmaceuticals	National Committee on Drugs and Pharmaceuticals, CII, 23, Institutional Area, Lodhi Road, New Delhi 110 003 Tel: 011-2462 9994 (4 lines) Fax: 011-2463 3168/2462 6149 E-mail: shalab@co.cii.ernet.in
National Committee on Oil and Gas Exploration-Production	National Committee on Oil and Gas Exploration-Production, CII, Gate No. 31, North Block, Jawaharlal Nehru Stadium, New Delhi – 110 003 Tel: 011-2436 6225/2436 6273/2436 6276/ 2436 6281 Fax: 011-2436 6271/2436 7844

Contd.

Major Manufacturing Associations *(Contd.)*

Name of firm	Address
National Committee on Petroleum	National Committee on Petroleum, CII, Gate No. 31, North Block, Jawaharlal Nehru Stadium, New Delhi 110 003 Tel: 011-2436 6225/2436 6273/2436 6276/2436 6281 Fax: 011-2436 6271/2436 7844
National Committee on Textiles	National Committee on Textiles, CII 23, Institutional Area, Lodhi Road, New Delhi 110 003 Tel: 011-462 9994 (4 Lines) Fax: 011-463 3168/462 6149 E-mail: hemant@co.cii.ernet.in
Oil and Gas Equipment Division	National Committee on Oil & Gas Equipment. Division, CII, Gate No. 31, North Block, Jawaharlal Nehru Stadium, New Delhi 110 003 Tel: 011-436 6225/463 6273/436 6273/436 6376/436 6281 Fax: 011-462 6271/464 7844
Oil and Gas Services Division	National Committee on Oil and Gas Services Division, CII, Gate No. 31, North Block, Jawaharlal Nehru Stadium, New Delhi 110 003 Tel: 011-436 6225/463 6273/436 6273/436 6376/436 6281 Fax: 011-462 6271/464 7844
Petrotech Society	Petrotech Secretrariate 601–603, Tolstoy House Tolstoy Marg, Connaught Place, New Delhi 110 001 Ph: +91 11 2335 4002-05 Fax : +91 11 2335 4001 E-mail: petrpotechscociety@vsnl.net, petrotechindia@touchteIndia.net Website: www.petrotech society.com
Pollution Monitoring and Control Equipment Division	The Secretary, Pollution Monitoring and Control Equipment Div. CII, 23, Institutional Area, Lodhi Road, New Delhi 110 003 Tel: 011-2462 9994 (4 Lines) Fax: 011-2463 3168/2462 6149 E-mail: shalab@co.cii.ernet.in

APPENDIX G

Cost of Various Crude Oils

Crude oils	Mar '13 ($/MT)	Crude oils	Mar '13 ($/MT)
Dubai	771	Iranian light	802
Brent	844	Iranian mix	793
Oman	784	Siri export	788
Tapis	917	Marib light	867
Arab heavy	743	Azeri light	851
Arab light	801	UMM shaif	855
Arab mix (50:50)	791	Arab mix (80:20)	787
Upper zakum	802	Essider	841
Basra light	760	Girassol	816
Forozon	769	Cabinda	800
Murban	852	Arab medium	780
Forcados	787	Rabi light	811
Escravos	820	N'kossa	859
Bonny light	852	Melittah	850
Masila blend	819	KG basin crude	916
Kuwait	770	CPC blend	835
BH	848	Zarzaitine blend	857
PY-3	919	Brega	875
Seriah light	879	COCO	807
Labuan	857	Sarir	840
Benchamas	853	Espior	887
Tantawan	852	Palanka	922
Saharan blend	887	Kissanje	873
Iranian heavy	768	Quaiboe	919

Courtesy: CPCL.

Approximate Cost of Petroleum products

Sl. No.	Product	Unit	Cost per ton
1.	Naphtha	MT (Metric tons)	65,000
2.	Furnace oil		41,000
3.	Superior kerosene	KL (1,000 L)	49,900
4.	Light diesel oil	KL	51,000
5.	Bitumen	MT	42,000
6.	LPG (commercial)	19 kg	15,600
7.	LPG (domestic) non-subsidized/subsidized	14.2 kg	920/425
8.	Automotive fuel	KL	74,000
9.	Aviation fuel	KL	74,000
10.	Sulfur	MT	12,000
11.	Motor sprit	L	73.00
12.	High speed diesel	L	65.00

Approximate Cost of Utilities

Sl. No.	Utility/product	Cost
1.	Steam • SHP • HP • MP • LP	₹ 2379/MT ₹ 2236/MT ₹ 2111/MT ₹ 1988/MT
2.	Power, ₹/kWh	5.44
3.	Cooling water, ₹/m^3	1.63
4.	Instrument air, ₹/m^3	1.3
5.	Plant air, ₹/m^3	1.3
6.	Nitrogen, ₹/Nm^3	₹ 0.82/Nm^3
7.	Service water, ₹/m^3	10.27

Index